Descartes in the Classroom

Medieval and Early Modern Philosophy and Science

The titles published in this series are listed at *brill.com/memps*

Descartes in the Classroom

*Teaching Cartesian Philosophy in the
Early Modern Age*

Edited by

Davide Cellamare
Mattia Mantovani

BRILL

LEIDEN | BOSTON

Cover illustration: Petrus Denique (engraver) and Jean-Joseph Havelange (student), *Pluralité des mondes*, MS., Louvain, MS. 308: *Physica* (1767), fol. 147ᵛ. Source: *Magister Dixit* Collection.

The Library of Congress Cataloging-in-Publication Data is available online at http://catalog.loc.gov
LC record available at http://lccn.loc.gov/2022038988

Typeface for the Latin, Greek, and Cyrillic scripts: "Brill". See and download: brill.com/brill-typeface.

ISSN 2468-6808
ISBN 978-90-04-52326-5 (hardback)
ISBN 978-90-04-52489-7 (e-book)

Contents

Figures and Tables

Tables

Abbreviations

[AT]   Charles Adam and Paul Tannery, eds., *Oeuvres de Descartes*, 11 vols. (Paris, 1897–1913).

[CSMK] John Cottingham, Robert Stoothoff, Dugald Murdoch and Anthony Kenny, transl., *The Philosophical Writings of Descartes*, 3 vols. (Cambridge, 1984–1991).

Introduction

Davide Cellamare and Mattia Mantovani

This volume studies the ways in which the philosophy of René Descartes (1596–1650) was diffused through teaching in the early modern age and, more specifically, between the 1640s, when Descartes's ideas started gaining ground, and the last decades of the eighteenth century, when the influence of Cartesianism was on the wane.

For over a century, Descartes's philosophy dominated the European landscape. Several of his works—most notably, the *Discourse on Method* (1637), the *Meditations on First Philosophy* (1641), the *Principles of Philosophy* (1644), and the *Passions of the Soul* (1649)—introduced momentous changes in the way in which learned persons conceived of the physical world and human nature.

A large body of scholarship is devoted to the complex historical trajectories along which Descartes's ideas rapidly spread, first in the United Provinces—where Descartes spent the greatest portion of his adult life—and then all over Europe. For instance, the diffusion of various forms of Cartesian philosophy has been recently analysed, in its "Dutch and French constructions," by Tad Schmaltz. Even more recently, the reception of Descartes's philosophy among its partisans and its critics has been one of the key themes of the nine-hundred-page *Oxford Handbook of Descartes and Cartesianism*.[1]

In explaining the rapid, international diffusion of Descartes's philosophy, the existing scholarship has oftentimes remarked that Descartes's ideas made their inroads in the European philosophical scene first and foremost through colleges and universities. Despite these important indications, no systematic study is available concerning the history of early modern Cartesianism from the perspective of the classroom and the ways in which Descartes's philosophy was taught and made its way through to early modern university curricula.

In this study, we do precisely this, and for a number of important reasons. Descartes's own attitude towards the teaching of his philosophy and its inclusion, later, within early modern teaching institutions, all of this seems to be pervaded by methodological, content-related, and institutional tensions. But what better way to explain these tensions than by looking at the actual way in which Cartesian philosophy was taught?

---

1 Schmaltz, *Early Modern Cartesianism*; Nadler, Schmaltz and Mahut, *The Oxford Handbook of Descartes and Cartesianism*.

Consider, for instance, Descartes's changing opinions about the traditional teaching that he received at his *alma mater*, the Jesuit college of La Flèche, as well as about the inclusion of his philosophy into university teaching.

In his *Discourse on Method*, Descartes recalled his feelings of relief on the day upon which he had reached "the age that permitted him to escape the command of his preceptors," and their uncertain sciences.[2] Yet, when asked by a correspondent where to enrol his son, Descartes recommended his old college, commenting that "there is no place in the world where philosophy is better taught."[3] Yet, before long, Descartes grew to utterly despise the teaching programmes that had been put in place, and to harbour the ambition of having his philosophy taught at the universities. Serious as this ambition was, it did not follow a linear trajectory.

Descartes was no academic philosopher. He never had to teach, and he expressly refused to do so when, in 1633, he declined the invitation to take the chair in theoretical medicine at the University of Bologna.[4] Moreover, in presenting his philosophy as a fresh start, Descartes notably criticized commentaries and academic disputations as sterile philosophical methods. His well-known plan to break away from traditional philosophy concerned not just the contents but also the method of philosophizing itself.

But despite his dismissive attitude towards scholastic philosophy, and the fact that he was never directly involved in teaching, Descartes invested a great deal of effort in having his ideas taught by others. It was also for this purpose that in 1644, Descartes reworked his metaphysics and physics into a textbook-like synthesis: the *Principles of Philosophy*. Moreover, Descartes famously associated with Dutch academics, notably his friend Henricus Reneri (1593–1639), and his one-time ally and later critic, the Utrecht professor Henricus Regius (1598–1679), through whom he sought to introduce his philosophy into the Dutch curricula.

Descartes's attempts to meddle with Regius's teaching have been well documented within the existing scholarship. Yet, Descartes's fluctuating attitude towards academic teaching—well beyond Utrecht—has not yet been given a full account. Importantly, Descartes's somewhat uneasy relationship with academic teaching makes one wonder about the policy of his followers working in academia. Did Cartesian teachers try to accommodate the content of Descartes's philosophy to the existing pedagogical practices, such as university

---

2   Descartes, *Discourse on Method* I, AT VI 9. Unless otherwise stated, all translations are our own.
3   Descartes to ***, 12 September 1638, AT II 378.
4   Manning, "Descartes and the Bologna Affair."

disputations? Or, did the teaching of Descartes's philosophy rather prompt—maybe even require—the emergence of new teaching methods?

In responding to these questions, this volume shows that the early diffusion of Descartes's philosophy had a specifically pedagogical dimension, which crossed paths with other ongoing attempts at curricular reform—notably, through Ramist education, for example. One should carefully consider the fact that professors both of a Cartesian persuasion and their anti-Cartesian colleagues primarily worked as *teachers*, and had therefore specific concerns and obligations, quite different from those of non-academic contemporaries like Pierre Gassendi (1592–1655), Blaise Pascal (1623–1662), John Locke (1632–1704) or like Nicolas Malebranche (1638–1715). But were these pedagogical concerns reflected in changes within early modern curricula? This question too calls for an in-depth investigation.

There are two further aspects that make the teaching of Descartes's philosophy a particularly interesting topic. In many ways, Descartes' ideas represented a very attractive and viable alternative to the Aristotelian system; one may think of the mechanist explanation of the natural world, for example. However, his philosophy had two important limits. First, in terms of the subjects it covered, the 'new' philosophy was not as complete as the 'old' one, which included an orderly system of knowledge and well-established pedagogical procedures with which to teach it. Secondly, Descartes's philosophy fitted uneasily with the overall goals of philosophical education in the university, which was often propaedeutic to the study of law, medicine, and theology.

One will remember that Descartes aspired to an all-encompassing system of philosophy, which in the 1647 French edition of the *Principles of Philosophy* he famously described as "a tree whose roots are metaphysics, whose trunk is physics, and whose branches, which issue from this trunk, are all the other sciences," which "reduce themselves to three principal ones: medicine, mechanics, and morals."[5] Nevertheless, as Roger Ariew has explained in detail, Descartes's philosophy covered only a small part of his philosophical tree; his overall contribution, at best, can be described as a general metaphysics and an incomplete physics. Let us not forget that the *Principles of Philosophy* were themselves the scaled-down version of what Descartes had originally envisaged as a complete *Summa philosophiae* in six books. Descartes scaled back his initial ambitions, in recognition of the fact that his own accounts of plants, animals, and (specifically) human animals were unsatisfactory.[6]

---

5   Descartes, *Principes de la philosophie*, *Préface*, AT IX-B 14.
6   Ariew, *Descartes and the First Cartesians*, chs. 9, 11, and *passim*.

While Descartes could allow himself to produce an abridged version of his initial *summa*, the pared-back nature of his project vexed many of his followers, especially those who were expected to teach the entire philosophical curriculum, including those disciplines that Descartes had not systematically treated, such as logic and ethics.

Studying the teaching of Descartes's philosophy is particularly interesting in this respect, as it enables us to watch how Descartes's supporters endeavoured to accomplish those things that he himself had failed to achieve. We have devoted particular attention to the investigation of whether early modern Cartesian teachers worked out systems of logic and ethics along Cartesian lines, or instead embraced a more eclectic stance, for example by juxtaposing existing views with Descartes's philosophy or by integrating different traditions into a synthesis. Several chapters of this volume attempt to do this by looking backward into the early modern classrooms and observing students taking notes or copying diagrams from Descartes's works. Other chapters consider the production of Cartesian textbooks about logic and ethics, and reveal exactly how Descartes's acolytes took over where their 'master' had left off.

Descartes's philosophy, incomplete as it was in comparison to the one then taught at early modern universities, faced also another challenge. As is well known, seventeenth- and eighteenth-century students of philosophy were enrolled in arts programmes, which were conceived as preparatory for the study of the so-called 'higher disciplines': law, medicine, and theology. Education in scholastic philosophy, which Cartesians set out to replace, was geared towards, and well-tested for, this purpose; Descartes's system was not, as contemporary critics often remarked. For example, the Utrecht professor in theology Gisbertus Voetius (1589–1676)—one of Descartes's first opponents—lamented the fact that the students trained in Descartes's philosophy made fools of themselves when reading the higher disciplines, as they had omitted to learn even the most basic technical jargon. The question, therefore, naturally arises as to whether and how Cartesian teachers managed to maintain the link between philosophy and the higher disciplines, after having rejected everything Aristotelian. The seriousness of this problem seems all the more acute when one considers the fact that Descartes's sole contribution to medicine was in the physiology that he sketched in the *Treatise on Man* (published only posthumously in 1662); this means that he left no pathology and no therapeutics. Descartes's claims in theological matters were few and far between; those that he did make, moreover, proved problematic to most of his contemporaries.

Given the paucity of Descartes's original contributions in these disciplines, the challenge for his academic followers was to fill in the gaps. How, for example, could they develop a new medical system based on Cartesian physiology? And how did they manage to successfully introduce their novel ideas about metaphysics and physics in a way that was in harmony with the purposes of teaching Christian theology? In this volume, we have taken particular care to address the hitherto less frequently studied teachings of medicine on the part of Cartesian philosophers. Moreover, by looking at the teaching documents, we disclose new perspectives on the well-known institutional tensions that characterized the early diffusion of Descartes's philosophy.

The great institutional resistance with which Descartes's philosophy was met from the very beginning is well documented—suffice it here to mention the seminal studies by George Monchamp, Caroline Louise Thijssen-Schoute, and Theo Verbeek.[7] The earliest official condemnation of Descartes's philosophy (Utrecht, 1642) was pronounced even before the *Principles of Philosophy* was published. Other Dutch universities, such as Leiden and Harderwijk, were soon to follow Utrecht's lead by issuing their own condemnations. In 1656, the States of Holland and West-Friesland proclaimed a decree which forbade the diffusion of Descartes's philosophy. This prohibition issued from the fact that the Cartesian influence was considered responsible for what was—from a Reformed point of view—an undesirable crossing of boundaries between philosophy and the reading of Scripture. But the struggles that Descartes's philosophy endured were not limited to the Reformed context of the United Provinces.

In the Southern Low Countries, the Catholic University of Louvain equally prohibited Cartesian doctrines in 1662. One year later, all of Descartes's works were placed on the 'Index of Forbidden Books,' by the Roman Congregation of the Index. Even twenty years after his death, Louis XIV banned the public teaching of Descartes's philosophy in 1671, explicitly on account of religious concerns. In sum, Descartes's philosophy was condemned in both the country of his birth and in his adopted home country, and the prohibitions on the teachings of his philosophy in both places persisted long after his death.

In spite of all the official resistance it encountered, Descartes's philosophy quickly attracted a substantial and enthusiastic following. How are we to explain this craze, that Dennis Des Chene dubbed *Cartesiomania*?[8] In this volume, we show that this question may best be answered when one observes the institutional frictions surrounding the teaching of Descartes from the perspective of the classroom.

---

7 Monchamp, *Cartésianisme en Belgique*; Thijssen-Schoute, *Nederlands Cartesianisme*; Verbeek, *La querelle d'Utrecht*; Verbeek, *Descartes and the Dutch*.

8 See Des Chene, "*Cartesiomania*: Early Receptions of Descartes." The term 'Kartesiomania' was used as early as 1654—with a negative connotation—by Jacobus Revius, in his *Kartesiomania, hoc est, furiosum nugamentum* (Leiden, 1654). On Revius, see Goudriaan, ed., *Jacobus Revius*.

Besides showing the intriguing tensions (methodological, content-related, and institutional) that characterized the early diffusion of Descartes's philosophy and its internal evolution, this volume also intends to contribute to a more general knowledge of the history of early modern teaching and universities. Several chapters focus on specific European universities and academic institutions, from the better-known (e.g., Cambridge, Leiden, Louvain) to those less well known, but no less relevant to the reception of Descartes's ideas (e.g., Breda, Frankfurt an der Oder, Nijmegen, and Uppsala). Furthermore, this volume furnishes several examples of what it meant to teach philosophy—and, in particular, a not-yet-established philosophy—in early modern European universities. In doing so, *Descartes in the Classroom* has a contribution to make to the reappraisal of the role of the university in the scientific revolution, for which John Gascoigne was already pleading in 1990.[9] The present study is therefore offered also to readers whose primary interest does not lie in Descartes's philosophy as such. In effect, the teaching of Descartes's philosophy is best studied as an affair that pertains to teaching theory and practice, without thereby forgetting its philosophical import and ambitions.

In compiling this volume, we have sought to find a balance between these different concerns—the same delicate exercise those university professors with a leaning towards Cartesianism had to practise in their classrooms. At the same time, the scope of our volume is not limited to university education, but, as we are about to explain, it takes into consideration a diverse range of teaching venues and practices.

Descartes in the Classroom is the strongly modified result of a three-day conference at Radboud University, Nijmegen in 2020. This conference was organized in the context of a research project that primarily studies the teaching of Cartesianism as it emerges from student notebooks produced in the Northern and the Southern Low Countries, between 1650 and 1750.[10]

During the conference and in preparing this volume, the following things became clear to us: first, the 'classroom' that one must consider in order to

---

9   Gascoigne, "A Reappraisal of the Role of the Universities in the Scientific Revolution." Since Gascoigne's article was published, the role of universities in the transformations of early modern philosophy and science has increasingly received more attention. Most recently, for example, the role of university disputations and dissertations has more specifically been treated by Friedenthal, Marti and Seidel, eds., *Early Modern Disputations and Dissertations.*

10   This research project (*The Secretive Diffusion of the New Philosophy in the Low Countries: Evidence on the Teaching of Cartesian Philosophy from Student Notebooks*) is carried out by the editors and is directed by Prof. Christoph Lüthy (RU, Nijmegen) and Prof. Jan Papy (KU Leuven). The project, the conference, and this volume are supported by the NWO (*Neder-*

understand the teaching of Cartesian philosophy is much bigger than the confines of the university classroom; secondly, the documents that need to be
studied by a history of the teaching of Cartesianism cannot be limited to student notebooks. By necessity, therefore, we have enlarged our scope so as to
include the widest possible array of teaching venues and practices, and an
equally diverse range of methods and documents. The 'classroom' in which
Descartes's philosophy was taught is more than just a classroom; the sites for
the diffusion of his philosophy included 'home' and 'distance learning' *ante litteram*, weekly conferences, learned discussions, and a much larger audience
than the academic audience alone. Cartesians themselves consciously pursued
this strategy, and they took pride in having their lectures attended by "people
of all stations and conditions: prelates, abbots, courtiers, doctors, physicians,
philosophers, surveyors, regents, schoolboys, provincials, foreigners, artisans—
in a word, people of every age, sex and profession."[11] The Cartesian classrooms
became much wider than the Schoolmen's.

1   Methods and Documents

Descartes conceived of the contents and the form of his philosophy as inextric-
ably intertwined. Yet, as Chapter 1 (Theo Verbeek) shows, Cartesian philosophy
could permeate and be preserved in university teaching precisely because
Cartesian authors managed to convey (parts of) its contents through methods
that fitted the existing pedagogies.

In the early modern age, the teaching of Descartes's ideas did not exclusively
take place through university lectures. Within academic teaching, as Chapter 8
(Andrea Strazzoni) shows, Cartesian philosophy was also conveyed through
experiments. Moreover, private lessons also played an important role, as shown
in Chapter 19 (Mihnea Dobre).

Not everyone who was instructed in Cartesian philosophy had access to
university lectures or to the experiments performed within university institutions. Women, for instance, were notoriously denied access to those venues. Nonetheless, as Chapter 12 (Sarah Hutton) explains, Anne Conway (1631–
1679) found her way to Descartes's philosophy through epistolary tutorials with

     landse Organisatie voor Wetenschappelijk Onderzoek) and the FWO (*Fonds Wetenschap-
     pelijk Onderzoek*) under grant number 326-69-001 (Samenwerking Vlaanderen-Neder-
     land). We wholeheartedly thank the institutions that made this project possible.
11   Rohault: *Oeuvres posthumes* (Paris, 1682), *Préface*, n.p.; translated in Roux, "Condemna-
      tions of Cartesian Natural Philosophy," 768–769 (modified).

Henry More (1614–1687)—a sort of 'distance learning' *ante litteram*. Moreover, as Chapter 18 (Marie-Frédérique Pellegrin) illustrates, French women between the seventeenth and eighteenth centuries—at least those wealthy and well connected—learned Descartes's ideas as they circulated in salons run by aristocratic women. As Chapter 5 (Henri Krop) explains, the translations of Descartes's works into Dutch was also an important means to make his ideas available to a non-academic audience in the United Provinces. In reflecting on these elements, we invited the contributors to consider the 'classroom' in a way broad enough as to encompass the diverse venues in which Descartes's philosophy was taught.

Going back to the university context, this volume covers a very diverse array of documents. University disputations, for instance, became an important battleground—sometimes quite literally so, given the occasional outbursts of violence—between Descartes's friends and foes. Several chapters study these documents from a variety of perspectives, such as the *materiae promotionis* and their authorship (for instance, in the case of disputations that were arranged by a teacher as together forming the different stages of a single argument). More specifically, Chapter 7 (Erik-Jan Bos) focuses on the role of Cartesian and anti-Cartesian 'corollaries,' or *impertinentia*, which students were allowed to add to the disputations they defended. This chapter, moreover, shows the utility of quantitative approaches in reconstructing long-term tendencies: in this specific case, the numerousness of the above-mentioned corollaries and their underlying logic.

Very central to our study are class dictations: these manuscript sources feature prominently in the chapters by Davide Cellamare, Steven Coesemans, Domenico Collacciani, Antonella Del Prete, Mattia Mantovani, Carla Rita Palmerino, and Christoph Sander. Besides revealing details of the teaching of Cartesian philosophy that would otherwise go unheeded, our consideration of student dictates contributes to the ongoing research on this type of document in general. *Dictata*—especially those related to scholastic philosophy—have in fact received increasing attention as a privileged point of observation when it comes to studying early modern intellectual history. Systematic research in this field is currently underway within the *Magister Dixit* Project at KU Leuven, which aims at digitizing and studying all extant students' lecture notes of the 'old' university of Louvain (1425–1797). This project has been a major source of support and inspiration throughout the different stages taken in the composition of this book.[12]

---

12   The manuscripts digitized by KU Leuven *Magister Dixit* Project, can be consulted at http://lectio.ghum.kuleuven.be/lectio/project_summary.

Among the various types of documents that are treated in this volume, printed textbooks produced for university teaching are studied in Chapter 13 (Roger
Ariew) and Chapter 20 (Tad Schmaltz); their circulation and reedition between
the United Provinces and England features prominently in Chapter 11 (Igor
Agostini). University regulations and the decrees by public authorities also
played a major role in the vicissitudes surrounding the teaching of Descartes's
philosophy: accordingly, they are duly studied in several places in this volume,
notably in Chapter 6 (Pietro Daniel Omodeo).

Virtually all contributions consider several of the above-mentioned types of
documents, thereby providing a nuanced picture of the early modern teaching
of Descartes's philosophy and exemplifying a number of strategies for studying
it. Indeed, we have framed this volume in such a way as to account for a wide
variety of documents, methods, venues, and geographic areas: this approach is,
we think, not just a crucial aspect of our work, but it is also apt for studying the
philosophy of a thinker who recommended travel to different countries so as
to get a wider and richer understanding of the world.

Before we take you, the reader, through the contents of this volume, it is
opportune to appreciate the importance of another document type through
which Cartesian philosophy was taught. Images played a crucial role in the diffusion of Descartes's philosophy. As Christoph Lüthy has explained, Descartes
made crucial use of illustrations to lend support to his corpuscular physics
and to account for phenomena as diverse as rainbows, waves, the revolutions
of planets and the circulation of blood.[13] Besides featuring in the published
works of his followers, the images used by Descartes also found their way into
students' notebooks produced in Louvain.[14] When it comes to understanding
the diffusion of Descartes's thought, Cartesian imagery is no less important
than Cartesian texts. While this point is often mentioned, a full and systematic
appreciation thereof—a history of Cartesian imagery—is still missing.

It is for this reason that we have decided to devote to this topic an essay
collection of its own, soon to appear in the present Brill series, under the
title: *Cartesian Images: Picturing Philosophy in the Early Modern Age*. The two
volumes are intended almost as twin publications, a combined attempt to
account for the complexity of early modern Cartesianism and its diffusion
through both word and image. With a few strictly-necessary exceptions, therefore, the present volume will forgo a systematic treatment of Cartesian imagery
among its already rich material.

---

13   Lüthy, "Where Logical Necessity Becomes Visual Persuasion."
14   The research concerning Cartesian images in Louvain textbooks owes much to the sem-
      inal study by Vanpaemel, "The Louvain Printers."

2   Themes and Subjects of This Volume

The impact of Descartes's ideas on seventeenth- and eighteenth-century philosophy is so pervasive and rich that many studies have been devoted to it,
and many more studies will no doubt continue to explore these themes in
the future. It is impossible to encapsulate in one book a definitive account
of the teaching of Cartesian philosophy, given the scale and complexity of
the international network of actors and institutions involved. We can make
no claim to have exhausted such a multifaceted and large-scale historical
phenomenon, and we look forward to future studies, amongst which will be
some—we hope—that have derived some inspiration from this present volume. This being said, we think that the present volume presents a rich and
diverse picture of the teaching of Descartes's philosophy in the early modern
age.

We have sought, especially, to cover as much as possible the whole teaching
curriculum (including topics that are often overlooked): from logic to ethics,
from physics and the study of the soul to metaphysics.

Given that the teaching of natural philosophy in the early modern age and
the eventual replacement of Descartes's physics by Newtonian physics have
been the object of several (also recent) studies, we have tried to enrich our
history with topics that have hitherto not received the attention they deserve:
for example, medicine, magnetism, and experimental hydraulics, as well as the
debates concerning the origin of life.[15] We have, moreover, documented the
introduction of a new discipline in seventeenth-century university teaching,
namely the history of philosophy. Sometimes done with the purpose of presenting Descartes's philosophy in a negative light (by comparison with the other
available philosophical schools), the history of philosophy ended up increasing the fame of Cartesian philosophy.

Little by little, the *Cartesiomania* of supporters and detractors alike contributed to turning Descartes into the leader of the *novatores* and the 'father of
modern philosophy.'

While we can hardly do justice, in these few pages, to the many topics treated
in this book, we trust that, taken together, these twenty chapters bring to you

---

15   On the teaching of natural philosophy in the early modern age, see at least, very recently,
        Feingold and Sangiacomo, eds., *Reshaping Natural Philosophy: Tradition and Innovation
        in the Early Modern Academic Milieu*. The remaining branch of Descartes's tree of philo-
        sophy, namely mechanics, will receive special attention in the above-mentioned volume
        Cartesian Images. There, we will show that, in fact, Descartes's mechanics (just like his
        mathematics) relied heavily on diagrams.

a good sense of what it meant to study the 'new philosophy' in the early modern age, and how entire generations of students first encountered Descartes's ideas.

The volume is roughly organized around geographical areas and, within each of them, a mainly chronological order. However, several figures, institutions, and periods are treated in different chapters. Our research, after all, also wishes to show that early modern Cartesianism and its teaching crossed the borders of nations and schools. Exceptions to the geographical and chronological criteria occur in those cases where the complex articulation of the diffusion of Cartesianism did not lend itself to rigid temporal and geographical categorizations. The twenty chapters of this volume should indeed be read as standing in dialogue with each other. To this end, we have inserted a rich system of cross-references, which should help the reader navigate the complex network of early-modern Cartesianism.

The diffusion of Descartes's philosophy in universities—and the condemnations it incurred—first took place in the United Provinces; accordingly, a fair number of the chapters of this volume is centred on that emerging nation. We have taken special care to examine the case of leading universities—such as Leiden and Utrecht—but also to consider institutions that have so far been largely neglected.

Chapter 1 (Theo Verbeek) reveals and explains structural frictions within Descartes's own attitude towards the teaching of his philosophy: Descartes was reluctant to have his ideas become part of academic pedagogy. However, he keenly attempted to influence the inroads of his philosophy into Dutch Universities, and in such a way as to engage with professional philosophy.

The next two chapters provide important information concerning the way in which Adriaan Heereboord (1613–1661) introduced Descartes's ideas in philosophical teaching at the University of Leiden. Chapter 2 (Howard Hotson) shows that Heereboord did so as part of a primarily pedagogical programme, which aimed to impart knowledge to his students about both the traditional Aristotelian views, as required by the university regulations, and Descartes's new philosophy. The case of Heereboord nicely illustrates how the eclectic and post-Ramist traditions, which Heereboord inherited from his teacher Franco Burgersdijk (1590–1645), provided the framework within which Cartesianism could first enter the *cursus philosophicus* in Leiden. Chapter 3 (Antonella Del Prete) compares Heereboord's teaching to that given by Johannes De Raey (1622–1702). By analysing extant *dictata* of their classes, this chapter sheds light on the different strategies adopted by both philosophers in order to teach Descartes's views, at a time when the existing regulations of Leiden University forbade the mention of his name.

A newly-discovered manuscript of De Raey's private *dictata* on the *Principles of Philosophy* is the subject of Chapter 4 (Domenico Collacciani): by means of a comparison between this manuscript and De Raey's published works, this chapter reassess the common understanding of De Raey as a *novantiquus* and uncovers the comparatively greater freedom with which De Raey's private teaching explains Descartes's philosophy.

The teaching of Descartes's philosophy in the United Provinces was not just a university affair. In fact, as Chapter 5 (Henri Krop) shows, members of Spinoza's circle, who were heavily influenced by Cartesian philosophy, strove to make "Descartes speak Dutch," and to make his ideas available to a larger readership. At heavy costs, given that they were not eligible for teaching positions, a group of Dutch Spinozists translated Descartes's works into Dutch, centuries before they could be read in English or German. This chapter describes these translations and explains the motivations behind their production and publication and explains their appeal to an audience of Dutch readers.

Cartesian philosophy was soon to make its way from the university to the even more rarefied milieu of the court. This is nicely illustrated by Chapter 6 (Pietro Daniel Omodeo), which documents two controversies on Cartesian philosophy at the University of Frankfurt an der Oder. This chapter shows how the mathematics professor Johannes Placentinus (1628–1688) successfully managed to teach Cartesian philosophy thanks to the patronage of Prince Friedrich Wilhelm of Brandenburg.

Taken together, the last two chapters enable us to appreciate the socio-political setting of early modern Cartesianism. The pressures exerted, from both above and below, on university professors are also shown here in contexts —such as the German-speaking lands—which feature but rarely in most histories of Cartesianism.[16]

The next four chapters of the volume consider a varied network of figures who worked between Utrecht, Leiden and Nijmegen, often crossing each other's paths. Chapter 7 (Erik-Jan Bos) looks at Utrecht University and more specifically at the—Cartesian or anti-Cartesian—corollaries included in the disputations defended there between 1650 and 1670. This chapter reveals the

---

16   One very important exception is the case of Johannes Clauberg (1622–1665), which has
      been treated in the seminal studies by Trevisani, *Descartes in Germania* and Verbeek, ed.,
      Johannes Clauberg. More recent scholarship includes Hamid, "Domesticating Descartes."
      Other geographical areas have received even less attention, such as the Iberian Peninsula
      and Scotland. Luckily, recent studies have started making up for this. For example, see
      Gellera, "The Scottish Faculties of Arts and Cartesianism"; Golvers, "Note on Descartes's
      Entrance at the Coimbra Colégio das Artes."

role played by corollaries in circumventing the above-mentioned prohibitions on mentioning Descartes's name and in allowing professors to test views that they might have wanted to publish in the future.

The next three chapters of the volume consider figures who worked between the universities of Leiden and Utrecht and the comparatively less well-known Academy of Nijmegen, and who often crossed each other's paths. Some of these figures will be familiar to Descartes scholars, such as Burchard de Volder (1643–1709), whose classes and empirical teaching through experimental practice are documented in Chapter 8 (Andrea Strazzoni). By looking at student notebooks and *dictata*, this chapter uncovers de Volder's experiments in pneumatics and hydraulics, as well as their integration of ideas taken from Descartes, Archimedes, Robert Boyle (1627–1691), and Simon Stevin (1548–1620). Other figures have hitherto received less scholarly attention, such as Theodoor Craanen (1633–1688) and Gerard de Vries (1648–1705). Craanen is taken into consideration in Chapter 9 (Davide Cellamare), which examines two types of documents: a series of university disputations on the conjunction of soul and body, presided over by Craanen at Nijmegen, and the class notes of Craanen's medical courses in Leiden. These documents reveal how Craanen used his teaching on medical pathology to explain and to try to prove the central points of Descartes's anthropology. Chapter 6 and 7 fit nicely with the recently increased interest in the empirical and the medical sides of early modern Cartesianism.

Gerard de Vries's teaching in Utrecht is treated in Chapter 10 (Daniel Garber). While his teaching was thoroughly anti-Cartesian, this chapter shows that his life-long attempt to refute the fundamental principles of Descartes's philosophy resulted in de Vries's Utrecht students having an in-depth knowledge of Descartes's ideas—if only in order to refute them. This chapter explains the vicissitudes of de Vries's teaching by looking at a series of university disputations he devised as a historical introduction to the philosophy of Descartes and its position within the existing philosophical landscape.

From the United Provinces, the teaching of Descartes's philosophy soon spread out to the rest of Europe, for instance, across the English Channel. Chapter 11 (Igor Agostini) illustrates this very well, by describing the teaching given by Johannes Schuler (1619–1674) at the less well-known Orange College of Breda and its transmission to England via Cambridge. This chapter describes Schuler's activity as inspired by the ideal of the *libertas philosophandi* as well as the influence of Henry More's theory of spiritual extension. This chapter moreover explains the English reedition of Schuler's works as partly motivated by an attempt to further support More's theories in Cambridge.

England was also the context of a very special way of teaching Cartesian philosophy, within the university and without it, and one in which More figures prominently. Chapter 12 (Sarah Hutton) addresses More's teaching of Descartes's philosophy by looking especially at two manuscript sources: a notebook produced by a Cambridge student, Thomas Clarke; and More's correspondence with Anne Conway. In both instances, Hutton sheds light on the presence and use of the Descartes-More correspondence in More's own classroom and in his remote—epistolary—teaching.

The teaching of Descartes's philosophy across the Channel also ensured that some of the branches of Descartes's philosophical tree started to bloom. As mentioned above, a complete course in philosophy required a module on ethics—a topic upon which Descartes had only briefly touched in a number of works and letters. Chapter 13 (Roger Ariew) considers an anonymous textbook (1685) which circulated in London and Cambridge, that tried to build a Cartesian system of ethics out of these scattered remarks, and discussed it alongside scholastic ethics—notably that of Eustachius a Sancto Paolo (1573–1640)—with the intention of making it suitable for teaching.

While it was making its important inroads into England, Descartes's philosophy rapidly penetrated many other institutions in continental Europe. We paid special attention to those institutional contexts and geographical areas that have hitherto been less well studied. The fact that the diffusion of Descartes's philosophy through its teaching crossed national and denominational boundaries emerges very clearly from Chapter 14 (Christoph Sander), which studies a varied array of documents—mostly, for the first time—concerning the teaching of Descartes's views on magnetism, from the Italian peninsula to Scandinavia.

Despite their acknowledged role, the Southern Low Countries are often a less studied context when it comes to the teaching of Cartesianism.[17] Nevertheless, the next three chapters of this volume show that the case of Louvain is of crucial importance for understanding the early phases—as well as the institutional consolidation—of the new philosophy. It suffices here to recall the fact that Louvain—a university that styled itself as a Catholic stronghold—was in fact the first Catholic university to condemn Descartes's philosophy.

Chapter 15 (Mattia Mantovani) sheds new light on the Louvain condemnation of 1662, by examining the earliest teaching of Descartes's philosophy, and the responses it elicited between 1637 and 1671. The notebooks and disputations at our disposal reveal an intense, almost obsessive interest in Descartes's the-

---

17   As far as Flemish Cartesians are concerned, Arnold Geulincx (1624–1669) has been studied
       by (among others) Van Ruler, Uhlmann and Wilson, eds., *Arnold Geulincx*, Ethics; Sangi-
       como, "Defect of Knowledge."

ory of perception, down to the minutest details of his physiology. This chapter shows how much the *fortuna* of Descartes's theory of perception reveals about Descartes's own philosophical agenda and the reasons behind its acceptance, or rejection, in the academic milieu.

Just like Descartes's theory of perception, his views in physics were the subject of much debate at Louvain. This point is nicely illustrated by Chapter 16 (Carla Rita Palmerino), which examines twenty-five *dictata*, preserved and digitized by the ongoing *Magister Dixit* project at KU Leuven. These *dictata*, which were produced in Louvain between 1674 and 1776, are shown to have used engravings and drawings representing the Cartesian theory of tides. Thanks to a critical assessment of this material, this chapter reconstructs the teaching of Cartesian physics in seventeenth- and eighteenth-century Louvain, up to the last decades of the eighteenth century.

But Louvain was also an important junction of the international routes along which Cartesian philosophy travelled. Just as they moved between the United Provinces and their neighbouring countries, Descartes's ideas moved also between the Flemish and the French contexts. An example of this is offered by Chapter 17 (Steven Coesemans). On the basis of students' *dictata*, this chapter illustrates another way in which the new philosophy percolated into the Louvain curriculum, namely through the teaching of the so-called *Port-Royal Logic* by Antoine Arnauld (1612–1694) and Pierre Nicole (1625–1695)— first published anonymously in nearby Paris in 1662. In studying documents belonging to logic, this chapter addresses a discipline that—*pace* Descartes, who deemed it unnecessary—remained the first course that first-year students were expected to take.

A substantial literature supplies a detailed picture of the reception of Descartes's philosophy in France. The present volume enriches this picture by looking at the variety of formats and genres employed by French Cartesians in order to convey Descartes's ideas to a varied audience: not only by means of the usual university lectures in Latin, but also through private tutorials, public conferences in French, and witty dialogues with larger social and political implications.

Just as in England, so too in France, being excluded from university teaching, women learnt about Descartes's philosophy in other venues. Chapter 18 (Marie-Frédérique Pellegrin) opens with an examination of some sixteenth-century debates about whether philosophy could find a place within the traditional 'education of ladies.' The chapter then focuses on François Poulain de la Barre (1647–1723), who employed Descartes's ideas in order to argue that women are as suited as men—if not in fact better suited—to studying philosophy, precisely because they lack a training in scholastic philosophy.

Chapter 19 (Mihnea Dobre) analyses Jacques Rohault's (1618–1672) private lessons and conferences regarding cosmology, and reconstructs their relationship with Rohault's published work. In so doing, this chapter uncovers the implicit use of Cartesian principles in Rohault's teachings on cosmology, as well as Rohault's general attitude towards the discipline.

Chapter 20 (Tad Schmaltz) examines two additional textbook sources: Pierre-Sylvain Regis's *Système de philosophie* (1690) and Edmond Pourchot's *Institutio philosophica* (1695). His chapter shows that the former document is linked to an attempt to popularize Cartesian philosophy outside French universities, while the latter was more properly associated with an academic type of Cartesianism and bears important links to disputes internal to the University of Paris.

The chapters of this volume show how useful the study of the teaching of Descartes's philosophy can be in understanding both historical episodes that we thought we already knew and less-known events, contexts, and figures who turn out not to be unimportant after all. The diffusion of Cartesian philosophy in the early modern age, the resistance it faced, its rise and fall—all of this happened to a large extent through teaching; and teaching practice, so we show, was in fact one crucial part of a central episode in the scientific revolution. Thanks to today's increased access to teaching documents and the combined efforts of many researchers, we can now narrate this episode from a privileged perspective and take you, the reader of this volume, into the 'classrooms' where Descartes's ideas were first heard.

Davide Cellamare and Mattia Mantovani
April 2022

CHAPTER 1

Descartes and the Classroom

Theo Verbeek

1 Introduction

After Descartes left La Flèche, the only socially acceptable way for him to be found in a classroom would probably be as a Jesuit teacher. Whereas to be a Jesuit was something noteworthy (before and after Descartes, several members of his family were Jesuits), lay teachers belonged by definition to an entirely different class from that of an upstart noble family like Descartes's. For Descartes himself, on the other hand, the figure of the professor or the teacher, and by implication that of the professional philosopher, easily coalesces with that of the *paedagogus*, the pedant.[1]

2 Professionalism and Pedantry

Descartes keeps the letters of Isaac Beeckman (1588–1637) in case he ever wants an illustration of the "silly vainglory of a pedant" (*la sotte gloire d'un pédant*).[2] Gisbertus Voetius (1589–1676) is, according to him, not only a "mean and miserable tyrant," but also "the most barefaced pedant on earth" (*le plus franc pédant de la terre*).[3] More generally, the ignorance of those who are in favour of occult qualities, that is, the ignorance of virtually all scholastic philosoph-

---

1   *Paedagogus* is Descartes's Latin word for *pedant*: "Cumque interea se valde doctos putent, propterea quod multa ex iis, quae ab aliis scripta sunt, memoria tenent, et iis credunt, hinc *insulsissimam arrogantiam et vere paedagogicam* acquirunt." *Epistola ad Voetium* IV, AT VIII-B, 43 (my emphasis). Most dictionaries agree that the word does not occur before the sixteenth century, that it was first used in French and Italian, and that its origin is presumably the post-classical *paedagogans* ('acting as a pedagogue'), perhaps by association with the Italian *pedante* ('foot-soldier'); *Oxford English Dictionary*, sub voce. Apart from words like *ineptus, putidus, tetricus, fastidiosus*, etc., there is no exact equivalent of *pedant* in classical Latin.
2   Descartes to Mersenne, 4 November 1630, AT I 172.
3   Descartes to Mersenne, 11 November 1640, AT III 231, CSMK III 156; Descartes to Regius, [3 or 4 February 1642], AT III 510, Bos, "Correspondence," 113. Bos's edition completes, for what concerns the correspondence with Regius, the standard reference to Adam and Tannery (AT) and, whenever possible, the translation of Cottingham, Stoothoff, Murdoch and Kenny (CSMK). The dates of the letters are those established by Bos.

ers, is "haughty and pedantic" (*superba et paedagogica*), because they assume
that what they do not understand cannot be known by any human being.[4]
To Franciscus Burman (1628–1679) Descartes confides that in his *Meditations*
he avoided certain questions in order not to irritate the "pedants" (*paedago-
gos*) but, according to the same source, pedants also forced him to include
rules of provisional morality, "because otherwise they would have said that he
[Descartes] is without religion, without faith."[5]

Entirely in the tradition of Michel de Montaigne (1533–1592), Descartes
sees the *pédant* as someone with a humble background, who learned a few
easy tricks (*artes*) by which to distinguish himself. Learning made him *doc-
tus* but also arrogant, obstinate and irascible, whereas the true scholar, that
is, the *eruditus*, became "better and wiser."[6] Although this is clearly a matter
of how and what they were taught, or, in other words, a matter of education
(*cultura*), nature and character (*natura*) play an important role. Both the *doc-
tus* and the *eruditus* are the product of a subtle interplay between individual
nature and education: "those who have an inclination to the good, become
better and wiser through learning," whereas those who are already inclined
to evil become even worse through learning. Someone with a low and vulgar
nature is interested in knowledge only insofar as it serves his worldly ambitions, whereas someone with a good and noble nature would be interested only
in knowledge that makes him a better human being.[7] In the Low Countries
probably more than in France, Descartes learned that, assisted by scholarships
and private sponsors, a poor but gifted young man could become a minister
or a professor. Especially as a minister such a man could have authority, and
dominate; he could influence the masses and be a person with whom politicians would have to reckon.[8] As a result, the temptation of pedantry would
be enormous, whereas, according to Descartes himself, moral perfection and
authority are ultimately matters of birth—although it would be difficult to

---

4   Descartes to Regius, [3 or 4 February 1642], AT III 507, Bos, "Correspondence," 116.

5   *Conversation with Burman*, AT V 155–156, 178; for the rules of provisional morality, see *Discours*
    III, AT VI 22–28, CSMK I 122–125.

6   *Epistola ad Voetium* IV, AT VIII-B 43–44: "Le guain de nostre estude, c'est en estre devenu meil-
    leur et plus sage." Montaigne, *Essais* I, 26, in Montaigne, *Oeuvres*, vol. 1, 26. The difference
    between *doctus* and *eruditus* is almost entirely of Descartes' own making and can be justified
    only by relying on the etymology of these words: a *doctus* has been taught (*doceo*), and so
    passively absorbed pieces of knowledge, whereas the *eruditus* was "polished" (*erudio*), and so
    acquired good manners. See Verbeek, "Descartes's *Letter to Voetius*."

7   *Epistola ad Voetium* IV, AT VIII-B 44.

8   Descartes makes himself an echo of critiques often voiced (especially by politicians and intel-
    lectuals) against Dutch ministers in the seventeenth century; see Groenhuis, *De predikanten*,
    especially ch. 5.

explain that in terms of his own philosophy. The ideal condition for generosity is *la bonne naissance*: "there is no virtue so dependent on good birth as
the virtue which causes us to esteem ourselves in accordance with our true
value."[9] Apart from the fact that it would not be fitting for a nobleman like
Descartes to serve in any other capacity than that of a soldier, a priest or a
magistrate, the teacher and the pedant have values that a nobleman should
reject. In sum, it is hard to imagine Descartes in the role of a professional philosopher.

3   Cartesians and Professional Philosophy

The paradox of Dutch Cartesianism is that, with the exception of Lambert
Van Velthuysen (1622–1685), all its representatives were exactly that: university
teachers and professional philosophers.

Obviously, much of what Descartes says about teachers and professional
philosophers can be written off as social snobbery and class prejudice. It
belongs to his biography rather than his philosophy, which in any case cannot explain how "the souls God puts into our bodies could not all be equally
noble and strong."[10] Generosity—literally nobility (*generositas*)—as it is officially defined by Descartes, presupposes a distinction between human beings,
all of whom are naturally free, and other animals, none of which are, but not
a distinction between humans and other humans.[11] Much also can be written off as the result of personal frustration and anger. Beeckman is a pedant mainly because Descartes' relations with him were upset by a duplicitous
monk, Marin Mersenne (1588–1648); Voetius, a saintly bore anyway, is a shame-
less pedant because he is opposed to Descartes' philosophy and managed to
have it banned in Utrecht University. On the other hand, if Descartes neither
wanted to be a professional philosopher nor meant his philosophy to be a professional philosophy—that is, a philosophy that can be taught—he would have
little reason to be angry with the Jesuits because they obstinately ignored his
Meteorology.[12] His long and hopeless quarrel with Pierre Bourdin (1595–1653)

---

9   *Passions de l'âme* III 161, AT XI 453–454, CSMK I 388.
10   "[...] que toutes les âmes que Dieu met en nos corps ne sont pas également nobles et for-
     tes." *Passions de l'âme* III 161, AT XI 453, CSMK I 388.
11   On generosity, see Verbeek, "Generosity."
12   Descartes to Fournet, [October 1637], AT I 455; Descartes to Mersenne, 27 July 1638,
     AT II 267, CSMK III 118 (*philosophie* here means, not the discipline, but the *classes de philo-
     sophie*, that is the highest forms of a college); *Epistola ad P. Dinet*, AT VII 573, CSMK II 386.

over the *Dioptrique* would be pointless.[13] He would have no reason for frustration when the University of Utrecht banned his philosophy in 1642, nor when, a few years later (1648), in a bid to depersonalize the Cartesian controversy, the administration of Leiden University decided that his name must no longer be mentioned in public lectures and disputations.[14] In fact, the picture of *'Descartes in the Classroom'* raises a host of difficult questions, like, whether it was ever Descartes's intention that his philosophy should or could be taught and professionalised; whether there is anything in it that could prevent it from giving rise to a new scholasticism and to new forms of pedantry; whether, insofar as they were professionals, Cartesians could remain faithful to Descartes's ideas and ideals; what adjustments were necessary if they were to do what they had to do as professional teachers and examiners—questions moreover that were raised right from the beginning of Descartes' public career, if not perhaps in a very explicit way by Descartes and the Cartesians themselves, then certainly by their adversaries.[15] While other chapters of this volume deal with Cartesians, would-be Cartesians and anti-Cartesians, I limit myself to Descartes himself.

4   Descartes and Professional Philosophy

Was it ever Descartes' intention that his philosophy be taught? At first, probably not. In the early stages, when he is still working on what became *Meteorology*, he sets out to explain rainbows, parhelia and other meteorological phenomena, as an *échantillon de sa philosophie* ("a specimen of his philosophy") in such a way that "those who understand only Latin will find it a pleasure to read"[16]—in other words, he reaches out to anybody who knows Latin, that is, all those who, without necessarily being professors or teachers, had enjoyed an academic education, like lawyers and physicians. However, barely a month later, it is already Descartes's intention to expand his explan-

---

13   Sortais, "Le cartésianisme chez les Jésuites français"; Sortais, "Descartes et la Compagnie de Jésus"; Collaciani and Roux, "La querelle optique de Bourdin et de Descartes."

14   For Utrecht, see Verbeek, *La querelle d'Utrecht*. For Leiden, see Verbeek, *Descartes and the Dutch*, 46–47.

15   Lack of professionalism is one of the main objections to Descartes by Martin Schoock (1614–1669) in his *Admiranda methodus* (Utrecht, 1643); French translation in Verbeek, *La querelle d'Utrecht*, 153–320.

16   "C'est une des plus belles matières que je saurais choisir, et je tâcherai de l'expliquer en sorte que tous ceux qui seulement entendront le latin, puissent prendre plaisir à le lire." Descartes to Mersenne, 8 October 1629, AT I 23–24, CSMK III 6–7.

ation of *parhelia* and *meteores* into an "explanation of all natural phenomena, that is, an entire physics." The fact that his main preoccupation is now that presentation and form must "satisfy some without giving others an opportunity to contradict," is indicative not only of a different audience but also of a certain apprehensiveness on his part with respect to the world of learning.[17] Unlike an *échantillon* or an *essai* written, like those of Montaigne or Francis Bacon (1561–1626), for the instruction and amusement of cultivated gentlemen, a complete physics (*toute la physique*) belongs to a professional tradition that has always adopted strict rules of presentation. And although Descartes seems to be ready to reject these, he knows that it is according to those rules that he will be judged. In other words, he addresses a general audience and dissociates himself from an academic tradition but is ready to engage with academic adversaries, that is, with professional philosophers. Although the early correspondence contains no indication about the language in which Descartes' "complete physics" is to be written, its first part, which we know as *The World*, was written in French and ready to be sent to Mersenne no more than three years later, at the end of 1633. As a result, it is reasonable to suppose that, from the beginning, Descartes intended his book to be in French, which is not the language of professional philosophers. In any case, with his physics, Descartes ultimately addressed an even wider audience than with the very first version of the *Meteorology*—not only those who, without being professional philosophers, understood Latin, but all cultivated Frenchmen. Finally, the distinctive form he already has in mind seems to be the *fable* of the new world, as presented in chapters 6 and 7 of *The World*– a literary device that deviates strongly from the stylistic conventions of professional philosophy, not only because it explicitly appeals to the imagination but also because it allows Descartes to remain vague and noncommittal with respect to his principles.[18] In sum, Descartes' stance in relation to professional philosophy at this early stage is a mixture of aversion and attraction: he is eager to deal with the problems of philosophy, not only those of physics but even of metaphysics, and is ready to engage with professional philosophers;

---

17 "Car depuis le temps que je vous avais écrit il y a un mois, je n'ai rien fait du tout qu'en tracer l'argument, et au lieu d'expliquer un phénomène seulement, je me suis résolu d'expliquer tous les phénomènes de la nature, c'est-à-dire toute la physique. Et le dessein que j'ai me contente plus qu'aucun autre que j'aie jamais eu, car je pense avoir trouvé un moyen pour exposer toutes mes pensées en sorte qu'elles satisferont à quelques-uns et que les autres n'auront pas occasion d'y contredire." Descartes to Mersenne, [13 November 1629], AT I 70, CSMK III 7–8.

18 *Le monde* VI, AT XI 31, CSMK I 90. On the literary aspects of Descartes's early works, see Gilby, *Descartes's Fictions*.

but, apart from reaching out at an audience of non-professionals, he certainly does not see himself—nor does he want to be seen—as a professional philosopher.[19]

This position remains virtually unchanged until 1638. In the *Discourse*, Descartes still believes it would be "unbecoming" (*il n'y aurait point d'apparence*) for "a private man" (*un particulier*) to want a reform of "the body of the sciences or [of] the established order of teaching them in the schools." Like the State and the Church, schools and universities belong to the public order and are better left alone:

> These large bodies are too difficult to raise up once overthrown or even to hold up once they begin to totter and their fall cannot but be a hard one. Moreover, any imperfections they may possess—and their very diversity suffices to ensure that many do possess them—have doubtless been much smoothed over by custom; and custom has even prevented or imperceptibly corrected many imperfections that prudence could not so well provide against. Finally, it is almost always easier to put up with their imperfections than to change them, just as it is much better to follow the main roads that wind through mountains, which have gradually become smooth and convenient through frequent use, than to try to take a more direct route by clambering over rocks and descending to the foot of precipices.[20]

Reform is left to "meddlesome and restless characters" (*humeurs brouillonnes et inquiètes*), with whom Descartes does not want to be associated, his own project being only "to try and reform my own thoughts and *build on land that is entirely my own*."[21] Still, the fact that schools and universities are built, not on a privately owned piece of land, but on common ground, does not mean that

---

19   Apart from the small treatise on metaphysics written at Franeker (a *commencement de métaphysique*, Descartes to Mersenne, [April 1637], AT I 350, CSMK III 53; cf. Descartes to Mersenne, 25 November 1630, AT I 182, CSMK III 29), there are several early letters that show an interest in metaphysics. However, with regard to metaphysics Descartes seems to be even more apprehensive than with regard to his physics: "[...] je crois que vous m'aviez déjà ouï parler auparavant que j'avais fait dessein d'en mettre quelque chose par écrit; mais je ne juge pas à propos de le faire, que je n'aie vu premièrement comment la physique sera reçue." Descartes to Mersenne, 15 April 1630, AT I 144, CSMK III 22.

20   *Discours* II, AT VI 14, CSMK I 118.

21   "Jamais mon dessein ne s'est étendu plus avant que de tâcher à réformer mes propres pensées, et de bâtir *dans un fonds qui est tout à moi*." *Discours* II, AT VI 14–15, CSMK I 118 (my emphasis).

they cannot change at all. If custom prevented and corrected their imperfections in the past, it is reasonable to suppose that it could prevent and correct them in the future. As a result, even professional philosophy could change, not totally and suddenly perhaps, but gradually and under the pressure of circumstances—as Descartes must have realized was already happening. As compared to Thomas Aquinas (1225–1274) and even Francisco Suárez (1548–1617), Dutch professors of philosophy (as opposed to many professors of theology) taught a much reduced version of Scholasticism, slimmed down to a few scholastic terms and concepts, and flexible enough to incorporate newish elements. A close and intimate friend of Descartes, Henricus Reneri (1593–1639), professor of philosophy at Deventer and at Utrecht, was more or less working in this way. Not only did he heap praise on Aristotle, "the prince of the philosophers" (*philosophorum princeps*), and on colleagues "in neighbouring towns" (Leiden and Amsterdam),[22] who taught Peripatetic philosophy, "thus giving themselves a great reputation, being much applauded by their audience, with a remarkable grace of telling and teaching, and in sum with huge success," he also proposed a programme of empirical research, which although it has Baconian overtones, does not really deviate from the usual principles and methods of professional philosophy, and, as his disputations show, leaves room for small innovations, for which, moreover, he never claims absolute certainty.[23]

Again, Descartes' stance in relation to professional philosophy seems to change in the summer of 1638, when thanks to the clever lobbying of Reneri and other Utrecht friends, Henricus Regius (1598–1679) was appointed professor of theoretical medicine at Utrecht University. Descartes must have been well aware that Regius was not cast in the same mould as Reneri. Still, his reaction is not that of someone who does not care:

> Those people [Pierre Fermat (1607–1665) and Gilles Roberval (1602–1675), who had been critical of Descartes' optics and geometry] may do and say and write whatever they wish but I am determined to despise them. And in the end, if the French do me no justice, *convertam me ad gentes*.[24] [...] I can tell you that this very week I received a letter from a doctor I

---

22   Those philosophers were Franco Burgersdijk (1590–1635) and Caspar Barlaeus (1584–1648) respectively. On Burgersdijk, see Chapters 2 and 9, in this volume.

23   Reneri, "De lectionibus ac exercitiis philosophicis" (Utrecht, 1634), n.p.. On Reneri in general see Buning, "Henricus Reneri."

24   "Then Paul and Barnabas waxed bold, and said, 'It was necessary that the word of God should first have been spoken to you: but seeing ye put it from you, and judge yourselves unworthy of everlasting life, lo, *we turn to the Gentiles*.'" Acts 13: 46.

have never seen nor known, who nonetheless thanks me very warmly for making him a professor in a university where I have neither friends nor influence; but I was told that having taught privately [*en particulier*] some of the things published by me to the students over there, this was so much to their liking that they asked the magistrate to give them that professor. There are others here who teach my geometry without being instructed in it by myself, and still others who comment on it. And I write this to you so you may know that, if truth can't find a home in France, it cannot fail to find one elsewhere and that I can't really bother.[25]

Descartes exaggerates and is not entirely truthful. To be sure, in the United Provinces there were a few people who understood Descartes's geometry and taught it to others. Not all of them were academics. To the contrary, apart from Jacob Golius (1596–1667), professor of oriental languages and mathematics at Leiden, they were either competent non-academic professionals, like Jacob Waesenaer (d. 1682), a surveyor, and Jean Gillot (ca. 1613–1657), who became a military engineer; or amateurs, like Alphonse Pollot (1602–1668) and Godefroot van Haestrecht (ca. 1592–1659). Frans van Schooten (1581–1646), professor in the Dutch School of Mathematics at Leiden, was competent enough to be one of the arbiters in Descartes' proxy conflict with Stampioen (1610–1653), but it was only his homonymous son (1615–1660) and successor in the Leiden chair of mathematics (1646), who, by producing a Latin translation of Descartes' *Geometry* (1637) and especially by completing it with commentaries and additional explanations, more than ten years after the letter cited above, made it accessible to the academic world.[26] The *Dioptrics* attracted much attention, not only from amateurs like Constantijn Huygens (1596–1687), but also from pro-

---

25 "En effet, que ces gens-là fassent ou disent ou écrivent tout ce qu'ils voudront, je suis résolu de les mépriser. Et au bout du compte, si les Français me font trop d'injustice, *convertam me ad gentes*. Je suis résolu de faire imprimer bientôt ma version latine pour ce sujet, et je vous dirai que j'ai reçu cette semaine même des lettres d'un docteur que je n'ai jamais vu ni connu, et qui néanmoins me remercie fort affectueusement de ce que je l'ai fait devenir professeur en une université où je n'ai ni amis ni pouvoir; mais j'apprends qu'ayant enseigné en particulier quelque chose de ce que j'ai fait imprimer, à des écoliers de ce lieu-là, ils y ont pris tel goût qu'ils ont tous prié le magistrat de leur donner ce professeur. Il y en a d'autres aussi qui enseignent ma géométrie, sans en avoir eu de moi aucune instruction, et d'autres qui la commentent. Ce que je vous écris, afin que vous sachiez que, si la vérité ne peut trouver place en France, elle ne laissera peut être pas d'en trouver ailleurs, et que je ne m'en mets point fort en peine." Descartes to Mersenne, 23 August 1638, AT II 334–335.

26 *Geometria à Renato Des Cartes anno 1637 Gallice edita, nunc autem cum notis* [...] *Francisci à Schooten* (Leiden, 1649). New editions were published in 1659, 1661 and 1683 (a German edition in 1695). See Van Otegem, *A Bibliography of the Works of Descartes*, vol. 1, 114–145.

fessionals like Golius and Franco Burgersdijk; a Latin translation, which made it more accessible to the wider academic world, would not be published before 1644, as part of a translation of the *Discourse* and the *Essays*, which eventually, together with the *Principles of Philosophy*—published in the same year and by the same publisher—would be widely used in lecture courses.[27] Moreover, Descartes did have friends at Utrecht, not only Reneri, but also Gijsbert van der Hoolck (d. 1680), one of the two burgomasters, who, more or less singlehandedly, ran the university. And although we do not know much about the students at Utrecht in this period, it was not the students who asked the magistrate to have Regius as a professor but the senate of the university, who probably used his success with the students as an argument in his favour. Finally, the fact that Regius became a professor and that Dutch academics were interested in Descartes' philosophy and mathematics obviously does not mean that truth found a definitive home in the United Provinces, as Descartes would experience himself a few years later: not only was Utrecht University the first in the world where Descartes' philosophy was taught; it was also the first where it was banned. Still, the fact that some of Descartes' ideas received some approval from some professors at some Dutch universities—unlike at French universities, which until the beginning of the eighteenth century remained closed to the new ideas—did entail a partial, albeit strictly local, reform of "the scholastic order" (*l'ordre établi dans les écoles*). Indeed, the very fact that in the United Provinces mathematics enjoyed academic respectability, which was a recent phenomenon, could be seen as a modest sign that scholastic reform was possible. Given the fact that the only application of physics was in medicine, the nomination of a Cartesian professor of medicine could easily be seen as the first indication of a more general reform of "the body of the sciences and the scholastic order." Descartes had the proofs of it under his eyes, not only in Utrecht, but also in Leiden, where in the same period students and professors in the medical faculty identified the circulation of the blood as a new medical paradigm.[28] Finally, Descartes' enthusiasm for what happened in

27   "Descartes is on familiar terms with Burgersdijk and Golius. The first claims that from (Descartes's) few pages on optics he learned more than from all the other books he had read. This Descartes is a rich nobleman, who prefers obscurity." Hartlib, *Ephemerides* (late 1634), in Hartlib, *The Hartlib Papers*, 29/2/44–55. For all details on the translation, see Corinna Vermeulen, "René Descartes, *Specimina philosophiae*: Introduction and Critical Edition."

28   Descartes to Regius, 24 May 1640, AT III 69, Bos, "Correspondence," 43–44. Descartes mentions the names of Franciscus de le Boë—Sylvius (1614–1672), then still a private lecturer at Leiden University, and Franciscus van der Schagen (ca. 1615–1673). Both were students of Johannes Walaeus (1604–1649), mentioned in the same letter (AT III 66, Bos, "Corres-

Utrecht and elsewhere in the United Provinces was undoubtedly also motivated by bitter disappointment over the reception of his ideas in France, where, again, Mersenne had managed to upset his relations with his fellow countrymen, particularly Fermat, and where the Jesuits showed themselves reluctant to embrace his ideas. However, neither Descartes's bitterness about what failed to materialize in France, nor his jubilation about breakthroughs at a few Dutch universities are fully explicable if he did not mean his philosophy to be taught and professionalised. In fact, his letters to Regius go far beyond expressions of gratitude and sympathy, but testify to an intense desire to meddle in Regius' teaching and to control his publications. To be sure, if Descartes found it a challenge, it was not only because he had to leave the defence of his ideas to someone of whose presentation he could not approve.[29] It was also because a professor has "the duty to have an opinion on whatever pertains to the subject of medicine," whereas, up to that point, he had been free to follow his own interests.[30] In any case, until the summer of 1641, that is, until the completion of the general part of Regius' *Physiologia*, Descartes reviewed, commented and corrected each and every disputation submitted by Regius and wanted to be sure that a correct version of his ideas was presented in the classroom.[31]

As far as we know, the episode with Regius is the only time Descartes has been actively involved, albeit at a distance, in university teaching. Despite the fact that he had been close with Reneri, and that Reneri was very active in propagating Descartes' fame, there is no evidence that Descartes ever tried to

---

pondence," 41), who, having criticized Harvey, later became convinced of the truth of his discovery by the demonstrations of Sylvius.

29 "[...] me multa legisse in tuo compendio physico, a vulgari opinione plane aliena, quae nude ibi proponuntur, nullis additis rationibus, quibus lectori probabiliora reddi possint ..." Descartes to Regius, [April 1641], AT IV 249, Bos, "Correspondence," 57.

30 "[...] longe enim difficilius est, de omnibus quae ad rem medicam pertinent suam sententiam exponere, quod docentis officium est, quam cognitu faciliora seligere, ac de reliquis prorsus tacere, quod ego in omnibus scientiis facere consuevi." Descartes to Regius, [November 1641], AT III 443, Bos, "Correspondence," 87. Cf. Descartes to Regius, [early May 1641], AT III 371, Bos, "Correspondence," 64: "[...] tantum video novum opus mihi imponi, quod nempe homines inde sint credituri, meas opiniones a vestris non dissentire, atque adeo ab iis quae asseruistis, pro viribus defendendis, me imposterum excusare non debeam; et tanto diligentius ea quae legenda misisti debeam examinare, ne quid in iis praetermittam, quod tueri recusem."

31 Regius, *Disputatio medico-physiologica pro sanguinis circulatione* (Utrecht, 1640); Regius, *Physiologia sive cognitio sanitatis, tribus disputationibus in Academia Ultrajectina publice proposita* (Utrecht, 1641); actually, six disputations, each of the three being divided into two parts. In September 1641 the *Physiologia* was continued, now as a series on disease. After a temporary suspension, it was completed in 1643; for an overview, see Bos, "Correspondence," 195–196.

intervene in his teaching or to play a role in editing his disputations.[32] The episode with Regius also seems to be the first time that Descartes realizes that a teacher, and by implication a professional philosopher, is not free to select his own problems; that he inherits a programme he cannot easily change; that there are things the public expects from him; in sum, that, whether he likes it or not, he is always, albeit partially, tributary to a tradition. It may have been the first time Descartes realized that it is impossible to do philosophy in a professional way without leaving one's own "piece of land."[33] In any case, even though he remains interested in what happens at Dutch universities, to the point of allowing a young man from Leiden to interrogate him on his work, we have no evidence of any further active involvement in university teaching. From now on, he seems more interested in the reception of his ideas by enlightened and cultivated 'amateurs,' like Claude Clerselier (1614–1684), Pierre Chanut (1601–1662), Princess Elizabeth (1618–1680), and Queen Christina (1626–1689).

5   Problems

Again, all this seems to belong to Descartes' biography rather than his philosophy, the more important question being, of course, whether, regardless of Descartes' private ambitions, his philosophy can be taught and professionalised at all, and, if so, whether there are any obstacles that must be surmounted, and whether anything essential was sacrificed by those who did teach it and thereby created a new professional tradition. Here we must make a distinction which, from Descartes' own point of view, would already be a sacrifice. Regius, for example, accepted and successfully taught much of Descartes' physics but, as we already saw, Descartes did not approve of his presentation. In other words, the example of Regius shows that, although Descartes certainly did not like it, a distinction could be made, and actually was made, between what for the sake of convenience I shall call 'contents' (concepts, theories,

---

32   There is some evidence that, occasionally, Reneri consulted Descartes; see, for example, Descartes to Reneri, 2 June 1631, AT I 205–208 (although at that point Reneri was not yet a professor); Descartes to Reneri, 2 July 1634, AT I 300–302. According to Schoock, Reneri was "the main herald of [Descartes's] fame" (*famae ejus praecipuo buccinatore*), *Admiranda methodus*, *Praefatio*, n.p.; Verbeek, *La querelle d'Utrecht*, 370. Posthumously, Reneri certainly contributed to Descartes' fame insofar as the funeral oration on his death turned into an elaborate eulogy to Descartes; Aemilius, *Oratio in obitum clarissimi et praestantissimi viri Henrici Renerii* (Utrecht, 1639), also in his *Orationes* (Utrecht, 1651), 105–124; cf. Descartes, *Lettre apologétique aux magistrats d'Utrecht*, AT VIII-B 203–204.

33   *Discours* II, AT VI 15 (see above).

explanations) and 'form' (presentation, method). Regius, in any case, but many other Cartesians as well, assumed that, by making that distinction, they did not lose anything essential. But is this assumption correct?

Even at an early stage, Descartes realised that his method, his general approach, his critical attitude towards tradition, not to mention his particular theories and explanations, could create problems, with respect not just to traditional philosophy but to theology as well. As he says in the *Letter to Dinet*, "the experience of many years has taught us that traditional and ordinary philosophy is consistent with theology, while it is uncertain whether this will be true of a new philosophy."[34] The examples of Cornelius Jansenius (1585–1638) and later Antoine Arnauld (1613–1694), showed that even to replace the authority of one Catholic saint, Thomas Aquinas, by that of another, Augustine, was not without danger. Moreover, from an early stage, Descartes had been acutely aware that in Roman Catholic countries his philosophy created problems with respect to the traditional conceptualization of the Eucharist and, although he was confident that he could solve them, we know that eventually these problems proved insurmountable.[35] In Reformed Calvinist countries like the United Provinces, the problems were different and of a greater diversity. In a way, Descartes arrived at the wrong moment and started in the wrong place. Unlike other Dutch universities, which were governed by a committee of the Provincial States, Utrecht University, the first where Descartes' philosophy was taught, was governed by the burgomasters of the town, who were probably more susceptible to local pressure than the administrators of Leiden University, for example, who were based in The Hague and only assembled in Leiden a few times a year to meet for business. Certainly on a local level, Voetius, Descartes' main opponent at Utrecht, enjoyed popular influence and, as a result, much informal power as a minister of the Church of Utrecht.[36] More importantly, the type of orthodox theology represented by Voetius, and in Leiden by Jacobus Revius (1586–1658), which relied heavily on Scholastic philosophy, reigned supreme only for a short period and, already in Descartes's

---

34   *Epistola ad P. Dinet*, AT VII 579, CSMK II 390.

35   Descartes realized that as early as 1630, when, working on a *discours* on "the nature of colours and light," that he had to explain "how the whiteness of the bread remains in the Holy Sacrament." Descartes to Mersenne, 25 November 1630, AT I 179, CSMK III 28. Probably as a result of his explanation of the Eucharist (in the *Fourth Replies* and the correspondence with Mesland), his works were placed on the *Index librorum prohibitorum* on 29 November 1663, "donec corrigantur."

36   In the *Epistola ad Voetium*, Descartes seeks to question the legitimacy of the use of this informal influence, emphasising the social and political effects of Voetius' preaching; see especially *Epistola ad Voetium* VII, AT VIII-B 107–135.

own lifetime, lost much of its influence to the 'federal theology' of Johannes Coccejus (1603–1669)—a theology which made little use of philosophy and allowed a strict separation between philosophy and theology.[37] It is important to emphasise that Coccejus and the Coccejans were perfectly orthodox insofar as they subscribed unreservedly to the confessional basis of the Dutch Reformed Churches as fixed by the Synod of Dordrecht (1619). It is no coincidence, therefore, that the majority of Dutch Cartesians, all of whom were orthodox Calvinists, embraced Coccejan theology. Still, despite their aversion to Scholasticism, even Coccejan theologians could be uneasy about certain aspects of Descartes' philosophy, particularly his metaphysics and method. The fact of the matter is that it proved to be perfectly possible to be a Cartesian while ignoring Descartes' metaphysics or paying minimal heed to it. In Leiden, Cartesians were even forced to do so because the administration of the university repeatedly proscribed the teaching of metaphysics in any form, whether Cartesian or Aristotelian.[38] Most teachers of Cartesian medicine, moreover, did not worry at all about metaphysics and very little about epistemology, but used the circulation of the blood as a new medical paradigm. Whereas Descartes still deplored the relative independence of his physiology, which ideally should have the same sort of foundation as his general physics, they simply used the idea that the animal body is a mechanical machine as a heuristic hypothesis, thus, paradoxically, detaching medicine from physics—questions like whether there is a void, or how secondary qualities relate to primary qualities, or, for that matter, what the exact relation is between mind and body evaporated, so to speak, insofar as they were not immediately relevant to physiology or medicine.[39] On the other hand, Cartesian physicians were free to incorporate non-

37   Johannes Coccejus (Johannes Coch) was professor of Biblical philology at Bremen (Germany) and of oriental languages at Franeker, before being professor of theology at Leiden (from 1650), where he died from the plague in 1669. On Coccejus' theology, see Van Asselt, *The Federal Theology of Johannes Coccejus*.

38   There were ways to circumvent those interdictions: Geulincx taught a semi-Cartesian metaphysics of his own, and Wittich must have given a seminar on Descartes' *Meditationes*; Wittich, *Annotationes ad Renati Des-Cartes* Meditationes (Dordrecht, 1688). In any case, the administration usually did not intervene unless they were forced to do so.

39   "De la description des corps inanimés et des plantes, je passai à celle des animaux et particulièrement à celle des hommes. Mais, *parce que je n'en avais pas encore assez de connaissance*, pour en parler du même style que du reste, c'est-à-dire, en démontrant les effets par les causes, et faisant voir de quelles semences, et en quelle façon, la nature les doit produire, *je me contentai* de supposer que Dieu formât le corps d'un homme, entièrement semblable à l'un des nôtres, tant en la figure extérieure de ses membres qu'en la conformation intérieure de ses organes, sans le composer d'autre matière que

mechanical explanations, taken from chemistry, for example, without being untrue to the idea that the animal body is a machine—in the Netherlands, at least, all Cartesian physicians were highly influenced by Sylvius (1614–1672), who leaned heavily on chemistry and was a Baconian in his theory of science. In sum, despite the fact that Descartes saw his philosophy as an indivisible whole, it proved perfectly possible to detach physics and physiology from their meta-physical basis.[40]

However, this did not solve what, for the sake of convenience, I called prob-lems of form, that is, problems of presentation and method. Teachers must have found these a challenge, also because, according to Descartes himself, present-ation and method are intimately related to contents. For not only does he see metaphysics as the foundation of physics, without which all certainty would be lost; metaphysics could also, in his view, be grasped and demonstrated only by what he calls—a bit inappropriately and perhaps misleadingly—*analysis*, which he contrasts with what he calls, also somewhat inappropriately and cer-tainly misleadingly, *synthesis*:

> Analysis shows the true way by means of which the thing in question was discovered methodically and as it were *a priori*, so that if the reader is will-ing to follow it and give sufficient attention to all points, he will make the thing his own and understand it just as perfectly as if he had discovered it for himself. But this method contains nothing to compel belief in an argumentative or inattentive reader; for if he fails to attend even to the smallest point, he will not see the necessity of the conclusion. [...] Syn-thesis by contrast, employs a directly opposite method where the search is, as it were, *a posteriori* (though the proof itself is often more *a priori* than it is in the analytic method). It demonstrates the conclusion clearly and employs a long series of definitions, postulates, axioms, theorems and problems, so that if anyone denies one of the conclusions it can be shown at once that it is contained in what has gone before, and hence the reader, however argumentative or stubborn he may be, is compelled to give his assent.[41]

---

de celle que j'avais décrite, et sans mettre en lui, au commencement, aucune âme raison-nable, ni aucune autre chose pour y servir d'âme végétante ou sensitive, sinon qu'il excitât en son coeur un de ces feux sans lumière ...." *Discours* V, AT VI 45–46, CSMK I 134 (my emphasis).

40   On the relationship between Cartesian physiology and medicine, as well as the combina-tion of Cartesian medicine and Sylvian chemistry, see Chapter 9, in this volume.

41   *Meditationes, Responsiones* II, AT VII 155–156, CSMK II 110–111.

According to Descartes, *synthesis* is a method suitable for geometry, because "the primary notions which are presupposed for the demonstration of geometrical truths are readily accepted by anyone, since they accord with the use of our sense," whereas in metaphysics "there is nothing which causes so much effort as making our perception of the primary notions clear and distinct." And this would be the reason why Descartes wrote "*meditations*, rather than *disputations* as the philosophers have done, or *theorems and problems* as the geometers would have done." The 'primary notions' of metaphysics would be such that, "if they were put forward in isolation, they could easily be denied by those who like to contradict just for the sake of it."[42] After all that has been said and written about these texts, and more particularly about the question whether in the *Meditations* Descartes actually uses the method he describes here as *analysis*, I may perhaps be forgiven for saying nothing on the subject and if I instead draw attention to the fact that what lurks behind the method here described as *synthesis*, is actually dialectic, and that one of the favourite forms of teaching—or in any case, the form in which the results of teaching were presented and tested—was precisely the disputation, the form of which Descartes was very critical indeed: "I never [...] observed that any previously unknown truth has been discovered by means of the disputations practised in the schools."[43] In fact, he claimed, nothing is as harmful for the right use of reason as precisely the kind of dialectical argument used during a disputation:

> Nor [...] does the true use of reason, in which all learning [*eruditio*], all good sense [*bona mens*], all human wisdom, is contained, consist in disconnected syllogisms, but only in a careful and accurate comprehension of whatever is required for the knowledge of the truths one is looking for. And given the fact that these can hardly be expressed in syllogisms, unless one takes a great number of them together, it is certain that those who use them in isolation almost always forget a part of the things that should be considered together and so become used to paying little attention and unlearn [*dediscere*] good sense [*bona mens*].[44]

---

42   *Meditationes, Responsiones* II, AT VII 156–157, CSMK II 111–112.

43   *Discours* VI, AT VI 69, CSMK I 146.

44   "Neque enim, ut scias, verus ille usus rationis, in quo omnis eruditio, omnis bona mens, omnis humana sapientia continetur, in disjunctis syllogismis consistit, sed tantum in circumspecta et accurata complexione eorum omnium, quae ad quaesitarum veritatum cognitionem requiruntur. Et cum haec vix unquam possint exprimi syllogismis, nisi multi simul nectantur, certum est eos, qui tantum disjunctis utuntur, fere semper aliquam partem eorum, quae simul spectanda sunt, omittere, sicque assuescere inconsiderantiae, ac dediscere bonam mentem." *Epistola ad Voetium* IV, AT VIII-B 45.

Accordingly, far from being only inappropriate as a method for metaphysics, the effect of synthesis, which is actually what Descartes is describing here, would be the destruction of good sense, *bona mens*, that is, the *bon sens*, or "the power of judging well and of distinguishing the true from the false," which according to the *Discourse* would be what makes us truly human.[45] In other words, one of the most common exercises of seventeenth-century schools and universities schools would, in the end, be utterly destructive of good sense and make students unfit to understand Cartesian philosophy.

The second favourite form of teaching, on the other hand, was the commentary—until the end of the eighteenth century (and even beyond) undoubtedly the form most used in lecture courses. But on this, Descartes is almost as dismissive as on disputations and dialectic. Commentators would be "like ivy, which never seeks to climb higher than the trees which support it." They

> take downward steps, or become somehow less knowledgeable than if they refrained from study, when, not content with knowing everything which is intelligibly explained in their author's writings, they wish in addition to find there the solution to many problems about which he says nothing and about which perhaps he never thought.[46]

If, *building on his own land*, Descartes tried to replace these traditional forms by new techniques, like meditation and doubt, his example was not much followed. Regius, for instance, completely ignores Descartes' metaphysics and certainly does not see it as the foundation of his physics. Like Reneri and Sylvius, and even to a certain extent Adriaan Heereboord (1613–1661), he seems to be more in favour of a vaguely Baconian theory of science.[47] As to the Cartesian method of doubt, while it perhaps worried some anti-Cartesians and certainly orthodox theologians, Cartesians were either silent about it, or trivialized and reinterpreted it. Indeed, it is hard to see how doubt and meditation could be practised in any seventeenth-century classroom.

6   Conclusion

My conclusion is threefold. First of all, despite the fact that Descartes saw his philosophy as an indivisible whole, Cartesians showed that it is possible to

---

45   *Discours* I, AT VI 1–2, CSMK I 111.
46   *Discours* VI, AT VI 70, CSMK I 147.
47   See Verbeek, "The Invention of Nature." On Heereboord, see Chapters 2 and 3, in this volume.

dismember it. Despite the *cogito* and the metaphysics, Descartes had, unwittingly, planted in his philosophy a centrifugal force, which made it possible to separate medicine and moral philosophy from physics, and physics from metaphysics. Secondly, the professionalization, and in fact the academic survival, of Cartesian philosophy, which I believe was the historical achievement of the Cartesians, was possible only because each of the parts of Cartesian philosophy could follow its own path regardless of the others. Thirdly, the only part (if one may call it a part) of Cartesian philosophy that could not be incorporated was its form. That problem was solved either by dropping Descartes' requirements altogether or by introducing a different didactic system.

CHAPTER 2

The Philosophical Fulcrum of Seventeenth-Century Leiden: Pedagogical Innovation and Philosophical Novelty in Adriaan Heereboord

Howard Hotson

Today, Adriaan Heereboord (1613–1661) is remembered primarily for his role in pioneering the introduction of Cartesianism into the Dutch universities.[1] The "academic breakthrough of Cartesianism in the Republic," to quote Wiep Van Bunge, centred above all on "the great concentration of Cartesians in Leiden"; and "what made the reception of Cartesianism in Leiden unique was mainly the person of Adriaan Heereboord."[2] Yet any survey of his *oeuvre* immediately reveals a paradox: the overwhelming majority of Heereboord's philosophical writings were devoted to expounding the old philosophy of Aristotle.[3] For those looking for a purely philosophical explanation of the advent of Leiden Cartesianism, this conservatism may seem puzzling. In order to understand it, two things must be acknowledged. The first is that even Heereboord's philosophically conservative early writings were pedagogically innovative (section 2). The second is that his early pedagogical innovation provided an indispensable preparation for his later embrace of philosophical novelty (section 3). A reappraisal of Heereboord must therefore begin by placing the familiar philosophical data in a less familiar pedagogical context (section 1).

1 Philosophical Pedagogy in the Pre-Cartesian Era

During the late sixteenth and early seventeenth centuries, the form in which philosophy should be transmitted was as contested in Protestant universities as

---

1 This paper derives substantially from Howard Hotson, *The Reformation of Common Learning: Post-Ramist Method and the Reception of the New Philosophy, 1618–c.1670* (Oxford, 2020). This material is presented here in modified form by permission of the Oxford University Press.

2 Van Bunge, "Philosophy," 306. See also Sassen, "Heereboord"; Verbeek, *Descartes and the Dutch*, 34, "the introduction of Cartesianism in Leiden was due to a professor of logic, Adriaan Heereboord." On Heereboord, see also Chapter 3, in this volume.

3 Van Bunge is of course well aware of this: "Philosophy," 301, citing Dibon, *La philosophie néerlandaise*, vol. 1, 116–119. Heereboord's only semi-Cartesian work, as we shall see, was published posthumously.

© HOWARD HOTSON, 2023 | DOI:10.1163/9789004524897_004

the substance of philosophy itself. The main pedagogical dispute of the period pitted the successors of Petrus Ramus (1515–1572) against more authentically humanist pedagogues. A brief rehearsal of this contest is required, because the tradition which emerged from Ramism has been consistently maligned, misrepresented, and misunderstood ever since the sixteenth century, including by its most influential twentieth-century interpreter, Walter J. Ong, S.J.[4]

What links Ramus with the tradition descending from him was the stress on educational efficiency and utility. The common aim which bound Ramus and the post-Ramist tradition together was the determination to transmit useful knowledge as quickly and effectively as possible. The principal means to that end was an evolving set of basic principles of exposition, tested and refined in a constant stream of new textbooks.[5] It is these principles, and the textbooks based on them, which set post-Ramist practice apart from more authentically humanist pedagogy. Although Ramism is rightly regarded as an offshoot of Renaissance humanism, the use of such textbooks contravened the fundamental principle of humanism: namely, the determination to teach the arts curriculum primarily by immersing students in the unmediated study of the ancient texts. As academic humanism reached the acme of its prestige in the final decades of the sixteenth century, Ramus's reforms were therefore rejected in most leading Protestant universities and academies in favour of the prestigious but painstaking practice of immersing students in the Greek text of Aristotle.[6] By 1623, no less a figure than Francis Bacon (1561–1626) could describe Ramism as "a kind of cloud that overshadowed knowledge for a while and blew over."[7] Many contemporary humanists triumphantly concluded that Ramism had been permanently banished from the republic of letters.[8]

Meanwhile, however, Ramism had mutated into a much more virulent pedagogical strain in an environment very different from Bacon's England. Within the fragmentary political and confessional landscape of the Holy Roman Empire, Ramus's streamlined pedagogy appealed strongly to sub-university institutions in two of the humbler strata of imperial polities; and it was these

---

4   Ong, *Ramus, Method and the Decay of Dialogue*. The following notes refer the reader to my own account of this prehistory, as well as some of the key pieces of literature upon which it draws.

5   Skalnik, *Ramus and Reform*; Hotson, *Commonplace Learning*, 38–51.

6   Hotson, *Commonplace Learning*, 51–68.

7   Bacon, *De augmentis scientiarum* (London, 1623), vol. 6, 2, in Bacon, *Works*, eds., Spedding, Ellis and Heath, vol. 4, 448–449: "Fuit enim nubecula quaedam doctrinae, quae cito transit."

8   For other examples, see Hotson, *Commonplace Learning*, 283.

institutions which drove the next two generations of educational innovation across the Protestant world. First, from around 1570 onward, teachers in the civic *gymnasia* of Hanseatic cities applied Ramist methods across the established arts curriculum.[9] Secondly, from the mid-1580s onward, the *gymnasia illustria* in mostly Reformed imperial counties introduced new ways of mixing Aristotelian philosophical content back into this Ramist pedagogical matrix and spread its application throughout the higher faculties as well.[10] Students graduating from these humble institutions then precipitated a crisis by demanding equally clear and useful instruction in full universities. The classic response to this crisis was formulated in Heidelberg and his native Danzig by the Reformed philosopher, Bartholomäus Keckermann (ca. 1572–1608).[11] In 1599, he derived—largely from the writings of the leading Paduan Aristotelian, Jacopo Zabarella (1533–1589)—a new "systematic" method of expounding essentially Aristotelian philosophical substance in quasi-Ramist pedagogical form. In the decade that remained to him before his premature death in 1609, he wrote a comprehensive series of textbooks—on logic and rhetoric, physics and metaphysics, geography and astronomy, ethics and politics—designed to replace the cumbersome 'textual Peripateticism' of the humanists with a much more efficient 'methodical Peripateticism' which retained the doctrine of Aristotle but abandoned his text as the basic vehicle of philosophical instruction.

The reception of Keckermann's pedagogical '*systema*', and the 'methodical Peripateticism' resulting from it, was immediate and almost universal: within a decade of his death, Keckermann's methods had become the standard method of teaching philosophy in most Protestant universities, Lutheran as well as Reformed.[12] But a few universities stood their ground, none more tenaciously than Leiden.[13] Three years after the arrival of Justus Lipsius (1547–1606) in 1579, the essentially Ramist curriculum originally proposed for Leiden was swept aside along with the use of textbooks for formal teaching in the arts fac-

---

9   An excellent example is Korbach, on which see Friedrich, *Die Gelehrtenschulen*; and, more generally, Hotson, *Commonplace Learning*, 25–37, 68–89.

10   The example *par excellence* is Herborn, which served the county of Nassau-Dillenburg in the first instance, and the broader network of Wetterau counties more generally. See Menk, *Die Hohe Schule Herborn*; Hotson, *Commonplace Learning*, 90–126.

11   Freedman, "The Career"; Hotson, *Commonplace Learning*, 135–165.

12   Hotson, *Commonplace Learning*, 157–161, 169–186; Hotson, *Reformation of Common Learning*, 32–41.

13   Another example is Basel, on which, see Hotson, *Reformation of Common Learning*, 153–155.

ulty. After Lipsius's departure for Louvain, Joseph Justus Scaliger (1540–1609) assumed the task of vetoing intermittent attempts to re-established Ramism and everything associated with it. The result was nearly a half-century in which humanist philosophical pedagogy reigned supreme in Leiden, despite the presence of Rudolph Snellius (1546–1613), who would have emerged as Europe's leading Ramist were it not for repeated humiliations by the university curators and the great scholars they appointed.[14]

In the years after 1618, four interrelated shockwaves temporarily loosened the grip of humanist pedagogy on Leiden's philosophical teaching.[15] The expulsion of Remonstrants deprived Leiden of some of its leading scholars. A reassessment of pedagogical priorities followed the Synod of Dordrecht (1618–1619), which gave rise to a new generation of textbooks for Holland's schools. Meanwhile, the destruction of the German Reformed academic world engulfed the Dutch universities in a tidal wave of central and east-central European students accustomed to Ramist and post-Ramist teaching methods. This brief period of radical instability also saw the appointment in 1620 of Franco Burgersdijk (1590–1635), who moved Leiden from the extreme margin to the centre of the unfolding post-Ramist tradition by applying Keckermann's method to a series of textbooks on logic, physics, and ethics which became standard throughout the Reformed world during the second quarter of the century.[16]

Given this widespread success, one might imagine that Burgersdijk's legacy was secure. But not in Leiden. In 1640, acting on the pretext that teaching standards had declined since his death in 1635, the university curators tasked a commission to determine the manner in which philosophy was henceforth to be taught in Leiden. Far from canonizing Burgersdijk's post-Ramist pedagogy, the commission completely repudiated it.[17] All philosophy lectures, they decreed, were henceforth to consist of *verbatim* commentaries on the Greek text of Aristotle, explained where necessary by recourse primarily to the ancient Greek commentators, without consideration of questions not derived from Aristotle's text. As textual support, the university printers were to provide affordable octavo editions of Aristotle's *Organon*, physics, and *Nicomachean Ethics*, with the original Greek and a Latin translation supplemented only by brief synopses of individual books and chapters.[18] Of the seven commissioners,

---

14   Hotson, *Reformation of Common Learning*, 30–59.

15   For the following, see Hotson, *Reformation of Common Learning*, 101–118.

16   Bos and Krop, *Burgersdijk*; Hotson, *Reformation of Common Learning*, 118–125.

17   On the changing use of Burgersdijk's works in Leiden, see Chapter 9, in this volume.

18   "Ordo, secundum quem deinceps in Academia Leidensi Philosophia docebitur," 8 August
      1641 (Molhuysen, *Bronnen*, vol. 2, 331–333*): "Generale esto praeceptum, ut ipse tex-

the only one to withhold his signature from the draft regulations of 8 August 1641 to this effect was the 27-year-old Heereboord, who had been appointed as Burgersdijk's indirect successor the previous year.

2   Heereboord and the New Pedagogy

On the cause of this dramatic reversal of educational methods, there is no consensus in the literature. Paul Dibon argued that the commission's reactionary proposals were intended to counteract "the penetration of the works and handbooks of Spanish neo-scholasticism," which worried Leiden's theologians; but this interpretation was purely speculative.[19] Edward G. Ruestow referred these proposals to the period "when the troubles [with Cartesianism] at Utrecht were mounting," but this seems premature: those troubles can be dated from the last of the second set of disputations in Utrecht by Henricus Regius (1598–1679), held between 24 November and 8 December 1641, several months *after* the commission's new regulations.[20] Heereboord himself blamed intrigue by the rector, Otto Heurnius (1577–1652), "a staunch Aristotelian," Theo Verbeek adds, "who bitterly opposed the theory of the circulation of the blood."[21] But none of these three explanations explains the key feature of these recommendations: to base philosophical instruction in Leiden exclusively, not merely on the *teaching* of Aristotle, but on the *text* of Aristotle as well, supplemented only by a small group of ancient Greek commentators.

The abrupt pedagogical volte-face recommended by the commissioners in August 1641 becomes perfectly intelligible, however, when contextualized in the history of the Ramist and post-Ramist tradition just outlined. Previously overlooked here is the unmistakable influence of the ranking humanist professor in contemporary Leiden: the distinguished philologist Daniel Heinsius (1580–1655). Alone amongst his Leiden contemporaries, Heinsius's published work consisted largely of editions, translations, paraphrases, commentaries,

---

tus Aristotelis praelegatur, et verbotenus exponatur; ac Aristoteles per Aristotelem et ex antiquis, praesertim Graecis, interpretibus illustretur .... nec alibi aliae quaestiones tractentur, nisi quam paucissimae, eaeque utiliores, et ex ipso textuque deriventur." Cf. Molhuysen, *Bronnen*, vol. 2, 259, 260–261; Bohatec, *Die cartesianische Scholastik*, 149–151; Dibon, *La philosophie néerlandaise*, vol. 1, 109–113; Verbeek, *Descartes and the Dutch*, 36.

19   Dibon, *La philosophie néerlandaise*, vol. 1, 113, "la pénétration des ouvrages et manuels de la néo-scolastique espagnole."

20   Ruestow, *Physics at Leiden*, 38; Verbeek, *Descartes and the Dutch*, 16–17, 37.

21   Verbeek, *Descartes and the Dutch*, 114, n. 17.

and studies of Aristotle's poetics, ethics, and politics, of the pseudo-Aristotelian *De mundo*, and of ancient Greek commentators on Aristotle such as pseudo-Andronicus of Rhodes.[22] In advocating the replacement of Burgersdijk's post-Ramist textbooks with *verbatim* commentaries on Aristotle's texts, Heinsius was merely reinstating the firm preferences of his own teacher, Scaliger, and Lipsius before him. In other words, Heinsius was merely attempting to perpetuate what Dibon called "one of the characteristic features of philosophical propaedeutics at the University of Leiden: the unceasing effort of the professors—from Bertius [taught 1593–1619] to Heereboord [taught 1640–1661]—to put their students in direct contact with the text of Aristotle, an effort which sometimes met with stiff resistance" from the professors of philosophy themselves, who preferred pedagogical practicality to philological perfectionism.[23] In short, the cause of the counter-revolution attempted in 1641 is not so much philosophical as pedagogical: it represents the final concerted attempt by leading high humanists to assert their hegemony over philosophy teaching just a few months before the controversies surrounding René Descartes (1596–1650) erupted.

Confronted with this retrograde recommendation, Heereboord was compelled during the subsequent four months to refight the campaign to replace textual with methodical Peripateticism which Keckermann had waged successfully in Heidelberg around 1600 and which Burgersdijk had rehearsed a quarter-century later.[24] In a letter to the university curators, Heereboord objected to these draft regulations, as Dibon pointed out, by "rehearsing in effect the arguments of Keckermann and Burgersdijk," and indeed those of Ramus as well.[25] The *Organon* was an inadequate basis for logical instruction, Heereboord advised, because it lacks a preliminary discussion of the definition, division, object, and end of logic (provided by Ramus and Keckermann's *praecognita*), an adequate discussion of definition and division (added by Burgersdijk), a separate treatment of method and order (pioneered by Ramus), and a method of teaching the application and use of logic (emphasized by all

22   Sellin, *Daniel Heinsius.*
23   Dibon, *La philosophie néerlandaise*, vol. 1, 15, "un des traits marquants de la propédeutique philosophique à l'Université de Leyde: l'effort que ne cesseront de tenter les professeurs—de Bertius à Heereboord—pour mettre leurs étudiants en contact direct avec le texte d'Aristote, effort qui se heurtera parfois à des sévères résistances ..."
24   Burgersdijk, *Institutionum logicarum libri duo* (Leiden, 1626), letter to the reader; cf. Hotson, *Reformation of Common Learning*, 118–120.
25   Dibon, *La philosophie néerlandaise*, vol. 1, 111, "Reprenant en effet les arguments de Keckermann et de Burgersdijk ..."

three), as well as several important technical subjects.[26] More generally, the
method of teaching philosophy through *verbatim* commentary on Aristotle's
text neither served the interests of students nor delivered the basic philosophical curriculum within the brief space of eighteen months, as the curators
demanded.[27] By early December, a compromise was laboriously negotiated:
Aristotle was to be expounded in lectures not strictly *verbotenus* but *secundum
ordinem contextus*. In other words, instruction was to begin with a brief analysis or compendium, focused on the clearest and most important passages of
Aristotle's text, and then briefly to treat all the necessary questions which arise
from a sequential reading of his text.[28] "In this way, the young professor safeguarded the heritage of logical reform of his teacher, Burgersdijk," and with it
the surprisingly tenuous hold of systematic pedagogical methods in Leiden.[29]
The price of this partial emancipation from the *form* of Aristotle's works was
Heereboord's agreement not to depart from the *substance* of Peripatetic teaching in his formal lectures.

Given this difficult beginning, it is perhaps unsurprising that the bulk of
Heereboord's public teaching was ostensibly devoted to revising Burgersdijk's
cycle of 'systematic' (i.e., post-Ramist) but essentially Peripatetic textbooks. In
1649 he published three *collegia* of physics, ethics, and logic which condensed-

---

26   Heereboord, *Meletemata philosophica* (Leiden, 1654), "Ad curatores epistola," 8: "I. Quod in
      organo Aristotelico deficiat ipsius artis Logicae definitio, partitio, objecti ac finis explica-
      tio, etc .... III. Non explicatur sufficienter nec suo loco ratio definiendi et dividendi modus
      .... V. Nihil traditur de methodo et ordine, de modo praxin Logicam instituendi, etc." Cf.
      Dibon, *La philosophie néerlandaise*, vol. 1, 111, 112 n. 96; Hotson, *Commonplace Learning*,
      43–44, 145–146; Hotson, *Reformation of Common Learning*, 118–125.

27   Heereboord, *Meletemata philosophica*, "Ad curatores epistola," 7: "Ego dixi, non me videre,
      qua ratione brevissimo temporis spatio, ad maximam Studiosorum utilitatem, Logica
      doceri publice queat, et D.D. Curatorum petitioni satisfieri, si constringar ipsius Aristotelis
      textum verbotenus explicare: longe facilius et felicius me id praestiturum utrumque, si lib-
      eram et non Aristotelico textui adstrictam servem docendi rationem."

28   Molhuysen, *Bronnen*, vol. 2, 337–339*: "Ratio ac modus, quo in posterum docenda in Aca-
      demia hac Philosophia, ex consensu iudicioque eorum quibus hoc commissum," 13 De-
      cember 1641. See also Dibon, *La philosophie néerlandaise*, vol. 1, 109–113.

29   Dibon, *La philosophie néerlandaise*, vol. 1, 113, "Le jeune professeur sauvegardait ainsi
      l'héritage de la réforme logique de son maître Burgersdijk." In fact, Dibon is being char-
      acteristically generous to Leiden here: Burgersdijk explicitly distinguished this kind of
      pedagogical approach—which he associated with two of his Leiden predecessors, Petrus
      Molinaeus (taught logic 1593–1596) and Petrus Bertius (taught 1593–1619)—from Kecker-
      mann's methodical Peripateticism, which he clearly judged superior and adopted himself.
      Cf. Burgersdijk, *Institutionum logicarum libri duo* (Leiden, 1626), letter to the reader; Hot-
      son, *Reformation of Common Learning*, 31, 118–119.

into short theses (of the sort sanctioned by the commission) the *Idea philosophiae naturalis*, the *Idea philosophiae moralis*, and the *Synopsis logicae* of Burgersdijk.[30] In 1654 he republished these three works together as a single, tripartite survey of *Philosophia naturalis, moralis, rationalis*.[31] The same year saw the first edition of his *Meletemata philosophica*, consisting principally of a survey of metaphysics by means of disputations. These two simultaneously published works, roughly complementary in subject matter, form, and scale, were sometimes bound together to form the outline of a basic *cursus philosophicus*.[32] In 1659, moreover, two years before Heereboord's death, a new edition appeared which integrated the tripartite *Philosophia* into the *Meletemata* to create the most complete philosophical *encyclopaedia* published within the Dutch golden age. The work saw three reprintings during the seventeenth century, the last of these in Amsterdam in 1680 for sale in London (see Table 2.1).[33] In all these works, Heereboord built on the framework erected by Burgersdijk and restricted himself to teaching philosophy "secundum Peripateticos, ... ex more Academiae nostrae."[34]

---

30   Burgersdijk, *Idea philosophiae naturalis* (Leiden 1622, 1627, 1635, 1640, 1645, 1648, 1652; Amsterdam 1657, 1648, 1657); *Idea philosophiae moralis* (Leiden 1623, 1629, 1635, 1640, 1644); for the *synopsis*, see Hotson, *Reformation of Common Learning*, 120 n. 58, 121 n. 62. Heereboord, *Collegium physicum: in quo tota philosophia naturalis aliquot disputationibus breviter & perspicuè per theses proponitur* (Leiden, 1649); *Collegium logicum: in quo tota philosophia rationalis [...] per theses proponitur* (Leiden, 1649); *Collegium ethicum: in quo tota philosophia moralis aliquot disputationibus [...] explicatur* (Leiden, 1649).

31   Cf. the descriptions of Dibon, *La philosophie néerlandaise*, vol. 1, 118; Ruestow, *Physics at Leiden*, 48.

32   The first edition—*Meletemata philosophica, maximam partem metaphysica* (Leiden, 1654)—was a quarto volume of 407 pp. ([6], 38, 362), but was published with the *Philosophia naturalis, moralis, rationalis* (Leiden, 1654) of 234 pp. ([6], 68, 23, [1], 136) and sometimes bound with it (e.g., Aberdeen University Library, sign. SB 102 Hee 1). The main texts of these and all final editions were published in two columns.

33   The second Leiden edition of the *Meletemata philosophica, in quibus pleraeque res Metaphysicae ventilantur, [...] Ethica [...] explicatur, [...] Physica [...] exponitur, summa rerum Logicarum [...] traditur* (1659) printed these two books as a single work and was also *sexaginta tribus [...] disputationibus auctior*: 869 pp. ([12], 382, 248, 136, 68, 23). The third Nijmegen edition of 1664/5 added the *Philosophia naturalis cum novis commentariis* (on which below) and a treatment of *Pneumatica* to make a volume in four parts of 982 pp. ([14] 382, [2]; 248; 136, [2]; 175, 23). The Amsterdam edition of 1665—in four parts and 1030 pp. ([52], 382, [2]; 248; 136, [2]; 112, 93, [3])—is mostly a reprint of the Nijmegen edition. The final edition (Amstelaedami: Sumptibus JHenrici Wetstenii; Prostant Londini: Apud Abelem Swalle, MDCLXXX) is a single volume of 1020 pp. ([4], 1007, [9]) including no new texts; *English Short-Title Catalogue*, R41241 (accessed 25 February 2020).

34   Heereboord, *Meletemata*, 2nd ed. (Leiden, 1659), 'Lectori.' Cf. Ruestow, *Physics at Leiden*, 48.

TABLE 2.1   Core philosophical textbooks in Leiden by Burgersdijk and Heereboord

BURGERSDIJK	HEEREBOORD			
Leiden 1622–1632	Leiden 1649	Leiden 1654	Leiden 1659	Nijmegen 1664–1665
Idea philosophiae naturalis (1622)	*Collegium physicum*	*Philosophia naturalis*	*Meletemata philosophica, in quibus pleraeque res Metaphysicae ventilantur, Ethica, explicatur, Physica exponitur, summa rerum Logicarum traditur*	With revised *Philosophia Naturalis cum novis commentariis* (1633) and *Pneumaticas*
moralis (1623)	*ethicum*	*moralis*		
Synopsis logicae (1632)	*logicum*	*rationalis*		
		Meletemata philosophica, max. partem metaphysica		

3   Heereboord and the New Philosophy

Behind this Peripatetic façade, however, Heereboord's personal philosophical interests ranged far more widely than his carefully regulated lectures. Johann Heinrich Alsted (1588–1638), the young Herborn professor who published the first edition of Keckermann's complete works in 1613, had immediately sought to go beyond the Danziger in using Keckermann's systematic method as a vessel for assembling a far more eclectic range of philosophical authorities.[35] In unmistakably similar fashion, Heereboord's philosophical interests rapidly spread beyond those of Burgersdijk. Shortly after the first foreign editions and translations of Bacon's main philosophical works appeared in Leiden and Amsterdam in the 1630s, "it was Heereboord who came forward as the most passionate advocate of Bacon in the Dutch Republic."[36] While Jan

---

35   Hotson, *Commonplace Learning*, ch. 6.i; Hotson, *Alsted*, ch. 2.
36   Van Bunge, "Philosophy," 288; see also Strazzoni, "The Dutch Fates of Bacon's Philosophy," which focuses principally on Heereboord and De Raey.

Amos Comenius (1592–1670) was still in London in January 1642, Heereboord charged an intermediary to obtain all of the editions of his books available in the city's bookshops, to inquire after the progress of his *Janua rerum*, and to urge him to visit Holland.[37] In Leiden, he acted as the Moravian's main host later that year, introduced him to Descartes, and maintained contact for at least five years thereafter.[38] In 1645, Heereboord published an edition of the *Metaphysica* dictated fifteen years earlier by Keckermann's former student, Joannes Maccovius (1588–1644), adding first an index of questions and then a *tractatus de anima separata* from a manuscript preserved by Johannes Polyander (1568–1646), before expanding it by another half-length again with his own annotations.[39] In 1647 he recommended the publication of the much reprinted quadripartite philosophical encyclopaedia of the French Cistercian, Eustache de Saint-Paul (1573–1640), as an accurate, concise, and easy introduction to the main branches of the received philosophy.[40] In 1654, Heereboord saw through the press a brief theological work which he had extricated from Alsted's brilliant protégé and son-in-law, Johann Heinrich Bisterfeld (d. 1655), whose philosophy had already begun to influence his own.[41] After Bisterfeld's

37   On the invitation, see Frijhoff, "Pieter de la Court."

38   On the visit, see Comenius, *Continuatio admonitionis fraternae*, § 55; Engl. transl. in *Comenius' självbiografi*; Young, *Comenius in England*, 44–45, 48, 50. For later contact, see Johann Raue to Comenius, 18 September 1642 (Hartlib, *The Hartlib Papers*, 7/93/1A–4B and 18/25/1A–4B); Bisterfeld to Comenius, 9 January 1643 (*The Hartlib Papers*, 7/63/1A–2A); Comenius to Johann von Wolzogen, 4 March 1647: Comenius, *Korrespondence*, ed. Patera, no. 109.

39   Maccovius/Makowsky, *Metaphysica: ad usum quaestionum in philosophia ac theologia adornata & applicata* (Leiden, 1645), with a dedication by Heereboord signed "pridie Kal. Octob. 1645," fols. *2ʳ–2ᵛ; new edition *accedit* [...] *tractatus de anima separata* (Leiden, 1650); *tertium edita, et explicata, vindicata, refutata per Adrianum Heereboord* (Leiden, 1658); also reprinted in Amsterdam, 1651; Bremen, 1668; see Van Bunge, Krop, Leeuwenburgh, Van Ruler, Schuurman and Wielema, eds., *Dictionary*, vol. 2, 661–665.

40   Eustachius a Sancto Paulo (1573–1640), *Summa philosophiae quadripartita, de rebus dialecticis, ethicis, physicis, et metaphysicis* (Leiden, 1647), with a letter from Heereboord to the reader (fols. *5ᵛ–6ᵛ) dated Kal. May 1647. I owe this reference to Mattia Mantovani. On the author, see Armogathe, "Eustachius."

41   Bisterfeld, *Scripturae sacrae divina eminentia et efficientia* [...] *denuo in duabus disputationibus. Accedit eiusdem authoris, ars concionandi* (Leiden, 1654), opens (fols. *2ʳ–3ᵛ) with a letter from Heereboord to Bisterfeld (dated simply 1654): "tuos tractatus, me obstetricante, lucem apud nos adspicere." See, for instance, Heereboord, *Meletemata philosophica*, 1st ed. (1654), "Disputatio de Encyclopaedia," 336–338. The opening thesis reads like a virtual paraphrase of Bisterfeld's view of the combinatorial *encyclopaedia* as a picture of the infinite web of reciprocal relations designated by the distinctive term "immeatio." See also De Dijn, "Heereboord," 61.

death in 1655, Heereboord printed one new work by Bisterfeld with a handful of reprints in two volumes, with prefatory material situating them within
the post-Ramist tradition.[42] Fragmentary surviving correspondence with Pierre
Gassendi (1592–1655), Constantijn Huygens (1596–1687), and Descartes finds
him deeply engaged in studying the mechanical philosophy in the mid-1640s.[43]
In 1646 Samuel Sorbière (1615–1670) reported to Thomas Hobbes (1588–1679)
that Heereboord was eagerly awaiting the publication of *De corpore* and *De
homine*.[44] The splendidly representative library auctioned after Heereboord's
death in 1661 leaves no doubt that his reading was omnivorous. Among the
174 "libri misc. in folio" listed are all the main landmarks of the post-Ramist
tradition: Ramus's *Professio regia*, Keckermann's *Opera omnia*, Alsted's definitive *Encyclopaedia*, Comenius's *Opera didactica omnia*, even the published
parts of his *Consultatio catholica*—not to mention the works of Bacon, Bisterfeld, Descartes, Gassendi, Robert Fludd (1574–1637), Athanasius Kircher (1602–
1680), the Jesuit *cursus philosophicus*, and a great deal else.[45]

It must have been as difficult to relate such wide reading to lectures framed
in strictly Aristotelian terms as it was frustrating to exclude it from lectures
altogether. Heereboord initially addressed this problem by allowing his students more liberty to pursue fresh approaches than he was allowed himself.
As he wrote in the letter to the Leiden governors which prefaces his *Meletemata*, "In those disputations of which I myself was author, I followed the principles of Aristotle, as is called for in the curricula of physics, ethics, and the
disputationes selectae," which were written by the professor rather than students. "In the theses and corollaries [...] prepared by the exertion and skill of

---

42   Bisterfeld, *Elementorum logicorum libri tres*: [...] *Phosphorus catholicus* [...] *Consilium de
      studiis feliciter instituendis* (Leiden, 1657). A letter to the reader by Heereboord (9 Decem-
      ber 1656, fols. *4ʳ–10ᵛ) situates Bisterfeld's logic within the tradition of Ramus, the Phi-
      lippo-Ramists, Keckermann, Alsted, and Burgersdijk. It is followed by a "Discrimen Aris-
      totelicorum et Rameorum. Ex Syntagmate Snellio-Rameo" (fols. *11ʳ–12ᵛ); see Bisterfeld,
      *Philosophiae primae seminarium ... editum ab Adriano Heereboord ... qui dissertationem
      praemisit de philosophiae primae existentia et usu* (Leiden, 1657). See further Viskolcz,
      Bisterfeld Bibliográfia, 38–40 and also 56, 63, 105. For further detail, see Hotson, *Reforma-
      tion of Common Learning*, 347–348.
43   Heereboord to Gassendi, 20 February 1644, in Mersenne, *Correspondance*, vol. 13, 27–28,
      no. 1252; Heereboord to Huygens, 24 July 1644, in Huygens, *Briefwisseling*, vol. 4, 11, no. 3634
      (on Descartes); Descartes to Heereboord, 19 April 1647, AT IV 632–636, with abundant
      annotation on his disputations *pro* and *contra* Descartes.
44   Sorbière to Hobbes, 11/21 May 1646, in Hobbes, *Correspondence*, vol. 1, 128–130.
45   *Catalogus* [...] *librorum* [...] *Adriani Heereboord*, "Libri misc. in folio," nos. 39, 151, 154, 156,
      167.

the students themselves," however, "I permitted the opinions and principles of other philosophers to be aired as well, so that their wits and mine together might be exercised and the direction in which reasoning might lead us might be discovered."[46] Thus, although the *disputationes selectae* are confined to the *philosophia recepta*, the *epimetra* and *paradoxa* appended to them include reference not only to Descartes, but also to Giovanni Pico della Mirandola (1463–1494), Bernardino Telesio (1509–1588), Francesco Patrizi da Cherso (1529–1597), Francis Bacon, Johannes Kepler (1571–1630), Tommaso Campanella (1568–1639), and others. By adopting this clever formula, Heereboord devised a means of teaching students the old and new philosophies together in a manner which respected the letter—if not perhaps the spirit—of the Leiden regulations of 1641. The academic genre of disputations—so much maligned by Bacon, Descartes, and the other proponents of the "new philosophy"—had come to the rescue of the *libertas philosophandi*.[47]

Even this compromise, however, failed to satisfy Heereboord's students: while their teacher had safeguarded the ability of more advanced students to debate contemporary philosophical issues in disputations, beginning students were still deprived of a more structured introduction to the new philosophies and their relationship to the old. In 1657, the students therefore began urging their teacher to revise his recently published survey of Peripatetic natural philosophy by incorporating into it the kind of modern views which he had clearly been discussing with them outside formal lectures. Heereboord responded during the last few years of his life but left the resulting work unpublished at his death in 1661—whether this was out of deference to the Leiden authorities is not known. The result, which only appeared posthumously in 1663, was an expanded edition of his physics, roughly double the length of the previous version.[48] In this new edition, the original Peripatetic

---

46 Heereboord, *Meletemata philosophica* (1654), "Ad curatores epistola," 9, cols. A–B: "*Quarum ipse* Disputationum *extiti author in iis* Aristotelis *principia fui secutus, quod* Physicarum, Ethicarum, selectarum Disputationum *docent curricula: in* Corollariis, *ut vocant,* & Thesibus Studiosorum *proprio Marte & arte confectis, aliorum etiam Philosophorum placita & principia ventilari, fui passus, ut meum simul & illorum exerceretur ingenium, ac, quo ratio nos ducere valeret, palam fieret.*" Transl. in Ruestow, *Physics at Leiden,* 58–59; see also Verbeek, "Tradition and Novelty," 183–188.

47 Cf. Descartes's views—in AT IV 69, CSMK I 146—with Heereboord, *Sermo extemporaneous de ratione philosophice disputandi* (Leiden, 1648), also included in the *Meletemata*; see Verbeek, "Tradition and Novelty," 187–188.

48 Heereboord, *Philosophia naturalis, cum commentariis peripateticis antehac edita: nunc vero hac posthuma editione mediam partem aucta, & novis commentariis, partim è Nob.*

commentaries on each precept were substantially augmented by further *commentaria* reflecting both the author's own views and those of three recent philosophers: Descartes, his early Dutch follower and later critic, Henricus Regius, and Claude Guillermet de Bérigard (1578–1663), who had challenged Aristotelian physics, not from a modern standpoint, but from that of the Ionian pre-Socratics.[49]

Figure 2.1 illustrates how the components of the expanded version were presented on the printed page. The Herborn printer, Christoph Corvinus (1552–1620), had enhanced the pedagogical clarity of Alsted's *Encyclopaedia* by distinguishing its *praecepta*, *regulae*, and *commentariola* from one another typographically.[50] In very similar fashion, the Leiden printer, Cornelius Driehuysen, printed the last two of this chapter's numbered Aristotelian theses in large Roman type, the briefest of commentaries on them in smaller italic type, and the more extended *commentaria* on them in two columns of smaller Roman type. Yet this typographical similarity masks a deeper pedagogical diversity. While Alsted's commentaries attempt to *reconcile* various philosophical opinions with one another, Heereboord's discriminate and critique the four philosophers *without* attempting to derive a unified philosophical system from them.

Confronted by this unique disposition of unreconciled material, historians of philosophy have offered a bewildering variety of interpretations of Heereboord and his philosophy. Already during the century before Dibon's classic survey, a multitude of incompatible characterizations accumulated: Heereboord was described variously as a neo-scholastic, an anti-scholastic, and a conciliator of old and new; as an able and zealous propagator of Cartesianism and as an incoherent eclectic; as a "progressive" critic of the old but a "fail-

---

D. Cartesio, Cl. Berigardo, H. Regio, aliisque praestantioribus philosophis, petitis, partim ex propria opinione dictatis, explicata (Leiden, 1663), esp. "Typographus lectori," which recounts the origins of the work. The text of the *Philosophia naturalis* of 1654 (in 68 two-column quarto pages) is reproduced virtually without change in 1663; but well over 100 of the 256 octavo pages in the 1663 edition are devoted to the new "sententia trium-viri," far more densely printed in two columns.

49   Regius, *Fundamenta physices* (Amsterdam, 1646). On the author and his relations with Descartes, see Van Bunge, Krop, Leeuwenburgh, Van Ruler, Schuurman and Wielema, eds., *Dictionary*, vol. 1, 818–821; Verbeek, *Descartes and the Dutch*, 13–19; Schmaltz, "Regius." On the work, most recently, see Verbeek, "Regius's *Fundamenta physices*"; Verbeek, "The Invention of Nature"; Claude Guillermet, seigneur de Bérigard, *Circulus pisanus* (Udine, 1643; Padua, 1661).

50   Hotson, *Commonplace Learning*, 184–186.

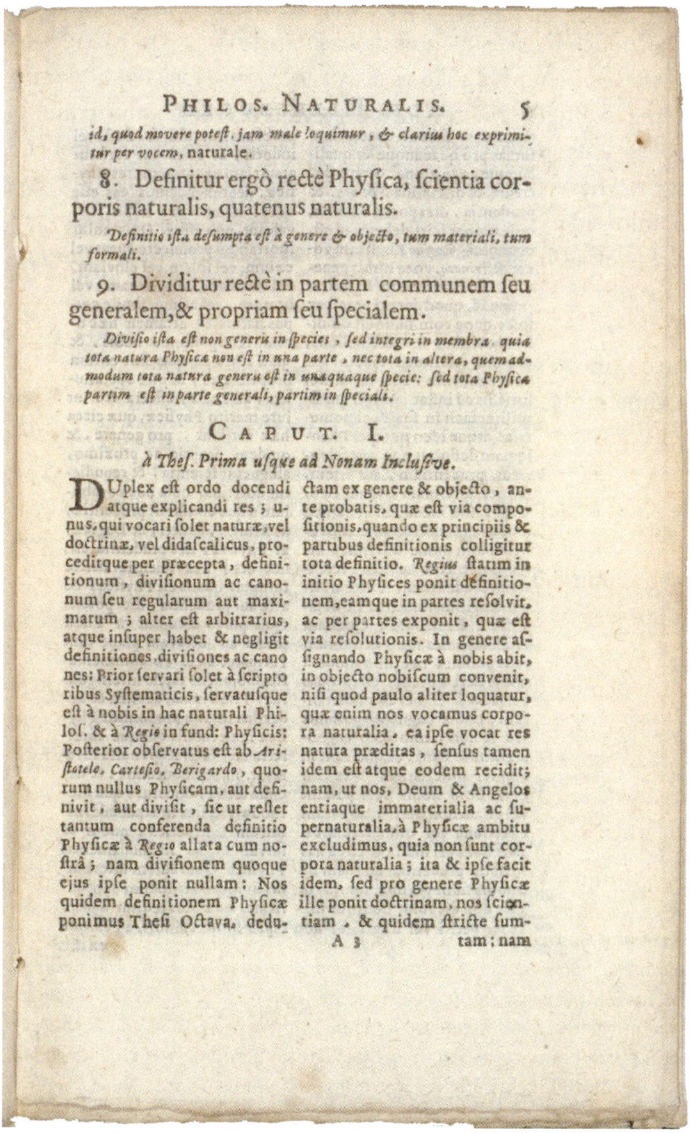

PHILOS. NATURALIS.   5

id, quod movere potest. jam male loquimur, & clarius hoc exprimitur per vocem, naturale.

8. Definitur ergò rectè Physica, scientia corporis naturalis, quatenus naturalis.

Definitio ista desumpta est à genere & objecto, tum materiali, tum formali.

9. Dividitur rectè in partem communem seu generalem, & propriam seu specialem.

Divisio ista est non generis in species, sed integri in membra. quia tota natura Physica non est in una parte, nec tota in altera, quemadmodum tota natura generis est in unaquaque specie: sed tota Physica partim est in parte generali, partim in specialis.

CAPUT. I.

à Thes. Prima usque ad Nonam Inclusive.

Duplex est ordo docendi atque explicandi res; unus, qui vocari solet naturæ, vel doctrinæ, vel didascalicus, proceditque per præcepta, definitionum, divisionum ac canonum seu regularum aut maximarum; alter est arbitrarius, atque insuper habet & negligit definitiones, divisiones ac canones: Prior servari solet à scripto ribus Systematicis, servatusque est à nobis in hac naturali Philos. & à *Regio* in fund: Physicis: Posterior observatus est ab *Aristotele, Cartesio, Berigardo*, quorum nullus Physicam, aut definivit, aut divisit, sic ut restet tantum conferenda definitio Physicæ à *Regio* allata cum nostrâ; nam divisionem quoque ejus ipse ponit nullam: Nos quidem definitionem Physicæ ponimus Thesi Octava, deduc-

ctam ex genere & objecto, ante probatis, quæ est via compositionis, quando ex principiis & partibus definitionis colligitur tota definitio. *Regius* statim in initio Physices ponit definitionem, eamque in partes resolvit, ac per partes exponit, quæ est via resolutionis. In genere assignando Physicæ à nobis abit, in objecto nobiscum convenit, nisi quod paulo aliter loquatur, quæ enim nos vocamus corpora naturalia, ea ipse vocat res natura prædita, sensus tamen idem est atque eodem recidit; nam, ut nos, Deum & Angelos entiaque immaterialia ac supernaturalia, à Physicæ ambitu excludimus, quia non sunt corpora naturalia; ita & ipse facit idem, sed pro genere Physicæ ille ponit doctrinam, nos scientiam. & quidem strictè sumtam: nam

A 3

FIGURE 2.1 This page, from the first chapter, contains (in large Roman type) the last two of the chapter's numbered Aristotelian theses (dealing, in this case, with definition and division), the briefest of commentaries on them (in smaller italic type), and finally the more extended *commentaria* (in two columns of smaller Roman type) comparing Aristotle with Descartes, Regius, and Bérigard (in this case, on the best method of teaching physics).

ure" as systematizer of the new.[51] A more recent account merely repeats an earlier verdict: "the French Cartesian Louis de La Forge [1632–1666] had it right when he referred to Heereboord in 1666 as a *sçavant peripatetic*, not a *sçavant cartésien*." According to others, "Heereboord limits himself to simply juxtaposing the opinions of various authors, without arriving at an original position"; he "lacked the energy to arrive at a new synthesis" and therefore "drifted anchorless in a sea of opinions and speculations that had been loosed by the loss of confidence in peripatetic thought."[52] Edward G. Ruestow articulated this critique at greatest length:

> The reliance on the older framework as well as the disparate character of the new commentaries reflected a failure on Heereboord's part to have worked out a consistent and systematic physics of his own. The 'progressive' character of his physics remained largely a diffuse and eclectic criticism of the older system, for which, however, since he could not accept Cartesian physics as a whole, he had no viable alternative. Though rejecting certain fundamental concepts of scholastic physics, he was unable to do without the systematization it provided.[53]

This retrospective judgment of Heereboord as a failure is difficult to reconcile with the avid reception of the work. During the two decades between 1663 and 1684, the updated *Philosophia naturalis* was printed eight times: five as a separate work and three within editions of the *Meletemata*. Nowhere was the work more enthusiastically received than in Restoration England. Between 1665 and

---

51   These are the views, respectively, of Dunin-Borkowsky, *Der junge Spinoza*; Van Vloten, *Ysselkout*; Dijksterhuis, Serrurier, and Dibon, eds., *Descartes et le cartésianisme hollandaise*; Boullier, *Histoire de la philosophie cartésienne*, vol. 1, 270: "Heereboord ne fut pas un moins habile et un moins zélé propagateur du cartésianisme"; Bohatec, *Die cartesianische Scholastik*, 19: "Er nahm verschiedene Elemente in seine Philosophie auf, ohne sie zu einer höheren Synthese verarbeitet zu haben. In seiner Naturphilosophie war er Cartesianer, in der Metaphysik und Ethik Aristoteliker, in der Logik, soweit er von seinem Lehrer, dem Aristoteliker Burgersdijk nicht abhängig war, Ramist .... Von einer *durchgreifenden* inneren Synthese, einem Einheitstrieb ist bei Heereboord fast nichts zu finden" (this passage is quoted also in Krop, "Scholam naturae ingrediamur," 4); as assembled in Dibon, *La philosophie néerlandaise*, vol. 1, 116–117: "il est bien difficile de faire le départ entre ce qui ressortit à la 'philosophia recepta' et à l'inspiration personnelle," exercising "une grande prudence pédagogique"; "la méthode de Heereboord est essentiellement celle du *disputator*. La dispute constitue pour lui ... l'instrument fondamental de la proédeutique philosophique."

52   Quoting, respectively, Schmaltz, *Early Modern Cartesianisms*, 72; Verbeek in Van Bunge, Krop, Leeuwenburgh, Van Ruler, Schuurman and Wielema, eds., *Dictionary*, vol. 1, 396; Verbeek, "Tradition and Novelty," 196; Ruestow, *Physics at Leiden*, 147.

53   Ruestow, *Physics at Leiden*, 48–49.

1676, the physics was reprinted three times in Oxford, where tutors continued to recommend the *Meletemata* into the early eighteenth century.[54] The final edition of the *Meletemata* was printed in Amsterdam in 1680 for sale in London, and the last edition of the physics appeared in the English capital four years later.[55] In New England, the *Meletemata* lasted even longer: it dominated philosophical instruction in the Harvard syllabus of 1723 and was still in use during the 1740s.[56] An adequate assessment of Heereboord's achievement must therefore reconcile his apparent "failure" as a systematizer with his success as a textbook author, above all in England and New England.

4   Pedagogical Innovation and the Reception of Philosophical Novelty

The first step toward such a reassessment is to return Heereboord to the proper professional category. Strictly speaking, he was not a professional philosopher: he was a professor of philosophy. His primary responsibility, in other words, was not to expound a unique philosophical 'system' of his own: it was to teach philosophy. This professional obligation, to be sure, was shared with the vast majority of his Dutch philosophical colleagues; but it was Heereboord's task to teach philosophy at a decisive moment of transition. His teacher, Burgersdijk, could work within the still consensual framework of Aristotle. His successors rapidly developed a Cartesian framework to replace it. Heereboord was the fulcrum on which the tradition tilted from the one to the other. It was during his career that the challenge which characterized his whole era was most acute: the challenge of teaching philosophy during the decades in which the old philosophical gods were dying, and the new ones were still being born.

Heereboord's strategy for meeting this challenge can best be appreciated by reading the *Philosophia naturalis* in the light of the *Consilium de ratione studendi philosophiae* appended to it.[57] There (as in several other texts) Heereboord divides philosophy into two categories. '*Philosophia vetus*' derives its principles principally from Aristotle. '*Philosophia nova*' derives its principles

54   Heereboord, *Philosophia naturalis* (Oxford, 1665, 1668, 1676): Madan, *Oxford Books*, vol. 3, nos. 2704, 2803, 3107. See Feingold, "The Humanities," 322; Feingold, "The Mathematical Sciences and New Philosophies," 403.
55   Heereboord, *Philosophia naturalis* (London, 1684): *English Short-Title Catalogue*, R28506. For the *Meletemata*, see n. 33 above.
56   Morison, *Harvard College in the Seventeenth Century*, vol. 1, 146–147, 233–234.
57   Heereboord, *Philosophia naturalis* (1663), fols. R1^r–3^v.

neither from Aristotle nor from any other authority, ancient or modern, but takes nature herself as the object of philosophical analysis. Although these definitions clearly reveal Heereboord's own preferences, he insists that "the old and Aristotelian philosophy must first be grasped as a whole, before you approach the new philosophy of one more recent author or another."[58]

Although not fully developed in the *Consilium*, the reasons for this advice, in the mid-seventeenth century, remained manifold and compelling. First and most basically, the old philosophy remained the indispensable preparation for studying the three higher faculties: Aristotelian natural philosophy was propae-deutical to Galenic medicine; moral philosophy was preliminary to the study of law; and metaphysics had once again become an essential part of the flourish-ing Reformed scholastic theology associated in the Netherlands especially (but not exclusively) with the school of Gisbertus Voetius (1589–1676) in Utrecht.[59] Secondly, familiarity with the old was scarcely less necessary for understanding the new: one could not fully understand Bacon, Descartes, or any of the other *novatores* without understanding what they were arguing against.

This need to master the old and a multiplicity of new philosophies simultan-eously posed an obvious, virtually unprecedented, and seemingly insurmount-able pedagogical problem. Ever since Ramus a century earlier, the difficulty of getting students through Aristotle's logical and philosophical corpus within a brief space of time had exercised the wits of Reformed Europe's most innov-ative pedagogues. Now the problem was radically exacerbated by the need to provide, within that same brief interval, some kind of comprehensible intro-duction to the new philosophy well as the old. To make matters worse, the '*philosophia nova*' was not centred around a single philosopher of the standing of Aristotle: instead, it involved a rapidly growing number of quite incompat-ible new philosophies in dispute with one another as well as with Aristotle. As a consequence, the generation which straddled the transition between old and new faced perhaps the most daunting pedagogical challenge in the entire his-tory of academic philosophy. It is this pedagogical problem, not the search for some new philosophical synthesis, which shaped Heereboord's revised physics and the encyclopaedic *Meletemata* into which it was eventually incorporated. The key purpose of his *Consilium* was to instruct students on how to use Heere-boord's works as a means of overcoming this daunting challenge.

For acquiring a grounding in the old philosophy of Aristotle, Heerebo ord's advice could scarcely be more familiar to a student of the post-Ramist tradi-

---

58   Heereboord, *Philosophia naturalis*, fol. R1ᵛ: "ex tota *Philosophia Vetus atque Aristotelica* prius est tenenda, antequam ad *novam* hujus vel illius recentioris accedatur."
59   A valuable rehearsal of these issues in the English context is Feingold, "Aristotle and the English Universities."

tion. At the outset of their philosophical education, students are advised to study the philosophical *praecognita* of Keckermann or Alsted, which define the genus, object, subject, and goal of any discipline, and treat its division into parts. Having done so, they should work their way back to the original texts through a graded sequence of authors: first the '*systematici*,' who express their precepts and commentaries in a manner accommodated to contemporary students and applicable to the uses of contemporary life; then the '*quaestionarii*,' who abstract key questions from the text of Aristotle and subject them to individual scrutiny; and finally the '*textuales*,' who comment directly on the texts themselves.[60] Heereboord's grounding in the post-Ramist tradition of his teacher becomes even more explicit in his recommendations of specific textbooks for individual disciplines.[61] For logic, to take the central case, the student should begin with Burgersdijk's brief *Synopsis* (expounded by means of Heereboord's own notes) and then master the fuller treatment contained in Keckermann's classic *Systema logicum minus et majus*. Under *quaestionarii*, Heereboord's principal recommendation is Clemens Timpler (1563–1624), Keckermann's teacher in Heidelberg, later professor in Steinfurt, who organized his textbooks around a series of questions on which a range of different opinions were collected and discussed.[62] Finally, "for gaining an understanding of the text of Aristotle's *Organon*," Heereboord notes, "Pacius suffices"—referring here to Giulio Pace (1550–1635), the Italian professor of philosophy and law in Heidelberg in the 1580s, who introduced the Renaissance Aristotelianism of Jacopo Zabarella, with whom he had studied in Padua, into Heidelberg in the 1580s, thereby laying the groundwork for Keckermann's subsequent pedagogical reforms there.[63] In short, Heereboord's sequence of Peripatetic textbooks lays out his own pedagogical lineage with crystal clarity: his own teacher, Burgersdijk, provides the students' point of departure; behind Burgersdijk lie the *praecognita* and *systemata* of Alsted and Keckermann; and before Keckermann the *quaestiones* of Timpler and the humanist Aristotelianism imported to Heidelberg by Pacius from Zabarella in Padua. Heereboord explicitly instructs his students to study the traditional philosophy, not in the textual manner—which had dominated Leiden pedagogy

---

60   Heereboord, *Philosophia naturalis*, fol. R1ᵛ. For the origin of these distinctions between 'praecognita' and 'systemata,' and between 'methodical' and 'textual' Peripateticism, see Hotson, *Commonplace Learning*, 145–148, 177–182.

61   Heereboord, *Philosophia naturalis*, fols. R2ᵛ–3ʳ.

62   Cf. Freedman, *Timpler*, 157–191; as summarized in Hotson, *Commonplace Learning*, 228–229.

63   Heereboord, *Philosophia naturalis*, fol. R1ᵛ: "Ad organi Aristotelici *textum* intelligendum suffecerit *Pacius*."

before Burgersdijk and which Heinsius and his colleagues had attempted to reimpose in 1641—but in the manner developed in Reformed central Europe in response to Ramism and semi-Ramism. His guidance for the remaining disciplines of the old philosophy is similar. Amongst systematic authors, for instance, he recommends Burgersdijk, Johann Combach (1585–1651), and Alsted for metaphysics; Keckermann, Combach, Burgersdijk, and Gilbert Jack / Jacchaeus (d. 1628) for physics; and for moral philosophy (ethics, politics, and *oeconomica*), Burgersdijk, Keckermann, Alsted, Marcus Friedrich Wendelin (1584–1652), and Johannes Althusius (1557–1638).

Having dealt with the ancients, Heereboord then divides modern philosophers into two categories depending on the thoroughness of their rejection of Aristotle. First are those who reject some but not all of the principles of ancient Peripatetic philosophy. These are led by a philosophical 'triumvirate': Juan Luis Vives (1493–1540), whose "golden book *de causis corruptarum artium*" was a major influence on Ramus; then Petrus Ramus himself, "whose two little books of logic in particular—those two golden nuggets, worth more in fact than all gold—one can never praise too highly, on account of their extraordinary usefulness in analysis and genesis"; and finally the historical critique of the Aristotelian tradition by Francesco Patrizi da Cherso.[64] Still more illustrious is the category of those moderns who overthrow Aristotelian principles altogether and base their investigations on entirely new foundations. This group includes Patrizi's bolder *Nova de universis philosophia* and those "illustrious heroes," Bacon's *De augmentis scientiarum* and *Instauratio magna* ("never too much esteemed"), Comenius's physics and *Pansophiae prodromos*, the eclectic physics of Kenelm Digby (1603–1665), and finally "the rising morning star of philosophy, René Descartes," who "has provided the true key of true philosophy."[65]

---

64   Heereboord, *Philosophia naturalis*, fol. R3ᵛ: "cujus inprimis nunquam satis commendavero aureolos, imo omni auro digniores, duos, inaestimabilis, ob singularem in *Analysi* & *Genesi* usum, *Logicae* libellos." The still more various group which follows betrays both interest in atomism and receptivity to Italian Renaissance philosophy: directly via Bernardino Telesio (1509–1588) and Tomasso Campanella (1569–1639), and indirectly via Sébastien Basson (1560–1621), David Gorlaeus (1591–1612), and the Boote brothers, Gerard (1604–1650) and Arnold (1606–1653).

65   Heereboord, *Philosophia naturalis*, fol. R3ᵛ: "& Philosophiae exorientis Phosphorus, *Renatus Des Cartes, in omnibus operibus Philosophicis*, in quibus editis, Philosophiae verae clavem veram exhibuit, naturae arcana non pauca reclusit, januam ad rerum immotam veritatem perveniendi aperuit: Atque utinam editis divini ingenii monumentis aliquando accedant inedita, quae alibi latent! in quo voto consiliium hoc meum finio." Very similar phraseology is found in *Meletemata philosophica* (1654), 10.

How then to introduce students to such a wide range of philosophies? Heereboord's solution was to adapt the methods devised for the old philosophy to teaching the new. The moderns, in other words, were to be studied initially, not via their own texts, but through a graduated series of *compendia*. "Descartes and Bérigard," he noted, "have not proceeded according to the standard method of teaching by means of definitions, divisions, and precepts"— that is, the method developed within the tradition deriving from Ramus and canonized by Keckermann's *systema*.[66] Bérigard, in keeping with his classical philosophical material, had adopted the dialogue form. Descartes, a gentleman philosopher with no teaching obligations, and one raised outside the Reformed educational world, employed continuous exposition, without the pedagogical structure of a post-Ramist textbook. Regius, on the other hand, "proceeds more properly" in this regard. Having studied Ramus's works and absorbed the expository method of Keckermann, Regius had sought to "methodize" Descartes: in the face of the Frenchman's strenuous objections, he had expounded a basically Cartesian physics within a dichotomously ordered sequence of definitions and rules of the kind familiar to his Utrecht students.[67] In short, Heereboord's procedure was two-fold. First, he teased out the definitions, divisions, and precepts more or less implicit in their texts in essentially the same manner that his predecessors, Keckermann and Burgersdijk, had done for Aristotle. Then, to facilitate an analytical comparison of old and new, he inserted brief introductions to these alternative philosophies within the framework of the old. Taken as a whole, the revised *Philosophia naturalis* offers a matrix of crucial and currently controversial topics arranged on the scaffolding of the received Aristotelian *cursus philosophicus*.

The essential procedure here was as old as the earliest stage of the post-Ramist tradition: from the foundation of Herborn onward, dialectic was taught by explicating Ramus "in comparison with Aristotle and others."[68] Similar com-

---

66   Heereboord, *Philosophia naturalis*, 3, "Quamvis autem Cartesius et Berigardus non processerint methodo doctrinae consueta, per definitiones, divisiones, similiave praecepta, uti quidem rectius hic procedit Regius, nos tamen studebimus ea, in hac comparatione instituendâ, ex continua oratione Cartesii, et dialogis Berigardi, quantum fieri potest, eruere, et definitionibus ac divisionibus commentariisque rerum Physicarum Aristotelicis, accommodare, et cum iis comparare." For background, see Hotson, *Commonplace Learning*, 148–150.

67   Descartes to Regius, July 1645, AT IV 239–240 and 248–249, CSMK III 181 and 254–255; Verbeek, *Descartes and the Dutch*, 52; Verbeek, "Notes on Ramism," 50–51. For context, see Bos, "Correspondence," 185–188.

68   Cf. *Leges scholae Nassovicae* (1609), §vi. vol. 3, "Tertius [professor philosophiae] dialecticam Rami explicato ... collatione cum Aristotele allisque"; repr. in Steubing, *Geschichte*

parison of old and new had then spread to other disciplines by employing the *collationes* or *collegia Gelliana* recommended by Alsed in at least a dozen places: in this method, each member of a private study group was tasked with studying and defending an authority with a different view of a controversial point, such as Ptolemy, Copernicus, and Tycho Brahe.[69] More specifically still, in revising his physics textbook between 1657 and 1661, Heereboord had employed essentially the same procedures used by Alsted in composing his physics nearly half a century earlier.[70] First, he analysed his authorities, ancient and modern, into individual philosophical topics; then he distributed discussions of their teaching into the previously established (and therefore Peripatetic) system of headings; and finally, he subjected the accumulated material to close critical and comparative scrutiny.

Yet Heereboord's approach differs from Alsted's in four significant ways. First, the statutory Peripateticism of Leiden initially forced Heereboord during the 1640s and 50s to suppress the non-Aristotelian material within the body of his textbooks in a way that Alsted was never required to do in Herborn in the 1610s. Second, the publication of the works of Bacon, Gassendi, Descartes, and others lent Heereboord's eclectic mixture an air of modernity unavailable to Alsted four decades earlier. Third, while Alsted's *Physicae systema harmonicum* juxtaposed four brief *treatises* on the four different schools of physics in his day, Heereboord went a step further: he broke his four schools of physics down into the same set of commonplace headings and then juxtaposed the competing doctrines on each commonplace, in a manner which facilitated comparison between the four competing systems. Fourth, Alsted, operating within the context of the *philosophia perennis*, then sought within his encyclopaedias of 1620 and 1630 to reconcile them with one another to create a *systema harmonicum*.[71] Heereboord's purpose was less philosophical and more thoroughly pedagogical: he sought to provide his students, not with a premature synthesis, but

---

      der Hohen Schule Herborn, 271–313, here 282–283; Hotson, *Commoplace Learning*, 101–106. In Harderwijk, the Herborn-educated German philosopher, Jodocus Höing from Unna (Mark), likewise taught "dialectica P. Rami cum collatione terminorum Aristotelicorum" from 1603 to 1637. De Haan, *Wijsgerig onderwijs* [...] *te Harderwijk*, 10–12.

69   Hotson, *Commonplace Learning*, 241–242.

70   Alsted, *Systema physicae harmonicae, quatuor libellis methodice propositum, in quorum I. Physica mosaica delineatur, II. Physica Hebraeorum, Rabbinica et Cabbalistica proponitur. III. Physica peripatetica ... plenius pertractatur. IV. Physica chemica perspicue et breviter adumbratur* (Herborn, 1612; rev. ed. 1616, 1642). For this paragraph, see Hotson, *Commonplace Learning*, 231–236.

71   Curiously, Krop, "*Scholam naturae ingrediamur*," also places Heereboord in the context of the *philosophia perennis*.

with an orientation to the bewildering range of philosophical options prolifer-ating rapidly in the mid-seventeenth century. Although Heereboord's physics represents a *systema* in the contemporary, pedagogical sense of a set of defini-tions, divisions, precepts, and commentaries designed to teach a discipline, it is not a 'system' in the derivative, modern sense of a self-contained presentation of a comprehensive and coherent set of philosophical doctrines: it is more like a *meletemata*, that is, a series of studies, essays, or meditations.

Better still, what Heereboord offers is a *vade mecum* which provides the stu-dent with both a point of departure at the outset of his philosophical studies and a constant point of reference in the course of it. Its first purpose is to intro-duce students to the Peripatetic framework which had structured academic philosophy for centuries and which every student in the mid-seventeenth cen-tury still needed to understand. Within this brief compass, only the most rudi-mentary, simplified presentation of physics could be provided. This compen-dium is not intended as a self-sufficient introduction to Aristotelian philo-sophy: it is to be followed by lengthier *compendia*, full-scale *systemata*, col-lections of controversial questions, and finally (for the devoted Peripatetic), a close study of Aristotle's original text and the ancient commentaries on it (in the manner preferred by Heinsius and his humanist predecessors). Having obtained some degree of familiarity with Aristotle, the student could return to read the continuous and more voluminous commentary on points of similarity and difference with three anti-Aristotelian philosophers, Bérigard, Descartes, and Regius; and this could be followed, in turn, by more direct engagement with their original works. The *Meletemata*'s core function, in other words, was precisely *not* to teach students a closed philosophical 'system'. Instead, it was designed to introduce students to contemporary philosophical debates in a manner which was both structured and open-ended. Rather than leading the student down a particular path to a predetermined destination, it provided a coherent introduction to a multitude of options—from a traditional Peripat-etic formation culminating with the Greek text of Aristotle, to the radically novel option of immersing oneself instead in the philosophical writings of Descartes.

When Heereboord's *Meletemata* is viewed, not as philosophy but as philo-sophical pedagogy, its critics appear ill-judged. Criticizing Heereboord for not becoming a thoroughgoing Cartesian implies that his only reasonable option, after emancipating himself from Aristotle, was to submit himself and his stu-dents to Descartes. Objecting to his "failure" as a systematizer likewise mistakes an ingenious pedagogical project for a failed philosophical one. Systematiz-ing the new philosophy into another idiosyncratic, premature, and therefore ephemeral philosophical "system" would arguably have served his students less

well than developing a means of introducing them quickly and relatively pain-
lessly into the complex philosophical debates of their time. The lively and
productive generation of philosophers which followed him clearly profited
from the liberating philosophical education he had devised for them: these
included not only the golden generation of "Dutch Cartesians"—Johannes De
Raey (1622–1702), Johann Clauberg (1622–1665), Christoph Wittich (1625–1687),
and the others—but arguably even Baruch Spinoza (1632–1677), whose introduction into philosophy came through Keckermann, Burgersdijk, and Heereboord.[72] When the subsequent generation retreated into a more thoroughly
"modern" Cartesian orthodoxy, the golden age of philosophy at the Dutch universities was arguably over.[73] From this perspective, the transition from Heereboord's deliberately agnostic introduction to the philosophical debates of his
day to the doctrinaire Cartesian orthodoxy of the next generation is not progress at all, but a regression from open-mindedness into sterile system building.

This contrast helps explain the longer shelf-life which Heereboord's work
enjoyed in Oxford than in Leiden. In the Dutch universities, philosophy was
taught primarily by professors, many of whom sought to distinguish themselves by producing philosophical 'systems' of their own. Indeed, it is in this
period that the term *systema* begins to transition from the primarily pedagogical meaning which Keckermann had given the term to the philosophical
meaning it has today. Regius (much to Descartes's annoyance) had led the way,
and Heereboord's own students—De Raey and Clauberg—had followed him
even before their teacher's death. In Oxford, despite the creation of chairs in
philosophy and mathematics between 1618 and 1626, the arts curriculum was
taught primarily by non-specialist tutors who had little interest in carving out
a philosophical reputation by producing works of their own. In that context,
the best preparation for the disputations in the arts and further work in the

---

72   When Spinoza first began formal philosophical studies in 1656, the key texts he studied
      were Keckermann's *Systema logicum*, Burgersdijk's *Institutiones logicae* and *metaphys-
      icae*, and Heereboord's *Meletemata*; and the influence of all of these texts and authors
      has been noted on fundamental aspects of his philosophy. "Spinoza was sympathetic to
      Heereboord's way of philosophizing, and he quotes directly from the *Meletemata* in the
      Metaphysical Thoughts"; Gabbey, "Spinoza's Natural Science and Methodology," esp. 144–
      145, 183–184.
73   Van Bunge, "Philosophy," 346: "Despite all the revolutionary fanfare that accompanied
      the ousting of the old philosophy, by the 1650s the 'innovators' were already creating
      an orthodoxy of their own [...]. One generation after the new philosophy had made its
      breakthrough in the Dutch universities [...], it became itself the greatest impediment to a
      further renewal of philosophy."

higher faculties was not to master the *systema* of some foreign professor but to survey the whole field of contemporary debates within the framework of the received philosophy. In the words of the most authoritative study, "Prescriptions of tutors leave little doubt that the undergraduate curriculum was consciously balanced in order to acquaint the student with the divergent versions of natural philosophy, both ancient and modern."[74] Heereboord's was not the only philosophical textbook adopted for this purpose, but it was one of the most influential during the 1660s and 1670s. In effect, Oxford remained relatively open to the adoption of experimental natural philosophy, and eventually Newtonianism, because its tutors had institutionalized neither Cartesianism nor the *philosophia novantiqua* but the open-minded, philosophical agnosticism epitomized by the pedagogy of Heereboord.

Failing to find a coherent philosophy within Heereboord's physics, historians of philosophy have tended to reserve their praise for Heereboord's championing of the *libertas philosophandi*:

> The stimulating disruptiveness of his presence at Leiden lay not in the modernity of his own philosophy but in his struggle to secure an open market place of ideas within the walls of the university [...]. The hallmark of his own thought and career was not a dedication to any philosophic system, but a resistance to the subjection of philosophic speculation and debate to tradition and authority.[75]

In this view, the hallmark of Heereboord's philosophy was to introduce into philosophical instruction in Leiden the kind of *libertas philosophandi* which Rudolph Goclenius the Elder (1547–1628) and Alsted been established in Marburg and Herborn a generation earlier. Many of the most striking features of Heereboord's philosophy were already present in the case of Alsted: the maintenance of Peripateticism as a structure within which to assemble material, the collection within that structure of highly heterogeneous material, the extreme eclecticism which resulted especially in natural philosophy, the consequent bewilderment of later seventeenth-century readers confronting this formally tidy but philosophically incoherent material, and the struggles of historians of philosophy to categorize the resulting mêlée of opinions.[76] The crucial difference is that Heereboord's commonplaces included a now celeb-

---

74   Feingold, "The Mathematical Sciences and New Philosophies," 404.
75   Ruestow, *Physics at Leiden*, 58, 59; a formulation adopted by Dibon, "Der Philosophieunterricht in den Niederlanden," 55.
76   Cf. Hotson, *Alsted*, 10–11; Hotson, *Common Learning*, esp. 200–246.

rated philosopher not available to the older encyclopaedist: René Descartes. Heereboord's presentation of natural philosophy is therefore best understood as post-Ramist eclecticism updated by a preliminary consideration of the "new philosophy."

5   Conclusion

Heereboord's work therefore documents with particular clarity the manner in which the post-Ramist tradition of philosophical eclecticism provided the matrix through which Cartesianism first percolated through the *cursus philosophicus* in select Dutch universities. The ultimate driver of this process was neither philosophical nor pedagogical but social: the need for the huge surge of students in post-Reformation universities to obtain a useful education as quickly and efficiently as possible.[77] This force gave irresistible traction to the claims—by the Ramists and semi-Ramists before 1600 and the methodical Peripatetics after Keckermann—that the best means of teaching philosophy was not via the Greek text of Aristotle or some mere compendium of it, but by a new generation of textbooks based on very carefully considered pedagogical principles. Despite tenacious rear-guard action by Leiden's humanists, by the time Cartesianism arrived on the academic scene, the old philosophy was no longer taught primarily by direct study of the text of Aristotle. Cartesianism was never taught primarily by direct study of Descartes's texts either. Instead, the first generations of Cartesian professors accommodated elements of Cartesianism into the framework already provided by the philosophical *systema* devised a half-century earlier by Keckermann. Pedagogical exigencies had first won for Reformed philosophers the freedom to depart from the text of Aristotle; and the resulting 'systematic' pedagogy then enhanced their freedom to depart incrementally from the teaching of Aristotle as well by using the framework of the post-Ramist textbook to add non-Aristotelian and indeed anti-Aristotelian doctrines to the mix. Heereboord marks both aspects of this transition in Leiden: the first phase of his career is devoted to rescuing systematic pedagogy from the final retrograde assault of the humanists; and his final, posthumous work shows how anti-Aristotelian and specifically Cartesian philosophical content could be incrementally introduced into the new systematic pedagogical framework.

---

77   This social dimension of the argument is fully developed in Hotson, *Reformation of Common Learning*, 1–18, 420–429.

Several methodological conclusions follow from this account. Most explicit in the foregoing discussion is the need to read early and mid-seventeenth-century philosophical texts in the milieus in which they were written. For the overwhelming majority of Dutch, English, and German philosophers, this means returning them to the academic context of the university. Since the professor of philosophy's foremost obligation was to teach, rather than to innovate or even to philosophize, this means that the great bulk of philosophical writing from this period needs to be read from an educational as well as a philosophical standpoint. This basic methodological starting point brings others in its train. If pedagogical considerations are re-prioritized, then individual works and writers need to be placed in the context of the era's evolving pedagogical traditions as well as philosophical schools. In that case, far more attention needs to be devoted to the most innovative and disruptive pedagogical tradition which swept through the Reformed and eventually also Lutheran worlds in the latter sixteenth and early seventeenth century: Ramism and its many ramifications. But this requires acknowledgement, in turn, that this tradition embraces much more than Ramism, narrowly construed. The Ramist and post-Ramist tradition was ultimately rooted in the Renaissance dialectic of Rudolph Agricola (1444–1485) of Groningen, received its most radical formulation by Pierre de la Ramée in Paris, but then gravitated back to the Rhineland after Ramus's death, and rapidly morphed into the 'methodical Pripateticism' of Keckermann in Heidelberg and Danzig, and the post-Ramist eclecticism of Alsted in Herborn before being imported into Leiden above all by Heereboord's teacher, Burgersdijk. Such an account clearly demands, in turn, that the history of Protestant philosophy in the post-Reformation period escape from the national categories which have constrained far too much historiography, and rediscovers the internationalism of the contemporary *peregrinatio academica* itself.

CHAPTER 3

Teaching Cartesian Philosophy in Leiden: Adriaan Heereboord (1613–1661) and Johannes De Raey (1622–1702)

Antonella Del Prete

1 Introduction

The Netherlands is a privileged context in which to study Descartes in the classroom. Thanks to the scholarship of some eminent commentators such as Paul Dibon, C. Louise Thijssen-Schoute and Theo Verbeek, we know that the diffusion of Descartes's philosophy was precocious in the Dutch Republic and that it centred mostly on the universities.[1] In Leiden, it is possible to compare different generations of Cartesians and to better evaluate the evolution of Cartesian philosophy.

The comparison between different generations of university teachers aimed at an investigation of the diffusion of Cartesian philosophy is a broad research field and deserves to be inquired systematically. I will narrow my focus down to two philosophy teachers—Adriaan Heereboord (1613–1661) and Johannes De Raey (1622–1702).[2] They both had personal contact with Descartes; both advocated for the introduction of his philosophy at Leiden University; and both were protagonists of what Theo Verbeek has called the 'Leiden Crisis'.[3]

In the available sources allowing us to reconstruct their teaching, there is an asymmetry: concerning Heereboord we must rely exclusively on printed texts, while some manuscripts come down to us from De Raey. The Heereboord texts do not always suit the purpose of our research: rarely can the original texts of the disputations be compared to those collected afterwards into volumes. After the first publication, and before being re-edited in a comprehensive volume in 1650, the disputations were often heavily reworked. This

---

1 McGahagan, *Cartesianism*; Thijssen-Schoute, *Nederlands Cartesianisme*; Dibon, *Regards*; Verbeek, *Descartes and the Dutch*.

2 Verbeek, "Heereboord"; Schuurman, "Raey, Johannes de." On Heereboord, see Chapter 2, in this volume. On De Raey, see Chapter 4, in this volume.

3 Verbeek, *Descartes and the Dutch*, 34–51; Van Bunge, *From Stevin to Spinoza*, 41–43.

precludes assessing the reliability of the charges filed by the supporters of traditional philosophy and determining the kind of Cartesianism embraced by
Heereboord, as we ignore the exact nature of the changes occurring between
the original disputations and their reprint. However, we do have access to a
copy of Heereboord's *Disputationes selectae*, different from any other: it consists of a collection of original disputations, arranged in the same order as the
above-mentioned volume printed in Leiden in 1650.[4] These particular disputations can therefore be dated and their paratext, including the corollaries, can be
read. This allows us to slightly anticipate the date of Heereboord's acquaintance
with Descartes's philosophy and to better specify the features and the steps of
his Cartesian commitment (see section 2 below).

For De Reay, by contrast, we have access to some manuscripts providing
his comments on Descartes's texts. Consequently, in his case, it is possible to
notice the differences between the lectures and the publications (see section 3
below). However, we should keep in mind that the manuscripts containing De
Raey's *dictata* refer, in all likelihood, to his private lectures. Leiden University
rules promulgated in December 1641 established that teaching philosophy consisted in reading and commenting on Aristotelian texts.[5] Academic authorities
continually reaffirmed Aristotle's centrality in the following years and in 1647
forbade commentary on Descartes's philosophy: the need to repeatedly insist
on silence in this regard indicates that breaches of the rule were rather common.

---

4 The copy of the *Disputationes selectae* that I found in Augsburg's library (Universitätsbib-
liothek Augsburg, 02/V.1.4.28) binds the original disputations in the same order they will be
published in the edition of 1650.

5 Molhuysen, *Bronnen*, vol. 2, 338*-339*: "In singulis Philosophiae partibus praelegetur Aristoteles: et quidem ordine, secundum contextum. Eo tamen modo ut sententiae ac verba,
quae prae ceteris mentem exprimunt Philosophi, praelegantur: cetera obiter modo pro-
ponantur. Ita ut ubique quantum fieri potest loci singuli, ex aliis, in quibus fusius ac magis
perspicue loquitur agitque, exponantur ea ratione ac modo, ut Graecorum quoque inter-
pretum ac philosophorum interpretatio addatur. Tum ut libris singulis capitibusque compendium praemittatur singulorum atque analysis: sed et quaestiones omnes necessariae,
quae ex contextu oriuntur, obiter discutiantur: postremo, ut breviter mens eorum, in quibus
cum reliquis philosophis disputât Aristoteles, proponatur. Ita visum, ita post seriam mat-
uramque deliberationem in Conventu ad hoc instituto, indicatum conclusumque a Mag-
nifico Rectore, Professoribusque qui a Nobilissimis Amplissimisque D. Academiae Curat-
oribus atque urbis Lugduno Batavae Consulibus ei rei delegati sunt D. Polyandro, D. Trigla-
dio, Scotano, Heinsio, L'empereur, Du Ban, Herebortio, eo loco quo Senatus Academicus
haberi ac convenire solet." On the Leiden 1641 regulation, see the Chapters 2 and 4, in
this volume. On the changes to this regulation effected in 1673, see the Chapter 9, in this
volume.

2   Adriaan Heereboord's Engagement with Cartesian Philosophy

Heereboord had a brilliant start to his career. He was appointed extraordinary professor of philosophy, with the chair in logic, in 1640, and he started giving lectures in 1641. The following year, he became vice-regent of the *Collegium theologicum*, having been preferred over two other candidates, one of whom was Martin Schoock (1614–1669), the man who was soon to establish a reputation for his strenuous opposition to Descartes's philosophy. From the very beginning, Heereboord challenged the academic rules and, as he claimed in his inaugural lecture, he aimed to free philosophy from the blind respect of Aristotelian and scholastic doctrines. He appealed to different sources to achieve his purpose: in his opening speech, humanist thinkers are praised above all, while in the speech he delivered upon his accession to the position of ethics professor in 1645, he mentioned Petrus Ramus (1515–1572), Giovanni Pico della Mirandola (1463–1494), Bernardino Telesio (1509–1588), Francesco Patrizi (1529–1597), Tommaso Campanella (1568–1639), Francis Bacon, and Descartes.[6] In the *De angelis* disputations, discussed between December 1644 and January 1645, some corollaries are clearly Cartesian while others are inspired by Johannes Kepler, Sébastien Basson (1573?–?), Bacon, Campanella, and Johann Sperling (1603–1658).[7] As Howard Hotson has so brilliantly demonstrated, the aims of such eclectic syncretism are varied: to stimulate students' and teachers' minds, presenting and discussing different philosophical traditions; to revive the tradition of debating the pros and cons of the same opinion; and, above all, to defend the *libertas philosophandi*.

Following the death of François Du Ban (1592–1643) in the previous year, Heereboord was appointed full professor in 1644; a year later he was to be granted a teaching position in ethics.[8] The upward trajectory of his career came to a halt, however, in 1644 when Adam Stuart (1591–1654) was appointed full professor and was given the priority to teach in the hours preceding those assigned to Heereboord. Moreover, in 1645 the Curators gave Stuart a chair in metaphysics and physics.[9] Is the end of Heereboord's ascent connected to the beginning of his adherence to Cartesianism? This is hard to tell, given that the timing of his full espousal of Cartesianism is difficult to date.

---

6   Heereboord, *Meletemata* (Leiden, 1654), 10; see also 28 (which includes a dithyrambic praise of Descartes). See Van Bunge, *From Stevin to Spinoza*, 48–50.

7   Heereboord, *Meletemata*, 10–11.

8   Molhuysen, *Bronnen*, vol 2, 276 and 178.

9   Molhuysen, *Bronnen*, vol. 2, 287, 290–291, 295–296, 343*-344*, 347*-348*.

Theo Verbeek has tried to determine when such a shift took place by examining contemporary records, and he has dated Heereboord's familiarity with Descartes's philosophy back to 1644.[10] Thanks to the copy of the above-mentioned *Disputationes selectae*, the disputation *De notitia Dei naturali*, containing Heereboord's first reference to Descartes, can be dated to 13 March 1643. Heereboord writes:

> Those wishing to know more, turn to Marin Mersenne's *In Genesin*, whose chap. 1 art. 2 proves with 35 arguments the natural notion of God, and to *Campanella* [who does so] with an entire book, which he has entitled *Atheismus triumphatus*. *Renée Descartes, in his prima Philosophia, meditatione tertia* and in the following *replies* to different *objections* against it, proves both notions of God dealt with in our thesis, the innate and the acquired. He employs reasoning and the idea of God himself introduced in us through creation, as if this idea was the imprint of the craftsman on his work.[11]

Since Descartes's name is quoted on March 1643, it can be assumed that Heereboord's acquaintance with Cartesian philosophy will have begun by this time and closely have preceded Adam Stuart's arrival in Leiden, which became effective in 1645 but was already planned in 1644. While Stuart was a traditional philosopher, Heereboord had a completely different view: he had studied with Franco Burgersdijk (1590–1635) as a teacher and can be defined as a post-Ramist.[12] Heereboord was aware of the polemics raised by Descartes's

---

10   Verbeek, *Descartes and the Dutch*, 37.

11   Heereboord (praes.) and Zouterius (resp.), *Disputationum ex philosophia selectarum tertia de notitia Dei naturali* (Leiden, 25 March 1643): "Qui plures desiderat, adeat *Mar. Mersen. in Genesin*, ubi cap. I. art. 2. probat argumentis 35. notitiam Dei naturalem, et *Campanellam*, integro libro, quem vocat *Atheismum triumphatum*. Et utramque Dei notitiam a nobis hisce thesibus pertractatam, tum insitam, tum acquisitam, ex ratiocinio et ipsius Dei idea per creationem nobis indita, tanquam nota artificis operi suo impressa, *subtilitissime* probat *Renatus Des-Cartes, in sua prima philosophia, meditatione tertia, et respons.* subjunctis *adversus* varias contra eam *objectiones*." This disputation was scheduled to be discussed once again in January 1648, with the addition of some corollaries treating Stuart's polemical replies to Descartes. The curators' action, however, would prevent its actual discussion. Heereboord replaced it with another, but he published it in the *Epistola ad curatores* in 1648.

12   Hotson, *Reformation of Common Learning*, and Chapter 2, in this volume, show that Heereboord's attitude toward philosophy is deeply rooted in his Ramist training, both in terms of the systematic imprint of his teaching and in terms of his tendency to introduce students to different philosophical traditions.

theories: he understood that any attempt to introduce Cartesian philosophy in Leiden would be met by strenuous opposition from Stuart, as his main academic rival. Establishing whether Heereboord's 'conversion' to Cartesianism caused the reaction of anti-Cartesians, or whether their reply to his first provocations guided him towards an increasingly fierce defence of Descartes, is therefore a complex question to settle.

During the following years, Heereboord, who did not seem initially to have wholeheartedly endorsed Descartes on an exclusive basis, examined a wide range of Cartesian allegations: the disputation supporting Gassendi quoted by Gijsbert Voetius (1589–1676),[13] for example, includes a summary of the *Meditationes* while the corollaries of the *De angelis* disputations (1644–1645) range from criticism of prejudices to clear and distinct ideas; from God and the soul to the objections to material forms, and the importance of doubt:

Metaphysics
1. It is necessary to eliminate all prejudices to search for the truth.
2. Without the knowledge of the Soul and of God, there is no certain knowledge of the Truth.
3. What we perceive clearly and distinctly is entirely true.
4. The cause of errors is the will which, since it is more extensive than the intellect, often expands to what the intellect does not clearly perceive.
5. Neither divine intervention nor the necessity of the will destroys freedom; on the contrary, it increases and corroborates it.

Physics
1. All material forms are only modes of matter.
2. The sky is corruptible and aerial in nature.
3. Stars move by their own motion through the liquid sky, like birds through the air.
4. Elements do not transmute into each other.[14]

---

13   Voetius, *Selectae disputationes* (Utrecht, 1669), v, 507–513; see Verbeek, *Le contexte historique*, 1–33.
14   Heereboord (praes.) and Du Pré (resp.), *Disputationum ex philosophia selectarum decima-quinta de angelis* (Leiden, 1644): "Metaphysica. 1. Omnium praejudiciorum eversio necessaria est ad veritatem indagandam. 2. Sine cognitione Animae, et Dei, nulla datur certa Veritatis cognitio. 3. Illud omne verum est, quod clare, et distincte percipimus. 4. Causa errorum est voluntas, quae, cum latius pateat, quam intellectus, saepe se extendit ad ea, quæ intellectus non clare percipit. 5. Neque concursus divinus, neque voluntatis necessitas, libertatem evertit; imo vero magis auget, et corroborat. Physica 1. Omnes formae

The introduction of Cartesianism into Leiden, therefore, can be compared to a tile in the overall mosaic: Descartes's works are not initially commented upon in the classroom but find room in the free discussion of the corollaries to the disputations.[15] The Cartesian text receiving most attention is the *Meditationes*, attesting to a change of focus compared to the Cartesianism of Henricus Regius (1598–1679), which is based on the *Discours* and the *Essais* thereon, and later, the *Principia*.

During following years, a progressive polarisation vis-à-vis his opponents' reactions can be noticed in Heereboord's writings. Between 1647 and 1648, the original disputations to which we have access show an increasingly marked opposition to Aristotelianism. Not only is scholastic philosophy the target of sharp polemics, as in his previous writings, but Aristotle himself is now targeted.[16] In Heereboord's justifications of these criticisms, a wide range of *auctoritates* are invoked, while the reference to specific Cartesian doctrines is absent most of the time. By no means does this imply a lack of appreciation of Descartes's philosophy, however: since 1647, as we have seen, any mention of Descartes and his opinions was forbidden. Heereboord was thus forced to speak indirectly, as we can see in the *De formis* arguments against Aristotle. Heereboord built them up around the text of an irreproachable English puritan, William Pemble (1592?-1623), and eliminated any references to Descartes, except with regard to the freedom of indifference (dealt with in the *Meditationes*) and the criticism of preconceived opinions.[17] This example shows that Heereboord's use of his sources was very sophisticated: as he could not mention Descartes, he presented his theses linking them to some other authors who could be safely quoted.

The second volume of the disputations, published in the *Meletemata* of 1654, mirror mostly Heereboord's teaching after 1649 and attests to a shift in Heereboord's agenda and his strategy. Contrary to the first volume, distinctly Cartesian theses feature in the disputations' text proper and not only in the commentaries. Sometimes Heereboord circumvents the prohibition to expli-

---

materiales sunt tantum modi materiae. 2. Coelum est corruptibile, et aereae naturae. 3. Stellae proprio motu per liquidum Coeli moventur, ut aves per aerem. 4. Elementa non transmutantur in se invicem." The *Epistola ad curatores* give us the reactions to these corollaries; Heereboord, *Meletemata*, 10.

15   On corollaries, see the Chapter 7, in this volume.

16   Heereboord (praes.) and Wassembergh (resp.), *Disputatio metaphysica ad Aristotelis l. 3 Metaphys.* (Leiden, 1646).

17   Heereboord (praes.) and Hardenbergh (resp.), *Disputationum ex philosophia selectarum trigesima-quinta quae est secunda de formis* (Leiden, 1648), thesis 11. See Ariew and Grene, "The Cartesian Destiny," 322.

citly mention Descartes by discussing (by name) "Cartesians" like Regius;[18] however, in other disputations, the reference to Descartes is quite explicit and concerns two disciplines, physics and metaphysics. Regarding the first, we can refer to a series of disputations dealing with the identity between quantity and matter, and the distinction between accidents and substance: they are especially significant and illustrate Heereboord's ability to switch between Aristotelian and Cartesian conceptual frameworks in his teaching. Concerning the first point, Heereboord claims that there is no real distinction between matter and quantity, in an analytical confrontation with the theories of Ruvius, Suárez, Hurtado de Mendoza, Arriaga and Pereira.[19] These same topics are taken up again in the *De rerum materialium natura* disputation, in a context more explicitly marked by Cartesian influences: the text consists of a paraphrasis of the first ten articles of the second part of the *Principia philosophiae*. Dealing with the distinction between substance and accidents, Heereboord links the refusal of the Aristotelian theory of qualities to the confutation of the existence of separate accidents. As a consequence, he gives a corpuscular explanation of light, sound, smells, tastes, and of hot and cold.[20]

By no means does the more explicit presence of Cartesian themes imply Heereboord's renunciation of other philosophical traditions: some disputations clearly refer to the Aristotelian philosophy, while the influence of the *novatores* is central to the disputations dealing with the number of the elements, the air, and atoms. We can conclude that in the 1650s Heereboord did not change his approach when dealing with physics: Descartes continues to be juxtaposed with other *novatores* and does not seem to be the favourite.

Moving on to metaphysics, the situation changes as Descartes seems to be the only alternative to scholastic philosophy. For example, the *Annexa* to the *De abusu philosophiae in theologia* disputation summarize Descartes's main theories, with a view to condemning scholastic philosophy as a whole: philosophy must be grounded on evident principles; what is perceived with clarity and distinction cannot be false; we have to set ourselves free from the prejudices of the senses acquired during childhood; the mind does not always need images (*phantasmata*) to perceive something; knowledge is drawn from ideas which are of three kinds (innate, acquired, and invented);—these are the only ways through which we can achieve authentic philosophical knowledge: "This is the

---

18   Heereboord, *Meletemata*, 207. Regius, alongside Claude Bérigard and Descartes, will be
      a protagonist in the commentary included in the posthumous edition of Heereboord's
      Physica; see Chapter 2, in this volume.
19   Heereboord, *Meletemata*, 192–195.
20   Heereboord, *Meletemata*, 205–207.

only, true, sound way to Philosophize and discover the truth, and to totally eradicate any doubt from the minds of Philosophers, and of all mankind."[21] Descartes thus becomes the example of how to properly set up the relationship between philosophy and theology, according with Luther's reform of religion and Ramus's transformation of logic.

Heereboord also participates in the internal debates within Dutch Cartesianism, as we can see in some of the disputations dedicated to doubt, such as the *De philosophandi ratione*.[22] He starts with a presentation of doubt which can profitably be compared to Johannes Clauberg's (1622–1675) interpretation of the same topic. In fact, both philosophers defend Descartes's use of doubt as a means to eliminate what is uncertain and provide human knowledge with a firm ground. However, the most scandalous aspects of Cartesian doubt are omitted: Heereboord does not mention mathematical truths, common notions, and God's existence among the beliefs which can be questioned and considered as false.[23] The aforementioned *Annexa* to the *De abusu philosophiae in theologia* show Cartesian influences which can be related to what De Raey wrote in the same years: doubt is useful in avoiding sense prejudices, but it must neither be applied to faith nor to everyday life.[24]

We can conclude that the references to Descartes intensify over time in Heereboord's works, moving from the corollaries to the actual text of the disputes and involving both physics and metaphysics. However, the presence of Cartesian arguments in Heereboord's works must be interpreted within the framework of his notion of *libertas philosophandi* and of his belief that his specific duty, as a university teacher, consisted in outlining to his students different schools of thought.

3   Johannes De Raey's Teaching in Leiden

Johannes De Raey offers a quite different scenario. A few years younger than Heereboord, De Raey came into contact with Cartesianism at first in Utrecht, and then in Leiden, where he immediately participated in the ongoing debates.[25] The University records give us further information: De Raey gave pri-

---

21   Heereboord, *Meletemata*, 222: "Haec unica est vera, germana, solida Philosophandi ac veritatis inveniendae via, et ratio omnem dubitationem funditus ex animis Philosophorum, hominumque omnium, eradicandi."
22   Heereboord, *Meletemata*, 329–332.
23   Heereboord, *Meletemata*, 341–343.
24   Heereboord, *Meletemata*, 221.
25   Heereboord quotes a remarkable episode: De Raey steps in as an opponent during a dispu-

vate lectures from 1648, and in March 1651 was entrusted with the teaching of
Aristotle's *Problemata*. Two years later, however, the entire philosophy course
was reconfigured: Stuart reclaimed the chair of metaphysics, Heereboord kept
that of logic, while De Raey was named extraordinary professor in physics, and
Henricus Bornius (ca. 1617–1675) in ethics.[26] De Raey's activity as a professor
did not raise so many controversies as his activities as a student did, but it is
still studded with setbacks and reprimands, so that he had to wait until 1661 for
full professorship.

Despite his constant commitment to Cartesian philosophy, De Raey is often
considered an exponent of the *philosophia novantiqua*. The title of his first printed work seems to justify this conclusion. Yet, one should not forget De Raey's
tactical reasons. His opening speech, pronounced on 25 March 1651, already
brings up the contrast between the *notitia vulgaris* and philosophy concerning
the knowledge of nature—a key concern in the evolution of De Raey's thought
which is based on the Cartesian dichotomy between sensory and intellectual
knowledge.[27]

In the following years, De Raey uses the distinction between *cognitio vulgaris* and *cognitio philosophica* to build a classification of the various kinds of
knowledge.[28] At the end of his life, the contrast between common and philosophical knowledge is meant to protect the other disciplines from the intrusion
of philosophy. This protection was deemed essential because of the printing of
Lodewijk Meyer's *Philosophia Sacrae Scripturae interpres* (1666), of Spinoza's
Tractatus theologico-politicus (1670), and the spread of a radical Cartesianism
at Franeker University.[29]

In this context, is it possible more precisely to define De Raey's Cartesianism? If we only take into account the *Clavis* of 1654, his concerns seem less
comprehensive compared to those of Heereboord. Moreover, the references
to the wide range of *novatores* and scholastic philosophers quoted in Heere-

---

tation chaired by Stuart and invites him to openly name the philosopher whose opinions the *praeses* is refuting. However, Stuart cannot name Descartes without breaking the
Curators' decree. When he orders De Raey to keep silent, the confrontation between the
two ends in a true uproar; Heereboord, *Meletemata*, 18–19. Cf. Verbeek, *Descartes and the
Dutch*, 48–49.

26   Molhuysen, *Bronnen*, vol. 3, 76.

27   De Raey, *Oratio inauguralis* (Leiden, 1651). The text was to be republished with some cuts
        and some minor changes in the *Clavis* of 1654.

28   De Raey (praes.) and Freherus (resp.), *Disputatio philosophica de constitutione logicae*
        (Leiden, 1668), [1–3].

29   See Bordoli, *Dio, ragione, verità*. De Raey comments on this developments of Dutch philo-
        sophy in his *Cogitata de interpretatione* (Amsterdam, 1692), 660–668.

boord's disputations are rarer in De Raey. He seems to be more compliant than Heereboord with the topics assigned by the Curators and the Academic Senate.

Some elements, however, invite us to broaden this picture. First, De Raey's caution in introducing Cartesian elements is slightly relaxed in the *Clavis* of 1654. The title announces a conciliation between Aristotle and Descartes, which in fact consists in employing Descartes's philosophy so as to establish what of Aristotle's philosophy is still relevant and what is outdated.[30] Second, we know that, in his private teachings, De Raey breaches the Curators' decree during the time spent in Leiden. Four manuscripts—one containing his *dictata* on Descartes's *Meditationes*, the others the *dictata* on *Discours* and *Principia*[31]—confirm his rule-breaking in this regard. The manuscript conserved at Leiden University library is a commentary on *Discours* and *Principia*. It belongs to the Petrus Van Musschenbroek collection and deserves further attention and comprehensive study. In this context, I can only offer some provisional considerations. The first issue raised by this manuscript is its dating. The title refers to De Raey as a teacher at Leiden University: therefore, it could be conjectured, as it has been until now, that the text was written between 1652 and 1668, as was noted by one punctilious librarian. However, the first pages contain a sort of bibliography, written by the same person who composed the text. It starts with Bacon's *Novum organum* and ends with Boyle's *De coloribus*, passing through a range of mainly scientific works. There are numerous references to Cartesian-minded authors (Regius, Clauberg, De Raey)—all works published before 1668. However, there is one potential exception: the list includes two editions of De Raey's *Clavis*, which was edited twice, in 1654 and 1677. This detail could prompt us to reconsider the dating of the manuscript. However, any conclusions drawn from a book list, which might have been added even long after the original text was written, must remain doubtful. It is more advisable to therefore date the manuscript on the basis of internal references.

The web of cross-references to Cartesian texts in the Leiden manuscript is large: quotations of the parallel excerpts of the *Meditationes*, the *Objectiones* and *Responsiones* are recurrent, as is the reference to the *Discours*, but sometimes we find some references to Descartes's correspondence. For example, in his commentary on the biographical facts emerging from the *Discours*,

---

30   De Raey, *Clavis philosophiae naturalis*, 1st ed. (Leiden, 1654), *Epistola dedicatoria*, n.p.. See Del Prete, *"Duplex intellectus,"* 161–174.

31   Verbeek, "Les *Principia*," 701–712; Collacciani, "Manuscrits cartésiens," 168–172.

De Raey spots some numerological similarities with Christ's life and explains Descartes's bias in favour of the Netherlands by quoting his letter to Guez de Balzac, where Descartes praised Amsterdam at the expense of Italy.[32] The *terminus post quem* for the manuscript can be established thanks to a marginal note at *Principia* I 57 referring to two disputations upheld by De Raey: "Vide Disputat nostras De formis et de vero et falso." I could find no disputation *De formis*. Yet, the table of contents of the *Cogitata de interpretatione* states that the disputations *De vero et falso* were debated in Leiden between 1667 and 1668. This manuscript would therefore attest to De Raey's last years in Leiden.

The manuscript is a testament to the relevance of De Raey's methodological concerns. We can identify some characteristic elements of De Raey's philosophical production, such as the therapeutic use of philosophy in eliminating prejudices. The distinction between disciplines is linked to the use of doubt, as we can see in the comment to *Principia* I 3. There, De Raey claims that philosophy cannot have a utilitarian aim since it is useless in everyday life. This singles out philosophy from the other disciplines: on De Raey's account, the quest for truth and the lack of utility seem to be directly proportional. Since it concerns the truth and not everyday life, philosophy can and must employ doubt.[33] In the Leiden manuscript as well as in the printed texts, more specifically, De Raey contrasts philosophy with three other disciplines—theology, law, and medicine:

> So too there may be some things in Medicine and Jurisprudence that exceed our capacities and yet we must accept them, because experience is the principle of Medicine, the will of the prince that of Jurisprudence, as divine revelation is the principle of Theology. But the one and only principle of Philosophy is reason: these things have not been well grasped so far.[34]

---

32   MS., Leiden, BPL 907: De Raey, *Analysis sive argumenta eorum, quae continentur in Cartesii Dissertatione de methodo et Principiis philosophiae* (*post* 1667), fol. 8ʳ.

33   MS., Leiden, BPL 907: De Raey, *Analysis sive argumenta eorum, quae continentur in Cartesii Dissertatione de methodo et Principiis philosophiae*, fol. 8ᵛ.

34   See MS., Leiden, BPL 907: De Raey, *Analysis sive Argumenta eorum, quae continentur in Cartesii Dissertatione de Methodo et Principiis philosophiae*, fol. 13ᵛ: "Sic etiam aliqua esse possunt in Medicina et Jurisprudentia quae excedant captum nostrum et tamen admittenda sunt, propter experientiam, quae Medicinae, et voluntatem principis, quae jurisprudentiae principium est, ut revelatio divina Theologiae, solius autem Philosophiae unicum et solitarium principium ratio est quae hactenus non bene animadversa sunt."

What about De Raey's university courses? Did De Raey respect the Curators' decree in his public lectures? To answer these questions we have to move on to the second edition of the *Clavis*, printed in 1677, and to the *Cogitata de interpretatione*, published in 1692. I was able to date most of the disputations included in these books: a number of them date back to De Raey's years in Leiden.[35] These works prove that he breaches the Curators' prohibitions in both his private and his university lectures and that he publicly taught Cartesian philosophy even before moving to Amsterdam in 1669.

The second edition of the *Clavis* and the *Cogitata* attest to a varied and intense activity, and are proof of De Raey's strong commitment to Descartes's metaphysics starting from the early 1660s.[36] For example, the *Pro vera metaphysica, quae de principiis humanae cognitionis tractat* starts by pointing out the distinction between the common and the true metaphysics: the first is Aristotelian metaphysics, the second is first philosophy which concerns the principles of human knowledge and is identified with the Cartesian metaphysics. The text continues by retracing the phases outlined in the first part of the *Principia*, making up their paraphrasis with some additions and omissions.[37] We are therefore able to compare De Raey's use of Cartesian philosophy with other kinds of Cartesianism and more generally with the developments in post-Cartesian philosophy. Some interesting connections emerge about the mind-body union, the definition of the soul, the relationship between immaterial beings and place, and the nature and use of doubt.

3.1   The Mind-Body Union

The topic of the mind-body union is not the subject of a specific treatment but emerges at the end of a long analysis of human knowledge, in the series of disputations discussed between 1662 and 1664 and collected in *De mentis humanae facultatibus et erroribus circa verum et bonum sectiones duae*. De Raey stresses the supremacy of intellectual knowledge over sense perception and openly questions the peripatetic saying *Nihil est in intellectu quod prius non*

---

35   See Appendix. The disputations available in online catalogues dealing, amongst other themes, with medicine can be added to this list: *De origo et natura ignis* (Leiden, 1654); *Disp. medicae de febribus* (Leiden, 1659); *Disp. medico-physiologica de calido nativo* (Leiden, 1659); *De sede animae* (Leiden, 1664); *De dubitatione* (Leiden, 1665).

36   Unlike Strazzoni, *Dutch Cartesianism*, 105–125, on the basis of the dates of the original disputations (see Appendix), I propose to antidate De Raey's interest in Cartesian metaphysics and to place it during his stay at Leiden University. On De Raey *Cogitata de interpretatione*, see also Strazzoni, "The Cartesian Philosophy of Language," 89–120.

37   De Raey, *Clavis philosophiae naturalis*, 2nd. ed (Leiden, 1677), 412–439. De Raey is more explicit than Descartes in the polemic against fatuous Scholastic verbiage.

fuerit in sensu.[38] Indeed, if our knowledge was to be based on senses and imagination, we would not be able to know God, since God is infinite and therefore impossible to grasp with these faculties. De Raey follows Descartes in his
distinction between the "comprehension" of infinity, which we do not possess, and its "intellection," which we do; following in Descartes's footsteps, he
explains the difference between the divine infinity and the indefiniteness of
extension.[39] Moreover, this disputation testifies to De Raey's involvement in
the wider post-Cartesian debate on the union between mind and body, and on
mind-body causality.

He uses an epistemological approach in line with the theme of the disputation: he starts with a discussion on sensible and intentional species, aiming
to criticize materialistic stances (there are several references to Epicurus and
Lucretius).[40] Having established the non-existence of material *species* travelling from external bodies to the mind via the senses, De Raey dismisses the
peripatetic theory of assimilation: neither the senses nor the intellect can in
any way become similar to the things perceived; there is no possible transition of the qualities from the objects to the senses, and from the senses to
the mind.[41] There is in fact a radical difference between the changes occurring
in the senses during a sensation, and the mind's perception of these changes:
alterations in our sensory organs do not represent the external objects whereas
mental species or ideas do (De Raey uses both terms). The connection between
sensations and intellectual perceptions, however, is due not to an occurrence
of efficient causality but to a sympathy and a consent relying on the union
between mind and body.[42] The issues De Raey addressed are inspired by the
complex Cartesian view on what it means that an idea is the image of a thing.
Descartes claims that an idea and the object it represents are in the same relationship as a word and the thing for which it stands.[43] De Raey goes further
than Descartes in this regard, just as his contemporaries do: although his opinions are close to Clauberg's on the union of mind and body, his thoughts can
also be linked to Louis De la Forge's in-depth analysis of human *conscience* in
the *Treatise of Man* (1666).[44] De Raey shares with the first the idea that the

---

38   De Raey, *Clavis*, 2nd ed. (1677), 297.
39   De Raey, *Clavis*, 2nd ed. (1677), 332–333.
40   De Raey, *Clavis*, 2nd ed. (1677), 363–374.
41   De Raey, *Clavis*, 2nd ed. (1677), 375–386.
42   De Raey, *Clavis*, 2nd ed. (1677), 386–396.
43   Descartes, *Le monde*, AT XI 3–4.
44   Clauberg, *Corporis et animae in homine conjunctio*, in Clauberg, *Opera omnia philosophica*
        (Amsterdam, 1691), vol. 1, 221; La Forge, *Traité de l'esprit de l'homme* (Paris, 1666), 101–128.

relationship between mind and body cannot be classified as a form of efficient causality; with the latter, the belief that the alterations of the body cannot be images, unlike the ideas of the mind.

3.2   *The Definition of the Soul*

The polemic against Aristotelianism seems to be a key theme of the *Clavis* of 1677. Let us consider the *De Aristotele et Aristotelicis* dissertation, which comes from a disputation discussed in Amsterdam in 1669. The claims with which the text starts are similar to the statements disseminated in the *Clavis* of 1654. The judgment on Aristotle's logic and metaphysics is, however, extremely negative: Aristotle has not sufficiently stressed the distinction between God and the world, between the soul and the body.[45] The same charge appears in the dissertation *De forma substantiali et anima hominis, ex Aristotele, contra aristotelicos*, which was debated in Leiden in 1665. Once again, the title might suggest Aristotle's separation from his disciples. However, the text decidedly lines up against the Aristotelian doctrine of the soul: De Raey blames Aristotle for identifying the soul with the body and making it mortal. This text is relevant because of its polemics against the definition of the soul as a substantial form. Not only does De Raey stand apart from Descartes, who still considered the human soul as the only substantial form, and from Heereboord, who endorsed Descartes's opinion, but he also drew closer to Clauberg and Wittich, who undoubtedly shared his concerns in this matter.[46] Needless to say, this is a very important evolution in Dutch Cartesianism: not by chance does Petrus Van Mastricht quote this work in his impressive *Novitatum cartesianorum gangraena*, comparing it to Wittich's writings concerning the relationship between philosophy and theology, and the rejection of the *physica mosaica*.[47]

3.3   *The Relationship between Immaterial Beings and Place*

The disputation *De loco*, debated in 1667 and included in the *Clavis* of 1677, reveals a significant convergence between De Raey, Lambert Van Velthuysen

---

45   De Raey, *Clavis*, 2nd ed. (1677), 201–218.

46   Descartes to Regius, January 1642, AT III 503 and 505; Heereboord (praes.), and Goethals (resp.), *Disputationum ex philosophia selectarum trigesima-quarta de formis*, (Leiden 1648); Heereboord (praes.) and Ouzeel (resp.), *Disputationum ex philosophia selectarum trigesimae-sextae pars prior, quae est tertia de formis*, (Leiden 1648); Heereboord, *Philosophia naturalis*, 27, 241–242; Clauberg, *Theoria corporum viventium* and *Corporis et animae in homine conjunctio*, in Clauberg, *Opera omnia philosophica*, vol. 1, 187–188 and 252, respectively; Wittich, *Theologia pacifica* (Leiden, 1671), 117–118.

47   Mosaic physics aims to ground natural philosophy on a literal interpretation of the Bible, which is supposed to be the primary source of our knowledge. On sacred, or Mosaic, physics see Ann Blair, "Mosaic Physics," 32–58; David S. Sytsma, "Calvin, Daneau, and *Physica Mosaica*," 457–476.

(1622–1685), Clauberg, Christophorus Wittich (1625–1687), and Frans Burman (1628–1679). These authors, referring to the definition of mind as a thinking substance and of the body as an extended substance, conclude that minds, angels, and God, being immaterial substances, can by no means occupy a particular place. Consequently, the scholastic distinctions between the *circumscriptive*, *definitive*, and *repletive* presence make no sense or, as claimed by De Raey, can only be applied to bodies. Immaterial substances can be ascribed to a place only through their operation, power, passive or active virtue (the lexical choices vary depending on the author).[48] From a philosophical point of view, these claims are connected to the separation of substances outlined by Descartes, to his denial of imaginary spaces, and to his correspondence with Henry More. From a theological point of view, they dissociate themselves from two theories: the Socinians' belief in a corporeal God and their opponents' defence of the presence of immaterial substances *per essentiam*.[49]

3.4   *The Nature and Use of Doubt*

The nature and the use of Cartesian doubt was a very controversial topic in the 1640s and the 1650s. De Raey addresses this issue in *De dubitatione*. In his opinion doubt must concern everything including those things we have very few reasons to question, as with the existence of our mind, God's existence, and the existence of all material objects. This kind of doubt, however, must be interpreted as a suspension of assent and not as the denial of the existence of our mind, God, or the external world: the Cartesian statement that everything we doubt must be considered as false is to be read in this way. In the pages that follow, the author lists the reasons why doubting the existence of mind, God, and external things is lawful.[50] In this disputation, the path of the *Meditationes* is fully delineated, in contrast to what happens in Heereboord's work and in other texts written by De Raey such as the *Pro*

---

48   De Raey (praes.) and Von Flammerdinge (resp.), *Disputatio philosophica de loco* (Leiden, 1667); Van Velthuysen: *Disputatio de finito et infinito, in qua defenditur sententia clarissimi Cartesii, de motu, spatio, et corpore* (Amsterdam, 1651), 72–80; Van Velthuysen, *Luculentior probatio* (Rotterdam, 1680), vol. 2, 1177–1183; Clauberg, *Corporis et animae in homine conjunctio*, in Clauberg, *Opera omnia philosophica*, vol. 1, 221–222; Wittich, *Consensus veritatis* (Nijmegen, 1659), 89–98; Wittich, *Theologia pacifica* (Leiden, 1671), 121–132, 144 and 167–184; Burman, *Synopsis Theologiae* (Amsterdam, 1699), vol. 1, 148–153.

49   Scribano, *Da Descartes a Spinoza*, 182–225; Del Prete, "Y-a-t-il une théologie (néerlandaise) cartésienne?" 89–106.

50   De Raey (praes.) and Duyrkant (resp.), *Disputatio philosophica de dubitatione* (Leiden, 1665).

vera metaphysica, where doubt is presented as a mere means to purifying the mind and allowing the establishment of metaphysics without specifying any examples of objects upon which doubt can and must dwell.[51] Cartesian doubt is explicitly opposed to sceptical doubt, and is introduced as a suspension of agreement and as a preliminary to the true philosophy, besides being clearly distinguished from concrete life and its needs, and the truths of faith. Therefore, it is strictly confined to those who are starting to pursue philosophical truth but have not yet discovered its foundations. De Raey followed Clauberg and the strategy he developed to answer the allegations of Jacob Revius (1586–1658) and Cyriacus Lentulus (ca. 1620–1678).[52] Here too, the interpretation of Cartesian doubt lends itself to a weakening and a normalisation,[53] somehow allowed by Descartes himself in the *Notae in programma quoddam* where he identifies the object of the *Meditationes* with the proof of God's existence and the path of doubt as a mere revival of the objections elaborated by the Sceptics.[54]

4   Conclusions: Comparing Two Generations of Professors

We can now draw some provisional conclusions. We have discovered a philological detail concerning Heereboord's acquaintance with Descartes's works, moving its date to early 1643. From a more general point of view, Heereboord's eclecticism is twofold. On the one hand, it corresponds to a philosophical and pedagogical ideal aimed to display different or even opposed points of view in order to challenge them. On the other, it was meant to bypass Leiden University rules concerning the use of Aristotle's works and the prohibition on making any references to Descartes. Similarly, De Raey's attempt to accommodate Aristotle and Descartes in the early 1650s can be assumed to have had a strategic aim: De Raey, indeed, would later abandon this approach. Lastly, some scholarly clichés about De Raey's Cartesianism can be rectified. His attempt to comment,

---

51   De Raey, *Clavis*, 2nd ed. (1677), 412–413.
52   Clauberg, *Defensio cartesiana*, in Clauberg, *Opera omnia philosophica*, vol. 2, 970–971; Clauberg, *Initiatio philosophi*, in Clauberg, *Opera omnia philosophica*, vol. 2, 1132–1139. For what concerns De Raey's communications with Clauberg, see the letters recently published by Strazzoni, "On Three Unpublished Letters," 66–103: an intellectual relationship continuous over time emerges, alongside a dense exchange of information aimed to coordinate the answers to be given to anti-Cartesians in the most heated moments of the polemics.
53   Borghero, "Discussioni," 1–25; Savini, *Johannes Clauberg*, 168–176.
54   Descartes, *Notae in programma quoddam*, AT VIII-B 366–368.

in his private and public lectures during his stay in Leiden, on Descartes's meta-physics attests that he did not share Regius' approach of rejecting Descartes's metaphysics.

What can be said about the differences between these two generations of professors, teaching Descartes's philosophy in Leiden? De Raey's Cartesian-ism presents common features with that of Heereboord, but there are also different elements. They share an enthusiasm for the *libertas philosophandi* and the building of a historical framework, allowing us to preserve part of Aristotle's thinking but to refuse scholasticism, and to open up to alternatives often belonging to Renaissance or Cartesian philosophy. One key difference, however, is seen in their choice of subject: Heereboord focuses on metaphys-ics and physics, rather than on logic and ethics, while De Raey, at least until 1654, abides by his teaching duties and principally deals with physics. Non-etheless, Leiden's manuscript BLP 970 provides us with evidence of a teaching activity completely centred on Cartesian texts, granting Descartes's wish to see the *Principia* adopted as a university textbook and focusing on the first two parts of this text, dealing with metaphysics and the foundation of phys-ics. Following the evolution of De Raey's thought through the works published after 1654, it can be noticed that these two features—putting Descartes at the centre of his teaching and giving room to metaphysics—are confirmed. While Heereboord, despite introducing Cartesian philosophy, never stops referring to a wide range of more or less traditional philosophers, De Raey adopts a critical stance towards Aristotle starting from the 1660s and seems to be fully immersed in the debates internal to Cartesianism. This commitment is con-firmed by De Raey's polemics with the eclectics: he clearly distinguishes the *libertas philosophandi* from eclecticism, which he often described as a current of Aristotelianism.[55]

Acknowledgments

I am indebted the participants in the Nijmegen conference *Descartes in the Classroom* and to the editors of this book: their comments were solid and very useful. A special thanks to the anonymous referees for their attentive and care-ful reading of the first draft of this chapter: their suggestions have substantially improved it.

---

55   De Raey, *Clavis*, 2nd ed. (1677), 201, 506, 544; De Raey, *Cogitata de interpretatione*, 435–436.

Appendix: List and datating of the disputations contained in De Raey's *Clavis philosophiae naturalis*, 2nd ed. (Leiden, 1677) and *Cogitata de interpretatione* (Amsterdam, 1692).

Clavis *1677*

De Aristotele et Aristotelicis *	Amsterdam 1669
De cognitione humana	Leiden 1660
De mentis humanae facultatibus et erroribus circa verum et bonum sectiones duae	Leiden 1662 and 1664
De origine erroris summatim *	Leiden 1666
Pro vera metaphysica, quae de principiis humanae cognitionis tractat	
De veritate et ordine humanae cognitionis *	Leiden 1665
De idea Dei	
De forma substantiali et anima hominis, ex Aristotele, contra aristotelicos	Leiden 1665
De mundi systemate et elementis	
De spiritu vitali in homine et animalibus	
De calore et frigore	Leiden 1665
De duro et fluido	
De humido et sicco	Leiden 1666
De loco	Leiden 1667
*De constitutione logicae**	Leiden 1668 / Amsterdam 1684
*De constitutione physicae**	Leiden 1668 / Amsterdam 1684
*De sapientia veterum**	Amsterdam 1669

* These disputations would be republished in the *Cogitata de interpretatione* of 1692.

Cogitata de interpretatione

Libertas et servitus universim atque etiam in philosophando	Leiden 1666 / Amsterdam 1684
De vero et falso	Leiden 1667 and 1668
Disputatio philosophica explicans quid nomina materia, corpus et spatium significent	Amsterdam 1686
Disputatio philosophica specimen exhibens modestiae et prudentiae in philosophando	Amsterdam 1687
Disputatio philosophica qua quaeritur quo pacto anima in corpore moveat et sentiat	Leiden 1663 / Amsterdam 1681

Reassessing Johannes De Raey's Aristotelian-Cartesian Synthesis: The Copenhagen Manuscript *Annotata in Principia philosophica* (1658)

Domenico Collacciani

1 Introduction

The study of the teaching of Cartesian philosophy in the early modern age involves problems of at least two different orders. In a narrow sense, the spread of Descartes's philosophy coincides with the history of the *institutionalization* of Cartesianism as a subject of study in universities. In a broader sense, however, the diffusion of Cartesian philosophy is also heavily marked by the many informal contexts of teaching and discussion—i.e., by its *oral transmission*—the importance of which cannot be underestimated.

In this chapter, I present a manuscript I discovered in Copenhagen Royal Library, containing student notes from a course by the Leiden professor Johannes De Raey (1621–1702). Cartesian scholars mostly know De Raey for having been involved in the so-called Leiden Crisis of 1647.[1] He most probably met Descartes in person, as suggested by some letters.[2] Although De Raey, to the best of our knowledge, never actively engaged in controversies, he is already known to scholars for his constant promotion of the new philosophy; for this, we have ample evidence from two other similar manuscripts: one from the period when he taught in Leiden (1651–1668), the other probably from the Amsterdam period (1669–1702).[3] De Raey's contribution to Cartesianism is especially worth studying since it can give us some clues about the status of Cartesian philosophy in the context of the controversies in the Low Countries.

De Raey is a representative of the *novantiqui* philosophers, that is of the thinkers and teachers who tried to mix old, Aristotelian-scholastic scholar-

---

1 Collacciani, "Manuscrits cartésiens"; The term "Leiden Crisis" refers to the clash between the proponents of Cartesianism and the Theologians (mainly Jacob Revius) that took place in 1647; see Verbeek *Descartes and the Dutch.*
2 Schuurman, "Raey, Johannes de."
3 Verbeek, "Les *Principia*"; Verbeek, "Descartes."

ship with Baconian and Cartesian new philosophy.[4] In his *Clavis philosophiae aristotelico-cartesiana* of 1654, De Raey strives to establish a synthesis in which Cartesian epistemology, interpreted as a doctrine of *praecognita*, i.e., the prerequisites of knowledge, is the foundation of a physical theory that complements and perfects Aristotelian science, rather than refuting it.[5]

I will argue that De Raey is much more careful and wary in his published books than he was in his lectures, wherein he felt free to expound on his interpretation of Descartes texts. While it is well known that the Leiden authorities controlled not only printed works but also teaching, yet De Raey seems to have been able to teach some notions in his classroom that he would not have been allowed to communicate to a wider audience. In the following pages, I will analyse the manuscript in three stages. First, I will explore De Raey's philological work on Descartes's work and life (section 2). I argue that his aim is to define a new eclectic philosophical canon that would also include the new philosophy. Next, I comment on the passages in which De Raey reconstructs the history of philosophy in order to explain how he understands the historical role of Cartesianism in relation to tradition (section 3). Finally (section 4), I will show that, for De Raey, Cartesian philosophy is a logic that deals only with mental signs and, as such, cannot come into conflict with either scholastic metaphysics or theology.

2   The *Annotata* Manuscript

The manuscript *Annotata ad* Principia philosophica *Rev. Des-Cartes, excerpta in collegio, habito sub Joh. De Raey, inchoato die 1. Maji 1658, finito die 20. Decembris* is an in-quarto volume bound in sixteenth-century parchment.[6] As the title suggests, it contains the notes taken by an unknown student at a semester course given by Johannes De Raey on Descartes's *Principles of Philosophy* (1644). The first part also contains a detailed commentary on the *Discourse on Method* (1637)—probably as a methodological introduction to the natural philosophy set out in the *Principles*. De Raey's pedagogical method is the dictation: in the two columns per page in which the unknown student wrote down De Raey's words, two kinds of text block can be clearly distinguished. The first type is the *analysis textus*, introduced by a short reference to Descartes's work (a short

4   Dibon, *Regards*; Tjissen-Schoute, *Nederlands cartesianisme*.
5   De Raey, *Clavis philosophiae naturalis* (Leiden, 1654); Strazzoni, *Dutch Cartesianism*, Strazzoni, "La filosofia"; Del Prete, *"Duplex intellectus."*
6   MS., Copenhagen, E don. var. 145 kvart: De Raey, *Annotata ad* Principa philosophica (1658).

quotation of two or three words that allows the reader to find the passage to which the comment refers). In this "explication of the text" the professor focuses more on linguistic and philological issues, while in the second kind of text block he undertakes a more philosophical analysis, often arguing by means of examples, and comparing Cartesian theses with more traditional ones.

The manuscript is structured as follows:

1) *Ad dissertationem de Methodo*, fols. 2^r–49^r;
2) *Annotata ad Principia I*, fols. 49^v–113^v;
3) *Annotata ad partem secundam*, fols. 114^r–159^v;
4) *Annotata ad partem tertiam*, fols. 158^r–224^v;
5) *Ad partem IV*, fols. 225^r–286^v.

The manuscript sheds light on a lesser-known period of De Raey's career—hence its importance. The course took place in 1658, four years after the publication of De Raey's main work—the *Clavis aristotelico-cartesiana*—and ten years before he took up a teaching post at the *Athenaeum illustrem* in Amsterdam.[7] In 1658, De Raey had recently taken up the post of lecturer in the *istitutiones medicinae*, having taught physics since 1651 as *professor extraordinarius*.[8] The fact that the course took place between May and September suggests that it was not a university assignment per se, and may instead have been conducted as a private course. The subject of the dictation, moreover—focusing on Descartes—strongly indicates a private course as it could not have been allowed between the walls of the university. Even as early as 1647, the curators of Leiden University had forbidden the philosopher to be named, and this prohibition was confirmed in 1651.[9]

Thus, the Copenhagen manuscript constitutes a vivid portrait of the opinions and the work of Cartesianism in the aftermath of the public debates initiated by Revius in 1647. In fact, the course coincided with efforts on the re-issuing of the Cartesian *Opera omnia* begun by De Raey himself and by Frans Van Schooten (1615–1660) with the publication in 1656 of the *Opera philosophica* and the translation of the letters in 1668.[10] The few references in the Copenhagen manuscript to page numbers follow the pagination of these editions. De Raey's commentary on Descartes's works might thus be understood as a set of philological and critical notes to his Elzevier edition.

With regard to the formal structure of the manuscript, it should be noted that De Raey's lesson does not differ, at least in its final form, from that of

7   Van Miert, *Humanism*.
8   Molhuysen, *Bronnen*, vol. 2, 159.
9   On the meaning of this prohibition, see Verbeek, "Dutch Cartesian Philosophy."
10   Descartes, *Opera philosophica*; Descartes, *Epistolae*; Van Otegem, *A Bibliography*.

the Göttingen manuscript known as *Conversation with Burman*.[11] Both are structured as text commentaries starting with a short quotation that allows the passage in question to be identified. Little is known about the manner in which the conversation between Descartes and the young Frans Burman (1628–1679) took place (the manuscript bears the date of 16 April 1648). Various hypotheses have been proposed: Burman may have edited the text under Descartes's close supervision or may have reworked some notes taken during the dialogue. The real question is indeed whether this "conversation" is a reliable testimony of Descartes's own views—one of the more far-reaching interpretations takes this as another text by Descartes himself—or a summary, written later by a student in theology.[12] If we consider that De Raey's course took place over a period of seven months, the commonly accepted hypothesis that the conversation with Burman lasted over just one lunch (testified by the immediacy of Descartes's reference to "this table") seems scarcely credible. Even if the dialogue form sometimes appears—i.e., when Burman asks some very pertinent questions—the Göttingen manuscript, like the Copenhagen manuscript, is more a detailed account of a lecture on Cartesian philosophy.

Such reflections should lead us to reconsider in a new light the question of the teaching of Cartesian philosophy. Before making its entrance into the university classroom, Descartes's philosophy was taught in informal settings. Descartes himself was the first teacher of Cartesian philosophy: it is well known how hard he worked to have his philosophy accepted into the teaching curricula of various institutions: first by the Jesuits, then the Dutch universities.[13] While all this may seem trivial—after all, every philosopher is by definition the first teacher of his own philosophy—one must consider that De Raey's teaching was a continuation of Descartes's own approach in the *Conversation with Burman*. This comparison between the two texts should not be judged as dictated solely by a similarity between the two manuscripts. More pertinently,

---

11   MS., Göttingen, MS. Philol. 264: *Responsiones Renati Des Cartes ad quasdam difficultates*; published in AT V 144–179 (*Conversation with Burman*), CSMK III, 332–354.

12   See the introduction in Descartes, *Conversation*; Ariew, "The Infinite in Descartes' Conversation"; Kieft, "Conversation with Burman."

13   Descartes conceived the *Principia philosophiae* by replicating the structure of the school textbook: "My intent is to write in order a textbook of my philosophy in the form of theses." Descartes to Mersenne, 11 November 1640, AT III 233, CSMK III 157. See Ariew, *Descartes among the Scholastics*. He was also actively involved in the preparation of Henricus Regius's courses at Utrecht University; see Verbeek, *Descartes et Regius*. For his critical assessment of the scholastic method of *disputationes*, see also Descartes, *Epistola ad Voetium*, AT VIII-B 120.

it is based on an important textual fact, namely that De Raey quotes in his own lectures the conversation Descartes had with Frans Burman in 1648. As a comment on the famous passage in the *Discourse on Method* in which Descartes mentions the "long chains of reasoning of the geometers," De Raey told his student that:[14]

> Our author was once asked if theological issues could be discovered by the same method. He answered that they can, except for those things that exceed the power of the intellect.[15]

The one who posed the question to the philosopher is undoubtedly Frans Burman since the quoted passage contains a clear reference to the paragraph of the Göttingen manuscript in which the same subject is discussed:

> but is it not the case that in theology too all the items are mutually related in the same sort of sequence and chain of reasoning? [Descartes] Undoubtedly, they are. But these are truths which depend on revelation, and so we cannot follow or understand their mutual connection in the same way.[16]

The fact that De Raey is familiar with this Cartesian argument and that he quotes it by explicitly referring to a spoken conversation is an important historical point because it provides an external, and therefore objective, confirmation of the authenticity and reliability of the Göttingen text.[17]

How did De Raey know what Descartes had said to Burman by way of a reply? There is a very high likelihood that this was related to him by Burman himself.[18] Or, maybe, we can suppose that Johann Clauberg (1622–1665) gave him

---

14   "Those long chains composed of very simple and easy reasonings, which geometers customarily use to arrive at their most difficult demonstrations, had given me occasion to suppose that all the things which can fall under human knowledge are interconnected in the same way." Descartes, *Discours* II, AT VI 19, CSMK I 120.

15   MS., Copenhagen, E don. var. 145 kvart: De Raey, *Annotata ad* Principa philosophica, fol. 21: "Rogatus aliquando Author noster an res Theologicae eadem Methodo inveniri possint Affirmavit excepto quod illae in nonnullis fugiant ingenii vim."

16   Descartes, *Conversation with Burman*, AT V 176, CSMK III 350.

17   It should be noted that passages from the *Conversation with Burman* were certainly also known to Wittich, who mentions them in his *Annotationes* but without explicitly referring to an oral colloquium, and to Clauberg; see Arndt's introduction to Descartes, *Gespräch*; Hübner "Descartes-Zitate."

18   On the Young Cartesian Circle near Burman, see Gootjes, "The *Collegie der Sçavanten*"; Thijssen-Schoute, *Nederlands Cartesianisme*.

his own copy of the *Conversation*. In the years between 1648 and the first edition of his *Logica vetus et nova* (1654), Clauberg studied Descartes's philosophy under De Raey,[19] which makes it reasonable to presume that they shared their sources. The third unknown copyist from whose pen was issued the Gottingen manuscript might even have been De Raey himself, but, in the absence of any other evidence, this remains only a hypothesis. I refer to the marginal note on the first page of the conversation manuscript which states that the surviving copy is the third version, copied from the original text at one remove: "Through Burman who transmitted it to Clauberg in Amsterdam on 20 April, from which manuscript I copied it in Dordrecht on 13 and 14 July."[20]

In addition to his very long and analytical commentary on Descartes's text, De Raey's manuscript is also important for some of the details it contains of Descartes's biography. These are given in De Raey's introductions which are furnished in order to put the philosophical discourse into the right context. As already mentioned above, De Raey knew Descartes, having acted as the *respondens* of some Regius's theses in Utrecht and having taken part in the Leiden Crisis.[21] Descartes's official biographer, Adrien Baillet, reveals that his requests for information whilst researching the *Vie de M. Descartes* were somewhat rudely rebuffed by De Raey, whom he quoted as giving this response: "*Vita Cartesii res est simplicissima, et Galli eam corrumperent*" ("Descartes's life is a very simple affair, and Frenchmen would corrupt it").[22] In so doing, De Raey will no doubt have deprived us of some details of the life of Descartes of which he alone may have been apprised.

A closer examination of De Raey's commentary, however, permits us to bring to light some new, all-important testimonies to Descartes's life, which are not to be found in any of the other sources—neither in the biographies by Lipstorpius (1653) or Borel (1656), nor in Baillet's "official" biography (1691).[23] Of course, this does not mean that everything that De Raey reports is true, and indeed some of his claims are quite bizarre. Perhaps most important of all is De Raey's claim that Descartes began his philosophical career much earlier than believed previously. In the introduction to his course, De Raey explains the word "meditation"

---

19   Verbeek, "Bio-Bibliographical Sketch."
20   Descartes, *Conversation with Burman*, AT V 146 (not translated in CSMK): "Per Burmannum qui 20 aprilis deinde communicavit Amstelod. Cum Claubergio, ex cujus Msto. Ipsemet descripsi, dordraci, ad 13 et 14 Julii."
21   Descartes to Regius, May 1641, AT III 371, CSMK III 181.
22   Baillet, *Vie de Descartes* (Paris, 1691), vol. 1, 30.
23   Lipstorp, *Specimina* (Leiden, 1653); Borel, *Vitae Renati Cartesii summi philosophi compendium* (Paris, 1656).

by presenting it as the third source of knowledge alongside the two mentioned by Descartes in the *Discourse*: books and the book of the world:[24]

> Here it must be observed that a man who has no knowledge whatsoever cannot meditate in himself; that is to say, he cannot conclude one thing from another by descending into himself. Therefore, a child can hardly acquire knowledge by meditating. So, our mind does not have to be a *tabula rasa*. Our author, being instructed in the multiple science mentioned above, could easily meditate. For the matter of meditation can be derived from any kind of knowledge, or from the study of letters, or from the book of the world, or from meditation itself, in such a way that one meditation gives birth to another. Our author [Descartes], who when he was fifteen years old discovered the most important things to be found in his *Dioptrics*, at the age of twenty began to reason about the earth's crusts and was learned in the threefold science. When he was fifteen-year-old he invented also algebra of his own, he solved some new mathematical problems proposed to him by the teacher only by looking into them, without the use of books.[25]

The chronology proposed by De Raey clearly contradicts the one which may be deduced from Baillet's work and from the works of the philosopher. If we take as true the claim that Descartes made the most important discoveries in the field of dioptrics at the age of fifteen, this would mean that, as early as 1611, he had already developed a research method which we previously thought had emerged no earlier than in the late 1610s. Certainly, there remains the problem of understanding what De Raey means by "the most important things" in *Dioptrics*; but in any case, whether we refer to the law of refraction or simply to the application of a physical-mathematical model to light, his dat-

---

24 Descartes, *Discours* I, AT VI 9–10, CSMK I 115.
25 MS., Copenhagen, E don. var. 145 kvart: De Raey, *Annotata ad* Principa philosophica, fol. 6ʳ: "Hic observandum hominem qui nullam plane habet scientiam, non posse in se ipso meditari, quod est descendendo in se ipso unum ex altero colligere. Hinc puer scientiam meditando vix acquirere potest. Non debet ergo mens nostra esse ut tabula rasa. Author noster multiplici instructus scientia supra memorata facile potuit meditari. Materia autem meditandi ex omni genere desumi potest, vel ex litterarum studio; vel ex mundi volumine; vel ex meditatione; ita ut meditatio una alteram pariat: atque hac 3plici materia instructus erat Author noster qui AEtatis 15 praecipuis ex iis, quae in dioptrica habentur invenit, AEt. 20 cogitare coepit de terrae corticibus. Algebram etiam 15 AEt. Ex proprio cerebro invenit, questiones mathematicas a Magistro sibi propositas hoc compendio resolvebat libros, ut novellas tantum inspexit."

ing largely precedes the known turning points of the philosopher's career.[26] De Raey believes that the physical explanation for the cohesion of the parts of the Earth's crust was known to Descartes long before he wrote *The World* in 1629–1633. This is a startling claim, given that the entire explanation relies on a theory of the movements of the fluid heavens and thus, ultimately, on the entire cosmology of the work.

Another noteworthy excerpt from the manuscript suggests that De Raey had a much deeper knowledge of Descartes's unpublished works than was previously thought. In the following quotation, he explicitly refers to the Descartes's notebook known as *Olympica*. The commentary refers to the introductory passage of the second part of the *Discourse*:

> Thus we see that buildings undertaken and completed by a single architect are usually more attractive and better planned than those which several have tried to patch up.[27]

De Raey's comment reads as follows:

> O very blessed invention! The author rightly congratulates himself on it, in some papers found after his death He alludes to a much more fruitful invention than anything he would later find in his life. Blessed is he who could know the causes of things![28]

De Raey certainly refers to the note dated 11 November 1620 in the *Olympica*: "11 November 1620, I began to understand the foundations of the marvellous invention."[29] By what means was he able to consult Descartes's notebook? At present, our poor knowledge of the circulation of Descartes's posthumous writings does not give us an answer to this question. The known extracts from the *Olympica* are available to us having been published by Baillet in his *Vie de M. Descartes*. They belong to the "French branch" of the dissemination of the philosopher's manuscripts that arrived in Paris from Sweden after his death. Clerselier had a key role in this dissemination. The writings that Descartes had

---

26   Schuster, *Descartes-Agonistes*; De Buzon, *La science cartésienne*.

27   Descartes, *Discours* II, AT VI 11, CSMK I 116.

28   MS., Copenhagen, E don. var. 145 kvart: De Raey, *Annotata ad* Principa philosophica, fol. 6ᵛ: "O felicissimum inventum iste quod Author merito gratulari sibi usus est, in chartis quibusdam post ejus mortem inventis; dum inventi alicujus longe felicissimi mentionem facit post omnibus quae in tota vita invenerat. Felix qui potuit rerum cognoscere causas."

29   Baillet, *Vie de Descartes*, vol. 1, 81: "XI. Novembris 1620, coepi intelligere fundamentum inventi mirabilis."

decided not to take to Sweden had been entrusted to his friend Cornelius Van Hogelande (1590–1662). We know that on 4 March 1650, De Raey attended the opening of the chest that Descartes had left behind in Leiden.[30] While he certainly had access to the unpublished writings contained therein, to the best of our knowledge, the *Olympica* notebook was not amongst the contents. The reference in our manuscript is evidence of a parallel circulation of the philosopher's early writings, which could have originated either from the chest left in the Netherlands or from another source. In fact, the only known source for the *inventum mirabilis* formula at the time of the making of the Copenhagen manuscript was the brief quotation in Borel's biography, which mentions it in the catalogue of posthumous writings found in Stockholm. However, the quote alone, without its context, does not give a measure of the significance of the formula. In his lecture, by contrast, De Raey's reference to the text shows that he fully understands its importance. For him, the *inventum* is undoubtedly the method. For this reason, he cites the passage in order to explain the metaphor of the *Discourse* in which the method is compared to a building.[31]

3   The Aristotelian-Cartesian Philosophy and the History of Philosophy

De Raey's course gives us an opportunity to understand some peculiarities of the first reception of Descartes's philosophy in the United Provinces. It should be noted here that, as in other similar cases, De Raey's so-called Cartesianism is nothing more than a convenient historiographic category that does not fully capture the complexity of the author's work. As Antonella Del Prete also shows in Chapter 3, in this volume, Johannes De Raey discusses Descartes's work from his own philosophical point of view, on the basis of which he judges and "makes use of" Descartes's thought. While it is true that he is a Cartesian insofar as he strove to defend the philosopher from the attacks of his adversaries, it should also be pointed out that De Raey considers Cartesian philosophy to be just one of the many options available to those who are freely disposed to reason.[32] In this section, I will attempt to shed light on two distinct problems: in the first

---

30   On the Stockholm inventory and the opening of the chest in Hogelande's possession, see Baillet, *Vie de Descartes*, vol. 2, 427–429.

31   Borel, *Compendium*, 17. For the most complete survey of the *Olympica* notebook, see Descartes, *Étude du bon sens*.

32   On Heereboord's "eclecticism," see Chapter 2, in this volume.

place, I will attempt to illustrate the defence strategies of Cartesian philosophy, and secondly, I will try to outline the main points of De Raey's own philosophy.

De Raey is the author of a series of academical *disputationes* collected in his two editions of the *Clavis philosophiae naturalis aristotelico-cartesiana* (1654 and 1677) and of *Cogitata de interpretatione* (1692). I will focus on the *Clavis*, as it originated from the same period as the *Annotata* manuscript, when De Raey was teaching physics in Leiden University. The manuscript may indeed shed additional light on the idea of an Aristotelian-Cartesian synthesis explored in the *Clavis* and therefore also on the meaning of the categories such "Aristotelian," "Scholastic" and also "Cartesian." It is important to recognize, however, that in the course on Descartes's philosophy, De Raey tackles the subject from a different perspective than that taken in the published dissertations. Instead of arguing in support of a philosophical syncretism, in these lectures he expounds Descartes's works in such a way as to stress the differences between Descartes and the rest of the philosophers. In short, the strategy required an understanding of these differences as shown in the *Annotata* which would then allow us to better elucidate the meaning of the Aristotelian-Cartesian synthesis. According to De Raey, Descartes's philosophy is the end of a long journey of human thought from antiquity up to the modern era. In a marginal note to the course, he presents this history of philosophy in brief, divided into five periods:

> It is not erroneous to divide philosophy into five periods, of which the first is the time of Democritus, Thales of Miletus, Leucippus, Anaxagoras and Hyppocrates, who neither had a school nor looked for disciples [...] the second period is that of those who founded schools, looked for disciples and created sects: such are, Socrates, Plato, Aristotle and, from then on, the Stoics, the Epicureans, etc. The third is at the time of the church fathers, their purpose was to guard the Christian religion. As it seemed to them that physics did not contribute anything to this, this part of philosophy was neglected. The fourth is scholastic philosophy, that is, during the darkness of the papacy. Its purpose was to defend its religion and mainly the Holy Mass. It lasted until 1500, that is, until the time of the Reformation, when Luther, Calvin and others began to reject the papal darkness. Their aim was to defend the reformed religion. This period extends until Descartes.[33]

---

33   MS., Copenhagen, E don. var. 145 kvart: De Raey, *Annotata ad* Principa philosophica, fol. 6ᵛ: "Philosophia non male dividitur in v periodos, quam 1 incidit saeculum Democriti, Thaletes Milesii, Leucippi, Anaxagorae, Hippocratis qui nec scholas habebant nec disciplos quaerebant [...] 2 eorum qui scholas erigebant, dicipulos quaerebant, sectas faciebant,

One of the most salient features of this brief five-step history of philosophy is in De Raey's provision of a precise definition of scholastic philosophy. In his view, the history of medieval and modern philosophy is in fact the history of how religion made use of philosophy to pursue its own ends, as shown in stages three to five. But De Raey suggests that even the second stage, the age of classical Greek philosophy, is characterized primarily by a peculiar social function resulting from its being structured into sects and schools. In the dedicatory epistle of the *Clavis*, one finds a similar, but much less critical reconstruction of the history of philosophy with regard to Aristotle.[34] As for scholasticism in the strict sense, De Raey argues that scholastic philosophy is only the medieval "papist" philosophy. If we apply De Raey's categories to an explanation of his own philosophy, we are forced to reassess the notion of Cartesian scholasticism in a different way. It is clear, at least, that De Raey excludes himself from this category, insofar as he does not consider himself a scholastic thinker. Of course, when scholars speak of Cartesian scholasticism they are thinking of the so-called second scholasticism, that is, the academic philosophy belonging to De Raey's fifth stage.[35] However, we must emphasize the fact that his notion of scholastic philosophy has primarily negative connotations as the opposite of "true philosophy." Therefore, we must clearly distinguish Aristotelians from Scholastics in order to better understand the nature of Aristotelian-Cartesian philosophy, since the short history of philosophy clearly entails a humanistic reaction against a background of medieval philosophy.

For De Raey, the real purpose of the Aristotelian-Cartesian synthesis is to rediscover true Aristotelian philosophy by getting rid of the dark age of papist scholastic philosophy. The thesis that Aristotle's physics and metaphysics were corrupted by later philosophers is even more radical than it might seem at first. Even before the Christian scholastic philosophy, the pagan philosophical schools had already accomplished the transformation of true philosophy into a system of dogmatic and abstract notions; and even before the medieval

---

Tales Socrates, Plato, Aristoteles, hinc Stoici, Epicuraei, etc. [...] 3 est circa tempus Patrum, horum scopus fuit tueri Religionem Christianam, ad id cum Physica ipsis nihil videretur conferre pars haec Philosophiae neglecta est. 4 est Philosophia Scholastica, in tenebris sc. Papatus, ejus scopus erat religionem suam et praecipue Missam defendere, duravit hac ad 1500, ad periodum sc. 5 quae sunt tempus reformationis cum Luterus, Calvinus alii tenebras pontificias cœperunt depellere, atque hujus scopus erat Reformatam religionem defendere: atque haec periodus extendit se ad Cartesium." On the role of the history of philosophy according to de Vries, see Chapter 10, in this volume.

34   *Epistola dedicatoria* in De Raey, *Clavis philosophiae naturalis*, [1–22].

35   Bohatec, *Die cartesianische Scholastik*; Dibon, *Regards*; Ariew, *Descartes among the Scholastics*; Ariew, *Descartes and the First Cartesians*.

scholastics, the "Peripatetics" or "Aristotelians" had already distanced themselves from the true teaching of the master. Who are "the philosophers," asks
De Raey, to whom Descartes refers when he states that:

> I have often noticed that philosophers make the mistake of employing
> logical definitions in an attempt to explain what was already very simple
> and self-evident; the result is that they only make matters more obscure.[36]

De Raey's answer is as follows:

> He is referring in particular to the Aristotelians and Scholastics who, lack-
> ing a science of the particular things themselves—which is indeed a hid-
> den and difficult science—employed all their subtlety about general and
> very simple notions which for the most part depend on abstractions and
> fictions, such as: entity, existence, power, necessity, and others in logic and
> in metaphysics.[37]

There is no doubt that, for De Raey, Aristotle himself should be distinguished,
and to some extent saved, from the distortions introduced into his philosophy
by his epigones. According to De Raey, the writings of Aristotle, which came
to Christianity through the Arabs, have been corrupted by errors in transcription and by a dogmatic interpretation which has obtusely denied the fallibil-
ity of the philosopher.[38] This runs contrary to the main teachings of Aristotle
which reflect his open attitude towards experiences and research on nature. It
is in this particular interpretation of the Stagirite's thought that the profound
significance of the synthesis between Aristotle and Descartes is therefore to
be found. According to De Raey, Descartes's anti-scholastic polemic combines
naturally with the humanist rediscovery of an Aristotle who is no longer the

---

36   Descartes, *Principia* I 10, AT VIII-A 8, CSM K I 195.
37   MS., Copenhagen, E don. var. 145 kvart: De Raey, *Annotata ad* Principa philosophica,
      fol. 12ᵛ: "Loquitur de Aristotelicis et Scholasticis, praecipue qui destituti rerum ipsarum
      et particularium scientia, quae abdita et abstrusa est, omnem sua subtilitatem exserunt
      circa notiones generales, et valde simplices, quae ut plurimum ab abstractionibus et fic-
      tionibus quibusdam pendent, qualia sunt ens, existentia, essentia, potentia, necessitas,
      aliaque in logica et metaphysica."
38   De Raey, *Clavis*, [7]: "Aliquot saeculis nemo vel Averrois vel Aristotelis autoritatem in
      dubium revocare, vel veritatem alibi quaerere ausus fuit vel potuit. Quotquot philo-
      sophabantur caeci imitators ac servum pecus errant" ("For several centuries no one ques-
      tioned the authority of Averroes or Aristotle or dared or was able to seek the truth else-
      where. All those who philosophized were blind imitators and enslaved flocks").

infallible founder of a sect, as the scholastics believed, but is rather the first one to have really interrogated the book of nature.

In the light of this, we might ask which school of philosophy De Raey believes himself to belong. He is certainly not a scholastic philosopher in the manner of the Medievals. Certainly, he teaches Aristotelian philosophy, as imposed by the curators of the University of Leiden to whom the *Clavis* is dedicated, but he reserves the right to criticize Aristotle where he judges him to be mistaken. Equally, he does not uncritically embrace Descartes's philosophy either, but instead makes use of it to reform the philosophy of nature transmitted to him. His only concern is with the study of nature itself, not with the books or the propositions of philosophers. In the introductory epistle to the *Clavis*, fidelity to Aristotelian texts is nonetheless recommended for reasons of prudence, as it is dictated by the need to maintain a traditional curriculum in a public university. From a strategic point of view, De Raey's edited work represents an attempt to bring an innovation into the school, albeit one that is still strongly tied to an ideal of science that has in the meantime become outdated by new scientific discoveries. In the private course, however, the author is more at liberty to devote himself exclusively to Descartes and can afford to expound more clearly his own views on philosophical research in general.

The programme of an Aristotelian-Cartesian philosophical synthesis is brought more sharply into focus if one reads it in the light of the aforementioned five distinct epochs into which De Raey divided his history of philosophy. One of the main theses of the *Clavis* is that Aristotelian prime matter has nothing to do with non-being and the abstractions of the scholastics. The true meaning of the Aristotelian prime matter, according to De Raey, is extension, in the sense in which Descartes discusses it in his natural philosophy: "It is clear that the mere extension in length, breadth and depth constitutes the substance of matter, which Aristotle says constitutes mass, i.e., the body."[39] From a historical point of view, a thesis like this implies that there is a direct link between the beginning of the history of philosophy and its end, namely the contemporary age; from the first age, and the start of the second until Aristotle, who marks the end of the golden age, through to the age of Descartes. De Raey's sketch of the history of philosophy is based on a well-defined epistemological structure on which all the historiographical categories so far seen depend.

---

39   De Raey, *Clavis*, 54: "Manifestum est, nudam et solam in longum, latum ac profundum extensionem, totam materiae substantiam, quae Aristoteli ὁ ὄγκος καὶ τὸ σῶμα dicitur constituere." I would like to thank Roberto Granieri for helping me with the transcription.

Here we can recall the twofold difference between philosophical knowledge and common knowledge. The first difference lies in the method
which is lacking in common knowledge, since it is a rough and chaotic
mass of notions accumulated from the beginning of life. It follows that
the common knowledge is not deduced from the first principles and
from the highest causes, but it only concerns the so-called τὸ ὅτι. For
example, people know that the sun is bright; philosophy asks why it is
bright and in what the nature of light consists. Indeed, the best way
to philosophize is to focus on what the knowledge of several things as
principles is dependent. For example, in physics, which studies motion,
[one must focus] on the figures and sizes of bodies so that through
them many things can be explained, while I affirm that whoever only
considers faculties, forms and qualities understands nothing. And therefore, it is necessary to ask why very subtle philosophers, both peripatetic and scholastic, have made so little advancement in the contemplation of nature. Concerning Democritus, Epicurus, and other ancients
who also focused on figures and motion: their principles are in part
the same as our author's, although they did not use the appropriate
method.[40]

This long text deserves quotation in its entirety because it makes explicit the
close link between epistemological theory and history of philosophy according
to De Raey. There are two modes of knowledge whose characteristics clearly
echo Descartes's definition of knowledge in the *Principles of Philosophy* and in
the *Discourse*. On the one hand, ordinary knowledge does nothing more than

---

40   MS., Copenhagen, E don. var. 145 kvart: De Raey, *Annotata ad* Principa philosophica:
     "Hinc colligi potest duplex differentia cognitioni Philosophica a populari. Una consistit
     in methodo, quae in vulgari cognitione nulla est, cum tantum sit rudis et indigesta moles
     variarum notionum quae sine ordine ab initio vitae accumulantur. Deinde vulgaris noti-
     tia deducta non est a primis principiis, et altissimis causis, sed tantum versatur circa
     τὸ ὅτι ut loquuntur. Ex. Gr. Vulgus novit Solem lucidum esse; Philosophia quaerit cur
     luceat, et in quo natura lucis consistat. Jam vero ea optima Philosophandi ratio est, in
     qua attendimus ad ea a quibus tanquam principiis rerum multarum cognitio pendet; ex.
     Gr. In Physica qui attendit ad motus, figuras et magnitudines corporum, multa inde expli-
     care potest, qui autem tantum de facultatibus, formis et qualitatibus cogitat, nihil inquam
     intelligit. Et hinc ratio debet peti, cur Philosophi subtilissimi, tum peripatetici, tum schol-
     astici, tam parum profecerint in natura contemplatione: quantum vero ad Democritum,
     Epicurum, aliosque antiquos; qui etiam ad figuras et motus attenderunt, adeoque iis-
     dem ex parte cum Authore nostro nisi fuere principiis illi recta methodo uti non fuer-
     unt."

passively accumulate sensible perceptions; on the other, philosophy seeks to go beyond appearances by aiming at principles and first causes. Of course, Descartes is neither the first nor the only one to highlight the importance of principles in epistemology. De Raey certainly has in mind sections both of Aristotle's *Organon* and Bacon's *Novum Organum* in which science is said to depend on the knowledge of first principles.[41] However, the definition of such principles as motion, figure, and magnitude is undoubtedly taken from Descartes.

Having made it clear that, for De Raey, the only true philosophy is corpuscular philosophy, we can better understand the way in which he interprets the philosophies of the past. At the dawn of Western philosophy, he argues, the philosophers of nature had correctly guessed what the correct principles of philosophical research were, but that, between this remote beginning and the contemporary rebirth, a long period of oblivion had intervened. In spite of the similarities between the old and the new philosophy, the latter is still by far the more successful, because it also includes the second distinctive element of true philosophy, namely method. It is clear that, for De Raey, only Cartesianism is synonymous with true philosophy, since it consists of both a corpuscular physics and the *felicissimum inventum* of the method. It remains to be understood, however, what role Aristotle was to play within a historical and theoretical framework so established. There is little space for reflections on Aristotle in the course devoted to the work of Descartes. But it is precisely the sincerity with which De Raey outlines his vision of philosophy that allows us to appreciate the nature of the Aristotelian-Cartesian synthesis he propounds. If it is true, on the one hand, that the first phase of atomistic philosophy comes to an end with Aristotle, on the other, it is also true that in the *Organon* he has instituted the philosophical method. The core thesis of the *Clavis* finally becomes clear. While Cartesian philosophy is the primary model, it is still possible to implement a corpuscular approach in Aristotelian physics by establishing an identity between prime matter and extension. To this end, one must accuse the scholastics of having corrupted the thought of the "true Aristotle," in order to comply with the curators' instructions, which limit teachers to Aristotle's philosophy alone. Into the traditional doctrine, one can thus introduce the two principles of true philosophy whose paradigm is provided by Descartes.[42]

---

41   Aristotle, *An. Post.* I, 6, 74ᵇ25; I, 9, 76ᵃ16; III, 3, 90ᵇ25–28. Bacon, *Novum Organum* I, CIV–CV, 160–162.

42   See Chapter 3, in this volume.

4   *Philosophia* as a Semiotic

The ideal presented here by De Raey is a far cry from the philosophy that was actually being taught in Dutch universities, even if it would seem that Descartes had merely achieved a synthesis of pre-existing elements that the history of philosophy had already encountered: namely, corpuscularism and the need for a method. In the course of his commentary on Descartes's *Principles*, as well as sketching out a history of philosophy and defining the structure of philosophy itself, the professor raises the question of how his project of reforming philosophical practice can be realized. The 'philosophers' (i.e., all those who oppose the new philosophy) hesitate to abandon the old knowledge because they have never seriously questioned the principles upon which it is based. A serious analysis of these so-called principles would show that they are only prejudices and far from being the basis of true science:

> In mathematics many things are obvious and known per se. Philosophers, believing their principles to be equally obvious, very often assume as principles absurd prejudices and the errors of childhood. The same principles, which must be sought after with effort and attention, shun the senses and common thought. For example, one principle of physics is that everything that moves is moved by something else. This is either completely missing or is denied by assuming the opposite principle, by the common and infantile errors, saying that all bodies have an internal natural principle of motion.[43]

De Raey's argument conflates two Cartesian theses that are not directly connected in the original. On the one hand, it addresses the issue of prejudices as obstacles hindering our access to a genuine philosophy. On the other, this theme is interpreted in the light of the Cartesian argument about principles in the second set of *Replies*. Prejudices do not merely hinder the understanding of truth: because of their simplicity, they also disguise themselves as absolute principles akin to those of mathematics. By contrast, it is a well-known

---

43   MS., Copenhagen, E don. var. 145 kvart: De Raey, *Annotata ad* Principa philosophica, fol. 39ʳ: "In mathematici pleraque sunt obvia et per se nota. Philosophi sic putantes obvia quoque eque sua Principia, absurda praejudicia et errores infantiae saepissime assumunt pro principiiis; ipsa autem principia quae studio et labore investigari debent, fugiunt sensum et cogitationem vulgi. Ex. Gr. Physicae principium est quicuid movetur ab alio movetur, hoc vel penitus latet, vel negatur per assumptionem principia contrarii, ex vulgi et infantiae erroribus, dicendo omnia corpora naturalia internum habere principium motus."

Cartesian tenet that the principles of metaphysics are not immediately access-ible but can only be accessed by means of long preliminary work. This is what the philosopher affirms in response to Mersenne who asked him for a synthetic demonstration of the meditations.[44]

The aim of the study of the different philosophical systems, therefore, is not so much to establish the truth of one philosophy against another. As said, Atomism and Aristotelian logic would both have good reasons for claiming to be true: the former, because of its corpuscular physics; the latter, because of its focus on method. However, the dark centuries of scholastic philosophy left such deep-rooted prejudices that it has become impossible to return to these original insights. According to De Raey, philosophical research must start all over again, seeking certainty by applying the rule of evidence and gradually destroying the false opinions (e.g., the substantial forms) that neither Descartes nor Aristotle ever supported. Research must therefore abandon the historical comparison of the truths of philosophers and must move to a higher level to investigate what are the unequivocal signs of truth.

> Sceptics deny that there is a criterion of truth and that there can be a certain sign of truth. Peripatetics, for their part, do not even seek a cer-tain criterion of truth, but hold it sufficient to say that what is true is that which conforms to things or, as Ramus says, states how things are. This is obviously useless and ridiculous because no one has ever doubted this, but the point is: from what we can know, we state the thing as it is and therefore our judgement conforms to things. Nor does it help to say that things are the measure of our knowledge and therefore we should know from things themselves when we speak the truth. For the mind perceives nothing closely except its ideas and opinions, and from these, as signs and measures, it must learn what is true and false.[45]

---

44   Descartes, *Meditationes, Responsiones* II, AT VII 157, CSMK II 111: "In metaphysics by con-trast there is nothing which causes so much effort as making our perception of the primary notions clear and distinct."

45   MS., Copenhagen, E don. var. 145 kvart: De Raey, *Annotata ad* Principa philosophica, fol. 32ʳ: "ideoque quod criterium esset et certum signum veritatis quod dari posse negant Sceptici. Peripatetici autem ne quidem inquirunt ullum certum criterium veritatis, sed satis esse putant si dixerint veram esse enunciationem quae rebus conformis est, seu ut Ramus loquitur, pronunciat uti res est, quod plane ineptum est et ridiculum cum nemo de eo dubitet, sed quaestio sit unde scire possemus, quod pronunciemus uti res est, adeoque judicium nostrum rebus conformis sit. Nec quicquam juvat dicere res esse mensuram no-strae cognitionis, adeoque ex rebus ipsis sciri debere quando veritatem dicamus […] Mens

According to De Raey, true philosophy must definitively abandon the illu-
sion of direct access to things themselves. Access to material reality is always
mediated by the modes of thought, and it is only within the analysis of ideas
that the notion of truth can find its justification. It is important to recall that the
cultural context of De Raey's analysis was shaped by the debate on the veracity
of Cartesian physics, which in Martin Schoock's *Admiranda methodus* (1643),
was accused of dealing with mere chimeras and abstractions that have nothing
to do with physical reality. More specifically, Schoock responded to Descartes
with the objection that the evidence of the "presentation" does not necessarily
imply the evidence of the contents that are represented. According to anti-
Cartesians, then, subjective certainty is no guarantor for the truth of knowledge
about things themselves.[46] Without entering here into the details of the debate,
I think it is important to point out that De Raey takes rather a radical defens-
ive position which consists in interpreting Descartes's theory of knowledge as
a mere system of signs that cannot in any way be confused with a method for
acquiring knowledge of things themselves. Since Descartes's method is a system of the selection of mental signs (the *repraesentamina*), it does not imply
any position regarding the nature of the known object, be it God or natural
phenomena. Only those who confuse the sign with that which is designated by
the sign can challenge the validity of the rule of evidence. The meaning of this
thesis is that any objection that attempts to refute Descartes by resorting to a
form of Aristotelian realism commits a logical fallacy.

The explicit target of this argument is the scholastics' substantial forms:

> So also, for the mind its own thoughts, from which it knows things by
> intuition, are better known than the things themselves which cannot be
> known in any other way. In this we fail miserably because we are accus-
> tomed to know only by a species and yet, unaware of this, we say that we
> know things themselves by a metonymy of the sign for the designated.
> Almost as if you say you see Peter when you see his portrait.[47]

---

autem proxime nihil percipit praeter ideas et opiniones suas, atque ex iis tanquam signis
et mensuris discere debet quid verum et falsum sit."

46   Verbeek, *La querelle d'Utrecht*, 216. For an extensive analysis of the Voetius / Schoock argu-
     ments, see Van Ruler, *The Crisis of Causality*; Savini, "La critique."

47   MS., Copenhagen, E don. var. 145 kvart: De Raey, *Annotata ad* Principa philosophica,
     fol. 14^v: "alias nobis incognitae, sic quoque notiores menti sunt cogitationes propriae, ex
     quarum intuitu res cognoscit, quam res illae ipsae quae alio nullo modo cognosci pos-
     sunt. Fallimur in eo misere quo assueti nihil cognoscere nisi per speciem, et id tamen non
     advertentes per metonimiam signi pro signato, dicamus nos res ipsas cognoscere; quasi
     dicas Petrum te videre cum vides ejus picturam."

Since things themselves cannot be known immediately, De Raey identifies true philosophy with logic, i.e., a semiotics that focuses on ideas as immediate signs that can give access to things by means of a mediation. From this point of view, it is possible to interpret the opposition between true and false philosophy in a new light. Scholastic philosophy misunderstands the relationship between sign-ideas and things by the surreptitious introduction of a rhetorical figure, namely metonymy, that abolishes the fundamental distinction between sign and object.

> For knowledge is nothing other than the sign (*repraesentamen*) of the thing in the intellect. Thus, clear and true knowledge means nothing other than the expression or representation in thought of those things which are outside thought in objects. E.g., I understand perspicuously that the circle is there where I represent it as a line equally distant from a central point. So insofar as we do not know something we do not represent what is in the thing and this nonrepresentation is a *quid negativum* which does not need God as its author.[48]

Why then, if Cartesian philosophy is nothing more than a logic of ideas, does it have so many opponents? The answer would seem to point to the fact that the reduction of Philosophy to a system of the intra-mental selection of representations is a double-edged sword. If, on the one hand, it saves Descartes from the attacks of scholastics, on the other, it sets up an ambiguity that risks proving impossible to resolve. The clarity and distinction of ideas is in fact a subjective sign of truth. Even atheists and heretics can claim to perceive their unorthodox theses as self-evident.

> Remove this criterion of truth and it will be the end of all human science. Not even Voetius dared to openly challenge this rule, even though he claimed that it favoured Socinians. <But> should one deny a truth that is misapplied by someone? No one in the grace of God has to lie. Even the plebs appeal to the evidence of the thing. It is objected that many others

---

48   MS., Copenhagen, E don. var. 145 kvart: De Raey, *Annotata ad* Principa philosophica, fol. 82ᵛ: "Cognitio autem nihil aliud est quam repraesentamen rei in intellectu: ergo cognitio perspicua clara et vera aliud nihil significat quam expressionem seu repraesentationem in cogitatione eorum quae sunt extra cogitationem in objectis, ex. Gr. Perspicue intelligo quid circulus sit ubi illus repraesento mihi tanquam lineam a puncto quodam medio equaliter distantem. Quatenus ergo aliquid non cognoscimus, etiam non repraesentamus nobis id quod in re est, atque hoc non repraesentare est quid negativum, quod proinde Deum Authorem non exigit."

say they perceive the thing clearly and distinctly, and yet are mistaken. <answer> The Holy Scripture is true. Even if someone does not use it correctly, one should not reject it. It is one thing to believe, quite another to be.[49]

In the lectures of 1658, De Raey already foresees that the first rule of the Cartesian method, which constitutes the strength of the new Cartesian Metaphysics, is also its weak point. Indeed, in the latter half of his career, he ended up acknowledging that Voetius was partly right. While some Cartesians, such as Spinoza and especially Lodewijck Meijer, employed the Cartesian method to counter the truths of theologians with scientific truths, thus challenging theological orthodoxy, in *Cogitata de interpretatione*, De Raey was led to a definitive separation between philosophical and scientific discourse.[50]

In conclusion, the manuscript of the *Annotata* is an important piece of evidence about the first reception of Descartes's work in the United Provinces. The private course classroom turns out to be the ideal place for De Raey to carry out his work of interpretation and integration of new and old philosophy. In sections 2 to 5 of this chapter, I have attempted to show that this operation takes place on three interconnected levels. In order to introduce Descartes into the canon of authors, De Raey deals first with the circulation of the texts and the author's biography, then with the significance of the new philosophy in relation to tradition, and lastly with its correct use.

---

49   MS., Copenhagen, E don. var. 145 kvart: De Raey, *Annotata ad* Principa philosophica, fol. 34ʳ: "Tolle hoc criterium veritatis, actum erit de omni humana scientia. Non hanc regulam aperte impugnare ausus est Voetius, tamen Socinianis eam favere propugnabat. <Sed, an E>. Veritas neganda, quod a quibus male applicetur? Ne quidem in gratia dei mentiendum est. Ipsa plebs provocat ad rei evidentiam. Objiciunt alii multos dicere se rem clare et distincte percipere, eos tamen falli. RSP. Sacr. Script. Vera est, jam si quis ea non recte utatur, an illa ergo rejicenda. Aliud est videri, aliud esse."

50   Del Prete, *"Duplex Intellectus."*

CHAPTER 5

"Let Descartes Speak Dutch": Spinoza's Circle Teaching Cartesianism

Henri Krop

1   Introduction

Descartes was much admired in the Dutch Republic, even during his lifetime. The luminary of Dutch science Christiaan Huygens (1629–1685), for example, upon hearing of the philosopher's death in Sweden, versified an *epithaphe*. It praised his wisdom, which revealed the hidden structure of the universe and invited nature to bemoan the loss of the great thinker:

> Nature, prends le deuil, viens plaindre la première
> Le Grand Descartes, et monstre ton désespoir;
> Quand il perdit le jour, tu perdis la lumière,
> Ce n'est qu'à ce flambeau que nous t'avons pu voir.[1]

In 1954, Thijssen-Schoute published her study of Dutch Cartesianism.[2] In this momentous study, she outlined the lives and works of a several dozen people inspired by Descartes's philosophy. The popularity of Descartes is also evident from the number of editions of his works published in the Netherlands. However, in the diverse landscape of reception, we may discern a specific undercurrent of people who make 'Descartes speak Dutch'. They wanted to reach a non-Latin reading public and, as such, their efforts are unprecedented in the seventeenth century. The translators and the publisher were part of Spinoza's network.

Descartes was a key influence on Spinoza. In his foreword to the *Opera Posthuma*, the Mennonite merchant Jarig Jelles (1620–1683), tells us that Spinoza studied theology during his adolescence, but afterwards "used his intellect for the study of nature," and that the "philosophical writings of the most noble and excellent philosopher René des Cartes helped him greatly."[3] And it was

---

1   Huygens, *Oeuvres de Christiaan Huygens*, vol. 1, 125.
2   Thijssen-Schoute, *Nederlands Cartesianisme*.
3   Spinoza, *Opera posthuma* (Amsterdam, 1677), * 2ʳ.

Johannes Colerus (1647–1707), one of Spinoza's first biographers, who called the French scholar Spinoza's "master teacher."[4]

Descartes was also an inspirational figure for Spinoza's circle. In the preface to the *Philosophia S. Scripturae Interpres*, Lodewijk Meyer (1629–1681) stated that the method of the "noble and incomparable Descartes" was needed to renew "philosophy and restore its natural splendour." By adopting his method, Meyer held out the hope that the boundaries of philosophy could be extended to deal with "God, the rational soul and man's highest felicity"—a clear reference to Spinoza's philosophical project as formulated in the *Short Treatise*—in such a way that all dissension in the churches might end.[5]

In the foreword to Spinoza's commentary on the *Principles of Cartesian Philosophy* (1663)—*Principia philosophiae cartesianae*—, Meyer states that Descartes took pity on the "wretched plight of philosophy and entered on an arduous new path." In mathematics "the brightest star of our age brought into light whatever was inaccessible to Ancients," and in philosophy the foundations of the sciences were laid by the mathematical method used by this "most noble and incomparable man." Spinoza, according to Meyer, had "thorough knowledge of Descartes's writings" and fulfilled the great task of rewriting Cartesian philosophy by means of the synthetic method.[6] It is obvious that Spinoza and Meyer owed a debt to Descartes's teachings and were prepared to pass these teachings on to others.

However, another member of Spinoza's circle, Adriaan Koerbagh (1634–1669), never refers to Descartes, neither in his work on rational theology—*Een ligt schijnende in duystere plaatsen*—nor in his dictionary—*Een Bloemhof*—both of which were written in 1668. Neither does Jarig Jelles (1620–1683) mention Descartes's name in his 1684 confession. In the preceding letter he states only that "Descartes's ideas will not lead to ancient Paganism and do not undermine Christian religion."[7] Philosophers of all creeds adopted them, so from a religious point of view, they are perfectly neutral. This observation seems to imply the irrelevance of Cartesian philosophy to religious observance. Meyer, likewise, does not really adopt the Cartesian method. To underline the ambiguity of the Bible, for example, he refers to the existing bulk of exegetical literature, and the ambiguity of all language is established by means of philology. Meyer's argument is a mixture of predominantly humanist-scholastic notions.[8]

4   Colerus, *The Life of Benedict de Spinosa* (London, 1706), 7: "at last having light[ed] upon the works of Descartes, he often declared that he had all his philosophical knowledge from him."

5   Meyer, *Philosophy as the Interpreter*, 25–26 and 240.

6   Spinoza, *Collected Works*, vol. 1, 226.

7   Spinoza, *Collected Works*, vol. 2, letter 48 A, 398. Jelles, *Belydenisse* (Amsterdam, 1684), *3ᵛ.

8   Krop, "The *Philosophia S. Scripturae interpres*," 105–108.

In the foreword to the already mentioned *Principia philosophiae cartesianae*, Meyer states that, according to Spinoza, some Cartesian doctrines are true, but many are false. Spinoza himself mentions Descartes's name only three times in the *Ethics*—and nowhere else in the *Opera posthuma* as a whole, with the notable exception of the Correspondence, where Descartes's ideas are frequently discussed. In the preface to part 3 of the *Ethics*, Spinoza ironically praises Descartes's ambition to deal with human emotions scientifically, even though he shared the prejudice of nearly all philosophers that human beings can place themselves outside nature. In the preface to part 5, Spinoza extensively quotes Descartes's *Passions of the Soul*. He comments that only "their subtlety" will convince people that such a "distinguished man" wrote these words; words that fall well below the level of intelligibility even of "obsolete" Scholasticism, which Descartes had told us to surpass.

Notwithstanding these oversights and sometimes overt criticism of Descartes, Spinoza's friends took great pains to propagate Cartesianism and teach this philosophy in their publications. They translated Descartes's works into Dutch "with incomparable labour and heavy costs."[9] It was the only option available to them for the dissemination of these ideas, since known Spinozists were not eligible for teaching positions. The second section of this chapter will describe these translations, which made the whole of Descartes's work available in Dutch, centuries before his works could be read in English or German. The third section discusses the motivation of the people involved—the two translators and the publishers—and the lay public they wanted to reach. Jan Rieuwertsz (1617–1686), Jan Hendrik Glazemaker (1619/20–1682), and Meyer had little apparent interest in the scientific significance of Cartesianism, which was falling out of fashion in the last part of the seventeenth century, but wanted to 'enlighten' those citizens who could only read Dutch. These translations were part of what we might call "a philosophy for the citizen of the world," as Kant would define it a century later, which combined a philosophical overview with historical writings, novels and travelogues.[10] In the seventeenth century, these categories of books, unlike philosophical works, were generally published in the vernacular.[11] To educate the citizens who lacked a knowledge of Latin, humanists had created a 'purist' Dutch vocabulary, which was then used for translating Descartes's works. This 'enlightenment' movement was inspired by a 'vernacular rationalism,' and speculated about God, man and mor-

---

9   *Relaas van de beroertens op Parnassus* (Amsterdam, 1690), 4.
10 See the preface of Kant's *Anthropologie in pragmatischer Hinsicht abgefasst* (Köningsberg, 1798).
11 Schaap, "Het fonds," 16.

ality, independently of any ecclesiastical dogma.[12] The Dutch translations of Descartes, accordingly, found their readers mainly among 'Christians without a Church'.

2   Teaching Cartesianism

2.1   *Outside the Classroom*

Two obvious ways to propagate a philosophy are formal teaching—as was the case with Aristotelianism in medieval schools, or Kantianism nowadays at our universities—and publishing (text)books. Spinozism, however, clearly overstepped the boundaries of academic freedom. In 1673 Spinoza refused a professorship at Heidelberg University, because "his teaching would surpass the limits of the freedom of philosophising."[13] Although this invitation, in the name of the Palatine Prince Elector, might have been prompted by the *Tractatus theologico-politicus* (1670),[14] the publication of this book, and the two other main texts of the Spinozist movement during the late 1660s, made Spinozism a clandestine philosophy and Spinozists *personae non gratae* for academic chairs.

In 1674, the States of Holland suppressed the *Tractatus theologico-politicus* together with Meyer's *Interpres*. Anticipating this prohibition, the three major texts of Spinozism had been published anonymously during the 1660s. The third of these texts, *Een Bloemhof van allerley lieflijkheyd* (A Garden with All Kinds of Sweetness, or a translation and explanation of all Hebrew, Greek, Latin, French and other foreign loan-words and expressions) was issued by Adriaan Koerbagh in 1668 under the pseudonym Vreederyk Warmond (Rich-of-Peace Mouth-of-Truth). The author argues that foreign words help to maintain pseudo-science, since people use these words without understanding them. Moreover, 'evil passions' motivate the clergy of all confessions, hiding their aims behind such words, and preventing ordinary citizens from judging for themselves. Under the entry 'Bible,' we read that the so-called Holy Writ is just a book, no more than legendary stories such as Reynaert the Fox and Till Eulenspiegel. Theologians hideously call it the Word of God, but the writers of the Old Testament books are unknown. An ecclesiastical assembly arbitrarily determined which books belong to the New Testament, but a later assembly

---

12   I borrowed this phrase from Buys, *Sparks of Reason*.
13   Spinoza, *Collected Works*, vol. 2, 397.
14   Cf. Hubbeling's comments on this letter in the Dutch version of Spinoza's *Tractatus theologico-politicus*.

might well change the canon.[15] In the same year, Koerbagh had only eleven quires set of his *Een ligt schijnende in duystere plaatsen* (A Light shining in Dark Places) before the printer reported him and he was arrested by the authorities after a long flight from the law. He was convicted on a charge of Socinianism, and sentenced to ten years hard labour in the Amsterdam Rasphuis.

Koerbagh's book shows that Spinozism was viewed as a radical movement which was duly supressed by the Dutch authorities. While Cartesianism was 'moderate' in comparison, it would have suffered from its association with Spinoza's circle, rendering it impossible for any members of Spinoza's circle to undertake a formal teaching of Descartes's philosophy at the Universities. To propagate Cartesian philosophy, they relied on books.

2.2   *Dutch Translations*

Between 1656 and 1661, Jan Hendrik Glazemaker translated nearly all of Descartes's works, and it is extraordinary that his translations into Dutch were available at such an early date.[16] The publisher was Jan Rieuwertsz, a lifelong friend of Spinoza. Although the Latin works were also edited in Amsterdam, they were published by other publishers.[17]

Of Descartes's main works, the *Discourse on Method* with its three essays attached—*Dioptrics, Meteorology* and the *Geometry*—was first printed in French in Leiden in 1637. It was then reprinted three times in Paris. The Latin version, however, was, as Van Otegem outlines, almost exclusively printed during the seventeenth century in Amsterdam, where twelve editions saw the light of day. Outside the Republic, there is only a 1664 Latin edition issued from London, whilst in Germany the first edition appeared in Frankfurt in 1692. The Dutch version was published under the title *Redenering van 't beleed* [*method*] *om zijn reden wel te beleiden*. Its first edition appeared in 1656 and spanned only 56 pages. In the margin of every page, we find French technical terms, which are translated by 'pure' Dutch word equivalents. On page three, for example: a. *bon sens* is translated as *goed verstand*, b. = a., c. *raison* (reden), d. *opinions* (meningen), e. *esprit* (vernuft), f. = e., g. *pensée* (gedachten), h. *mémoire* (geheugenis), and i. *imagination* (inbeelding). There is much repetition in these marginal notes and technical terms are nearly always translated by the same Dutch word. *Esprit(s)*, for example, is given twenty times in the margins of the *Redenering*, even though it is translated only in three different ways. The word is translated nine times as *vernuft*, five times as *verstand*, and otherwise as *geest*, mostly if

---

15   Koerbagh, *Een Bloemhof* (Amsterdam, 1668), preface and 95–97.
16   Thijssen-Schoute, "Glazemaker," 235–240, 242–246, 257–259; Keyser, ed., *Glazemaker*, 1–11.
17   All data in this section are taken from Van Otegem, *A Bibliography*.

combined with *animaux*. These variations are exceptional.[18] In the margin we find also a few phrases, for example, *douceurs tres ravissantes* (p. 8).

The second edition of the translation of the *Discourse* was entitled *Proeven der wijsbegeerte of Redenering van de middel om de reden wel te beleiden, en de waarheit in de wetenschappen te zoeken: de verregezichtkunde, verhevelingen, en meetkunst*, (Philosophical Essays or Discourse on the Method of Rightly Conducting One's Reason and of Seeking the Truth in the Sciences: Optics, Meteorology, and Geometry) and was based on the Latin version, which was published in Amsterdam in 1644. The change to the title exemplifies how contemporaries regarded the Cartesian method as a device to practise science. It appeared in 1659 with a foreword by the translator in which he outlined the different parts of the book, and thanked his friend, a trained mathematician, who helped him to solve some of the problems involved in doing the translations. Glazemaker also announced "a version of the Letters in order to propagate them among the Dutch."[19] This version included an extensive table of contents, which would enable the reader to navigate the text more easily. Glazemaker reworked his translation, but the technical terms, now quoted in Latin, were translated with the same Dutch word. On page one: a. *bona mens* translated by *goed verstand* as in the first edition, b. = a., c. *ratio* (reden), d. *opiniones* (meningen), e. *ingenium* (vernuft), and on page 2 of the second edition, a. *cogitationes* (gedachten), c. *memoria* (geheugenis), and b. *imaginatio* (inbeelding).

The last edition is of 1692, and is at the same time the third part of the collected works of Descartes published in Dutch. Of the original three essays included in the second edition, only the *Dioptrics* and the *Geometry* were retained, while other essays were added. These included *On Man* (Verhandelinge des menschelijken lighaams) and the *First Thoughts on the Generation of Animals*. These were not translated by Glazemaker, who died in 1682, but by Stephan Blankaert (1650–1704), the author of a Latin-Dutch oeuvre on Cartesian medicine and botany. The Latin original of the *First Thoughts* was not published until 1704. The printer was Jan Claesz ten Hoorn (1639–1714), who in 1687 unsuccessfully attempted to print the *Tractatus theologico-politicus*

---

18   See, with respect to his translation of the *Ethics*, Akkerman, *Studies*, 134–137. His conclusion, however, is "that Glazemaker limits his conscious variation to harmless spheres, and translates the truly philosophical information as consistently as possible"; *Vernuft*: p. 3 (twice), 4 (twice), 8, 9, 13, 29 and 55; *Verstand*: pp. 5, 43, 48, 51, 52 (also the translation of *entendement* and *sens*); *Geest*: pp. 6, 7, 13, 40 (twice), 41, 44, 45, 51.

19   Descartes, *Redenering van 't beleed om zijn reden wel te beleiden*, transl. Glazemaker (Amsterdam, 1656), Voorreeden, *3ᵛ: "Terwijl wy bezich zijn met de Brieven van deze Schrijver in twee delen bestaande te vertalen en aan de Nederlanders door druk gemeen te maken."

in a Dutch version—this was finally published in 1693.[20] Ten Hoorn printed scientific literature in Dutch, travel books, and the books of the basket-maker and radical Cartesian Willem Deurhoff (1650–1717). Like Rieuwertsz previously, he combined an interest in Spinoza's ideas with translating Descartes's works. His younger brother Timotheus ten Hoorn (1644–1715), also a bookseller, had a Dutch version of the *Tractatus theologico-politicus* for sale in his inventory. In 1695, some men—who later became witnesses for the prosecution—bought in Timotheus ten Hoorn's shop "seeker boek in quarto gebonden in een gemormert papier sijnde geintituleerd de naegelate Schriften van Benedictus de Spinosa" (a certain book in 4° bound in marbled paper entitled the Bequeathed Writings of Benedictus de Spinosa).[21] Timotheus ten Hoorn was summoned to appear before the magistrate, together with the son of Spinoza's publisher, Jan Rieuwertsz, but no further action seems to have been taken. In a defamatory pamphlet, both Ten Hoorn brothers were accused of having acquired Glazemaker's translation dishonestly, due to the negligence of Rieuwertsz junior.[22] Of the three editions of the Dutch version of the *Discourse*, some 35 copies are preserved, mainly in Dutch libraries.

The *Meditations on First Philosophy* were published in 1641 in Paris, and then again (in a revised edition) in 1642 in Amsterdam. Thirteen editions had appeared in the original Latin by 1692, mostly in Amsterdam and only three outside the Dutch Republic. The French version was first published in 1647 and received far less attention than the Latin original; it was merely reprinted twice. In 1656, the first Dutch version appeared without the objections and responses, but with a short message from the translator announcing these would be included in a second edition, which "will appear in due course."[23] However, the first Dutch edition does include three chapters on Cartesian logic abstracted from "Descartes's works by Jacques du Roure."[24] This section also reproduces the four rules Descartes outlined in the second part of the *Discourse*. Glazemaker translated this part from *La philosophie divisée en toutes ses parties, tirés principalement des péripatéticiens et de Descartes*, a manual which, in the first part, aligned Aristotelian logic with the Cartesian method. This translation was reprinted in 1683 together with Johannes Clauberg's comparison of scholastic

---

20   Peeters, "Timotheus ten Hoorn," 242–244.
21   Peeters, "Leven en bedrijf van Timotheus ten Hoorn," 27.
22   Peeters, "Leven en bedrijf van Timotheus ten Hoorn," 26.
23   Glazemaker, "Bericht," in Descartes, *Bedenkingen van d'eerste wysbegeerte*, transl. Glazemaker (Amsterdam, 1656).
24   Title page: *Beneffens de Redenkonst uit de schriften van Renatus Descartes door Jaques du Roure getrokken en in de Fransche taal gstelt.*

and Cartesian philosophy—probably not by Glazemaker, since it is far less purist—and Antoine Dilly's *De l'âme des betes* (Lyon, 1676).[25] Such a miscellany shows that there was an interest, among the Dutch reading public, in a new logic which would replace the traditional dialectics. As usual, the margins of the *Meditations* contain Latin technical terms. The second edition of the Dutch version, which appeared in 1657, included the *Objectiones* and *Responsiones*. A note by Glazemaker announced that the translation of the *Principles of Philosophy* was nearly finished and that he wanted to continue until all of Descartes's works were translated, including the Cartesian *Correspondence*, which had "recently appeared." Here Glazemaker refers to the French edition by Claude Clerselier of the same year. Of the two editions of the translated *Meditations*, 29 copies are known to be preserved. The last Cartesian text translated by Glazemaker was the *Passions of the Soul* in 1656. He used the original French version, and hence the marginalia were in French. We here find the same repetition of terms as in the Dutch versions of the *Discourse*. With the exception of 'passion,' for which Glazemaker sometimes gave two-word translations (e.g., lydingen of tochten), he consistently translated a French technical term with a single Dutch word. In the 1659 edition, the technical terms in the margin were in Latin, although the title remained *Les passions de l'ame, of de lydingen van de ziel, door Renatus des Cartes in de Fransche taal beschreven.*

In 1687, after Glazemaker's death, a new translation of the *Meditations* appeared, which, however, was to a large extent a reprint of Glazemaker's version, albeit in a different arrangement. It was entitled *Mediationes de prima philosophia of Bedenkingen van d'eerste wysbegeerte: met eenige verandering gebracht tot vijf door W.D.H.* These initials refer to Willem Deurhoff, a shopkeeper and a radical Cartesian who had been expelled from the Reformed Church. His foreword praises Descartes, who provided us with knowledge of the breadth of things that mankind will be able to know. He is contrasted with Aristotle, who having been driven by the senses was unable to acknowledge truth, and with Spinoza, who "undermines virtue and true religion."[26] Descartes opens "the only way to the highest truths," and the theologians, it is claimed, criticize Descartes out of fear of losing their privileges. A mind, emancipating itself from the senses, will be able to acknowledge the Divine truth with the help of Descartes, and, according to the poem at the end of the preface, will

---

25   Thijssen-Schoute, "Glazemaker," 237.
26   Descartes, *Mediationes de prima philosophia, of Bedenkingen van d'eerste wysbegeerte,* transl. Deurhoff (Amsterdam, 1687), "Voorreden," 4ᵛ. As far as I can see, Van Otegem does not mention this edition.

even be able to correct Descartes, if necessary.[27] Deurhoff himself accepted the help of Descartes in clarifying theological concepts and willingly left the Reformed Church. Of this translation, only two copies have been located: in the university libraries of Leiden and Tilburg. Deurhoff's foreword shows how Descartes was taught and used as an instrument to criticize both scholasticism and theology, and to stimulate people's capacity to think for themselves about God.

Of the *Principles of Philosophy*, Descartes's main work, fourteen editions were published during the seventeenth century: twelve in the Dutch Republic and only two outside of it. According to Van Otegem, the French version went into seven editions. The Dutch translation, with the programmatic subtitle: *A Work Enabling a Diligent Student to Climb to the Highest Stage of Wisdom and Science That Can Be Reached by the Human Intellect and to Arrive at the True Causes of the Effects of All Material Things,* (een werk, bequaam voor des zelfs naerstige betrachter, om door klare gronden tot de hoogste trap van wijsheit en wetenschap, die van 't menschelijk verstant bereikt kan worden op te klimmen en tot de ware oorzaken der uitwerkingen aller stoffelijke dingen te geraken) was published in 1657, the same year as the complete version of the *Meditations*. It combined, as the title stated, a philosophical interest in wisdom with more mundane scientific aspirations. It was made after both the Latin original and the authorized Picot translation, and included a version of Descartes's letter to Picot. In the margin we find the usual single terms in Latin and their fixed Dutch translations. On the first page we also find two phrases: *integrum usus rationis nostrae* (het geheel gebruik van onze reden) and *minima suspicio incertitudinis* ('t minste vermoeden van onzekerheit).

The last separate work to be mentioned is Descartes's *Correspondence*, which, unlike modern editions, was ordered thematically. As was the case for Spinoza's *Correspondence* and as the title of the translation states—*grotelijks tot verlichting der andere werken dienen* (contributing to the clarification of the other works)—its intended goal was to lend thematic clarifications to the works of the philosopher. The first volume appeared at Paris and Leiden in 1657, as the title page has it. It contained letters Descartes wrote in French and Latin, and some responses from others. The second volume was published in 1659 and the third in 1667. The Dutch version of the first two volumes was published by Rieuwertsz in 1661 in a single tome. In a break with the conventions used in the other translations, marginal notes referring to Latin technical terms are rare, and often we find captions, which (in the manner of a running header) refer to

---

27   Descartes, *Mediationes de prima philosophia*, transl. Willem Deurhoff, "Voorreden," ** 4ᵛ.

the title or theme of a section. On page one, for example: "some observations about the highest good" and on page two: "the nature of the highest good."

The third volume of the *Correspondence* appeared after Glazemaker's death in 1684, and was appended to the *Rules for the Direction of the Mind*, the *Search for Truth* (both incomplete texts), and *The World* (translated as *Van het licht* [Of Light]). The introduction is not made by Glazemaker. This we know, because it uses different Dutch words, such as *deurzichtkunde* instead of *verregezichtkunde* in translating *dioptrica*; *philosophie* instead of *wijsbegeerte*; *disputatie* instead of *twistgesprek*; whilst *phantasia* is translated as *verbeelding* instead of *inbeelding*. However, in all these works, the notes in the margins mention technical terms only in Latin. Of the 1661 edition, fifteen copies are preserved; of the 1684 edition, eleven; and of the 1692 collected works, also eleven. The first Latin edition of the first two volumes of the *Correspondence* came out in 1668, after the Dutch version, and that of the third volume in 1683.

In 1661 Rieuwertsz added a title page to all these previously translated works, which stated, "Alle de werken van Renatus des Cartes," and lists "all the works of Descartes," available in separate editions. Besides the five works already mentioned, the *Compendium of Music* and letters Descartes sent to the Utrecht magistrate and to Gisbertus Voet (1589–1676) were included. Rieuwertsz addresses the reader and expresses his esteem for Descartes, who did not echo the words of others, but was prepared to think independently and establish his ideas on clear evidence. Rieuwertsz was confident that all readers would share his view and would read the Cartesian works without prejudice. If so, they would "reap the fruits to be found in these works."[28] Here, once again we find the call to intellectual autonomy being presented as the core of Descartes's teaching.

A new edition of the complete works appeared between 1690 and 1692. It has some works not translated by Glazemaker, for example, the *Treatise of Man* and consists of four volumes: the first contains the *Principles of Philosophy*; the second, the *Meditations*, the *Rules for the Direction of the Mind* and the *Passions of the Soul*; the third, the *Discourse*, the *Geometry* and *Of Man* (*De homine*); and the fourth volume contained the *Correspondence*. Bookseller Jan ten Hoorn states in his preface that the previous edition was sold out and "although the printing was very costly due to the many clarifying illustrations, he gave way to the constant pressure of the many devotees."[29] We do not know how many

---

28   Descartes, *Alle de werken* (s.l., 1692), "Voorreden," ** 4ᵛ: "Geniet de vruchten, die in de Werken van deze Schrijver te vinden zijn."

29   "De boekverkoper aan de lezer," in Descartes, *Alle de werken*, n.p. (printed before the table of contents): "en door het sterk aandringen van veele Liefhebberen. Maar overdenkende

copies Rieuwertsz printed or sold but, according to the printing list of 1712, Jan ten Hoorn and his son Nicolaas ten Hoorn (1674–1728), also a bookseller, still had copies of this edition in stock.[30]

In order to get an impression of what else was on offer for the potential buyers of Descartes's works in Dutch, we find a three-volume history of the United Provinces in this catalogue next to a history of the Reformed Churches in France and their persecution, and a book in folio on naval architecture. In quarto we also find Hartsoeker's *Optics* and *Physics* in Dutch translations, and A. De Graaf's mathematical works. Moreover, some travel journals were for sale, as well as a double biography of Johan and Cornelis De Witt. In octavo we find the works of Willem Deurhoff, the satirical poetry of Salomon Rusting, and the works of Frederik Van Leenhof, the Swol minister who was inspired by Spinozist ideas. The second part of this catalogue is completely devoted to medicine and contains a few works in Latin. This catalogue of 1712 included the Dutch versions of Descartes's works in a publishing programme which offers 'radical' philosophy and worldviews to the enlightened Dutch citizens. Of the two editions of Descartes's complete works in Dutch, 31 copies are known to be preserved.

These efforts of Rieuwertsz to propagate Descartes's works to a non-Latin reading Dutch public included translations of some Cartesians. In 1658 the Amsterdam company published a logic by Clauberg under the title *Redenkonst, Het menschelyk verstandt in de dingen te beghrijpen, oordelen en onthouden stierende*—apparently not a translation of one of the two logics to be found in the Amsterdam *Opera* of Clauberg, of which the translator is probably Glaze-maker. After Glazemaker's death, his translation of Clauberg's *Paraphrasis* of the *Meditations* was published by Jan ten Hoorn. In the foreword, Glazemaker praised the eminence of Clauberg's intellect, of which this *Paraphrasis* gives proof, and his diligence for the purgation of "true science." Clauberg is like a Hercules, who cleansed the world of "age-old wrong opinions and wrong habits."[31] This translation is a device for propagating philosophy as such, and "due to his own inclination for this science," Glazemaker takes it to be his duty to give it "a Dutch cloth." In this substantial foreword, Glazemaker gives an impression of the audience of his translations: people who did not learn Latin, but were interested in scholarly questions and wanted to assess philo-

---

de groote onkosten van een ontallijke menichte der Figuren die in dit werk bevat warden ... heb ik het dan eyndelijk aangevaart om die te drukken."

30   *Catalogus van de boeken die gedrukt of in meerder getal te bekomen zijn* (s.a., s.l.), fol. A^v.

31   Glazemaker, "Foreword," in Clauberg, *Nadere uitbreiding*, transl. Glazemaker (Amsterdam, 1683), fols. * 3^r–3^v.

sophical problems independently. Rieuwertsz also published Lodewijk Meyer's *Philosophiae S. Scripturae interpres* in 1666, which the printer re-edited in 1673 and in 1674 in one volume with Spinoza's *Tractatus theologico-politicus*. It is highly possible that Meyer himself translated his own *Interpres* into Dutch in 1667.[32]

Glazemaker gave Dutch readers access to the works of Spinoza as well. Spinoza's first published work, his commentary on Descartes, was translated by the Mennonite Pieter Balling in 1664 under the title *Renatus Des Cartes Beginzelen der wysbegeerte, I en II deel, na de meetkonstige wijze beweezen, door Benedictus de Spinoza, Amsterdammer* [...] *Alles uit 't Latijn vertaalt door P.D.* (*Principles of Philosophy* parts I and II demonstrated in a geometrical manner by Benedictus de Spinoza, Amsterdammer, all translated from the Latin by P.D.). Between 1663 and 1665, Balling began with translating the *Ethics*. Jan Rieuwertsz published all of these translations and—unlike with the works of Descartes—he published them also with the original Latin.

3   "A Laborious Work Done with Great Costs": The Motives behind the Enterprise

3.1   *Glazemaker, the Translator*

The previous section outlined all translations of Descartes's works into Dutch. In this section I will try to answer the following questions: what motived Glazemaker to take on this laborious work? Why did Rieuwertsz publish them notwithstanding their costly nature? And, how were the translator and publisher related to Spinoza's circle? A simple—and correct—answer to the first two questions is that they were commissioned by his friend, the publisher Rieuwertsz and perhaps paid for by Jelles.[33] However, Glazemaker must have had a real interest in the Cartesian texts, since, as a well-to-do citizen, he had no financial need to commit to this kind of work. They were part of a whole of his translating activities, which, as we will show, were aimed at widening the intellectual horizons of citizens who, as merchants, were unable to read them in the original Latin versions. Glazemaker himself is an example of this group of citizens aspiring to education but unable to read Latin. In what follows, I present a short biography and bibliography.

---

32   Van Hardeveld, *Lodewijk Meijer*, 53.
33   Akkerman, "Glazemaker," 27. Piet Visser suggests that Rieuwertsz commissioned many translations; see Visser, "Blasphemous and Pernicious," 316–317.

Jan Hendrik Glazemaker was born as Jan Hendriksz Van Abcoude in 1619 or 1620 to Mennonite parents in Amsterdam.[34] His father was a cloth worker. His mother became a widow a few years after giving birth; in 1626 she remarried Wybrandt Beyniers, a Mennonite glazier. There is no evidence of any schooling. Jan Hendriksz adopted the art of his stepfather and became a craftsman. At the time of his marriage to a Mennonite girl in 1651, he still practised this mechanical art. Jan Hendrick adopted as a family name the Dutch word for the professional designation glazier—*glaze-maker*. In 1656 Rieuwertsz started to publish Glazemaker's translations. Of the 58 titles Glazemaker translated, 28 were published with both Glazemaker's and Rieuwertsz's names on the title page. The 58 titles listed by Thijssen-Schoute include the works of Descartes and Spinoza; eight travelogues, often in combination with other publishers; Homer (numbers 17 and 27 in the list); some essays and the *Annotations on the New Testament* of Erasmus (nr. 10 and 38); the Qur'an (nr. 30); Epictetus' *Discourses* (nr. 29); Montaigne's *Essays* (nr. 57); and the *Disquisitiones politicae* of Boxhorn, professor of history and politics at Leiden (nr. 45).[35]

During the 1660s, Glazemaker became involved with Pieter Balling and his publisher in internal Mennonite controversies. In 1683 Glazemaker's stepfather made a will appointing the publisher Jan Rieuwertsz as an executor. It shows that Rieuwertsz and Glazemaker belonged to a network of Mennonite families. Spinoza also belonged to this free-thinking Mennonite network, but there is no hard proof that Glazemaker ever met Spinoza. Letter 44, asking Jelles to stop the publication process of the Dutch version, suggests that Spinoza simply did not know that Glazemaker was translating the *Tractatus theologico-politicus*.[36] On the other hand, Glazemaker might be the addressee of letter 84 on the *Tractatus politicus*.[37] Over the course of his career, Glazemaker became a man of substantial financial means: his only daughter died in 1699, leaving a considerable fortune. Rieuwertsz auctioned off his very rich library in 1683. Glazemaker was also a man of immense erudition, at least in the second half of his life, and was—like Spinoza—an autodidact.

In 1643 Glazemaker began his arduous translating activity with the French book *L'honnete femme* by Jacques de Bosc, one of the most important feminist

---

34   Thijssen-Schoute, "Glazemaker," 206–225.
35   The numbers in the text are taken from Thijssen-Schoute's list. The 1982 exhibition catalogue gives 123 numbers: I. Philosophy 23 (20 Descartes); II. Christian morality 31, III. Religion 9; IV Political Theory 10; V. Travelogues 17; VI. History 19 and VII. Fiction 12 and VIII. Other works 2.
36   Spinoza, *Collected Works*, vol. 2, 190.
37   Klever, "Glazemaker," 25–31.

works of the seventeenth century, which advances the argument that women possess the same virtues as men. The translation is dedicated to several women, most of whom were Mennonite. The second translation of the same year is a collection of stories called *Toonneel der werreltsche veranderingen* written by several authors and also translated from the French. This book outlines the lives of people from all classes; as Glazemaker observes in the introduction, it aims to be a mirror of human life driven by passions in a changing world ruled by fate. It was reprinted in 1648 and 1651. The third book Glazemaker translated is a political allegory, originally written in Latin by John Barclay. He retranslated it in 1680 because, according to his own self-critical admission, he was completely ignorant of Latin in 1643. Having improved his Latin in the interim, he translated Livy's *Roman History* in 1646. In all, Glazemaker is presumed to have translated sixty-eight titles. The most translated author is Descartes, with eleven titles in twenty editions. Besides those previously mentioned, the other philosophers he translated include Lessius, Seneca, Plutarch and Marcus Aurelius: nine titles in all from authors dealing with morality and not with science, nor philosophy in a technical sense. Political writings that he translated include works of Lipsius, John Barclay and Boxhorn (9 editions). Other categories are history—Curtius Rufus, Livius and Bentivolgio (19 editions)—and travel books (17 editions). Glazemaker's translation of the Qur'an from a French version had five printings, only one of which was published during his lifetime. Nearly all these works were translated from French or Latin. Notably absent are translations of theological works; in the confessional society of the seventeenth century, these would not have had a unanimous acceptance amongst the reading public. Physics, too, would have excluded some parts of the general readership, and so it is only parts of the *Corpus cartesianum* that pass for translations of scientific writings.

Glazemaker's library inventory only partially confirms this profile of his translating activities, because he owned theological works as well, whilst non-Latin books were in a minority.[38] According to the auction catalogue, his "very instructive library" contained all kinds of books, "but mainly of theology and history." It contained no fewer than 3,655 numbered titles of bound books— whereas Spinoza had only about 160. Of the folio books, 105 were classified as theology and 331 on other subjects. The section of sizes smaller than octavo mentions 252 bound books in French, followed by 156 in Italian, 64 in French again, 74 in Dutch and the last twelve items in Italian again: 560 in all. If we glance through the many pages of this auction catalogue, it is striking that

---

38   *Catalogus instructissimae bibliothecae Joannis Henrici Glazemaker* (Amsterdam, 1683).

Glazemaker had relatively few books in Dutch: 39 in folio, 93 in quarto, 75 in octavo and 50 in duodecimo; 257 in all, that is to say, a mere 7 per cent of all of his books. Considering his lack of formal education, it is impressive that Glazemaker had so many Italian books, and was apparently completely at ease reading French and Latin, the languages which predominate in his library. The absence of Greek and Hebrew books is the only clear sign that the translator had no academic education. Glazemaker possessed editions of reformed orthodox theologians such as John Calvin (1509–1564) (for example, f. 34 a Dutch version of the *Institutio* and f. 53–55), Theodorus Beza (1519–1605) (f. 46–47), Voet (31 8°) and Johannes Hoornbeek (1617–1666) (74 8°). Unsurprisingly, he possessed the works of many heterodox theologians, like Dirck Volkertsz Coornhert (1522–1590) (f. 64), Jacobus Arminius (1560–1609) (8° 18, 22), Hugo Grotius (1563–1645) (f. 61), and the Socinians Johannes Crell (1590–1633) (f. 41 and 52), Conrad Vorstius (1569–1622) (4° 56–61) and Faustus Socinus (1539–1604) (8° 96). He also collected Erasmus (f. 8, 32, 73–83, 4° 20–49, 8° 40–52): for example, he had six editions of Erasmus's annotations of the New Testament, one of them his own Dutch version of 1663. He possessed Descartes in several editions and languages as well: the original Latin editions, the French versions and his own translations (unbounded books *sub littera* D.). He owned a copy of Spinoza's *Opera posthuma* (4° 203), and his own Dutch version, the *Nagelate Schriften* (4° 759). A copy of the *Tractatus theologico-politicus* in the Latin original seems to be missing, but he had a copy of the French version. The catalogue does not refer to any works of Koerbagh, but Glazemaker had Meyer's *Interpres* (4° 180) and several copies of the works of De La Court. He also possessed the Dutch translation of Hobbes's *Leviathan*. When it came to the philosophy of antiquity, Glazemaker seemed to prefer Stoic authors like Cicero and Seneca, but he had in his possession also Aristotle, for example an edition with Averroes's commentary published in Venice in 1503 (f. 197), and Lucretius. With this focus on practical rationality, Glazemaker's library reflects the reading interests of "the rationalistic citizen" of the Low Countries in the second half of the seventeenth century. It included theology, but the works of orthodox authors were studied together with the writings of freethinkers. So, Spinoza and his Mennonite friends were not exclusively interested in teaching Cartesianism; Descartes's philosophy was included as part of a broader programme to enlighten Dutch society.

3.2 *Jan Rieuwertsz, the Publisher*

The second question section 3 attempts to answer is why Rieuwertsz published this unique set of translations and how they fit into the rest of his publication activities. The unconventional nature of the books published by Jan Rieuwertsz is underlined in the secondary literature.[39] His bookshop was a well-known meeting place of "Socinians" and of "all kinds of scoffers."[40] He was born in 1616/17. In 1640 he became a member of the St. Lucas guild, which brought together painters and artisans in the book-trade. In 1649 he married for the first time and in 1653 for a second time. Like Jelles, Balling and Glazemaker, he was member of the Flemish Mennonite congregation, and he associated with them in the so-called 'Lammerenkrijgh' (War of the Lambs, the name refers to the place where the congregation gathered). They pleaded for free speech in the congregation, the irrelevance of confessions, and for a Socinian Unitarianism, aspirations outlined in the confession published by Jelles in 1684.

In 1640 Rieuwertsz started to work as bookbinder, after which he became a printer and ultimately commenced his activities as a publisher and bookseller. In publishing books, he collaborated with other Mennonite booksellers and printers. In 1675 he became municipal printer, and in 1685 he was succeeded as such by his son of the same name, who distinguished himself from his father by adding 'de Jonge' (junior) to the family name. Rieuwertsz senior died in 1687.

One hundred and ninety printers were active in Amsterdam between 1650 and 1674.[41] Together they published some 1,750 books—mostly in Latin—divided among the following categories:

Dutch books	Latin books
264 were devoted to theology and piety	521 Theology and philosophy
118 History	295 Classics
66 Travelogues and maps	189 Jurisprudence and politics
58 Mathematics and physics	134 Medicine
45 Medicine	78 History and geography
18 Jurisprudence	68 Mathematics and physics[a]

a   Schaap, "Het fonds," 27.

Comparing his publications with the figures given above, it is clear that, besides the works of Spinoza, Meyer, and Cuffeler, Rieuwertsz hardly published any other books in Latin. There are only three more titles to be found: two disputations he printed as a municipal printer, and a defence of Remonstrantism by

---

39   Visser, "Blasphemous and Pernicious," 313–315; Manusov-Verhage, "Jan Rieuwertsz," 238–245.
40   Luyken, *Ondersoeck over den inhoud van twee boecxkens* (Amsterdam, 1655), fol. B2ᵛ.
41   Schaap, "Het fonds," 26.

Poelenburgh (1655).[42] The remaining Dutch publications were mainly of a religious and pious nature, as was usual, although they were almost exclusively written by 'dissident' authors. Rieuwertsz, for example, published works by the Arminian minister Gerard Brandt (1626–1685), such as histories of the Reformation and devotional poems.[43] Another Arminian author he had in his inventory was the minister and poet Dirck Camphuysen (1586–1627).[44] He also published Mennonite authors and many pamphlets written by representatives of both sides during the War of the Lambs (some 15 titles). Between 1680 and 1683 he edited the works of the Roman Catholic mystic Antoinette Bourignon (1616–1680), who established a non-confessional religious movement.[45] These consisted of nine titles that he also edited in the original French and German versions, amounting to almost forty editions in all. To this category of religious and devotional works belongs Glazemaker's translation of the Qur'an and Bouwmeester's translation of *Philosophus autodidactus, sive Epistola Abi Jaafar, ebn Tophail de Hai ebn Yokdhan.* In the foreword to the latter, Bouwmeester stated that it was the writer's aim to show how the human intellect will arrive "at the knowledge of God and His worship" by making use of "philosophical freedom."[46] In the margin we find Latin terms, as in the translations of Descartes's works. Rieuwertsz established the same link between rational religion and freedom in the foreword to Camphuysen's theological works, where he celebrates the printing press as being extremely useful to humankind. It enables the reading public to buy books at a low price, offering them beatitude and clear insights into Scripture. He urges the readers to make use of Christian freedom and judge for themselves: "Given the diversity of sentiments, even discords, we should avoid making our own little puppet an idol."[47] Riewertsz

---

42   Poelenburgh, *Epistola* (Amsterdam, 1655); de Bie, *Disputatio logica de qualitate* (Amsterdam, 1679); De Raey, *Miscellanea philosophica* (Amsterdam, 1685).

43   Brandt, *On-partydig chronyxken* (Rotterdam, 1658); Brandt, *Verhaal van de Reformatie* (Amsterdam, 1663); Brandt, *Historie der Reformatie* (s.l., 1668); Brandt, *Stichtelijke gedichten* (s.l., 1665).

44   Camphuysen, *Stichtelyke rymen,* 20th ed. (Amsterdam, 1685); Camphuysen, *Theologische werken,* 3rd ed. (Amsterdam, 1682).

45   He published her books in collaboration with another Mennonite publisher, Pieter Aremysz; see De Baar, *'Ik moet spreken,'* 550. As dissenters they probably sympathized with her mission without being her disciples, but this affinity nevertheless goes some way towards explaining why Rieuwertsz published so many works of this mystic. Bourignon sought her readers beyond confessional boundaries.

46   Bouwmeester, "Voorreede," in Ibn Tufail, *Philosophus autodidactus,* transl. Bouwmeester (Amsterdam, 1672), *2–3.

47   Rieuwertsz, "De boeck-verkooper aan den Christelijcken leser," in Camphuysen, *Theologische werken,* 1st ed. (Amsterdam, 1661), *2–3: "Wie kan in soo grooten verscheydenheyt,

published mainly religious works—Schaap gives 58 titles in this category—but no books on dogmatic or controversial theology.

In terms of the number of books, the second largest category is philosophy: 25 titles. Besides contemporary authors like Descartes, Spinoza, Meyer, Balling, and Jelles, he published books by Epictetus, Erasmus, Petrus Ramus (1515–1572) and Omer Talaeus (1510–1562). These somewhat theologically sensitive books he published alone and not in collaboration with other publishers. Of Meyer he also published the *Nederlandsche Woordenschat*.

The outline so far shows that the guiding principles of Rieuwertsz's publications were an open mind and rationalism. In the foreword of *Zederymen* (moralistic poetry, 1655), written by his fellow Mennonite Anthony Jansen Van ter Goes, Rieuwertsz outlined the motivations behind his intention to start publishing Descartes's works.[48] Until now, he writes, he had published works on Christian religion and theology. He also gave the public pleasure by informing them about unusual and miraculous things by means of travelogues, but soon he will give them the "scherpzinnige filosofy van den vermaarden Descartes" (acute philosophy of Descartes), which will be of "aangenamen dienst" (pleasant service) to the accurate examiner. In the foreword to his edition of the complete works of Descartes, he explains that this philosophy proceeds from the axioms of the mind and uses clear arguments, and does not, therefore, get bogged down in parroting the opinions of others. According to his assurances, it will be seminal to the reader with respect to the knowledge of God and the nature of the soul.[49] For Rieuwertsz, Cartesianism is basically a God-centred philosophy.

The third largest category is travelogues, with eleven titles, which also exemplifies his open mind. The next largest sections consist of books on history (9) and politics (7). It should be noted that Rieuwertsz published only three scientific works: an edition of Abraham De Graaf's *De beginselen van de algebra of stelkonst, volgens de manier van Renatus Des Cartes* (The Principles of Algebra dealt with in a Cartesian Manner); more practical, a collection of medical aphorisms by Santorio Santorio (1561–1636) translated as *De ontdekte doorwaasseming of de leidstar der genees-heeren* (Transpiration Discovered, or the Guide of the Physicians); and a pamphlet on the controversy surrounding the delivery of unborn children in case of need.

(op dat wy niet seggen verdeeltheyt, waar in elk sijn popjen schier tot sijn Afgod maakt) het alle man van passen schrijven."

48   Rieuwertsz, "Den drucker tot den lezer," in Van ter Goes, *Zederymen* (Amsterdam, 1656), *2.

49   Rieuwertsz, "Den drucker tot den lezer," in Van ter Goes, *Zederymen*, *2.

As a bookseller and—as he explicitly described himself on the title pages of his books—the municipal printer, Rieuwertsz did not turn himself away from the world, as had Jelles and Spinoza. The list of his publications shows, however, that he "realized that money and material goods would not make his soul happy."[50] Jelles craved wisdom and, without knowledge of languages other than his mother tongue, he asked for translations of philosophical literature. To satisfy the spiritual needs of Jelles and other dissenters, Rieuwertsz published the Dutch versions of the works of Descartes and other philosophers.

3.3   *Purism*

After this outline of the translation and publication activities of Glazemaker and Rieuwertsz, which betrayed their motives, I will discuss now the nature of the Descartes translations. In two papers, Fokke Akkerman analyzed Glazemaker's translations. He concluded that they were undertaken as rather mechanical exercises, without refinement, free from interpretation, and that they respected the technicalities of the original. Glazemaker worked very quickly. He translated Spinoza's *Tractatus theologico-politicus* between autumn 1669 or the beginning of 1670 and 17 February 1671, at which time Spinoza heard that it had been translated. Between 1668 and 1671 he also translated 22 other titles, while the translations of Descartes's writings, minus the *Correspondence*, took him just two years, namely 1656 and 1657. In his translations of philosophical works, he used the 'purist' vocabulary created by the Dutch engineer and mathematician Simon Stevin (1548–1620), who converted every scientific term into a purely Dutch word and had the original term put in the margin. In 1589 Stevin wrote a programmatic essay, "Uytspraeck van de Weerdicheyt der Duystche Tael" (*On the dignity of Dutch language*), included in *Weeghconst*. Here he provided three arguments for the superiority of Dutch as a tool for science. The abundance of monosyllables facilitates the correspondence between sounds and things, and shows that the Dutch language is directly linked with the Adamic language used before the Fall. Moreover, the facility for forming composites strongly contributes to its clearness and unambiguity, because such words are in fact abbreviated definitions. Technical terms in Dutch, therefore, are clear and easy to learn and remember. The last argument for the Dutch language is its emotional force. In his mathematical and logical works written in Dutch, Stevin formed new words which should replace the traditional Latin and Greek ones. In these cases, he put the original terms in the margin. Dijksterhuis lists some hundred words which Stevin coined to correspond with terms of

---

50   Jelles, *Belydenisse*, 162.

scholastic discourse.[51] These terms in the margins are to be found in the translations of Descartes, Spinoza and Meyer, but not in the translations of Seneca,
Epictetus, Boxhoorn and Montaigne.

I created a list of 'philosophical' terms included in the margins of Glazemaker's translation of the second edition of the *Discourse on Method*, which
makes this 'purism' obvious. After the original term you will occasionally find
between brackets the number of times some of these terms were used, and if
a word is in italics, it is still in use with the same meaning. Of the 182 words in
the margin only *natura* is borrowed from a Romance language.

accidentia   toeval

actio   doening (bedrijf)

algebra   stelreegel

analysis   ontknoping

anima rationalis, sensitiva, vegetativa
   ziel, redelijke, levende, gevoellijke

arithmetica   *rekenkunst*

axioma   *grontreegel*

Bonus   *goed*

causa   *oorzaak*

certitudo   verzekering

cogitatio   *gedachte* / denking

cognitio   *kennis*

compositio   samenzetting

conclusio   besluit

conformatio   *gelijkvormigheit*

conjectura   *gissing*

consequentia   gevolg

cupiditas   begeerlijkheid

demonstratio   betoging

differentia   *verschil*

dispositio   gesteltenis

disputatio   twistreeden

distinctio   *onderscheiding*

effectus   uitgewerkte

ens   wezen

esse   wezen

essentia   wezentheit

ethica   *zedekunst*

evidentia   klaarblijkelijkheit

existentia   wezentlijkheid

experientia(e)   ervarentheit(-nisse)

fantasia   *inbeelding*

figura   gestalte

forma   *vorm*

generatio   *voortbrenging*

gloria   *roem*

geometria   *meetkunst*

habitus   hebbelijkheid

idea   *denkbeelt*

imaginatio   *inbeelding*

immediate   *onmiddellijk*

indefinite   *onbepaaldelijk*

indifferens   onverschillend

individuum   ondeelbare

ingenium   vernuft

intellectus   *verstant*

intelligentia   verstandelijkheit

intelligibile   verstandelijk

inventio   *vinding* / vonst

logica   redenkunst

materia   *stoffe*

mathematica   *wiskunde*

mens   *verstant*

memoria   *geheugenis*

---

51   Dijksterhuis, *Simon Stevin*, 307–312.

metaphysica   overnatuurkunde

methodus   beleed

modus   *wijze*

natura   *natuur*

nihil   *niets*

notio   kundigheit

notitia   *kennis*

objectio   *tegenwerping*

objectum   *voorwerp*

obscuritas   *duisterheit*

observatio   *waarneming*

opinio   *mening / gevoelen*

oppositio   *tegenstelling*

passio   tocht / lijding

philosophandi   wijsheitbetrachten

philosophia [12]   *wijsbegeerte*

physica   *natuurkunde*

potentia   *vermogen*

praeceptum   leerreegel

praesupponere   *vooronderstellen*

principium [20]   *beginsel*

probabile   *waarschijnelijk*

propositio   *voorstelling*

qualitas   *hoedanigheid*

ratio   *reden*

ratiocinatio   *redenering*

repugnatio   *tegenstrijdigheit*

res   *ding*

scientia [9]   *wetenschap*

sensus   verstant / zin

spatium   *ruimte*

speculativus   *beschouwelijk*

speculatio   bespiegeling

spiritus animalis   dierlijke geest

subjectum   *onderwerp*

substantia   *zelfstandigheid*

suppositio   onderstelling

syllogismus   *sluitreeden*

theologia   *godgeleertheit*

verisimilitudo   *waarschijnelijkheit*

veritas   *waarheit*

voluntas   *wil*

In the margins of the second edition of the *Discourse on Method*, we find 450 terms in all, that is to say, an average of 8.4 words per page. However, they are unequally distributed: on pages seven and nineteen only two terms appear, while on page 25 there are no fewer than 31. The words in the previous list are all in the singular, even if the margin sometimes also gives them in the plural. Moreover, some terms are regularly repeated: *bona mens* on page 1, *scientia* on p. 3, *philosphia* on p. 6, *demonstratio* on p. 13, *principia* on pp. 15, 44 and 48, *idea* on pp. 24 and 27, *ens perfectum* and *imaginatio* on p. 25, *arteria* on pp. 31–37, *venum* on pp. 32–36, *organum* on p. 39, *habitus* on p. 49. Crapulli counted 299 notes in the margin of the *Regulae*, not counting repetitions. He noticed that many of these notes were not those of the translator and that they were not included in the copy that was handed to the setter.[52] The use of repetition is difficult to explain, but its aim might have been to stimulate the reader to memorize the terminology.

---

52   Crapulli, *Le note marginali latine*, 116 reproduces the foreword to the second edition of the Dutch translation of Spinoza's *Tractatus theologico-politicus* (1694). The publisher

This purist vocabulary is a feature we also find in the margins of Meyers's translation of the *Philosophiae Scripturae interpres*. Below the 22 words with A:

absolutus volstreckt
acceptio beteeckenis
actio doeningh / werckingh
actus (signatus / exercitus) (gheteek-
  ent / gheoeffent) bedrijf
aequivocatio dubbelsinnigheydt
adverbium byword
affectio aendoening
ambiguitas dubbelsinnigheydt /twijf-
  felzinnigheydt
amphibolia twijffelsinnigheydt

analogia ghlyckvormigheydt
analytica ontknoopingh
anthropopatheia mensch-gelijck-
  vormigheyt
apodictice betoghelijck
appellatio benamingh
appellative benoemighlijck
argumentum bewys
attributum toe-eigheningh
axioma geloofsspreuck

These marginal notes are also unequally distributed throughout the text. The margins of the *Interpres* contain 478 words, some of which are of a very technical nature. However, on the first 40 pages we find 344 terms—an average of 8.5 per page—and in the next 80 pages 86 terms, that is, an average of only 1 per page. In the last 20 pages, the number increases slightly to 48, that is, an average of 2.5 terms. The precise policy behind this distribution is unclear to me. However, this translation of the *Interpres* meant the end of Meyer's translating activity which dated well before his introduction to Cartesianism. The purist impetus behind his translations is testified by the forewords.

3.4 *The Underlying Aims of These Translations to Dutch*

Another translator of philosophical and scholarly work, connected to Spinoza's circle is Lodewijk Meyer. His motives are akin to Glazemaker's. His biography makes obvious that their translation activities originated in an earlier tradition. Meyer was born in Amsterdam in 1629.[53] His half-brother was the famous translator of scholarly work, Allart Kók, who died in 1653 at the age of 37. Between 1654 and 1660, Lodewijk studied at Leiden University, first philosophy, and after 1658, at least officially, also medicine. In 1660 he graduated in both disciplines. His thesis in philosophy, on matter and motion, gives evidence of Cartesian influences. After his studies he returned to Amsterdam, where he set up a prac-

---

observed that Glazemaker's manuscript had no marginal notes. This is confirmed by the Vatican manuscript of the *Ethica*.

53   We find Meyer's biography in Van Hardeveld, *Lodewijk Meijer*, 11–93.

tice as a physician and became part of Spinoza's circle of friends. After 1667 he returned to the literary pursuits of his pre-university days, but maintained his contact with Spinoza.

Meyer's first publication is a poem in praise of his half-brother, and was printed alongside the Dutch version of Burgersdijk's logic, which Meyer also translated.[54] Meyer outlined the advantages for Dutch society of philosophizing and doing scholarly work in the vernacular. Translating science into Dutch is to be recommended because "Dutch is clearer than Latin, which is tainted by Greek." In the preface of the *Nederlandsche Woordenschat*, he invokes the "Batavian myth," linking political and cultural freedom. In ancient times the Dutch people fearlessly defended their "mother tongue" and liberty.[55] Meyer refers in the last lines of his laudatory poem to the enlightenment of the "Dutch people" as the objective of Kók's translating activities.

There is also a pragmatic argument. Doing science in Dutch could free the schools from the onerous duty of having to teach the pupils unfamiliar words, which overload their memory.[56] This argument returned in the *Nederlandsche Woordenschat*. Meyer urges that these years should rather be available for the exploration of things unknown until now. Young students, in particular, will be able to study the principles of mathematics directly, which will be beneficial to the study of the other sciences.[57] Two obstacles hinder the correct use of 'external reason' (that is to say speech) by the public. The first is the use of foreign words—*barbarismus*—for which Meyer suggested *ontaal* as a translation, that is, dysfunctional language; the second is the violation of linguistic rules, *solaecismus*, that is, bad speech. In order to cure the Dutch language of this 'cancer,' he lists such foreign words. These words, taken from other languages, are mostly used in a Dutch form, but in some cases are pure French. They have a general meaning and are primarily current in the fields of jurisprudence, medicine and the art of warfare. In the relevant entry the meanings are given and explained.

These objectives were anything but new and had already been formulated by his half-brother. In the programmatic texts included in the six Dutch versions of Burgersdijk's philosophical manuals published between 1646 and 1648, Kók

---

54   Meyer: "Op de oeffening der reden-konst van A.L. Kók," in Burgersdijk, *Logica practica, oft Oeffening der reden-konst*, transl. Meyer (Amsterdam, 1648), fol. xxx4: "Zo steekt de Batavier met ruime schreeden / den Romer na de Kroon der weetenheden / en stelt 't gheen hy met Grieksch be-morste taal / nauw zegghen kan in duitscher spreek-ghe-praal."

55   Meyer, "Voorreeden: Den Nederduitschen Taallieveren geluk en voorspoedt," in Meyer, *Nederlandsche Woordenschat* (Amsterdam, 1650), xv4.

56   Meyer, "Op de oeffening der reden-konst van A.L. Kók," fol. xxx4v.

57   Meyer, "Voorreeden," in Meyer, *Nederlandsche Woordenschat*, xxr.

addresses his compatriots who want instruction in the learned sciences.[58] By giving the liberal arts a Dutch form, the nation is best served and the fatherland appropriately cultivated. Like Meyer, Kók invokes the Batavian myth, observing that in ancient times "we" shed "the unjust yoke of criminal tyranny by kings."[59] Kók also suggested that he undertook his translations at the direct behest of the Dutch public, thus contravening the efforts of scholars to hinder access to the sciences. Such hideous efforts can only be motivated by envy or avarice.[60]

These arguments outline a strategy to enlighten the Dutch citizen. After the death of his half-brother in 1653, Meyer completed the translation of Wendelinus's *Christian Theology* that Kók had left unfinished.[61] In the dedication he stated that after translating works on the mechanical arts, he continued by translating treatises in philosophy itself, in order to give people access to the most "delightful science," i.e., theology, which discipline develops the higher faculties of man. In this manner, Meyer explained the link between scholarly works in the vernacular and the ability to pursue one's own happiness. "Arrogant" scholars stress the impossibility of translating scholarly works into Dutch. Moreover, some theologians object to the translation of theological works for fear of the intellectual emancipation of the common believer. Meyer adds, that if the efforts of these theologians will be successful, "Popish ignorance" will drag on and the Reformation will be in vain. Moreover, their attitude is irreligious, because the apostle Paul himself both urges the Christian to examine all things by himself and to practise charity, which primarily implies the duty to provide for the spiritual needs of our neighbour.[62] According to Meyer, theology should be part of the programme to make "the sciences speak Dutch," and in 1656 he translated, of his own accord, William Ames's *Marrow of Theology*.[63] The work is of a perfect orthodoxy; the Franeker divine is called the "flonkerstar onzer Kerken" (twinkling star of our churches), and the book an "kostelijke puikstaal" (excellent specimen) of our theology. In the preface Meyer once more underlines the moral importance of all theology, which leads people to their utmost happiness. He calls logic and metaphysics the most sublime sciences, which are like "twee vleughelen, waar mede men tot de Ghodtgheleerdtheidt, de ver-

58   In Kók, *Ontwerp der Nederduitsche letter-konst*, pxv-xxi and 74–93.
59   "het on-ghe-rechtigh juk en bal-daadighe dwing-landy der Koningen." Kók, "Op-dracht aan Mijn Heeren de Burgher-meesteren der ver-maarde Stadt Amstelredam," xxx3.
60   Kók, "Den ghoedt-hartighen Leezer," in Meyer, *Nederlandsche Woordenschat*, A2ʳ; Dibbets, "Koks Burgersdijk," 21.
61   Wendelinus, *De christlijke ghódt-ghe-leertheidt* (Amsterdam, 1655).
62   Meyer, "Den Lief-hebbers der Ghódt-ghe-leertheijdt kennis en zaligheid," in Wendelinus, *De christlijke ghódt-ghe-leertheidt*, xx2r-xx2v.
63   Meyer, "De vertaaler an den Leezer," in Meyer, *Nederlandsche Woordenschat*, x3.

heevenste der Weetenschappen opsteighert" (two wings one may use to ascend to theology, the most eminent science).[64] Meyer called translating Amesius's *Medulla theologiae* a difficult job, but necessary. It prepared the way for his compatriots to find their supreme bliss.[65] The *Interpres* itself was also translated, because of the excellence of its theology which "shows the mortals the way to live good and happy lives and is able to guide them to the eternal salvation."[66] By means of translations, the public could be enlightened about religion as well.

Meyer's own translation of the *Interpres* is his last translation into Dutch, but other members of Spinoza's circle—Balling, Bouwmeester and especially Glazemaker—continued his work. Akkerman drew particular attention to the exceptional number of Cartesian works that Glazemaker translated. They were intended for a lay reading public, who had a taste for scholarship and looked for moral education. If we are looking for an exemplar, in terms of people whom Glazemaker and Rieuwertsz wanted to teach, Jarig Jelles may well be considered a prototypical reader of Cartesian works in Dutch.[67]

Members of Spinoza's circle made Descartes speak Dutch and taught Cartesian philosophy by making an exceptional set of translations. These translations used a carefully conceived 'purist' terminology, which made them easily understandable to all literate citizens of the Dutch Republic, even if they had no schooling beyond the elementary school. Glazemaker and Meyer joined in with an earlier movement of Humanist scholars, who started to translate learned works into the vernacular. These translations became part of "a philosophy for the citizen of the world."

64 Meyer, "De vertaaler an den Leezer," in Meyer, *Nederlandsche Woordenschat*, x5.
65 Meyer, "De vertaaler an den Leezer," in Meyer, *Nederlandsche Woordenschat*, x3.
66 Meyer, *Philosophia Scripturae interpres*, Prologus, A2^{v}.
67 Akkerman, "Glazemaker," 26–27.

CHAPTER 6

Patronage as a Means to End a University Controversy: The Conclusion of Two Cartesian Disputes at Frankfurt an der Oder (1656 and 1660)

Pietro Daniel Omodeo

1   Introduction

In 1660, the Polish Cartesian of Bohemian origin, Johannes Placentinus—the Latinized form of Jan Kołaczek (Leszno ca. 1629—Frankfurt an der Oder 1683)—summarized the new philosophy of Descartes, which he supported, in a booklet he published in Frankfurt an der Oder under the title *Dissertatio philosophica, exhibens modum praecavendi errorem in veritati philosophicae, imprimis naturalis inquisitione atque dijudicatione, iuxta principia Renati des Cartes* (Philosophical dissertation, showing the manner to avoid error in philosophical truth, especially in the investigation and judgement of nature in accordance with the principles of René Descartes). [*1] The author indicated in the frontispiece that he was the dean of the Philosophical Faculty at the Brandenburg Electoral University of Frankfurt an der Oder (*Alma Electoralis Viadrina*). However, the dedication to two prestigious patrons pointed to a courtly milieu rather than a university readership for the publication. The most important of the two was Count Raimondo Montecuccoli (1609–1680), the imperial field-marshal who conducted the war against the Swedes, in which Placentinus's local patron, Prince Friedrich Wilhelm of Brandenburg (1620–1688), was also directly involved. The second dedicatee was the imperial infantry commander, Count Strozzi. Placentinus, who was both a university professor of mathematics and court mathematician to Friedrich Wilhelm since 1656, addressed the two "chiliarchoi" (or military commanders) as his patrons (*Viris sapientissimis, Patronis et Dominis suis gratiosissimis*). Such a dedication signalled that his introduction to Cartesian philosophy responded to the curiosity of some of the most powerful men of the imperial elite. This gesture was an exhibi-

---

1   [*] A revised version of this essay is included in the monograph Omodeo, *Defending Descartes in Brandenburg-Prussia: The University of Frankfurt an der Oder in the Seventeenth Century* (2022) as part of a larger chapter on Scholastic-Cartesian controversies over the origin of life which brought Frankfurt an der Oder and Wittenberg into opposition in 1659–1660.

© PIETRO DANIEL OMODEO, 2023 | DOI:10.1163/9789004524897_008

tion of symbolic power aimed at securing the legitimacy of his philosophical views.[2] At that time, courtly sociability and patronage were considered by men of science to be better suited to their intellectual ambitions than teaching in institutions such as universities or gymnasia.[3] The life and science trajectories of Galileo Galilei (1564–1642) and Johannes Kepler (1571–1630) witness to the fact that a mathematician (or a mathematically-minded natural philosopher) could achieve more independence in a court like Medici Florence or Habsburg Prague than at traditional cultural institutions—despite the dramatic end of Galileo's efforts to propagate the Copernican system in Italy.[4] Beginning in Placentinus's time, during the seventeenth century a new generation of scientists benefitted from royal support in France and England through newly founded scientific societies, which transformed the Renaissance modes of patronage and academic sociability by transferring them to more stabilized institutional frameworks.[5] Yet, it should be remarked that, at least in the case of Brandenburg, university teaching and courtly science were not too far apart. Frankfurt an der Oder is one instance of a Lutheran University with a Reformed suzerain, or *Landesherr*. Since the *Kurfürst* (the Electoral Prince) of Brandenburg was the political authority of reference, Frankfurt university also fell within the compass of courtly cultural politics.[6] In this essay, I discuss the political entanglements and the role of patronage as an important asset for Placentinus which allowed him to bring to a successful conclusion the Cartesian controversies in which he was involved. Through political support, he could legitimately teach and disseminate Cartesian ideas in his classrooms.

I here take into consideration two university controversies over the legitimacy of Cartesian philosophy. Both ended through an external appeal to powerful patrons. The first one pitted Placentinus against Frankfurt philosophers and theologians who disdained his Cartesian convictions. It particularly concerned cosmological issues, but was also extended to a more general conflict over the

---

2   On Placentinus's legitimation strategies through the mobilization of his scholarly network, see my essay "Asymmetries of Symbolic Capital."

3   On the concept of court society, cf. Elias, *Die höfische Gesellschaft*. The reference work on how a Renaissance scientist's individual ambitions were magnified by the court is Biagioli, *Galileo, Courtier*. Jürgen Renn and myself considered the socio-political setting of such early modern cultural politics in Omodeo and Renn, *Science in Court Society*.

4   See Biagioli, "The Social Status of Italian Mathematicians," 41–95; Henry, "'Mathematics Made No Contribution to the Public Weal,'" 193–220. On the German context, see Moran, "German Prince-Practitioners," 253–274; Moran, *Patronage and Institutions*.

5   Biagioli, "Le prince et les savants," 1417–1453. See also Giannini and Feingold, *The Institutionalization of Science*.

6   Cf. Appold, "Academic Life and Teaching," 65–116.

reconcilability of the Aristotelian legacy with Cartesian innovations. It ended in 1656 with Placentinus's appointment as court mathematician. The second controversy originated at a distance. It was sparked off by attacks from professors and students of the Saxony University of Wittenberg against Placentinus's pseudo-Cartesian conceptions of the origin of life from the stars. The controversy unfolded through disputations which were printed and defended at Frankfurt and Wittenberg. Placentinus's publication of the *Dissertatio philosophica* of 1660, dedicated to Montecuccoli and Strozzi, marked the conclusion of this polemic.

Printed university disputations are the principal sources for my reconstruction, as they provide evidence of the arguments and themes that were scrutinized. Disputations were a fundamental educational instrument of medieval and early modern universities, although their practice was profoundly transformed in the Gutenberg era, in the transition from oral culture to written culture that especially characterized Protestant universities.[7] However, I also refer to archival documents for more contextual information. I specifically consider the controversies of Placentinus and his opponents as a window into the teaching of Cartesian philosophy in seventeenth-century German classrooms. At the same time, these controversies indicate important elements of the broader European circulation of Cartesian philosophy.

2   The First Act of the Cartesian-Placentinian Disputations: Cosmology and the Problem of the Reconciliation with Aristotle

The first Cartesian controversy (I should say, polemic) took place at Frankfurt an der Oder in the years 1653–1656. Having already reconstructed its details elsewhere, I will now limit myself to summarizing it and dealing with the specific theme that is of relevance to this essay, namely the introduction of Descartes into German classrooms and the means of cultural-political legitimation of this curricular innovation that came from outside the university, in a context of early modern patronage.[8]

---

7   On university disputations in the long history of universities, see Friedenthal, Marti and Seidel, eds., *Early Modern Disputations*, 233–254; Weijers, *A Scholar's Paradise*, ch. 8, 121–138 ("The omnipresent disputation"). On the passage from orality to written disputations in early modernity, especially in Protestant contexts, see Clark, *Academic Charisma*; Chang, "From Oral Disputation to Written Text," 129–187. On university disputations and Cartesianism, see Chapter 7, in this volume.

8   Omodeo, "Central European Polemics over Descartes," 29–64.

Placentinus was appointed professor of mathematics at Frankfurt in 1653, after he successfully defended a rather generic *Disputatio mathematica* on 3 December under the presidency of the Greek professor, Georg Mellman.[9] Placentinus's disputation was not original but proved his proficiency in the areas of his future teaching. Yet, the *themata ex universa mathesi* (themes from all mathematics) that he defended comprised astronomical theses stemming from Copernican astronomy. This became a hotly disputed issue directly after his appointment, as he endeavoured to teach the physical tenability of heliocentrism on the basis of Descartes's principles of nature.

Placentinus's education was that of a wandering scholar. Most likely, he had received his first education in Johannes Comenius's Gymnasium at Leszno, the place where he was born. He studied in Gdańsk (1648) and Königsberg (1649), and then in Groningen (1651) and Leiden (1652). In the Netherlands, he became acquainted with Descartes's thought in the version that was imparted to students by Tobias Andreae and Johannes De Raey.[10] He enthusiastically embraced this novel philosophy and took upon himself the task of disseminating and teaching it.

At Frankfurt, the local philosophers sensed his bias towards Cartesian ideas. This they considered to be in contradiction with the peripatetic tradition, which they followed in accordance with Philipp Melanchthon's reform of protestant universities.[11] At the moment of his appointment, his colleagues gave him clear instructions to keep these Cartesian tendencies in check. As one reads from a later denunciation against him, it was expected that

> he should abandon all strange and abnormal opinions differing from the genre of philosophy that is accepted in our [university]. In the case that he adheres to some new doctrines, he should keep them to himself and should certainly renounce disseminating them among the students.[12]

---

9   Placentinus (praes.) and Mellemanus (resp.), *Disputatio mathematica exhibens themata ex universa mathesi* (Frankfurt an der Oder, 3 December 1653).

10   On the teaching of Descartes in the Netherlands, in particular De Raey's contribution to scholasticizing Cartesianism, see Strazzoni, *Dutch Cartesianism*.

11   On Melanchthonian educational and scientific contexts, see Bauer, ed., *Melanchthon und die Marburger Professoren*; Kathe, *Die Wittenberger philosophische Fakultät*; Darge, *Der Aristotelismus an den europäischen Universitäten der frühen Neuzeit*; Omodeo and Wels, eds., *Natural Knowledge and Aristotelianism*.

12   MS., Berlin, GR, I HA Rep.51, Nr. 94: *Acta*, fol. 56ᵛ (*Kurze Relation*, 26 Jan. 1655): "Daß er alle frembde, ungewöhnliche, und vom *Philosophandi genere apud nos recepto* abweichende *opiniones* fahren laßen solte. Hette er etliche newe *opiniones*, so solte Er dieselben für sich behalten, mit nichten aber unter die studiosos proseminiren."

In spite of this injunction, Placentinus began to disseminate a new physics and a new cosmology in his classes of mathematics. In the teaching programme (*programma inauguralis*) for his first classes, he announced (on 27 December 1653) that he would present Cartesian doctrines about the heavens to his students:

> In the solar body, spots are sometimes generated, sometimes corrupted, sometimes converted into flames [*faculas*]. Fixed stars are often generated all of a sudden and gradually disappear so that they degenerate into comets.[13]

This programme transgressed the disciplinary boundaries that separated mathematics and natural philosophy, mentioned theories that were in direct conflict with Aristotle's teaching of the heavens' perfection and incorruptibility, and introduced a new method of natural inquiry that was at odds with the well-established Peripatetic mindset. In the mid-seventeenth century, such ideas still looked scandalous to philosophers who had been raised according to an Aristotelian *Denkstil*. In January 1654, the professor of Greek, Melmann; the professor of logic, Johannes Walter Lesle; and the professor of physics, Philipp Beckmann, all tried to prevent Placentinus from disputing a series of Copernican-Cartesian theses *de Terra*. Placentinus argued that the Earth is a heavenly body that moves along with the other planets (and comets) through space, in accordance with a new elemental theory, namely that of the three Cartesian elements. His colleagues demanded that he follow the university statutes to which he swore an oath of obedience. They were especially opposed to the definition of body (*corpus*) as extension (*res extensa*). Such a doctrine contravened Aristotelian hylomorphism. Additionally, they did not accept the validity of his geometrical approach to physics—his way of thinking *more geometrico*, as they called it. In fact, Placentinus argued alongside Descartes that the Earth's *delatio* (or transportation) through a material vortex was reconcilable with the thesis of its immobility, understood as the absence of self-produced *motus localis* (local motion). Such a claim looked like sophistry to his Peripatetic censors and still constitutes a controversial issue of Cartesian studies. Yet, his defence of the Earth's 'immobile mobility' was not unique in those days but rather in line with another Cartesio-Copernican Placentinus knew

---

13   MS., Berlin, GR, I HA Rep.51, Nr. 94: *Acta*, fols. 57ʳ–57ᵛ: "In corpore solari maculae mox generantur, mox corrumpuntur, mox in faculas convertuntur. Stellae fixae saepius ex improviso apparent, ac paulatim disparent, sicque in cometas degenerant."

well. Daniel Lipstorp (1631–1684) of Lübeck authored a work entitled *Coperni-cus redivivus, seu de vero mvndi systemate, liber singularis* (Revived Copernicus, or one book on the true system of the world) (1653), which embraced the same Cartesian conception of a *delatio*.[14]

In November 1654, Placentinus's adversaries blocked the publication of his disputation entitled *De delatione Terrae* (On the Earth's transportation) when it had already been sent to the printer. As we read in a letter of protest that Placentinus sent to the Prince of Brandenburg, the theologian Friedrich Beck-mann, who was then the university rector, threatened the printer with physical punishment if he went ahead with the production of Placentinus's scandalous-looking theses. Beckmann then hung the offending disputation at the entrance to the main church of Frankfurt and tore it apart during the mass.[15]

As the polemic was escalating, Friedrich Wilhelm intervened in person. In January 1655, he wrote to the Frankfurt polemicists to defend the rights of pro-fessor Placentinus. He was allowed to freely dispute and print the theses linked to the mathematical truths he was expected to teach. Soon thereafter, Pla-centinus defended and printed a Copernican disputation, while, on 24 Febru-ary 1655, his student, Samuel Kaldenbach disputed *An Terra moveatur?* (Whether the Earth moves).[16] In this context, Placentinus quoted Galileo Gali-lei's *Dialogo sopra i due massimi sistemi del mondo* (Dialogue on the Two Chief World Systems):

> In the first dialogue *On the System of the World* [...] Galileo Galilei wisely says: 'Philosophy itself will greatly benefit from our disputations. In fact, if we propose true [theses], it [philosophy] will grow through the addition of new [knowledge]. By contrast, if we propose something false, the old doctrines [*sententiae*] will be reinforced through the refutation of these false propositions'.[17]

---

14   Lipstorp, *Copernicus redivivus* (Leiden, 1653). Cf. Vermij, *The Calvinist Copernicans*, 142–146. Placentinus mentions Lipstorp as one of his Cartesian acquaintances in the preface to Placentinus, *Renatus Des-Cartes triumphans* (Frankfurt an der Oder, 1655).

15   MS., Berlin, GR, I HA Rep.51, Nr. 94: *Acta*, fol. 64r.

16   The Earth's motion had been disputed by German mathematicians in university set-tings for a long time, at least judging by Johannes Regiomontanus's disputation on the motion of the Earth, entitled *An Terra moveatur an quiescat ... disputatio* (Disputation on whether the Earth moves or rests); Johannes Schönberg printed it in Nuremberg in 1533. See Omodeo and Bardi, "The Disputational Culture of Renaissance Astronomy," 233–254.

17   Placentinus (praes.) and Kaldenbach (resp.), *Discussio mathematica erotematis, an Terra moveatur?*, A2r: "Scite *Galilaeus Galilaei* Dial. I de mund. Syst. p. 29 *Ipsa philosophia*, inquit,

In this disputation, Placentinus reaffirmed the thesis of a terrestrial *delatio*. In a syncretistic vein, he even made the gesture of accepting Scholastic theses on celestial motion by *de facto* assimilating their spheres into Descartes's vortices. He presented the latter's vortex theory as an actualized version of the celestial movers that were once supposed to transport heavenly bodies through the heavens.[18]

Placentinus's counterattack went much further. He gathered several students in his house and let them dispute all of Descartes's natural conceptions following the order of the *Principles of Philosophy*. The meetings took place at a regular pace, one every week, for ten weeks in a row, and resulted in the publication of ten disputations, which were eventually collected in one printed booklet under the combative title *Renatus Des-Cartes thriumphans* (Triumphant René Descartes). They were dedicated to the Prince of Brandenburg and eventually distributed like a pamphlet in front of the main church by a student on a November Sunday of 1655. Placentinus's opponents saw this as an act of provocation and vociferously protested. Friedrich Wilhelm intervened again and invited both parties to moderation rather than continuing the dispute.

Another polemical front concerned institutional roles and university status. In October 1655, Placentinus expected to be elected dean of the Philosophical Faculty, according to an automatic principle of turnover. Yet, in the light of the philosophical positions he promoted, his opponents orchestrated his exclusion from the post. As Placentinus put it in an ensuing letter of protest to the Prince,

> this affair happened for this reason: because I privately taught Cartesian philosophy, thus fulfilling the order of His Princely Serenity [...]. Actually, they gave me a condition under which I could become dean, that I should abandon the Cartesian philosophy. I replied that I should obey His Princely Serenity and not them. Did they want to count more than the Serene Prince?[19]

---

nostris e disputationibus nonnisi beneficium recipit. Nam si vera proponimus, nova ad eam accessio fiet: sin falsa, refutatione eorum sententiae priores tanto magis stabilientur."

18   On ancient and Scholastic views on the heavenly spheres, see Lerner, *Le monde des spheres*; Grant, *Planets, Stars, and Orbs*. On the sixteenth-century crisis of the astronomy of celestial spheres, see especially Granada, *Sfere solide e cielo fluido*.

19   MS., Berlin, GR, I HA Rep.51, Nr. 94: *Acta*, fols. 50^r–50^v: "Diese aber praetention ist darumb geschehen, weil ich *Philosophiam Cartesianam privatim* dociret, und mandato E. Churfl. D. welches durch Ihr Gnaden hern von Wolzogen freiherr clementissime mir ist angedeutet worden, obtemperiret habe. Sie schlugen mir zwar vor ein condition danlich wenn ich also bald *Philosophiam Cartesianam* fahren ließe, so solte ich *Decanus* werden: aber ich

At the Prince's request, rector Brunnemann, who was a jurisprudent and therefore neutral in relation to the philosophers' conflicts, wrote the following report:

> Concerning the dean's election, Mr. Magister Mellman advanced that Mr. Placentinus first of all infringed the Statutes of the Philosophical Faculty, since he privately defended Cartesian philosophy in [a series of] disputations after he rejected the Aristotelian philosophy in some disputations. By contrast, Mr. Magister Placentinus objected that his predecessors already defended the thesis of the motion of the Earth. Moreover, [he declared that] His Princely Serenity had graciously conceded him the permission to present and defend Cartesian philosophy, partly in written form, through a letter a copy of which he showed us, and partly in oral form [...].[20]

The problem of reconciling Descartes with the Peripatetic tradition, which was longstanding in the Philosophical Faculty, was readdressed in a letter of March 1656, in which the anti-Cartesian party explained the reasons for their aversion to Placentinus:

> [...] We always directed ourselves according to the Statutes that we swore to, and we never desired anything but what they express, that is: *None shall deviate from the philosophy that has been transmitted to us*; and: *We will defend unanimity on the Peripatetic doctrine*.[21]

---

antwortete darauf E. Churfl. D. müßte und wolt ich pariren und nicht ihnen, sie würden ja nicht wollen mehr seyn, alß E. Churfl. D."

20   Rector Brunnemannus to Friedrich Wilhelm of Brandenburg, Frankfurt an der Oder, 27 October 1655, in MS., Berlin, GR, I HA Rep.51, Nr. 94: *Acta*, fol. 41ᵛ: "Hingegen hatt H.M. Mollemannus *pro electione Decani* angezogen, das erstlich H.M. Placentinus *contra statuta Philosophicae Facultatis* gehandelt in dem er *relicta Philosophia Aristotelica* in etlichen *disputationibus* die *Philosophiam Carthesianam* [sic!] privatim *in disputationibus* defendiete. Dagegen H. Placentinus eingewand, das die *hypothesin de Terrae motu* auch seine *antecessores* defendiret hätten und von E. Churfl. Durchl. auch die *Philosophiam Carthesianam* [sic!] zum proponiren und zum defendiren gnädigste Concession habe, theils schrifttlich wie Er uns dan eine *Copiam Rescripti* gezeiget, theils mündlich durch herren Ludwig von Wolzowen [sic!] Freiherren dem E. Churfl. Durchl. solches gnädigst mündlich befohlen hätte."

21   MS., Berlin, GR, I HA Rep.51, Nr. 94: *Acta*, fols. 10ʳ–10ᵛ: "Wir haben unß iederzeit nach Unseren beschwornen *Statutis* gerichtet, und nichts anders, alß was die im munde führen, begehret, nemblich: *Nemo a recepto apud nos philosophandi genere recedat. Item: Tueamur consensum doctrinae Peripateticae*."

It is clear from these exchanges that teaching Descartes became an issue of authority, both intellectual and political. A professor who did not acknowledge Aristotle's *auctoritas* should not be given a position of responsibility and prestige in the university. Still, the highest political and institutional authority at university was the Prince himself. An intervention from such an exalted position could put an end to the controversy. In the spring of 1656, Friedrich Wilhelm bestowed upon Placentinus an increase of salary and conferred upon him the honorific title of court mathematician (*mathematicus Electoralis*). Directly thereafter, Placentinus obtained the position he desired in his Faculty; he became its dean.

Among his first acts as dean, he addressed the theme of Cartesian-Aristotelian concord. On 19 July 1656, he defended a thesis aimed at establishing, as the title goes, the "Syncretism between René Descartes and Aristotle" (*Syncretismus philosophicus inter Renatus Des-Cartes et Aristotelem institutus*). Rather than a concession to his adversaries, this was an affront—as Cartesian philosophy was presented as striving for the truth that any right philosophy should aim at, including Aristotle's. But the university polemic was stilled, at least for a while, thanks to the Prince's intervention. The will of the Electoral Prince could dispel all doubts about the legitimacy of the teaching of Cartesian philosophy. Moreover, by formally embracing a Cartesian-Aristotelian 'concordism,' Placentinus could pretend that he was not infringing the statutory norms.

3   Second Act: Wittenberg versus Frankfurt on the Heavenly Origins of Life according to Cartesian Principles

Thanks to the ducal support, Placentinus could continue spreading Cartesian ideas at Frankfurt. He even occupied the position of university rector twice, in 1658 and 1665. During his first mandate, a second Cartesian controversy erupted. The cause, this time, was a *Dissertatio philosophica* in which Placentinus argued for the origin of life from the starry heavens (1659). Unlike the previous controversy, the attack was not internal, but originated in Saxony. Placentinus's main critic was Georg Kaspar Kirchmayer (ca. 1635–1700), who was an eclectic professor of eloquence at the Electoral University of Saxony, Wittenberg, with a keen interest in natural philosophy.[22] The controversy took place at a distance of some 180 kilometres, with a back-and-forth of

---

22   Kirchmayer is best known for his studies on mineralogy and phosphorescence, for which he was renamed "Phosphorus" by his fellow members of the *Academia naturae curiosorum*. See, among others, his *Commentatio epistolica de phosphoris* (Wittenberg, 1680) and

many disputes and publications, both in Wittenberg and Frankfurt. Students were involved as disputants in both institutions. It might be supposed that there were connections between the anti-Cartesian (anti-Placentinian) fronts in both universities but, for the time being, I have no evidence to support this supposition. In what follows, I only touch upon the most significant moments and issues regarding this controversy, leaving the details for another occasion.[23]

The title of Placentinus's *dissertatio* of 1659 summarizes the thesis that proved most controversial: *Philosophical dissertation proving that the natural heat and motion of the limbs in the human body, even life, do not stem from the soul—which is unique, namely rational—but rather from the very subtle matter of the heavens, which is analogous to the element of the Sun and the fixed stars.*[24] Although the focus of the disputation was the cosmic source of human life, the implications were quite problematic from a Peripatetic perspective. In fact, Placentinus decoupled the explanation of life from the action of the soul, and redefined the soul as *res cogitans*. As such, it was separated from the body and, in particular, its functions. As Placentinus stated: "It would be absurd to attribute to our soul that which we have in common with beasts."[25] The soul's control of the body, as one reads in Placentinus's disputation, is exerted from the centre of the brain, according to the Cartesian theory of the pineal gland.[26]

It should be remarked that the theory of the cosmic origin of life was more in line with Renaissance medical Platonism—as represented by authors such as Girolamo Cardano, Jean Fernel and Daniel Sennert—than with anything that could be directly derived from the letter of Descartes's physiology. In the course of the controversy, Placentinus clearly indicated that he took inspiration from earlier animist theories:

> The most celebrated Sennert assumes a heavenly matter, or spirit, that is inside the four elements albeit distinct, and analogous to the heaven [...]. Jean Fernel, in his *Universal Medicine*, solidly writes among other things, that 'everything that has life, is irrigated, contained and supported by a

---

Institutiones metallicae (Wittenberg, 1687); cf. *Allgemeine Deutsche Biographie*, vol. 16, *sub voce*.

23   I offer a more detailed reconstruction of the controversy in Omodeo, "*An vita hominis procedat.*"

24   Placentinus, *Dissertatio philosophica, probans, calorem et motum membrorum naturalem, in humano corpore* (Frankfurt an der Oder, 1659).

25   Placentinus, *Dissertatio philosophica, probans, calorem*, B3^v: "Absurdum foret id quod nobis commune est cum bestiis, animae nostrae tribuere."

26   Placentinus, *Dissertatio philosophica, probans, calorem*, A4^r, thesis VI.

beneficent heat. This heat, which is a quality, is fully divine and heavenly; it consists in ethereal spirit'.[27]

Drawing on such authors, Placentinus could even argue that the doctrine of an innate heat of heavenly origin, which functions to initiate and sustain the organic processes, is a conception that can ultimately be brought back to Aristotle and Hippocrates.[28] However, the reference to Sennert was strategically important to Placentinus's argument from authority, as the former physician had been a reputed professor at Wittenberg, whence opposition to his theses originated.

Placentinus claimed his cosmic theory of life could be traced back to Renaissance forerunners and the most authoritative among classical sources, but he in fact transformed these older doctrines by giving them a new Cartesian foundation. In the fifteenth thesis of his *Dissertatio philosophica* of 1659, he argued that

> the natural heat and the motion in the body do not stem from the soul, but rather from the heavenly matter, which is very subtle, rapidly agitated, and analogous to the matter of the Sun and the fixed stars. It is the constituent of light [*lumen*], the magnet and of fire (the one which is warm and lucid, as well as the one which is only lucid, or only warm). René Descartes calls it 'first element,' out of which the bodies that emit light are constituted.[29]

---

27 Placentinus (praes.) and Reinhold von Boyen (resp.), *Dissertatio physica, refutans* (Frankfurt an der Oder, 1660), thesis XIII: "Hanc materiam coelestem, sive spiritus, insitum a quatuor elementis, distinctum, et coeli analogum, [...] celeberrimus Sennertus concedit [...]. Solide quoque Johannes Fernelius, in *Universa Medicina*, inter alia [...] scribit: 'quicquid vitam agit, salutari calore perfusum continetur et regitur, atque is calor, qualitas cum fit, totus tamen divinus est atque coelestis, et in aethereo spiritu subsistit'." For the Fernel quotation, see Fernel, *Universa Medicina* (Geneva, 1577), 73.

28 Placentinus (praes.) and Tinctorius (resp.), *De origine caloris et motus membrorum naturalis* (Frankfurt an der Oder, 1660), A5ʳ. According to Hiro Hirai's learned reconstruction, Renaissance cosmic conceptions of generation like those of Sennert and Fernel were to be found earlier in Aristotle's *De generatione animalium* II 3 and Hippocrates' *De carnibus*, authoritative sources to argue for a connection between the life-generating heat of the seed and the heavenly element. Cf. Hirai, *Le concept de semence*; Hirai, *Medical Humanism and Natural Philosophy*; Hirai, "Il calore cosmico," 71–83; see also Clericuzio, *Elements, Principles and Corpuscles*, ch. 1.

29 Placentinus, *Dissertatio philosophica, probans, calorem*, A7ᵛ, thesis XIV: "Atque sic calor naturalis et motus in corpore non procedit ab anima, sed a materia coelesti, subtilissima, celerrime agitata, analoga materiae Solis et stellarum fixarum, in qua natura luminis, magnetis, et ignis tam calidi et lucidi, quam lucidi solum, atque calidi tantum consistit, et

If life originates from heavenly matter, as Placentinus contended, this matter has to coincide with the subtlest element of Descartes's physics. This, in turn, is perceived by our senses as light. Placentinus ascribed this interpretation to his Leiden professor De Raey, to whom he referred on this as on many other occasions.[30]

The reaction came from Kirchmayer's student, Arnold Lendericus of Utrecht who defended a counter-disputation in Wittenberg on 13 August 1659. Although the general theme of the disputation is the Aristotelian doctrine of the heavens, as is indicated by the title, *Disputatio de coelo*, it explicitly criticized Placentinus's Cartesian theory of heavenly matter and the origin of life. The two main Placentinian issues are discussed in the form of 'problemata':

> Problem One: What is the matter of the heavens? Is it elemental or a special one which is different from the elements?[31]

And

> Problem Two: Does man's life stem from his soul or from the very subtle matter of the heavens which is analogous to the element of the Sun and the fixed stars?[32]

For the first problem, Kirchmayer rejects the basic idea of a cosmological homogeneity like that of Descartes's *Principles of Philosophy*, according to which the same matter and laws of physics apply throughout the cosmos. In accordance with Peripatetic philosophy, Kirchmayer acknowledges that first matter (*materia prima*) is the same everywhere.[33] Nonetheless, he contends that the heavens show a behaviour different to the elementary realm beneath

---

quam Renatus des Cartes primum elementum, ex quo corpora lucem emittentia conflata sunt, appellitat."

30   Placentinus, *Dissertatio philosophica, probans, calorem*, A6ʳ-A6ᵛ, thesis XVI. On De Raey, see Chapter 4, in this volume.

31   Kirchmayer (praes.) and Lendericus (resp.), *Ex physicis disputatio publica de coelo* (Wittenberg, 1659), B1ᵛ, *Problema* I: *Quaenam coeli sit materia, elementarisne, an peculiaris et ab elementis diversa?*

32   Kirchmayer (praes.) and Lendericus (resp.), *Ex physicis disputationem publicam de coelo*, B2ᵛ, *Problema* II: *Utrum vita hominis procedat ab anima eiusdem, an vero a materia coelesti subtilissima, analoga elemento Solis et stellarum fixarum?*

33   The idea of the 'two physics,' one beneath and one above the Moon, is ubiquitous in Scholastic philosophy, just like debates about matter. For a classical treatment of these topics, see Grant, *Much Ado about Nothing* and Grant, *Planets, Stars, and Orbs*.

the moon. As a consequence of the 'formal' difference of the heavenly and
sublunary phenomena—summarized by the opposition between the perfectly
circular motions of the heavenly bodies and the imperfect linear upward and
downward motions of the elements—one has to assume a material difference,
too. The principle is that "another form, requires another matter" (*alia forma,
aliam requirit materiam*).[34]

As for the second problem, Kirchmayer rejects the thesis of the heavenly origin of life and argues that life exclusively derives from the soul on the basis of
the following Scholastic argument:

> We affirm [the following theses]: Life, seen from its origin, immediately
> stems from the soul, actually it is in reality the human soul itself. *I say 'in
> reality' not 'formally' nor 'modally'*. In fact, considered in relation to sub-
> stance, life is the first act, the form itself of a thing. *Formally*, life consists
> of the union of the soul and the body. *Materially*, it consists of innate
> warmth and vital spirits. These spirits move from the blood in the solid
> [parts of the body]. They are not communicated from the celestial mat-
> ter.[35]

A theological concern emerges, too. Kirchmayer believes that God imparted the
necessary heat to the Earth at the moment of its creation. For him, this makes
the starry causation of terrestrial life superfluous.

On 3 September 1659, Placentinus responded to his Wittenberg adversaries through the *Disputatio philosophica … refutans quatuor opposita argumenta
in Universitatis Wittenbergensi* (Philosophical dispute opposing the four arguments of the University of Wittenberg). It was another student, Johannes Reinholdus, who defended the theses aimed to reaffirm the Cartesian "truth."[36]

The first issue Placentinus and his student addressed was terminological.
Their main concern here was with the definition and, as a consequence, the
comprehension of the soul. According to Placentinus, Kirchmayer did not perceive the importance of the adjective "rational" (*rationalis*), as used in relation

---

34   Kirchmayer (praes.) and Lendericus (resp.), *Ex physicis disputationem publicam de coelo*,
      B3r.
35   Kirchmayer (praes.) and Lendericus (resp.), *Ex physicis disputationem publicam de coelo*,
      B4v: "Nos dicimus: Vitam originaliter spectatam, esse immediate ab anima, imo realiter
      ipsamet animam humanam esse. *Realiter inquam, non formaliter aut modaliter*. Vita enim,
      substantialiter considerata, actus primus est, ipsa rei forma est. *Formaliter* consistit in uni-
      one animae et corporis, vita. *Materialiter* vero in calido innato spiritibusque vitalibus. Qui
      spiritus a sanguine in solidum proficiscuntur. Non a materia coelesti communicantur."
36   Placentinus (praes.), *Disputatio philosophica, defendens, calorem*, A1r.

to considerations of the soul. For a Cartesian, this attribute was crucial as it pertained to the comprehension of the very essence of the soul as thought (*cogitatio*), reflection and rationality. Another issue related to the definition of matter was extension. The concept of *res extensa* was linked to the idea of a unification of the terrestrial and extra-terrestrial realms of the world. Through these terminological distinctions, Placentinus shifted the focus from the origin of organic life to the philosophical foundations of Cartesian cosmology and physics.

His adversaries welcomed the radicalization of the debate. In an *Exercitatio physica responsoria* (Physical Exercitation, a Response) that was held in Wittenberg on 7 January 1660, Kirchmayer argued for the necessity of moving from "consequences" (*erga*) to "assumptions" (*parerga*); he agreed with his adversary that the controversy over the origin of life should move to the clarification of the epistemological and natural principles underlying the definition of the soul and nature. On this occasion, he (or rather, the student acting as disputant) addressed and dismissed the fundaments of Cartesian philosophy by dealing with and rejecting crucial theses such as methodological doubt, the hypothesis that God could deceive the cognizant mind, scepticism concerning the reliability of the senses, the argument that the *cogito* could be the initial point of philosophizing, the three-elements physics, the rejection of forms in favour of material mechanistic explanations, light theory, post-Copernican astronomy, and the understanding of the Sun as a star. The disputation repeated well-known doubts and controversies over Cartesian thought, those Descartes himself discussed at the moment of the publication of his *Meditations*. In view of the many philosophical errors that he imputed to Cartesianism, Kirchmayer concluded, with a reference to the Dutch anti-Cartesian theologian Gisbert Voetius, that Descartes did not qualify for university teaching as he was an extra-academic autodidact who did not understand philosophy and subverted its tenets owing to his irredeemable ignorance.[37]

Of all these criticisms, Placentinus took most seriously those concerning God's benevolence, as he was accused of endorsing, together with Descartes, the radical doubts of sceptics and atheists. In a *Dissertatio physica, refutans propositam in Academia Wittenbergensi* (Physical dissertations refuting the one that was presented at the University of Wittenberg) dated 24 March 1660, Placentinus responded that doubt is not an end in itself in Cartesian philosophy, but is a methodological means to achieve a better comprehension of truth and

---

37 Placentinus (praes.), *Disputatio philosophica, defendens, calorem*, fols. B3^v-B4^r. On the Dutch controversies that involved Voët and Descartes in person, cf. Verbeek, *Descartes and the Dutch*.

a compelling demonstration of God's existence. As for the idea of a deceptive God, he argued that the problem has more to do with the validity of knowledge and the question of the reliability of the natural instinct (*naturalis instinctus*) that we received from God. A philosopher should be aware that we cannot base our knowledge on natural instinct—as the latter often deceives us—without first undertaking a preliminary assessment of our faculties and a cautious suspension of all that we think we know.[38] As for the senses, to cast doubt on them does not affect our life in practice but only in speculation. In knowledge theory, senses cannot constitute a firm ground for higher truths. Moreover, Placentinus cautioned that methodological (and temporary) suspension of what we claim to know should not be regarded as applicable to theological claims calling into question the goodness of God.

4   The Courtly Dissertation of 1660 as the Conclusion of the Second
       Controversy

The controversy over Cartesian philosophy between Frankfurt and Wittenberg continued for some time, and included interventions from other scholars. Among them was Samuel Hentschel, a lecturer at the philosophical faculty of Wittenberg who originally came from the same town as Placentinus, Leszno, and presided over six anti-Cartesian disputations directed against the Frankfurt professor. This series of disputations, which bore the title *De vita hominis* (On man's life), took place between the autumn of 1659 and 1661.[39] They addressed the mind-body problem as well as physiological and psychological aspects of Cartesian philosophy.[40] In the course of these further developments, the debate grew to encompass more than just the specific problem of the stellar origin of life—which, in fact, was not specifically Cartesian, if Cartesian at all—to include a more general discussion of the foundations of philosophy, namely the natural, cosmological, physiological and epistemological pillars of Descartes's philosophy. It seems to have unfolded as a 'clash of paradigms' which could not be solved in the framework of standard university practices. For the second time, Placentinus resorted to external support through a courtly publication, his *Dissertatio philosophica, exhibens modum praecavendi errorem in veritatis philosophicae, inprimis naturalis inquisitione atque dijudicatione,*

---

38   Placentinus (praes.), *Dissertatio physica, refutans*, B6ʳ.
39   Hentschel, *De vita hominis disputationes*, I–VI.
40   On Descartes's medical philosophy, see, amongst others, Aucante, *La philosophie médicale de Descartes*.

iuxta Principia Renati des Cartes (Philosophical dissertation, showing the manner to avoid error in the investigation and evaluation of philosophical truth, especially natural [truth], according to René Descartes's principles) of 1660, which sought the protection of high-ranking imperial commanders.

Following the Wittenberg-Frankfurt polemic's tendency to go deeper into the premises, Placentinus's *Dissertatio philosophica* especially tackled the epistemological problem of the origin of error and its solution according to Descartes's method. In the *prefatiuncula* (short preface), Placentinus began with a profession of a mechanistic creed. He dismissed the *formalism* of Aristotelian philosophy, for it remains at the surface of natural problems, and praised an investigation of the material structures capable of looking beyond appearances:

> As far as the contemplation of nature is concerned, a philosopher's task is not to register the external differences nor the forms of things (i.e., he should not persist with crusts and the most superficial wrap of things). Rather, he should go deep into the interiority and crack the structure of the smallest parts, their intervals, connections and hidden motions. Thence, he should joyfully discover all of nature's artifice and usefully extract the causes of its most hidden secrets.[41]

In the *Dissertatio*, Placentinus referred to Descartes's *Discourse on Method, Meditations on First Philosophy, Principles of Philosophy*, and *Passions of the Soul*. Additionally, he mentioned a disputation that he himself defended at Groningen under the presidency of Tobias Andreae (1604–1676), entitled *Dissertatio de methodis recte regendae rationis* (Dissertation on the method of rightly using reason).[42]

In his philosophical dissertation, Placentinus clarifies that the kind of error he is dealing with is not practical but theoretical, for it concerns the distinction between truth and falsehood. He explicitly writes that this concern ought to be separated from the ethical and religious inquiries into error and sin, that is, from the problem of good and evil in the conduct of life and faith.[43] However,

---

41 Placentinus, *Dissertatio philosophica, exhibens modum* (Frankfurt an der Oder, 1660), A3ʳ-A3ᵛ: "Circa naturae contemplationem officium Philosophi in eo consistit, ut non tam externas rerum differentias formasque, observet, non circa crustas, et prima rerum involucra haereat; quam ad interiora descendat, minutissimarum partium structuram, intervalla, nexum, atque latens motus aperiat, totum denique naturae artificium jucunde retegat, et eiusdem abditas rerum causas utiliter eruat."

42 Placentinus, *Dissertatio philosophica, exhibens modum*, C1ʳ.

43 Placentinus, *Dissertatio philosophica, exhibens modum*, A6ʳ-A6ᵛ.

since methodological doubt drew accusations of impiety, Placentinus stresses the importance of God's existence as the theoretical cornerstone of Cartesian epistemology. Placentinus infers God's existence from error itself. In fact, as he argues, any judgement about error presupposes a perfect yardstick of reference. Similarly, finitude presupposes infinity and our sense of limitation calls for the existence of a perfect infinite God as the source of our contingent and finite existence.[44]

Placentinus further affirms that God cannot be the *positive* cause of human errors.[45] However, some Scriptural passages suggest that He sometimes acted in a manner that could appear deceptive to humans. For instance, Jonah was sent to Nineveh to announce that the city would be destroyed in forty days but this did not happen. In this example, as Placentinus explains, God's assertion should not be seen as mendacious. In fact, destruction was to be visited upon the city unless the Ninevehans met God's condition, that is, to change their behaviour. As this happened, Nineveh was eventually spared from Divine wrath. Another example given was the Pharaoh's blindness to divine truth. Although this can only be explained through God's withholding from the Pharaoh the grace of faith, nonetheless this only amounts to a *negative* form, as it were, of 'deception'.[46]

Error itself is nothing 'real' but rather a privation or a negation of truth. It presupposes the finitude of our intellect.[47] Placentinus apologetically asserts that it is our fault, and not God's deception, that causes us to make a wrong usage of our faculties and incur error.[48] The scope of our will is broader than that of our intellect, hence our will can overtake the latter in the drive toward the fulfilment of its desires. Error specifically derives from misguided freedom.[49] Placentinus adds that the source of error should not be brought back to our faculties (intellection and will) in themselves but in dependence of the way we operate (*operatio*) with them. Hence, he argues for the necessity of a method of inquiry as an instrument that helps us channel freedom. He specifies that the mistakes we commit should not be ascribed to Divine Providence because they depend on our freedom. This was a line of thought that possibly ran counter to crucial doctrines of the Protestant theologies of *servum arbitrium*.[50]

---

44   Placentinus, *Dissertatio philosophica, exhibens modum*, A7ʳ.
45   Placentinus, *Dissertatio philosophica, exhibens modum*, A8ᵛ.
46   Placentinus, *Dissertatio philosophica, exhibens modum*, B1ʳ-B1ᵛ.
47   Placentinus, *Dissertatio philosophica, exhibens modum*, B2ʳ.
48   Placentinus, *Dissertatio philosophica, exhibens modum*, B2ᵛ-B3ʳ.
49   Placentinus, *Dissertatio philosophica, exhibens modum*, B5ʳ.
50   Placentinus, *Dissertatio philosophica, exhibens modum*, B6ᵛ. On the theological implica-

Placentinus explained that the criterion for establishing truthfulness, the *regula veri*, is to give our assent only to that which appears clear and distinct to us.[51] Mathematics offers the best instances of such truths. In this case, too, Placentinus stresses the fundamental theoretical importance of God's epistemological assistance. An atheist, he writes, could never truly believe that the sum of the angles of a triangle is equal to two right angles, because he could not have any certitude concerning the ground on which the validity of the evidence of the geometrical demonstration rests.[52] Similarly, the certitude of sensorial perceptions can only be secured through God.[53] These statements seem designed to overturn the allegations of impiety by arguing that Descartes's methodological doubt and the notion of a deceptive God are actually arguments against atheism.

Yet, the senses cannot be naïvely followed in accordance with our habitual trust of their objectivity. The qualities of worldly things are not intrinsic to material reality but depend on perception. The principle is simple: "The senses do not exhibit what a thing is in absolute terms but only in what manner it is in relation to us."[54] Reasoning and mathematics must guide judgements which are connected with sensory experience, otherwise one follows prejudices that derive from a childish use of perception. As an example, Placentinus offers the motion of the Earth, the evidence for which depends on reason and by no means on the senses alone.[55]

Placentinus eventually claims that Cartesian meditations have an almost Stoical component because they enable the soul to reach a state of tranquillity. Once one accepts that cogitations are the only thing in our absolute

---

tions of Descartes's *Discourse on Method*, see especially Janowski, *Cartesian Theodicy*. Placentinus addressed the problem of the relation between Scripture and nature in an early disputation of his, which he defended under the presidency of the French theologian Samuel Desmarets, *De duplici volumine Naturae et Scripturae, ex quo notitia Dei arcessenda est* (On the Twofold Book of Nature and Scripture, from Which a Notion of God Can Be Obtained) (1652). In it, Placentinus argued for the complementarity of natural philosophy and biblical exegesis, seen as two paths to the divine. Later, in Frankfurt, he clashed with his Melanchthonian and Aristotelian colleagues and this led him to rather emphasize the difference of the two approaches to the natural in order to guarantee a space of autonomy for his Cartesian investigation of nature. I discuss the details in Omodeo, *Defending Descartes in Brandenburg-Prussia*.

51   Placentinus, *Dissertatio philosophica, exhibens modum*, B7ᵛ.
52   Placentinus, *Dissertatio philosophica, exhibens modum*, C1ᵛ-C2ʳ.
53   Placentinus, *Dissertatio philosophica, exhibens modum*, C2ʳ.
54   Placentinus, *Dissertatio philosophica, exhibens modum*, C4ʳ: "Sensus non exhibent quid res in se absolute sit, sed quomodo ad nos relata se habeat."
55   Placentinus, *Dissertatio philosophica, exhibens modum*, C6ʳ.

power (*absolute esse in nostra potestate*), one can cultivate a form of philosophical detachment from the world:

> If we accustom our spirit [*animus*] to look at all things that are outside of us in this manner, thanks to a very long exercitation and often repeated meditations (something which is very difficult but not impossible), then we will not be bent by the desire of health, if we get sick, nor by that of freedom, if we are detained in prison; nor will we regret that our bodies are not as resistant to corruption as diamond (and other similar cases). It is our will's nature that it is never conducted for any reason to desire something, unless our intellect represents something as somehow possible to [the will]. Therefore, it [the spirit] is not upset if some desire is not fulfilled which is out of reach and is conceived as impossible for us.[56]

5   Concluding Remarks

In this chapter, I considered two Cartesian controversies that took place after the appointment of a staunch defender of Cartesian natural philosophy, Placentinus, as a professor of mathematics at Frankfurt an der Oder. The first was an institutional and intellectual controversy, in which Peripatetic philosophers tried to hinder his teaching of new views on nature, in connection with a mathematical method at odds with the Aristotelian tradition. Cosmology took centre stage in this controversy, as a Cartesian foundation of heliocentric astronomy, cometary theory and a redefinition of the elements confounded philosophers who had been trained in a different intellectual tradition and who wished to continue in the footsteps of their Peripatetic forerunners. Placentinus, the Cartesian mathematician, could not prevail over his adversaries through standard university means, such as his students' defence of Cartesian theses through disputations. It was only thanks to the backing of the Electoral

---

56   Placentinus, *Dissertatio philosophica, exhibens modum*, D2^v-D3^r: "Quod si et nos longissima exercitatione et meditatione saepissime iterata, animum nostrum, ad res omnes extra nos posita ita spectandas assuefaciamus (quod licet perdifficile, non tamen impossibile est) tunc non torquebimur sanitatis desiderio, si aegrotemus, nec libertatis, si carcere detineamur; nec tristabimur, quod nostra corpora non sint tam parum corruptioni obnoxia, quam est adamas, et sic in caeteris etc. Ea enim est voluntatis nostrae natura, ut erga nullam rem unquam feratur, illam optando, nisi quam illi noster intellectus, ut aliquo modo possibilem repraesentat; adeoque minime turbetur, si eorum non fiamus compotes, quae extra nos posita sunt et tanquam aequaliter nobis impossibilia spectantur."

Prince of Brandenburg, Friedrich Wilhelm, that Placentinus received permission to continue teaching Cartesian philosophy.

The second controversy was a university polemic, too. It unfolded through disputes and counter-disputes between students of Wittenberg and Placentinus's students in Frankfurt. A striking aspect of this controversy is that it originated with a topic seemingly far removed from the central preoccupations of Cartesian theory: the doctrine of the stellar origin of human life. Yet, both Placentinus and his main Wittenberg opponent Georg Kirchmayer found the controversial thesis to be properly Cartesian. As Kirchmayer put it in one of his disputations:

> Are the origin and principle of human life the rational soul, or some heavenly and very subtle matter which is analogous to the element of the Sun and the fixed stars? We [Aristotelians] support the first position, while Cartesians defend the latter.[57]

As my reconstruction has shown, it is not the thesis of the starry origin of life that was Cartesian, but rather the underlying natural and methodological assumptions of Placentinus's argument in its favour. One important point concerned the identification of the life-giving light of the stars with the subtlest element of the three elements of Cartesian physics. The doctrine of the stellar origin of life stemmed from Renaissance medical philosophers such as Cardano, Fernel and Sennert. In Kirchmayer's view this infringed against the Aristotelian conception of the soul as the source of life of an organic body without considerations of stellar agency. Another crucial issue concerned the definition of the soul as the rational realm of the *cogito*, on the basis of which Placentinus rejected Aristotelian hylomorphism and thus incurred the displeasure of his Peripatetic opponents.

At the end of the first controversy, Placentinus felt so confident of his position as court mathematician to the Prince of Brandenburg as to boast that he was promoting a form of Aristotelian-Cartesian syncretism. Since truth is eternal, he posited that there cannot be any contrast between Aristotelian and Cartesian truths. However, the very possibility of such a merging of perspectives proved increasingly remote over the course of the Frankfurt-Wittenberg *Streit* concerning the origin of life. The polemic did not stop at the specific

---

57 Kirchmayer (praes.) and Lendericus (resp.), *De origine vitae*, A2ᵛ: "Utrum vitae humanae origo et principium sit anima rationalis, an vero materia quaedam coelestis subtilissima, analoga elemento Solis et stellarum fixarum? Pro primo nos, pro posteriore membro militant Cartesiani."

'astro-biological' problem but became so broad as to encompass the fundamental differences of approaches, natural philosophy and epistemology. This
time, Placentinus sought a patron higher than the authorities of Brandenburg
and Saxony. With an appeal to the royal court that repeated his earlier strategy,
he penned a work of Cartesian epistemology, which dealt with the problem of
error and the piety of Descartes's theory of knowledge, dedicating it to powerful members of the imperial aristocracy in the tumultuous years of war with
Sweden.

The case of Placentinus is interesting not only because it permits us to trace
the diffusion of Cartesian ideas in German universities, but also because it
helps us to assess the proximity between the classroom and the court. A socially
sensitive history of seventeenth-century Cartesianism has to consider not only
university sociability and its institutional forms but also the cultural-political
settings that affected university dynamics. The question arises as to why the
Prince of Brandenburg, Friedrich Wilhelm, would have supported a heterodox
Cartesian like Placentinus. It seems to me that—not unlike in the Dutch controversies over Cartesian philosophy—the political authorities backed innov-
ative philosophers and defended their academic liberty in order to limit the
clerks' influence.[58] Such a cultural politics was all the more urgent for a Calvinist prince like Friedrich Wilhelm, as there was a certain tension between his
own confession and that of his Lutheran subjects. As a young man, the prince
spent several years of his schooling in the Dutch States, from 1634 to 1648, making him receptive to ideas that circulated in the Low Countries, not least under
the influence of his cousin, Elisabeth of the Palatinate, the patron of philosophy
who corresponded with Descartes and other prominent thinkers.[59]

As for the specific form of Cartesianism defended by Placentinus, he conceived and 'normalized' post-Copernican astronomical theses and vitalistic
conceptions—specifically cosmo-vitalistic ones—and translated non-Cartesian doctrines into the elements of a physics and cosmology of Cartesian inspiration. According to his creative reworking, the corpuscular light emanating
from the stars operates on Earth, thus enabling the possibility of organic life.
The innate heat, on which physicians of the Renaissance era like Fernel had
speculated within a framework of medical neo-Platonism, was transplanted
into a very different philosophical terrain. The soul, which had always played a

---

58   Verbeek, *Descartes and the Dutch*.
59   Elisabeth also corresponded with the physician Franciscus Mercurius van Helmont and
      the philosophers Nicolas Malebranche and Gottfried Wilhelm Leibniz. Placentinus dedic-
      ated to her a work on geodesy, *Geotomia sive Terrae sectio exhibens* (Frankfurt an der Oder,
      1657). Elisabeth's relationship with Descartes is described by Gaukroger, *Descartes*, ch. 10.

crucial explanatory role in theories of the origin of life, became irrelevant. As *res cogitans*, in fact, the soul received an exclusively intellectual character with no bearing on vital operations.

More generally, Placentinus's case raises fundamental historiographical questions. Can we speak of Cartesianism in the singular? To be sure, today we are in a good position to assess the transformative power of university teaching, benefitting from in-depth studies on early modern Aristotelianisms and Scholastic Cartesianism.[60] It has already been argued that the unitary concept of Cartesianism—first introduced by the Cambridge Platonist, Henry More in 1662—is an oversimplification which is not apt to capture the wide variety of philosophical and scientific strands that were directly inspired by Descartes's work, of which the French and the Dutch variants are the most relevant.[61] Yet, we need to go beyond the French-Dutch dualism and consider a multicentric legacy which comprises German and other local paths to Cartesianism with specific cultural, confessional, political, and national connotations. Hence, it seems that we historians of the early reception of Descartes's ideas have to place a greater emphasis on the variability of the ideas that were circulated under his name. Placentinus's cosmology of organic life is certainly one additional witness to the eclectic transformation of the Cartesian legacy in university and courtly settings.

Acknowledgments

I am thankful to the European Research Council for supporting this research, which is part of the Early Modern Cosmology endeavour (funded by the European Union's Horizon 2020 Research and Innovation Programme, Grant n. 725883). I am also thankful to the Max Planck Institute for the History of Science in Berlin, at which I began to first investigate German Cartesianism. For their support and suggestions, I am thankful to the ERC research group in Venice, and the editors of this volume, Davide Cellamare and Mattia Mantovani.

---

60   On 'Aristotelianisms,' see, among others, Schmitt, *Studies in Renaissance Philosophy and Science*; Bianchi, *Le verità dissonanti*; Martin, *Subverting Aristotle*; Sgarbi, "What Does a Renaissance Aristotelian Look Like?," 226–245. On Scholastic Cartesianism, see especially Bohatec, *Die cartesianische Scholastik*; Verbeek, *Descartes and the Dutch*; Trevisani, *Descartes in Germania*; Ariew, *Descartes among the Scholastics*; Strazzoni, *Dutch Cartesianism*.

61   Schmaltz, *Early Modern Cartesianisms*.

CHAPTER 7

Cartesian and Anti-Cartesian Disputations and Corollaries at Utrecht University, 1650–1670

Erik-Jan Bos

1   Introduction

In a short article published in 1995, Koert Van der Horst, curator of manuscripts at Utrecht University Library from 1975 to 2007, describes the limited sources available for the study of the university curriculum at Utrecht in the seventeenth century.[1] He points out that his library features just 33 Utrecht lecture notes from this period, one third being class notes from courses given by the professor of eloquence, Johannes Georgius Graevius (1632–1703). As to philosophy proper—all disciplines except Theology, Law and Medicine, belonged to the Faculty of Philosophy—there is only one such notebook dating to before 1675, a dictated commentary by Daniel Voet (1629–1660).[2] Van der Horst furthermore emphasizes that many of the publications by Utrecht professors on which they would often have built their lessons are not present in the university library and not even in other Dutch libraries. Finally, only a small number of the countless academic disputations have been preserved, and these are hard to find.

Since 1995, the situation has dramatically improved, primarily due to the internet. Nowadays books that once seemed rare or untraceable can be found and, in many cases, even downloaded in the blink of an eye. The same advantages exist for locating disputations, although this usually involves some more effort, because both the author and the title can be sources of confusion: for the former, is it the presiding professor or the student?; for the latter, is the name of the disputation series part of the title? As to student notebooks, a recent research project at KU Leuven and Radboud University Nijmegen promises to fill this gap in our knowledge.[3]

---

1   Van der Horst, "De twee vroegste *Series lectionum*."

2   See below. Utrecht University Library owns over 2,000 lecture notes; see Van der Horst, *Catalogus*. See also Van der Horst, "Collegedictaten." I thank Bart Jaski, curator of manuscripts, Utrecht University Library, for the information that there are no seventeenth-century lecture notes among the acquisitions since 1994.

3   As part of the KU Leuven *Magister Dixit* Project, Davide Cellamare (Nijmegen) and Mattia

© ERIK-JAN BOS, 2023 | DOI:10.1163/9789004524897_009

This chapter investigates the reception and teaching of Cartesianism at Utrecht University between 1650 and 1670, based upon the disputations for and against Descartes, with special attention to the corollaries. Corollaries, usually a list of short propositions below the text of the disputation, may relate to the topic discussed in the disputation but do not necessarily relate to the said topic. Sometimes they address contemporary issues in philosophy and science, such as the Copernican system, Harvey's discovery of blood circulation, or Cartesian philosophy. Corollaries can supply information on current debates within the academy, even when the texts of the disputations themselves are silent on these topics.

According to Descartes's classical biographer, Utrecht University was "born Cartesian".[4] After the founding of the university in 1636, its first professor of philosophy was Henry Reneri (1593–1639), a close friend and follower of Descartes. Apparently, he discussed the *Discourse* and the *Essays* with his students, but his disputations show no trace of Cartesianism.[5] In stark contrast, his colleague Henricus Regius (1598–1679), professor of medicine since 1638, openly taught Descartes's philosophy, and had it defended during ten public disputations in 1640 and 1641.[6] It elicited a hostile response not just from the theologians, as should be underscored, but also from professors of philosophy and medicine. The subsequent ban on Cartesian philosophy in 1642 and the censure directed at Regius seemed to smother the introduction of Descartes's ideas at Utrecht University, but these were only temporary setbacks, for the New Philosophy made a resurgence in Utrecht disputations in the 1650s which culminated in the next decade.

The main characters to whom we will turn our attention are the Cartesian Johannes De Bruyn (1620–1675) and the neo-Aristotelian Daniel Voet. In the same Faculty of Philosophy, they both struggled, on the one hand, with the philosophical tradition and the regulations of the academic senate and the *Vroedschap*, and, on the other, with the startling discoveries in science and medicine, as well as the rise of Cartesianism. On top of that, being mutually antagonistic, they struggled against each other, with De Bruyn in an underdog position at first, because he faced an anti-Cartesian majority at the university, led by Gisbertus Voet (Voetius) and Voet's sons Paul and Daniel.

Mantovani (Leuven) are collecting and analysing the lecture notes by Cartesian professors in the seventeenth-century Dutch Republic and Leuven University. At present, the only articles published on Dutch Cartesian lecture notes are Verbeek, "*Principia*," and Cellamare, "A Theologian Teaching Descartes."

4   Baillet, *Vie de Descartes*, vol. 2, 2.
5   On Reneri, see Buning, "Reneri."
6   On Regius, see Clarke and Bos, "Regius," with further references to secondary literature.

Before turning to the Utrecht disputations, some preliminary words are required on the main characters De Bruyn and Voet, and on the phenomenon of disputations and the university curriculum at Utrecht. After that, we will deal with the disputations and corollaries themselves.

2   Johannes De Bruyn and Daniel Voet

Johannes De Bruyn studied at the Illustrious School of 's-Hertogenbosch in the early 1640s. Here he defended theological disputations under Samuel Maresius (1599–1673), who later became professor of Theology at Groningen University, and philosophical disputations under Florentius Schuyl (1619–1669), who published the Latin translation of Descartes's *L'Homme* (*De homine*, 1662) and became professor of Medicine in Leiden.[7] In 1643 De Bruyn enrolled at Utrecht University, graduating in philosophy the next year. He then moved to Leiden to study theology. At the recommendation of his philosophy teacher in Utrecht, Jacob Ravensberg (1615–1650), he was appointed Ravensberg's successor and on 3 June 1652 was inaugurated professor *extra ordinem* to teach physics and mathematics. His first known disputation was held on 20 October 1652.[8] In March 1656 he became full professor. De Bruyn never published a textbook, and hence his disputations are an important source of his views and his philosophical development. Next to disputations, he published a defence of the Cartesian theory of light against Isaac Vossius (1618–1689) (*Epistola ad Vossium*, 1663), and a defence (*Defensio*, 1670) of Cartesian method and metaphysics against the attacks by a professor of Theology at 's-Hertogenbosch, Reinier Vogelsang (ca. 1610–1679).

Daniel Voet matriculated at Utrecht University in 1644, where his father occupied the first chair in theology and his brother a chair in philosophy.[9] He took his degree in philosophy at the University of Harderwijk (1648) and became Doctor of Medicine at the same university in 1651. On 9 February 1653, he inaugurated as professor *extra ordinem* at Utrecht to teach logic and metaphysics, replacing Paul Voet (who took up a chair in Law). His first known disputation was held on 29 June 1653. He became full professor in April 1656. Except

---

7   On De Bruyn, see Van Bunge, Krop, Leeuwenburgh, Van Ruler, Schuurman and Wielema, eds., *Dictionary*, vol. 1., 175–176; Sassen, *Studenten*, 32–34.

8   De Bruyn (praes.) and Taeispil (resp.), *Disputationum physicarum prima de philosophia in genere* (Utrecht, 20 October 1652), the first of eighteen *Disputationes physicae* (1652–1660).

9   On Daniel Voet, see Van Bunge, Krop, Leeuwenburgh, Van Ruler, Schuurman and Wielema, eds., *Dictionary*, vol. 2, 1028–1029, and Dieckhöfer, *Daniel Voet*.

for his (very!) many disputations, he did not publish anything during his lifetime. After his death, his father and brother published several of what appear to be notebooks that he used for his classes in metaphysics and physics, the most important one being *Physiologia*, which was reprinted several times. A heavily annotated edition by Gerard de Vries (1648–1705), professor of Logic and Metaphysics at Utrecht, appeared in 1678; a third and enlarged edition in 1688 (reprinted in 1694).[10] The *Physiologia* shows that although Voet had to teach metaphysics and logic, his real interest was in physics and medicine.

3   Disputations and the University Curriculum at Utrecht

Disputations played an important role in the academic curriculum.[11] Much more than the static performances on obligatory subjects that usually constituted public lectures, public disputations showcased the current interests of the professors, and were more often than not lively events. There were, generally speaking, two kinds of disputations: *pro gradu*, submitted in order to obtain a doctoral degree, and *exercitii gratia*, to practise the skills of students. The ordinary disputations were submitted *sub praesidio* of the student's professor, who drew up the texts and was responsible for its contents. Disputations that were part of a series, e.g., De Bruyn's *Disputationum physicarum prima-decima-octava* (1652–1660), were always the intellectual work of the professor. In the first decades of Utrecht University, this kind of disputation was the most common one in the Faculties of Arts, Medicine and Theology. In case of disputations *pro gradu* or *inauguralis*, the candidate had to defend the theses, which, if he is mentioned as author, he had composed himself to a greater or lesser extent, but the professor's approval was still needed. A hybrid form also occurred from time to time, when, in a *sub praesidio* disputation, the student would describe himself on the dedication page as *author et respondens*. The examples I have seen are never part of a series of disputations, and are often

---

10   Utrecht University Library owns three notebooks (MSS. 716–718) dictated by de Vries on Daniel Voet's posthumous works.

11   For the phenomenon of Dutch academic disputations, see Dibon, *Enseignement philosophique*, 33–49; Ahsmann, *Collegia*, 274–341 (German translation: Ahsmann, *Collegium*). For an outline of the situation at Utrecht University between 1636 and 1815, see Kernkamp, *Utrechtse Academie*, 145–170. Dirk Van Miert wrote a fine article on the role of disputations at Dutch Illustrious Schools and universities and their value as a source for intellectual history, with references to recent literature: Van Miert, "Disputation Hall." In the three following paragraphs I largely follow the text from my doctoral dissertation, Bos, "Correspondence," li–lii.

longer than the standard eight or twelve pages, which fact suggests that (much of) the text was indeed devised by the student. The presiding professor was however still considered responsible for the contents.

At the end of the text of the disputation there was often some space left for corollaries, in most cases announced as 'corollaries of the respondent.' They offered the student the opportunity to personalise the disputation, because the texts they had to defend were normally those of the professor. Yet the professor remained accountable, and ordinarily would not allow corollaries expressing views contrary to his own.[12] Some corollaries to the disputations by Voet and De Bruyn reoccur over and over again, and I suppose that students from time to time just borrowed propositions from earlier disputations, which they themselves liked or which they knew would make for a lively debate, and perhaps the professor had a list of corollaries at hand.

Because no records were held of the senate's meetings during the first five years, we have no detailed description of the rules on disputations in Utrecht.[13] In later years, the subject was brought up again in the *acta* of the senate and in the resolutions of the *Vroedschap*, which provide us with the following picture. Disputations *pro gradu* took place in the Dom Church, in public and *sine praesidio*, that is, without a presiding professor who might come to the rescue. A *moderator*, either the rector or the promotor, supervised the proceedings. These inaugural disputations began at nine o'clock, and lasted until ten or a quarter past ten. Students took care of the first round of opposition, followed by the graduates. Priority was given to those in whose faculty the graduation took place.[14] Disputations *sub praesidio* were submitted in one of the auditoria. Theological disputations were scheduled on Saturdays, juridical and medical disputations on Wednesdays, and philosophical disputations took place on Wednesdays or Saturdays. There were to be no two disputations at the same time, unless they were juridical and medical. Disputations should be announced two weeks in advance to the rector. Of each disputation, 200 copies were printed; the printer should distribute 130 among the students, the *praeses* received 20 copies, the respondent 30, and the beadle

---

12   Rienk Vermij noted a corollary to one of De Bruyn's disputations defending that "it is taught in Holy Scripture that the stars are rotating all round, and the earth is resting in the centre." De Bruyn (praes.) and Shepheard (resp.), *Disputationis physicae de alitura pars quinta* (Utrecht, 22 September 1660). The respondent, Nicolaus Shepheard, remarked that he added the corollary "with kind permission of the President, who upholds the opposite." Vermij, *Calvinist Copernicans*, 172.

13   Kernkamp, *Acta*, 102.

14   Wijnne, *Resolutiën*, 58–59, 64; Kernkamp, *Acta*, 160, 174, 220, 241, 526.

(*pedellus*) delivered the remaining 30 copies to the professors and members of the *Vroedschap*. The copies should be ready three days in advance, and the registrar would nail the title pages *ad valvas academiae*. The professors could publish up to ten disputations of eight pages at the expense of the municipality.[15] For the professors in theology and philosophy, this number was raised to twelve in August 1655, and to fourteen disputations of twelve pages in February 1669.[16]

Despite regulations on the ways in which a disputation should be conducted, it was more often than not a noisy happening. In 1648, the Leiden professor Adriaan Heereboord (1613–1661) stated that during a disputation the public should not "shout, laugh, pull faces, bleat, whistle, stamp, or make fun of the proceedings," which implies that this kind of behaviour was in fact the order of the day.[17] Both the Utrecht municipality and the senate tried to suppress any mischief by imposing restrictions—for example, no drinking before a disputation, or no arms in the classroom—and in 1661 they even decided to place a fence around the respondent's chair to prevent serious misconduct.

4   Philosophy in the *Series lectionum* of 1656, 1659, 1663, 1668, and 1672

In 1995 Van der Horst made a most welcome addition to the limited sources available at the time: two *series lectionum* from the seventeenth century, the first ever to resurface from this period.[18] The first *series* gives the programme for the second semester (February to July) of 1656, the second *series* for the second semester of 1672. They show that professors were required to lecture in public four times per week. The philosophers lectured on Monday, Tuesday, Thursday and Friday, and held their private courses and disputations on Wednesday and Saturday. Typical for Utrecht, apparently, are the details supplied on private courses. We highlight below the information that these *series lectionum* provide regarding the philosophy professors, and we offer information about three additional *series lectionum* (1659, 1663, and 1668).

---

15   Wijnne, *Resolutiën*, 58; Kernkamp, *Acta*, 177–178, 220.
16   Wijnne, *Resolutiën*, 84, Kernkamp, *Acta*, 337; Wijnne, *Resolutiën*, 105, Kernkamp, *Acta*, 479.
17   Heereboord, *Sermo*; cited from Verbeek, *Descartes and the Dutch*, 65. On Heereboord, see Chapters 2 and 3, in this volume.
18   Van der Horst, "De twee vroegste *Series lectionum*."

4.1   Second Semester, 1656

Public lectures: De Bruyn will teach physics on Monday and Tuesday; on Thursday and Friday he will teach astronomy using Willem Blaeu's *Institutio astronomica* (Amsterdam, 1634; many reprints during the seventeenth century). Voet will teach logic on Monday and Friday, metaphysics on Wednesday and Saturday. Private courses: these are listed in the second section of the *series*, headed by the somewhat misleading title *Exercitia publica et privata*. Whereas this section contains many details regarding the theological programme, as to philosophy they merely state that disputations will be held on Wednesday and Saturday, and the courses in logic, metaphysics, physics, mathematics and ethics will be continued.[19]

4.2   Second Semester, 1659

The *series lectionum* of the second semester of 1659 is lost, but the acts of the university senate and the Reformed Church council offer interesting information. In February 1659, Paul Voet called for a meeting of the senate, because of De Bruyn's announcement on the published curriculum (*series lectionum*) for the second semester to teach Descartes's *Meteors* and *Passions of the Soul*. Paul Voet strongly objected, pointing to the university decree banning the philosophy of Descartes. De Bruyn, however, did not give in, and even refused to accept the suggestion at least not to mention Descartes's name. The senate decided to inform the *Vroedschap*, but the burgomaster responsible for academic affairs, took the rector aside for a moment, urging him not to consult the *Vroedschap*, but to again warn De Bruyn not to mention Descartes by name.[20] The appearance of Descartes's name and works on the *series lectionum* was also discussed in a meeting of the Reformed Church council in March 1659. The council appointed a committee to take the matter up with the *Vroedschap*. A week later they reported back that the burgomasters had already talked to De Bruyn, who had promised not to mention Descartes's name anymore unless with the *Vroedschap*'s consent. To one of the members of the council—possibly

---

19   Van der Horst supposes that the lecture notes of Daniel Voet were written in this period
        (MS., Utrecht, MS. 715). These lecture notes, written by the student Johannes Rauwers, who
        matriculated in February 1655, bear the title "Dictata in Phisicam Senguerdi[i]," a com-
        mentary on Arnoldus Senguerdius's *Introductio ad physicam libri sex* (Amsterdam, 1653).
        See Van der Horst, "De twee vroegste *Series lectionum*," 272. Rauwers acted as respondent in
        a disputation by Voet in 1657; Daniel Voet (praes.) and Rauwers (resp.), *Disputationis meta-
        physicae, de formali ratione libertatis pars quarta* (1657), at the University Library Edin-
        burgh. That Voet did not confine his teaching to logic and metaphysics is clear from his dis-
        putations: in 1654 and 1655 he held various disputations on vacuum, motion, and gravity.
20   Kernkamp, *Acta*, 336–337; cf. De Vrijer, *Regius*, 47.

Voetius, who in his capacity as minister also sat in the council—the burgo-masters ensured De Bruyn would not get permission to do so. In February the following year, a burgomaster informed the church council that he would see to it that the upcoming *series lectionum* did not mention a course on any of Descartes's writings.[21] We may conclude that as long as De Bruyn did not explicitly mention Descartes's name or his writings, he could discuss and teach Cartesian philosophy.

4.3   First Semester, 1663

To the copies published by Van der Horst we can add a third *series lectionum*: the English naturalist John Ray (1627–1705) visited Utrecht and copied the programme of the first semester of 1663 in his travel journal. Unfortunately, he only mentions the public lectures.[22] Public lectures: Regnerus Van Mansvelt—a Cartesian who succeeded Daniel Voet in 1660[23]—will teach logic and meta-physical questions on Monday and Tuesday, on Thursday and Friday he will teach natural theology. De Bruyn lectures on physics on Monday and Tuesday, on Thursday and Friday he will continue his explanation of the foundations of mechanics.

4.4   Second Semester, 1668

The reverse situation of March 1659 occurred in March 1668. The professors of philosophy protested against the announcement in the *series lectionem* by the theological professors to submit disputations on 'theologico-philosophical' subjects. The senate being divided, the question was put before the burgomas-ters, who subsequently agreed with the philosophers, forbidding the theolo-gians to touch upon philosophical matters.[24]

---

21   *Het Utrechts Archief*, 746: Kerkeraad van de Nederlandse Hervormde gemeente te Utrecht, Acta 1658 mei–1660 oktober. (The dates of meetings mentioned are 21 and 28 March 1659, and 13 February 1660).

22   Ray, *Observations* (London, 1673), 45–47.

23   Regnerus Van Mansvelt (1639–1671) studied in Leiden and in his hometown of Utrecht. He defended philosophical theses under Daniel Voet—Daniel Voet (praes.) and Van Mans-velt (resp.), *Disputatio philosophica continens quaestiones illustres philosophicas* (Utrecht, 1656); British Library (536.e.12.(25), not seen by me)—and theological theses under Gis-bertus Voet. He took his degree in philosophy with Daniel Voet in September 1658; *Album promotorum*, 14. Like De Bruyn, he was a staunch proponent of Cartesianism and wrote a confutation of Spinoza's *Tractatus theologico-politicus*, and he was involved in a polemic with Maresius concerning the latter's *De abusu philosophiae Cartesianae* (Groningen, 1670). See Van Bunge, Krop, Leeuwenburgh, Van Ruler, Schuurman and Wielema, eds., *Dic-tionary*, vol. 2, 672–674.

24   Kernkamp, *Acta*, 465–466; cf. De Vrijer, *Regius*, 60.

4.5   *Second Semester, 1672*

De Bruyn lectures on physics on Monday and Tuesday, and on Thursday and Friday he explains the principles of astronomy. There are no public lectures in logic and metaphysics; Van Mansveld died in May 1671, and his chair remained vacant until de Vries's appointment in 1674. The deficiency is however remedied by De Bruyn in his private courses, who will teach in every philosophical discipline: logic, metaphysics, pneumatica, mathematics, physics, ethics, and politics, and he will read Grotius's *De iure belli ac pacis* (1625), and other authors at the request of the students. Significant for the history of the teaching of Cartesianism is the announcement of a public *Philosophicum Collegium Disputantium* on the *Principles of philosophy* of Descartes. This is a clear sign that the teaching of Descartes at Utrecht University had reached its summit. Decline was imminent, however: de Vries was a vehement anti-Cartesian, and De Bruyn's successor, Johannes Luyts, was no less opposed to Cartesianism.[25]

5   Resurfacing Disputations by De Bruyn and Voet

A stupendous number of disputations has been preserved in libraries all over Europe and North America, but, as Van der Horst pointed out, in 1995 it was still difficult to find them.[26] Disputations could be catalogued under the name of the student instead of that of the professor; disputations collected in a single volume received a general title (e.g., 'Dutch disputations, 1642–1674'), or they simply remained uncatalogued. In this respect, much progress has been made over the last decades: library catalogues have become accessible online, and many libraries have started to catalogue their disputations properly. Moreover, libraries have begun to digitise less attractive items, such as disputations, and to make them freely available on the internet. This enterprise was boosted by Google Books, digitising indiscriminately whatever is on the shelves of important libraries.

Thanks to these efforts, I was able to download no less than 89 disputations by Johannes De Bruyn, ranging from 1652, the year he was appointed professor of philosophy at Utrecht University, until 1670, five years before his death. All

---

25   For de Vries, see Chapters 9 and 10, in this volume.

26   When Van der Horst indicated that only a small number of the countless academic disputations have been preserved, he presumably referred to the disputations preserved in Dutch libraries. It is a fact that many Dutch disputations are owned by libraries outside the Netherlands, due, probably, to foreign students collecting and cherishing these rare imprints.

89 disputations come from the same collection in the Austrian National Library.[27] The British Library has 22 disputations defended under De Bruyn, which are all present in the Austrian collection as well. The Staatsbibliothek zu Berlin has 27 disputations by De Bruyn, one of which is not found anywhere else but which I, unfortunately, was unable to consult.[28] In contrast to these repositories, the Dutch libraries are quite poorly endowed: according to the *Short Title Catalogue Netherlands* the university libraries of Leiden and Utrecht have two disputations by De Bruyn apiece.

Is the number of 90 disputations close to the total number of disputations over which De Bruyn presided? Presumably not, because a philosophy professor at Utrecht could publish ten to fourteen disputations a year at the expense of the municipality (see above); De Bruyn, having taught at the university for more than 20 years, might, in principle, have delivered around 250 disputations. The 90 disputations we have so far cover the years 1652–1670; disregarding the last five years, we would still expect there to have been twice as much disputations as we presently have. As is shown in Figure 7.1 below, the distribution of the 90 disputations is uneven. There is a serious falling off in the number of disputations available from the years between 1658 and 1662, with each year having just three disputations or less (just one disputation in 1659 and 1661). During the last two years, 1669 and 1670, there is a decline in the number again. I have not found an explanation for these low numbers. Koert Van der Horst points out that a sharp decline in student numbers occurred between 1657 and 1662: the *Album studiosorum* lists just sixteen matriculations for 1657, whereas the yearly average for the previous five years was close to 200.[29] Despite this decline, no concerns about dramatically low student numbers are voiced in the acts of the senate or in the resolutions of the *Vroedschap*, and numbers of matriculations moreover seemed to surge back to normal levels shortly thereafter. In March 1661 the rector noted that during the last year (March 1660-March 1661) the university saw a record in the number of graduations.[30] In his history of Utrecht University, Kernkamp supplies the

27   Monika Kiegler-Griensteidl, specialist in manuscripts and rare books at the Austrian National Library, kindly informed me that these disputations are part of the library's historical holdings, which means they most likely came to the library close to the time they were published. The exact provenance, however, is unclear.

28   De Bruyn (praes.) and Van Rhee (resp.), *Disputatio philosophiae aliquot nobilissimae materiae ex physica et mathesi selectae continens* (Utrecht, 11 December 1661). Cf. Hoogendoorn, *Bibliography*, 760.

29   Van der Horst, "De twee vroegste *Series lectionum*," 261, n. 1.

30   Kernkamp, *Acta*, 349. Between March 1660 and March 1661, 30 graduations took place.

explanation: early in 1657, the *Vroedschap* revoked the tax exemption on beer
and wine for students as fraud had risen to unprecedented levels. As a result,
weighing the costs, many students decided not to pay the registration fee.[31] And
indeed, in 1660 and 1662 we encounter several students who defended theses
without being registered in the album.[32]

Returning to the question of the number of De Bruyn's disputations, the acts
of the senate mention no absence or illness of De Bruyn, nor of any other circumstance that may explain the low amount of disputations during 1658–1662
and 1669–1670. It thus seems safe to say that there must have been more disputations, although it is impossible to give an estimation of how many were lost
or may yet resurface.[33]

The New College Library of Edinburgh University owns a collection of 78 disputations by Daniel Voet, ranging from 1655 to 1660. I was able to acquire copies
of 24 disputations.[34] Other libraries have seven disputations that are not found
in Edinburgh.[35] To these 85 disputations that have survived in their original
format, we can add 26 disputations of which the original publication is missing
so far, but which are reprinted in Voet's posthumous works.[36] The reprinted text
supplies the date of the disputation and the name of the student, but corollaries
are omitted. The total number of known disputations by Voet is then 111, which
is much more than the 90 or so disputations the municipality would have paid
for; who it was that took charge of these additional expenses remains unclear.

31   Kernkamp, *Utrechtsche Academie*, 86–90.

32   Christiaan Melder, who became professor of mathematics at Leiden University, acted as
       a *respondens* in disputations by both De Bruyn, and Voet in 1660. Petrus Clerquius and
       Martinus Martens for De Bruyn, in resp. 1660 and 1662.

33   One possible way to shed light on the question is the study of the expense accounts of the
       town treasurer who paid out the university printers. Cf. Monna, "Gedichten," 260.

34   I thank Denise Anderson (Curatorial Assistant, Special Collections), and Elizabeth
       Quarmby Lawrence (Rare Books Librarian) at the Centre for Research Collections, Edin-
       burgh University Library, for their kind assistance and information.

35   The British Library has nine disputations by Daniel Voet, four of which are not present
       in Edinburgh. The Staatsbibliothek zu Berlin has a single disputation, not available in
       Edinburgh. The Herzogin Anna Amalia Bibliothek at Weimar, and the Universitäts- und
       Landesbibliothek Bonn both have a unique disputation as well. The Austrian National
       Library appears to have no disputations by Voet, nor does any Dutch library, according to
       Short Title Catalogue Netherlands.

36   These disputations are found in Daniel Voet, *Physiologia*, edited by Paul Voet (adding 14
       disputations of his own), and *Compendium pneumaticae*, edited by Gisbertus Voet. More
       disputations are to be found in an extremely rare work, of which only one copy (not
       seen by me) seems to survive in the Bibliothèque publique et universitaire de Neuchâ-
       tel: *Meletemata philosophica*.

Finally, Aza Goudriaan (Vrije Universiteit Amsterdam) has been so kind as to provide me with a reproduction of 19 disputations by Andreas Essenius (1618–1677), professor of theology at Utrecht from 1653 to 1677.[37]

6   What Is Cartesian or Anti-Cartesian?

Before turning to the data-analysis of the Utrecht disputations and their corollaries, and investigating to what extent they were either Cartesian or anti-Cartesian, some remarks are in order as to what exactly defines these texts as either Cartesian or anti-Cartesian? Obviously, there is no clear-cut answer to that question, although the matter is somewhat more straightforward in those instances where Descartes is mentioned by name. Some examples of such corollaries are:

> Does Descartes offer a valid proof of the equality of the angle of incidence and reflection? Denied.

> Whether we have a clear and distinct idea of God, in the way Descartes explains it? Affirmed.

> That Cartesian doubt is legitimate and does not lead to atheism has been proven very often in the disputations of our presiding professor.[38]

Very similar corollaries can be found where the only difference is that Descartes's name is not mentioned, but where the determination as to the category in which they belong is, again, simple:

---

37   Essenius (praes.), *Disputationis practicae de conscientia* [*pars prima-decima nona*] (Utrecht, 1666–1668). The original disputations are kept at Pitts Theology Library, Emory University, Atlanta.

38   Daniel Voet (praes.) and Ketelaer (resp.), *Disputationum selectarum octava continens quaestiones duas de evidentia et certitudine pars tertia* (Utrecht, 23 February 1656): "An bene demonstret Cartesius aequalitatem anguli incidentiae et reflexionis? Neg."; De Bruyn (praes.) and Van Swanevelt (resp.), *Disputatio philosophica de libero hominis arbitrio* (Utrecht, 25 April 1657): "Num Dei in nobis sit quaedam idea clara, et distincta, ut Cartesius explicat? Aff."; De Bruyn (praes.) and Witkint (resp.), *Disputationum physicarum decimaquinta de motu pars sexta* (Utrecht, 6 November 1658): "Dubitationem Cartesianam legitimam est neque ducere ad Atheismum sapissime in disputationibus D.C. Viri praesidis demonstratum est."

Can doubt be a metaphysical principle? Denied.

The axiom saying that we must doubt everything until we find principles that are certain, does not pave the way to atheism but rather to wisdom.[39]

Indeed, any corollary advancing the method of doubt, or claiming that animal behaviour is automatic, that we have an innate idea of God, or that the essence of the soul is thought—in short, hardcore Cartesian propositions, or their opposites—pose no problem. As we may assume that any student of philosophy in the 1650s would be well aware with whom to connect the concept of 'doubt,' we can predicate 'Cartesian' to any corollary speaking about doubt even when it sounds innocent enough, such as in "A learned man doubts anything of which he is not most certain."[40]

"Atoms exist" (*Dantur atomi*); is that an anti-Cartesian statement? Descartes denied the possibility of atoms, and he was upset by the accusation that he was an atomist, but it is a corollary that is found several times in De Bruyn's early disputations. De Bruyn, defended atomism in various disputations during the mid-1650s, but from 1658 onwards the contrary corollary is seen: "There are no atoms," indicating De Bruyn had abandoned the idea of atomism. In any case, I do not count the corollary "Atoms exist" as anti-Cartesian.

Many corollaries are of course neither Cartesian nor anti-Cartesian, like "The Moon does not produce light," and so on. However, when going over the numerous corollaries, it might be easy to miss specific Cartesian or anti-Cartesian statements. Contrary corollaries are a strong indication that something of that binary dynamic is at hand. Take for example the contrary corollaries "There is only one world" and "It is not contradictory that there is more than one world," seen in various disputations in the 1650s. The thesis that only one world or universe exists, is not a very well-known Cartesian statement, but Descartes did make that claim in his *Principles of Philosophy* (II 22). For this particular corollary and several others, a curious work by Essenius, published under

---

39   Daniel Voet (praes.) and Vossius (resp.), *Disputationum selectarum sexta continens quae-stiones duas de evidentia et certitudine pars prior* (Utrecht, 1 December 1655): "An dubitatio possit esse principium metaphysicum? N."; De Bruyn (praes.) and Michielzon (resp.), *Disputatio physica de mari* (Utrecht, 14 May 1656): "Axioma quod dicit de omnibus esse dubitandum, donec certa principia invenimus, non sternit viam ad Atheismum sed potius ad sapientiam."

40   De Bruyn (praes.) and Camp (resp.), *Disputationum physicarum septima de corporis loco et modo existendi in loco* (Utrecht, 14 December 1655): "Viri docti est de quacunqu[e] re dubitare, quarum certissimam cognitionem non habet."

a pseudonym in 1656, is a convenient tool. The *Specimen of Cartesian philosophy expressed in several theses* lists 52 Cartesian propositions followed by the remark "Denied."[41] Remarkably, the propositions are not followed by their refutation, but by quotations from Descartes's works and letters where he made these claims. Presumably the intent here was to muster the students with the material to rebuke any proponent claiming Descartes had not said it as such or was misinterpreted.

Essenius's compilation also helps to identify De Bruyn's earliest disputation expressing core views of Descartes's philosophy. It lists as proposition 29: "Are there three classes of matter, as three elements of the world?" which view De Bruyn defended in his disputation of 14 May 1653 (thesis iv).[42] A closer study of his pre-1656 disputations may yield more Cartesian-inspired views.

7   De Bruyn's Cartesian Disputations and Corollaries in Numbers

The next four figures display the number of disputations submitted by De Bruyn and Voet per year. Additional information, such as the numbers of Cartesian or anti-Cartesian disputations among them, is visible in the columns. Every chart is followed by an explanation and an analysis.

Figure 7.1 shows how the 90 disputations by De Bruyn are distributed over the years 1652–1670. The vertical extent of the year columns gives the number of disputations that year, which number is also seen inside the column. Columns in blue denote 'neutral' disputations, that is without clearly recognisable Cartesian content; columns in orange indicate the number of disputations that do have Cartesian content. In 1652 De Bruyn submitted six disputations that were neutral with regard to Cartesian philosophy; in 1667 nine of the disputations defended by his students were clearly Cartesian. When a column is divided into both blue and orange sections, the total height is the sum total of disputations that year, and each coloured part has a number corresponding to the disputation type. 1656, for example, knew five neutral disputations and three Cartesian disputations. The grey column of 1661 represents the one disputation I have not seen; as I have explained above, it is the only known disputation from that year.

---

41   [Essenius], *Specimen philosophiae cartesianae, thesibus aliquot expressum* (Utrecht, 1656). The proposition "An proinde nec plures mundi esse possint?" is found on p. 19.

42   "An sint tria genera materiae; ut tria mundi elementa? Neg."; [Essenius], *Specimen*, 15, referring to Descartes's *Principia* III 52, AT VIII-A 105. See also Vermij, *Calvinist Copernicans*, 169–171, discussing De Bruyn's disputations and corollaries regarding heliocentrism.

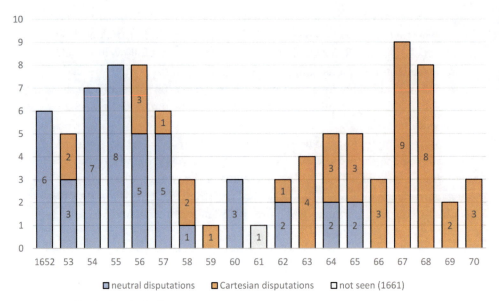

FIGURE 7.1 Disputations by De Bruyn,

Of the total of 89 disputations that were seen, 45 can be labelled Cartesian (51 per cent). De Bruyn's first Cartesian disputation was defended by Gerard Bornius on 14 May 1653. Gerard Bornius was a brother of Henricus Bornius (1617–1675), professor of philosophy at the Illustrious School of Breda (1646) and at Leiden University (1651). In *De mundo* the student explained that the universe consists of three types of matter: that of the Sun and the stars, that of the Earth, the planets and moons, and finally the aether that fills the space between these bodies. A son of Gisbertus Voetius, Nicolaas, one of Gerard's fellow-students in theology, adorned the disputation with a laudatory poem. In the second Cartesian disputation, on 10 December 1653 (first part), a student proposed that we must, for "once" (*semel*), ascertain ourselves of the foundations of philosophy, and call into doubt everything of which we do not have certain knowledge (theses IX and X).[43] Undeniably Cartesian as this sounds, it paves the way not for a Cartesian physics, but for an atomist worldview.[44] This eclectic approach may to some extent account for the fact that, in the

---

43   A clear reference to the opening sentences of Descartes's *Principia philosophiae* (1644), AT VIII-A 5, CSMK I 193: "It seems that the only way of freeing ourselves from these opinions is to make the effort, once in the course of our life [*semel in vita*], to doubt everything which we find to contain even the smallest suspicion of uncertainty." The expression is also found in the *Mediationes* (1641), AT VII 17, CSM II 12.

44   I will deal with De Bruyn's atomism in a future article.

subsequent two years, no overtly Cartesian ideas are found in De Bruyn's disputations. In 1656 three Cartesian disputations were submitted, the first having the same contents as the disputation defended by Bornius in 1653 (19 April 1656), the second on the tides (14 May 1656), and finally, the third disputation (2 July 1656) which happens to be the first in which Descartes's name is mentioned in the text.[45] In defence of the Cartesian philosophical method, it offers a long quote from Descartes's *Notae in programma quoddam* (1648) that fills almost an entire page.[46]

I have already discussed the uneven distribution of the occurrence of De Bruyn's disputations, for which I found no explanation. It seems likely that there were more disputations during the years 1658–1662 and 1669–1670, some of which may yet resurface. While De Bruyn did encounter opposition to his plan in February and March 1659 to discuss Descartes's *Meteors* and *Passions of the Soul* (and presumably to deal with them in disputations), it appears that he was authorized to continue if he refrained from mentioning Descartes by name. In the 25 disputations from 1655–1658, Descartes is mentioned just five times within the texts and/or corollaries. The first reoccurrence after 1658 is in 1662, so De Bruyn did, for the time being, comply with the demands of the senate and the church council. Remarkably, Descartes's name never appears in any disputations or corollaries between the years 1668–1670.

From 1662 onwards, the majority of the disputations—and during the five years up to 1670, all disputations—are strongly Cartesian. Whereas De Bruyn had previously been the only Cartesian professor among five opponents of Cartesianism (Voetius, Daniel and Paul Voet, Johannes Hoornbeeck, Andreas Essenius, and Matthias Nethenus), the Cartesians won ground in the 1660s: first

---

45 De Bruyn (praes.) and Van Swanevelt (resp.), *Disputatio philosophica miscellanea* (Utrecht, 2 July 1656). The respondent dedicated the disputation to Adriaan Heereboord. Swanevelt also defended theses of De Bruyn in 1655, dedicated to De Bruyn and the Amsterdam professors Arnoldus Senguerd and Alexander de Bie, and in 1657, dedicated exclusively to Jan Amos Comenius: De Bruyn (praes.) and Van Swanevelt (resp.), *Disputationum physicarum quarta de corporis divisibilitate* (Utrecht, 13 October 1655); De Bruyn (praes.) and Van Swanevelt (resp.), *Disputatio philosophica de libero hominis arbitrio* (Utrecht, 25 April 1657).

46 Descartes, *Notae in programma quoddam*, AT VIII-B 367–368, CSMK I 309, starting with "Secondly, I have never even taught that 'God is to be denied, or that he can deceive us, or that everything should be doubted, or that we should entirely withdraw our confidence in the senses, or that we should not distinguish between being asleep and being awake', and other things of that sort—doctrines of which I am sometimes accused by ignorant detractors. I have explicitly disavowed all such views, and refuted them with very strong arguments—stronger, I venture to add, than any that anyone before me has employed in refuting them."

in the Faculty of Philosophy with the appointments of Van Mansvelt in 1660
and Graevius the next year, and with the appointments in Theology of Franciscus Burman (1628–1679) in 1662 and Louis Wolzogen (1633–1690) in 1664. The
number of opponents was reduced to three: Voetius, Essenius, and Paul Voet.
The other professor of Theology, Johannes Leusden (1624–1699), often sided
with Voetius.

In the mid-1660s the Cartesian professors began to meet weekly, joined by
the well-known Utrecht physician, philosopher and theologian Lambertus Van
Velthuysen (1622–1685). This scholarly society, nicknamed the 'College of savants', discussed scientific and philosophical topics, and promoted Cartesian
philosophy.[47] They will have also undoubtedly discussed all of the latest academic turmoil arising from the disputations. Three students were also alleged
members of the society: Johannes Fuyck (1647–1703), who defended a disputation for De Bruyn on 18 May 1667, Antonius Van Schayck (d. 1714), respondent
during three disputations of De Bruyn (15 December 1666, 27 April 1667, and
3 October 1668), and a further student for whom we have no further details of
identification beyond the fact that he was named Specht.

Figure 7.2 shows the number of De Bruyn's disputations that have 'neutral'
and/or Cartesian corollaries. Columns that are (partially) blue denote disputations with 'neutral' corollaries (without clearly recognisable Cartesian content), (partially) orange columns indicate the number of disputations having
Cartesian corollaries. In 1656, for example, De Bruyn submitted four disputations with neutral corollaries and four disputations with one or more Cartesian
corollaries. The grey column of 1661 represents the disputation I have not seen
(see above).

Of the total of 89 disputations that were seen, 56 have one or more Cartesian
corollaries (63 per cent). The first disputation having a striking Cartesian corollary was defended on 12 October 1653 ("A wise man doubts anything, until he
has found principles that are certain, which establish its truth").[48] This recourse
to the method of doubt in a corollary precedes by two months the first occurrence of such an appeal in the text of a disputation (10 December 1653).

All disputations after 1661 have Cartesian corollaries, each successively bolder than the one before. They testify that at Utrecht University in the 1660s,
the prevailing attitudes had changed in favour of Descartes. One disputation

---

47   On the College of Savants, see the definitive article by Gootjes, "*Collegie der Sçavanten.*"
48   De Bruyn (praes.) and Van Sypesteyn (resp.), *Diputatio physica de causis descensus gravium*
      (Utrecht, 12 October 1653): "Viri prudentis est de quâcunque re dubitare, donec certa prin-
      cipia invenerit, quibus veritas ejus innititur."

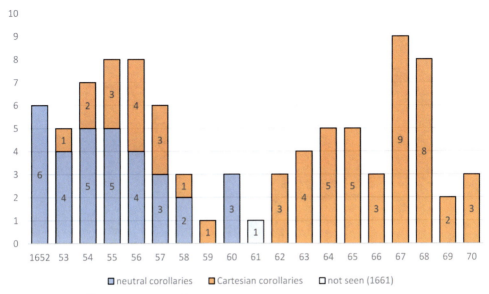

FIGURE 7.2   Corollaries to De Bruyn,'s disputations

from 1665 stands out, however, for being neither overtly Cartesian, nor having overtly Cartesian corollaries. It was defended by Thomas Bolwerck (1645–1711), who dedicated the disputation exclusively to Arnold Senguerd (1610–1667), an Aristotelian philosopher who had taught at Utrecht University but moved to the Athenaeum Illustre in Amsterdam in 1648. Bolwerck, born in Amsterdam, honours his former professor, who was also his uncle.[49]

8   Voet's Anti-Cartesian Disputations and Corollaries in Numbers

We now turn to the disputations by Daniel Voet. During the eight years of his professorship, he submitted at least 111 disputations, of which 85 survive in their original format, and 26 are known via their reprints in posthumous publications. I was able to consult 24 of the surviving disputations; three disputations

---

[49]   De Bruyn (praes.) and Bolwerck (resp.), *Disputatio physica de specialibus quibusdam motus effectibus* (Utrecht, 7 June 1665). The first corollary, "Caelum non esse animatum, nec ejus materia a sublunari distincta," can be labelled Cartesian, but compared to the far more candidly Cartesian corollaries in the other disputations at the time, the labelling as such seems quite tenuous. Bolwerck defended a disputation under Essenius (*De conscientia pars quarta*, Utrecht, 1666), with the same dedication to Senguerd. Bolwerck became minister in Beusichem.

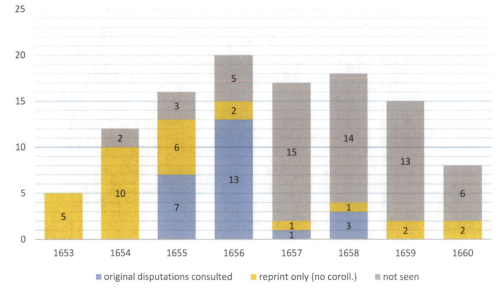

FIGURE 7.3 Disputations by Voet

that survive, but which I could not consult, are available as reprints; adding to these the 26 disputations solely known as reprint, we arrive at 53 out of 111 disputations (48 per cent), which I was able to access. This entails that no firm conclusions can be drawn, and that any interpretation remains provisory.

Figure 7.3 shows how the disputations by Voet are distributed over the years 1653–1660. The original disputations I have seen are coloured blue, those I have not seen are coloured grey. The reprinted disputations are in yellow, including the three disputations that were available to me in reprint (1654: 1; 1656: 2), while their original imprint is kept in New College Library, Edinburgh. One further caveat has to be made: the series *Disputationes ex theologia naturali selectae* (not consulted) comprises 39 disputations submitted between 1657 and 1660, but the catalogue of New College Library does not specify the year of publication of each disputation. I allotted the first seven disputations, all entitled *De natura sprituum*, to 1657, the next thirteen disputations (eight to twenty) to 1658, the next thirteen again (twenty-one to thirty-three) to 1659, and the last six disputations to 1660.

Figure 7.4 displays the disputations and corollaries by Daniel Voet directed against Cartesian philosophy, excluding obviously the 58 disputations not consulted. Each year has three or four columns. The first one (blue) gives the total number of disputations submitted that year. The second (yellow) column gives the number of disputations for which I had to use the texts reprinted in the posthumous works. The third (red) indicates how many dis-

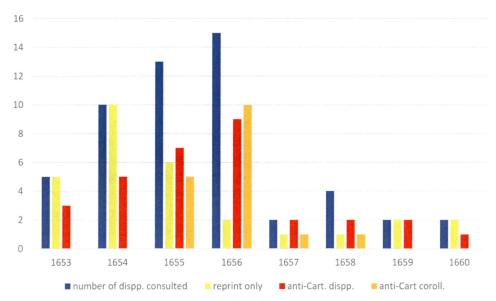

FIGURE 7.4 Voet's anti-Cartesian disputations and corollaries

putations contain anti-Cartesian content. Finally, the fourth (orange) column gives the number of disputations having one or more anti-Cartesian corollaries. Note that the corollaries are omitted in the posthumous works, which is why disputations from the years 1653–1654 and 1659–1660 in Figure 7.4 lack such corollaries. We may, however, safely assume there were corollaries against Descartes added to these disputations, first, because the years in question knew anti-Cartesian disputations and will therefore have had similar corollaries as well, and secondly, because since the 1640s the Voetian faction habitually added corollaries against Descartes and/or Regius. Paul Voet, for example, added the following corollary, clearly against Regius, to one of his disputations of 1651: "Is the human soul an accident or an attribute, or a mode of the body, and is it only in the brain? Denied."[50] The disputation has an additional set of six corollaries specifically announced as *Anti-Cartesiana*, of which the fifth reads "Is the first principle: I think therefore I am? The answer is: Nonsense."[51]

---

50 Paul Voet (praes.) and Van Ryssen (resp.), *Disputatio metaphysica pro novitate essendi* (Utrecht, 15 November 1651): "An anima humana sit accidens seu attributum vel modus corporis, et an in cerebro tantum existat? Neg."

51 Paul Voet (praes.) and Van Ryssen (resp.), *Disputatio metaphysica pro novitate essendi* (Utrecht, 15 November 1651): "An primum principium sit cogito, Ergo sum? Resp. nugae sunt."

Daniel Voet took up where his brother had left off. In 1653, the first year of his professorship, he submitted three (out of five) disputations aimed against Descartes and Regius (see below). Voet particularly criticizes the attribution of a role to the pineal gland as the seat of the soul and of common sense (*sensus communis*).[52] He also attacks the Cartesian work published by Lambertus Van Velthuysen, *De finito et infinito* (1650), which attack he continued in 1654.[53] Voet's general anti-Cartesian stance did not prevent him from adopting various elements of the New Philosophy and other scientific results. Whereas the rector of the university had tried to prevent Regius from publicly discussing blood circulation in 1640, Voet mentions (and accepts) the discovery by Harvey in passing (14 December 1653). He agrees with Descartes and Regius on the central role in sensation of the nervous system, accepts the existence of valves in nerves, and in the main accepts Regius' explanation of muscle movement as well (especially 24 May 1656). While rejecting the pineal gland as the seat of the common sense, Voet agrees that such a locus is located in the brain, namely in the fourth brain ventricle, where, according to Voet, all nerves come together.[54] Unlike Harvey, Descartes and Regius are not credited with having contributed anything of value, but Voet comes close to saying something positive about Descartes when, in comparing views opposite to his own theory of motion, he writes that Descartes's argument is at least better than that of Gassendi (15 November 1654).

If the years 1653–1656 are representative, Voet attacked Cartesian philosophy in half of his disputations. The number of anti-Cartesian corollaries will have matched the number of anti-Cartesian disputations. The first extant corollary directed against Descartes reads "The idea of God is God himself" (19 May 1655). According to Reformed orthodox theology, the idea of God comprises completely his essence and attributes, and is thus God himself. Consequently, and contrary to Descartes's view, humans cannot have an idea of God. The opposing corollary, that we do have an idea of God, is found in various disputations by De Bruyn. Later corollaries defended under Voet attack the method of doubt, the Cogito as a metaphysical principle, or state that Cartesianism leads to atheism, and so on. By contrast with De Bruyn, Voet had no apparent hesitation in mentioning Descartes by name: in almost half of his disputations from 1653 and

---

52 Daniel Voet (praes.) and Niepoort (resp.), *De ubi spirituum* (Utrecht, 29 June 1653); Daniel Voet (praes.) and Beets (resp.), *De ubi spirituum, pars quarta* (Utrecht, 9 December 1653); De Bruyn (praes.) and Camp (resp.), *Disputationum physicarum septima de corporis loco et modo existendi in loco* (Utrecht, 14 December 1655).

53 Daniel Voet (praes.), *De ubi spirituum* (Utrecht, 29 June 1653, 15 March 1654, 28 June 1654).

54 Daniel Voet, *Physiologia* (Utrecht, 1661), VI, III, 12, 86.

1654 the name appears, but some caution is recommended here, as we do not have the original disputations, but only the reprinted texts edited by Gisbertus Voetius and Paul Voet. In original disputations, Descartes's name figures in five disputations from 1656 and 1657.

9   De Bruyn vs. Voet

On 25 April 1657, De Bruyn presided over a disputation on free will. Right at the beginning, the respondent reminded the audience that the question of human free will is discussed by both theologians and philosophers; the disputation will, however, deal with this question from a philosophical point of view only, leaving aside any theological issues. In what follows, De Bruyn offered a Cartesian explanation of free will, with many references to Descartes's works and letters (the first volume of the correspondence edited by Claude Clerselier in Paris had just become available). At the end, De Bruyn explained that the whole disputation had been aimed "at informing the students entrusted to my care about the opinion of Descartes, and to arm them against the scruples with which they may be injected by its adversaries."[55] One of these adversaries is mentioned in the disputation, namely the author of a pamphlet entitled *Den overtuyghden Cartesiaen* (1656) who hid behind the pseudonym Suetonius Tranquillus. Perhaps De Bruyn did not know that Voetius was the author, but he will certainly have recognised the pamphlet as Voetian. De Bruyn devotes just one of the ten theses of the disputation to refute a difficulty raised in *Den overtuyghden Cartesiaen*, so clearly the more substantial adversaries were to be identified elsewhere. One such adversary might have been Daniel Voet, who, in 1657, devoted six disputations to the question of free will and determinism (*Disputationis metaphysicae, De formali ratione libertatis*). Regrettably, I have not seen these texts.

Except for Suetonius Tranquillus, De Bruyn did not mention his opponent(s) by name, nor did he ever explicitly name Voet in his disputations; from his side, Voet adopted the same policy. Indeed, the university senate had decided, during the Utrecht Crisis over Cartesian philosophy in the early 1640s,

---

55   De Bruyn (praes.) and Van Swanevelt (resp.), *Disputatio philosophica de libero hominis arbitrio* (Utrecht, 25 April 1657): "Plura de libero arbitrio dici possent, verum haec impraesentiarum sufficiant, cum praecipue hic animus fuerit, *studiosos curae nostrae commissos, hîc informare circa sententiam Cartesii, eosque munire adversus scrupulos, qui ipsis ab adversariis possent injici*; atque hoc sufficienter factum esse putamus." (my emphasis). The translation is taken from Vermij, *Calvinist Copernicans*, 170.

that colleagues ought not to be attacked openly. And this decree was well observed, at first. But from 1655 onwards, as De Bruyn and Voet explained or refuted Cartesian philosophy in their disputations, their addition of corollaries contradicting corollaries previously added by their opponents made these attacks somewhat more pointed.[56] That these debates were animated events is shown by a corollary added by one of Voet's students (namely the son of professor Essenius) to a disputation defended in December 1656: "As far as I know, nobody has said that the new philosophers are magicians, and has attacked them using that argument."[57] Imbedded in eleven aggressively anti-Cartesian corollaries, it is clear that 'the new philosophers' refer to De Bruyn and his supporting students. As it happened, one of De Bruyn's students had added two corollaries on magicians in a previous disputation: "Magicians cannot change themselves into cats or other animals" and "Magicians do not go to their assemblies sitting on pitchforks and brooms."[58] In all likelihood, the discussion of these two corollaries occasioned much amusement amongst the opposing students, especially when some audacious student compared Descartes and Cartesians to magicians—echoing an accusation already voiced by Maarten Schoock in *Admiranda methodus* (1643). The outcry caused by these corollaries was enough for Voet's student to state no such magical things had happened, which, as we may expect, was enough to prolong the amusement.

It is remarkable that De Bruyn and Voet were able to continue their battle for so long without any complaints being raised in the senate. Apparently, both professors were quite happy to ventilate their opinions undisturbed. In December 1658, however, the senate finally stepped in, deciding that the professors of philosophy should, from that time onwards, abstain from tirelessly provoking and attacking each other in disputations and corollaries, and that if, on occasion, they had to differ in opinion, they should refrain from the use of cruel invective.[59]

---

56   As already observed in Verbeek, *Descartes and the Dutch*, 88.

57   Daniel Voet (praes.) and Essenius A.F. (resp.), *Disputationum selectarum decima-quarta de errore, secunda* (Utrecht, 22 December 1656): "Nemo quod sciam dixit neotericos esse magos, eoque argumento eos oppugnavit."

58   De Bruyn (praes.) and Scharbach (resp.), *Disputationum physicarum duodecima de motu pars tertia* (Utrecht, 29 October 1656): "Magi non possunt se transformare in faeles nec alia animalia."

59   13 December 1658, Kernkamp, *Acta*, 335–336. The immediate cause for this directive were "certain philosophical corollaries," but it is not clear which corollaries exactly or who voiced them.

10   De Bruyn and Craanen on the Souls of Beasts

In 1654, Theodor Craanen (ca. 1633–1688), who later became a Cartesian professor at Nijmegen and then at Leiden University, proposed the following provocative corollaries to one of De Bruyn's disputations:

> The souls of beasts are also immaterial.
> The souls of beasts are as susceptible of immortality as are the souls of humans.[60]

The council of the Reformed Church reacted with great indignation, and De Bruyn had to explain and defend himself during several meetings with the ministers. De Bruyn told them that the student had written and inserted the corollaries, but that he nonetheless accepted them as his own. He claimed not to know that they might be regarded by some as scandalous or a danger to religion, but refused to take them back or to apologise. When the council voiced their dissatisfaction and the affair escalated, De Bruyn decided to leave the Reformed Church, to become a member of the Walloon Church, whose minister in Utrecht was Louis Wolzogen. Although Daniel Voet did not immediately jump on this affair in his disputations, a student repeated Craanen's second corollary with the qualification "Denied".[61] It is probably no coincidence that in 1658 Voet presided over two disputations on beasts, dealing with questions such as whether animals can think, feel pain, and so on, although Craanen's corollaries are not mentioned (10 and 24 March 1658).

In 1666, Essenius, professor of theology and also one of the ministers who had dealt with the 'souls of beasts' affair twelve years earlier, listed the 'beastly corollaries' among the manifold absurdities of Cartesian philosophy (10 November 1666). De Bruyn wrote a lengthy and furious reply, which he appended to his next disputation (12 December 1666). He gave a detailed account of what

---

60   De Bruyn (praes.) and Craanen (resp.), *Disputatio mechanico-mathematica de trochlea* (Utrecht, 21 June 1654): "Animae brutorum etiam sunt immateriales. Animae brutorum aequè sunt immortalitatis capaces ac hominum." The Cartesian stance of professor and student is shown by the first corollary, advocating the principle of doubt. Craanen's corollaries were 'prepared' in a corollary to an earlier disputation, stating that there are no substantial forms except in beasts and humans; De Bruyn (praes.) and Clemens (resp.), *Disputationis physicae de alitura pars prima* (Utrecht, 29 April 1654). For Craanen, see Chapter 9, in this volume.

61   Daniel Voet (praes.) and Junius (resp.), *Disputationum selectarum undecima de quaestionibus de nomine pars prior* (Utrecht, 19 March 1656).

exactly had happened at the time, not sparing Essenius for his deceitful behaviour. But, surprisingly, De Bruyn now conceded that his defence of the corollaries was a mistake. Explaining himself, he said that he had erred because of
the Aristotelian philosophy he had had to learn when he was a student; now he
accepted the account given by Descartes.[62] Essenius is thus mistaken, according to De Bruyn, when he attributes these absurdities to Descartes; turning
the tables, De Bruyn further states that if anyone accepts that animals have
substantial forms, the conclusion that animal souls are immortal must also be
theirs.

De Bruyn's extraordinary intervention probably aroused new interest in
the 'souls of beasts' affair. Just like Essenius, the Voetian theologian Herman
Wits (1636–1708) numbered De Bruyn's corollary on the immortality of animal
souls among pernicious new opinions, claiming he had been present when it
was being defended.[63] The minister and philosopher Petrus Allinga (d. 1692),
entangled in a polemic with Wits, came to De Bruyn's defence. In his rejoinder, Allinga admits that in themselves the words are very offensive; however,
they should be understood in their proper context. The proposition was uttered
in an academic philosophical disputation, when philosophers are prone to
advance provocative theses, which they usually solve with a distinction. It
is generally known, Allinga continues, that philosophers often come up with
extraordinary theses, which do not reflect their own views, only to test the
skills of their students, or to allow themselves some space to explore specific points, or to discover difficulties in the opinions of their opponents.[64]
Allinga then reveals that he too had been present at the notorious disputation:

> I was also present at this disputation, and I still cannot forget the rude
> impoliteness of a certain student against the presiding professor, who,

---

62   Indeed, his disputation of 12 Dec. 1663 has the corollary "Omnes actiones brutorem sint
      automatice."
63   Wits, *Twist*, 276. Herman Wits became minister in Leeuwarden in 1668, and subsequently
      held chairs in theology at the universities of Franeker (1675), Utrecht (1680), and Leiden
      (1698). He enrolled at Utrecht Universty in 1651, at Groningen University in 1654, and again
      at Utrecht in 1655. That he actually witnessed De Bruyn's disputation from 1654 is not
      impossible, but strangely he does not mention De Bruyn, but refers to a disputation by
      Mansvelt from 1669. This explains why Thijssen-Schoute (*Nederlands Cartesianisme*, 44)
      assumes the corollary was Mansvelt's.
64   Allinga, *Verdedigingh*, 60. Petrus Allinga, a Cartesian-Cocceian theologian, matriculated
      at Utrecht University in 1651, and became minister at Wijdenes. On him, see Van Bunge,
      Krop, Leeuwenburgh, Van Ruler, Schuurman and Wielema, eds., *Dictionary*, vol. 1., 6–8.

being unlearned and unable to use reasonable arguments, used slander instead, yelling that this paved the way to atheism and denied the immortality of the human soul. I do not recall how the professor or the respondent explained the thesis, but I do remember very well that the professor neutralised these allegations, and called upon a prominent Utrecht preacher, also present, to attest to his religious orthodoxy and to the innocence of his explanation of the corollary. Our author [Wits] ought to have remarked this as well.[65]

In a quiet and moderate way, Allinga analyses the ramifications of the corollary, ultimately reaching the same conclusion as De Bruyn, that the thesis is the logical conclusion for everyone claiming that animal souls are substantial forms.

Evaluating the reports of De Bruyn himself and of his ally Allinga, it seems likely that the corollaries in question were invented during De Bruyn's private lectures, and that Craanen was audacious enough to defend them in public. Allinga states that philosophers often advance paradoxical ideas during their disputations, but the corollaries in question are unlike anything De Bruyn had proposed before. To me it seems that De Bruyn allowed the corollaries, only to demonstrate the untenability of the theory of substantial forms, or, as Allinga put it, to discover difficulties in the opinions of their opponents. That he was sincere in defending the corollaries before the church council in 1654 seems doubtful. His unyielding attitude towards the church council is reminiscent of Regius's behaviour during the Utrecht Crisis which was characterized by a determination never to give in. De Bruyn's stubbornness was, moreover, amply demonstrated in his conflict with Paul Voet in 1659.

---

65 "Ick hebbe oock dit dispuyt by-ghewoont, en kan als noch niet vergeten de onbeschofte onbeleeftheyt van een seker Student tegen de *praesiderende* Heer *Professor*, welcke door sijn ongeleertheyt geen reden vindende om dit met reden tegen te spreecken, sich met lasteren socht te behelpen, roepende dat hier mede wierde gebaent een wegh tot *Atheisterye*, en geloochent de onsterffelijckheyt van des menschen ziele; het is my ontgaen op wat wijse den Heer *Professor* of den *defendent* die stellingh heeft uytgeleyt: doch dit is my in goede geheugenisse, dat dien Heer *Professor* tegen die lasteringh sich selven voor alle suyverde, en tot getuyge van sijn rechtsinnigheyt en onnosele uytleggingh over dat *Corollarium* versocht een deftich *Leeraer* van *Utrecht*, welcke op die tijdt in dat *dispuyt* tegenwoordigh was. Dit behoorde onsen *Auteur* oock aengeteeckent te hebben." Allinga, *Verdedigingh*, 61. The "prominent Utrecht preacher (*Leeraer*)" is presumably Louis Wolzogen.

11   Conclusion

I offer three observations by way of conclusion. After reviewing the practice of the teaching of philosophy at Utrecht, my first observation is that—at least in the 1650s—the categories Cartesian and Anti-Cartesian were not completely black and white. De Bruyn developed from being an eclectic Cartesian in the early 1650s into an orthodox Cartesian in the 1660s. It seems that he initially he explored the possibility of integrating the concept of atomism into Descartes's system. Looking at the Figures 7.1 and 7.2 presented above, we get the impression that his turn towards orthodox Cartesianism was made shortly after 1656. It is presumably no coincidence that the turn occurred after the intense pamphlet dispute over Cartesianism that raged in the Dutch Republic between 1653 and 1656, inducing Cartesians to group together under the single aegis of the teachings of Descartes himself. The picture of Daniel Voet as anti-Cartesian is no more black and white, unlike that of his father. Voet, who in many respects was wholly anti-Cartesian, nevertheless accepted and endorsed the Regian/Cartesian explanation of muscle movement, which is not a minor issue in physiology, and also showed great interest in the nervous system and the brain.

The second observation is that corollaries preceded disputations. It is important to remember that before De Bruyn defended the broad spectrum of Cartesian philosophy in his disputations, students added corollaries covering the metaphysical and physical aspects of Cartesianism to the disputations. The corollary "Atoms exist" also appeared before De Bruyn discussed atomism in his disputations. This suggests that in corollaries we find an indication of the subjects that professor De Bruyn discussed with his students in his classroom but not yet in public. By allowing or perhaps encouraging students to append such corollaries, De Bruyn could test the waters. It also shows that students seemed eager to ventilate Cartesian positions; and when they added to a single disputation 15 or more radically Cartesian corollaries, and that happened in the 1660s time and again, one can really speak of a Cartesian frenzy.

My third and final observation is that however orthodox the Reformed theologians were at Utrecht, the Cartesian philosophers presented themselves in the 1660s as no less part of an orthodoxy. Because I have not studied all of De Bruyn's disputations in detail—let alone those of Van Mansvelt and Burman—it is presumably premature to say that De Bruyn and his colleagues developed no interesting amplifications or adjustments to Descartes's philosophy, as had the Leiden Cartesians De Raey and Geulincx. However, their teaching influenced an entire generation of philosophers, physicians, and theologians. Their names, and the persons to whom they dedicated the disputations, are known, as are the exact dates of submission, and the persons who supported the students with laudatory poetry. Combining these data with those that can be retrieved from other sources, such as the various *alba studiosorum,* and the numerous Dutch disputations besides the 200 of De Bruyn, and Voet, and adding what is known of the careers of the students after leaving university, will give us a wealth of information on the scientific and social networks of the intelligentsia educated at Dutch universities. *Une mer à boire* perhaps, but as so much progress has been made since Van der Horst's article was published in 1995, such a project, in many ways, does not seem infeasible at all.[66]

---

66   As a first step, on my webpage at the Erasmus School of Philosophy, Erasmus University Rotterdam, I will make available my database in a spreadsheet format, listing about 15 different types of metadata for each disputation by De Bruyn, and Voet.

CHAPTER 8

Between Descartes and Boyle: Burchard de Volder's Experimental Lectures at Leiden, 1676–1678

Andrea Strazzoni

1   Introduction

Burchard de Volder (1643–1709), professor of philosophy and mathematics at Leiden from 1670 to 1705, has long been studied from a variety of perspectives. His significance in the history of philosophy and science has been recognized, if only for his role as founder of the Leiden experimental *theatrum* in 1675.[1] He has been viewed as a 'discontent' Cartesian who allegedly converted to Newtonianism.[2] He has also attracted attention as a correspondent of Leibniz (1646–1716).[3] Moreover, historians have debated his supposed 'crypto-Spinozism'.[4] As Tammy Nyden has put it,

> what makes de Volder an interesting and valuable object of study is precisely that he does not fit neatly into the categories of Cartesian, Newtonian, and Spinozist, or perhaps we should say, his case indicates how untidy these categories actually were in the seventeenth century.[5]

In particular, it is his peculiar approach to teaching a natural philosophy essentially inspired by the ideas of René Descartes (1656–1650) by means of an experimental agenda based on that of Robert Boyle (1627–1691) that renders de Volder an interesting figure for understanding how Cartesianism was disseminated in the course of the seventeenth century. He figures as part of

---

1   This has been analysed in De Pater, "Experimental Physics," though with no attention for de Volder's overall natural-philosophical ideas.

2   Sassen, "The Intellectual Climate." This claim has been later disproved: see Krop, "Medicine and Philosophy"; see also Ruestow, *Physics*, 89–112.

3   Russell, "The Correspondence"; Hall, "Further Newton Correspondence"; Lodge, ed., *The Leibniz-De Volder Correspondence*; Rey, "L'ambivalence de la notion d'action" (first and second parts).

4   Sustained in Klever, "Burchardus de Volder"; criticized in Lodge, "Burchard de Volder."

5   Nyden, "De Volder's Cartesian Physics," 228–229. Nyden, in particular, has analysed de Volder's epistemology.

what has recently been described as 'Cartesian Empiricism'.[6] Though experiments played an important role in the development of the ideas of Descartes himself, he did not set forth an experimental agenda. This was developed later in the seventeenth century, in particular in France, by Robert Desgabets (1610–1678), Jacques Rohault (1618–1672), and Pierre-Sylvain Régis (1631–1707).[7] In the Netherlands—the first country in which Cartesianism entered at the university—it was Henricus Regius (1598–1679) who first developed an empirical approach to Cartesianism, rejecting innatism as a source of knowledge. However, only de Volder eventually provided a teaching by means of experiments, based on sources other than Descartes yet aimed at demonstrating the validity of his philosophy, which de Volder continued to teach in the same years, by dictating commentaries (*dictata*) on his *Principles of Philosophy* (1644).

In fact, given the lack of any systematic treatise by de Volder, as his printed works were limited to academic orations and disputations,[8] in order to capture his thought one needs to look at sources different from the ones usually considered by historians—learned correspondences, printed texts, etc.—and to focus instead on the specific types of document that provide us with insights into the contents of his academic teaching: students' notebooks and academic textual commentaries on the works of Descartes, extant to us as handwritten *dictata*. By considering such sources, in this chapter, I shed light on the ways Cartesian ideas were taught and discussed at Leiden in the last quarter of the seventeenth century, and how these became entangled with the early modern experimental philosophy. After an overview of de Volder's life and works (section 2), I will detail and discuss a selection of contents of his experimental lectures taking place at Leiden in 1676–1678 (based mostly on Boyle, though with important references to Simon Stevin, 1548–1620) (section 3), and I will analyse the contents of his theoretical physics (based mostly on Descartes and on Archimedes' theory of floatation) (section 4). I will discuss how his experimental and theoretical physics interrelated, by considering de Volder's explanation of the lack of the sensation of pressure under water, for which he relied mostly on Descartes and on a Cartesian interpretation of the theory of floatation of Archimedes (section 5). As a conclusion, I argue that de Volder failed to capture an essential conceptual shift in the treatment of hydraulic and pneumatic pressure in the seventeenth century: namely, the passage, high-

---

6   Dobre and Nyden, eds., *Cartesian Empiricisms.*
7   Roux, "Was There a Cartesian Experimentalism in 1660s France?"
8   Wiesenfeldt, "Academic Writings."

lighted by Alan Chalmers, from a 'common' interpretation of pressure to a more 'technical' meaning, where the "former relates to forces on bounding surfaces between media whereas the latter refers to forces within the body of media."[9] This ultimately led to a divergence between de Volder's experimental and theoretical physics.

2   De Volder's Life and Works

Born in Amsterdam on 26 July 1643, de Volder studied philosophy at the Amsterdam Athenaeum illustre from 1657 onwards, following the lectures of Arnold Senguerd (1610–1667) and Alexander de Bie (1623–1690).[10] Subsequently, he matriculated at the University of Utrecht, graduating on 18 October 1660 as *magister artium* under the Cartesian professor Johannes De Bruyn (1620–1675). Then, he moved to the University of Leiden, where he graduated in medicine with a disputation *On Nature* (3 July 1664), dedicated to Franciscus Sylvius (1614–1672) and Johannes Hudde (1628–1704). After some years spent in Amsterdam as town physician, he was able to obtain a chair in logic at the University of Leiden in September 1670, through the recommendation of Hudde himself. de Volder started his teaching activities by lecturing on the logic of Franco Burgersdijk (1590–1635), and, after a few weeks, was allowed also to teach natural philosophy.[11] At this point, he had aligned himself with the 'Cartesian faction' at Leiden, of which, after the departure of Johannes De Raey (1620/1622–1702) and the death in 1669 of Arnold Geulincx (b. 1624), he was a representative alongside Theodoor Craanen (ca. 1633–1688) (appointed in June 1670) and the theologians Christoph Wittich (1625–1687) and Abraham Heidanus (1597–1678). It was with Wittich and Heidanus that de Volder was actively involved in the defence of Cartesianism following the departure from the University of the Aristotelian philosopher Gerard de Vries (1648–1705) in 1674, as a consequence of student disturbances during his lessons and disputations.[12] Indeed, Jean Le Clerc (1657–1736) reported that de Volder, in a now lost manuscript, explained

---

9   Chalmers, *One Hundred Years of Pressure*, 6.

10   Biographical information on de Volder is mostly provided in Le Clerc, "Éloge"; Gronovius, *Burcheri De Volder laudatio*; Niceron, "Burcher de Volder."

11   Concerning the practice of teaching Burgersdijk's works at Leiden, see the Chapters 2 and 9, in this volume.

12   Concerning Craanen and de Vries's tumultuous departure from Leiden, see Chapter 9, in this volume. Concerning de Vries's later activity and position concerning Cartesianism, see Chapter 10, in this volume.

how he himself had attempted, in June 1674, to convince the Grand Pensionary of Holland, Gaspar Fagel (1634–1688), that Cartesian philosophy did not pose any danger; and that, to the contrary, it had already inspired learned institutions such as the Royal Society.[13]

Between July and August of the same year, provided with a presentation letter by Philipp Van Limborch (1633–1712), de Volder travelled to England, where he visited Cambridge and met Isaac Newton (1642–1726).[14] Having returned to Leiden, in December 1674, he asked the Curators of Leiden University to fund the establishment of a *Theatrum physicum*, or *Auditorium philosophiae experimentalis*, which was to be a classroom wherein "to teach and to point, through experiments, to the truth and the certainty of the principles and doctrines that the students had learnt in *physica theoretica*."[15] The *theatrum*, whose main tool was an air-pump built by Samuel Van Musschenbroek (1640–1681), based on the model by Robert Hooke (1635–1703) and now extant at the Boerhaave Museum at Leiden,[16] opened the following year.[17] In 1674, de Volder also commenced his activities as a full-fledged academic philosopher, presiding over a series of disputations: in particular, his *On the Principles of Natural Things* (1674–1676), which is a defence of the use of the ideas of matter, movement, size, figure and disposition as first principles in natural philosophy; his *On the Weight of the Air* (1676–1678), giving some insights into his experimental lectures; his *Against the Atheists* (1680–1681), providing Cartesian-inspired demonstrations of the existence of God; and his *Philosophical Exercises* (1690–1693), a criticism of the *Censorship of Cartesian Philosophy* (1689) of Pierre-Daniel Huet (1630–1721). In the meantime, he also assumed the chair of mathematics and the directorship of the Leiden astronomical observatory (1682). Upon its publication, he carefully studied Newton's *Mathematical Principles of Natural Philosophy* (1687),

---

13   Le Clerc, "Éloge," 356–359. On the 'Cartesian' character of the Royal Society, see Jalobeanu, "The Cartesians of the Royal Society." In 1676 de Volder also partially authored a defence of Cartesian theses which appeared as Heidanus, *Consideratien, over eenige saecken onlangs voorgevallen in de Universiteyt binnen Leyden* (Leiden, 1676), and which caused the dismissal of Heidanus—the only official author of the book—from his academic post.

14   See Des Amorie Van der Hoeven, "De Philippo a Limborch," 39; Hall, "Further Newton Correspondence."

15   Molhuysen, *Bronnen*, vol. 3, 298: "by experimenten moghten werden gedoceert en aengewesen De waerheyt ende seekerheyt van die stellingen ende leeren, die in Physica theoretica De studenten werden voorgehouden." Unless taken from an already translated primary source, all translations are by the author.

16   De Clercq, *The Leiden Cabinet*, 67–68.

17   Molhuysen, *Bronnen*, vol. 3, 301–302. Experimental lectures were nonetheless already being given, from 1672, at Altdorf University by Johann Christoph Sturm (1653-1703).

finding its contents *veritables* and sharing insights on his reading of Newton's
text with Christiaan Huygens (1629–1695).[18] The late 1690s saw de Volder distance himself from Cartesianism, when, according to Le Clerc, he became dis-
satisfied with teaching Descartes's *Meditations on First Philosophy* (1641) and
with Rohault's *Treatise on Physics* (1671).[19] He started a long correspondence
(1698–1706) with Leibniz, from whom he requested a demonstration of the
activity of material substance, deeming the recourse to God as a 'universal
mover', typical of Descartes and his followers, "not worthy of a philosopher."[20]
Eventually, he retired from Academic teaching in 1705, leaving a well-equipped
academic *theatrum*, and died in 1709.[21]

While it is clear that de Volder defended Cartesian ideas during his career
and at least into the 1690s, it is less clear how these relate to his experimental lectures, which started in the mid-1670s.[22] On these, indeed, de Volder
provided some evidence in his aforementioned *On the Weight of the Air*, which
offers only a limited view of the actual contents of his experimental teaching, however, being a text (as I will show in section 4) dedicated mostly to
the natural-philosophical interpretation of a limited number of experimental
observations. More insights are afforded, on the other hand, by a document first
brought to attention by Adriaan De Hoog in his unpublished doctoral dissertation (1974), and then discussed by Gerhard Wiesenfeldt (2002), who has considered it in the context of the history of experimental practices at Leiden, and
which I aim to discuss in the light of de Volder's broader natural-philosophical
positions.[23] The handwritten report, namely, of de Volder's experimental lectures in the years 1676–1677, was provided by an English student of his, Christopher Love Morley (1645/1646–1702) under the title *Experimenta philosophica naturalia*. Written mostly in English and containing relatively free annota-

---

18   Le Clerc, "Éloge," 379–380.

19   Le Clerc, "Éloge," 398. There are traces of a lost series of academic *dictata* of de Volder
      on the *Treatise on Physics* of Rohault, dated 1698 and 1699, listed in the auction catalogue
      of a private library of 1752: "Annotata in Jacobi Rohaulti, Tractatum physicum, a clariss. et
      nobil. Professore Burchero de Volder, 1698. MS," and "Annotata in Rohaultii, tertiam Partem
      De rebus terrestribus a clar. et nobili Professore Buchero de Volder, A° 1699," Scheurleer,
      Bibliotheca martiniana (The Hague, 1752), 116 and 427.

20   Leibniz to de Volder, 19 November 1703, in Lodge, ed., *The Leibniz-De Volder Correspond-
      ence*, 279.

21   An inventory of the instruments of the *theatrum* is given in Molhuysen, *Bronnen*, vol. 4,
      104*.

22   As to de Volder's Cartesianism, especially in metaphysics, see Strazzoni, *Dutch Cartesian-
      ism*, ch. 6.

23   De Hoog, *Some Currents of Thought*; Wiesenfeldt, *Leerer Raum*, 54–64 and 99–132.

tions on lectures rather than literal *dictata*, this notebook—together with many other reports by Morley, who collected some 40 volumes of lecture notes, mostly on chemistry and medicine—can be accessed at the British Library.[24] Moreover, a further report by a student of de Volder, Hermann Lufneu (1657–1744), appeared in the *Nouvelles de la république des lettres* in 1685–1687, providing evidence of de Volder's activities for the years 1677–1678. Additionally, insights into de Volder's explanations of experimental results are provided in his dictated commentaries on Descartes's *Principles of Philosophy*, now extant at various libraries. These are the main sources to be considered in an analysis of de Volder's experimental approach to natural philosophy and its connection with his broader theoretical physics.

3   De Volder's Experimental Lectures in 1676–1678

Morley's *Experimenta philosophica naturalia* reports 28 experiments which took place from 12 March 1676 to 25 March 1677. For the present purposes, it suffices to consider a few of these experiments. In particular, in experiments 1–2 (the same experiment, repeated once), de Volder made two cylinders of marble, one of which was appended to a support, sticking together at their flat surfaces, in open air (such as in Figure 8.1, from Morley's notebook, and in Figure 8.2, from a sale catalogue (1700) of Johannes Joosten van Musschenbroek (1660–1707), successor of Samuel).[25] From here, he appended some weight to the lower cylinder, until they fell apart.[26] The cohesion of bodies along their flat surfaces was described as early as in Lucretius' *On the Nature of Things*.[27] In early modern times, such an experiment was attempted by Boyle and described in his *Defence of the Doctrine Touching the Spring and Weight of the Air* (1662): Boyle, as remarked by de Volder himself, could not, however, append to the cylinders so much weight as de Volder did.[28] In fact, Boyle had attempted such

---

24   MS., London, MS. Sloane 1292: De Volder (professor) and Morley (student), *Experimenta philosophica naturalia* (1676–1677). Morley also took notes of courses given by Craanen, one of which is considered in Chapter 9, in this volume.

25   Figures 8.1 and 8.2 are respectively in MS., London, MS. Sloane 1292: De Volder (professor) and Morley (student), *Experimenta*, fol. 78ᵛ and Van Musschenbroek, *Descriptio antliae*, plate 3, figure 8.2.

26   MS., London, MS. Sloane 1292: De Volder (professor) and Morley (student), *Experimenta*, fols. 78ʳ–78ᵛ.

27   Lucretius, *De rerum natura* I, 384–398; VI, 1087–1089.

28   MS., London, MS. Sloane 1292: De Volder (professor) and Morley (student), *Experimenta*, fol. 80ᵛ; Boyle, *A Defence*, 84–86.

FIGURE 8.1
Cohering marbles used by de Volder. MS., London, MS. Sloane 1292: De Volder (professor) and Morley (student), *Experimenta philosophica naturalia* (1676–1677), fol. 78ᵛ.
BRITISH LIBRARY

an experiment both in open air and in a vacuum (as in experiment 31 of his *New Experiments Physico-Mechanicall*, 1660), in order to show the role of air pressure in the cohesion of the two pieces of marble (which do not cohere in a vacuum).[29] This was also one of the aims of de Volder in performing the experiment, who however did not perform it in a vacuum, arguing that air pressure kept them together through a dismissal of the idea, to which the Aristotelians (like Gaspar Schott, 1608–1666) usually reverted, that cohesion in situations like this is due to the fear of a vacuum—as in fact it is nonetheless possible to pull the cylinders apart.[30] Also, the experiment served de Volder to shed light on a further phenomenon, namely the lack of the sensation of air pressure or water pressure on our bodies: de Volder argues that we do not feel such a pressure because of its uniformity, as it "compresses equally on all sides."[31] This theory, as I discuss further in section 5, can be traced back to Stevin and was

29   Boyle, *New Experiments Physico-Mechanicall*, 229–233.
30   MS., London, MS. Sloane 1292: De Volder (professor) and Morley (student), *Experimenta*, fols. 81ʳ–81ᵛ; Schott, *Mechanica*, 25–26.
31   MS., London, MS. Sloane 1292: De Volder (professor) and Morley (student), *Experimenta*, fol. 81ʳ.

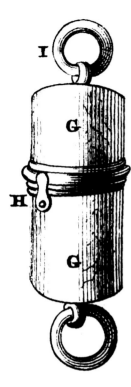

FIGURE 8.2
Cohering marbles sold by van Musschenbroek. Van Musschenbroek, *Descriptio antliae pneumaticae et instrumentorum ad*
eam inprimis pertinentium (s.l., 1700), plate 3, figure 2.
UTRECHT UNIVERSITY LIBRARY

later embraced by Boyle. It is crucial to a correct understanding of de Volder's
approach to hydrostatics as well as of the ways in which his experimental and
theoretical physics interrelate, as in commenting upon Descartes's *Principles*
of Philosophy he was to criticize it. Moreover, de Volder used the experiment to
argue for a theory alternative to Descartes's explanation of cohesion as due to
the parts of a body being at rest, which for de Volder has to be rejected, since a
hard body can easily be moved without breaking it. Instead, atmospheric pressure is a better cause of cohesion, as maintained also by Boyle in his *History of*
Fluidity and Firmness (1661).[32]

Other experiments involve the use of the air pump, e.g., experiment 3, in
which one Magdeburg hemisphere (built for de Volder by Samuel van Musschenbroek (1653–1703) and now extant at the Boerhaave Museum),[33] closed
by a plate, was emptied of air by means of the air pump, and weights were pro-
gressively appended to the plate until they were pulled apart. The same weight

32   MS., London, MS. Sloane 1292: De Volder (professor) and Morley (student), *Experimenta*,
     fols. 81ᵛ–83ʳ; Boyle, "The History of Fluidity and Firmness," 213. Descartes's theory of cohe-
     sion is given in *Principia philosophiae* II 54–55 and 63, AT VIII-A 70–71 and 77–78.
33   De Clercq, *The Leiden Cabinet*, 78.

could be appended to the apparatus when the plate was replaced by a second hemisphere. This is a variant of the demonstrations of Otto Von Guericke (1602–1686) with his hemispheres (see Figure 8.3).[34] It allowed de Volder to show that air "did only press per lineas perpendiculares," as the weight one can append to the lower hemisphere or to the plate depends only on the diameter of the hemisphere itself, not on its volume or external surface.[35] If we want to make sense of this claim (which contradicts his aforementioned statement that water or air "compresses equally on all sides"), we can articulate it as follows: what matters in the cohesion of the two hemispheres is only the pressure exerted perpendicularly with respect to the line of cohesion between the hemisphere and the plate or of the two hemispheres, not the oblique or lateral pressure exerted by air. As a matter of fact, not even de Volder himself could ignore the fact that air and water exert a pressure also in oblique directions, and not only in those directions perpendicular to the horizon (as is illustrated by the fact that if one rotates the hemispheres these do not lose their cohesion). However, as I discuss in sections 5 and 6, his strict reliance on an Archimedean model for hydrostatics and pneumatics did not provide him with the conceptual means to account for phenomena like lateral pressure.

Besides experiments involving air-related phenomena, de Volder performed a number of experiments in hydrostatics. In particular, he showed (experiment 23), contrary to the opinion of some Aristotelians (like Schott), that elements gravitate or press on themselves.[36] For instance, that water presses on water is, for him, shown by the fact that when tubes containing different liquids (as oils or mercury) are dipped in a basin containing water, their level will vary in accordance to the depth the tubes reach, i.e. in accordance to the different pressure exerted by water at different levels.[37] This is an experiment certainly drawn from Boyle's *Hydrostatical Paradoxes* (1666), where it is widely treated in Paradox 1.

Moreover, de Volder attempted to demonstrate, in experiment 25, the same tenet illustrated in experiment 3, this time with regard to water. Namely, that "water [...] presses only by straight lines, in no way by lateral, transverse, or oblique [lines]." This is shown by a very simple experiment too, namely, by

---

34　Von Guericke, *Experimenta nova* (Amsterdam, 1672), 106. See book 3, chapter 25, devoted to such demonstrations.

35　MS., London, MS. Sloane 1292: De Volder (professor) and Morley (student), *Experimenta*, fols. 84^r–86^v.

36　Schott, *Magia universalis naturae et artis* (Würzburg, 1657–1659), vol. 3, 430–438; criticized in Boyle, *Hydrostatical Paradoxes*, Appendix 1.

37　MS., London, MS. Sloane 1292: De Volder (professor) and Morley (student), *Experimenta*, fols. 127^v–128^v.

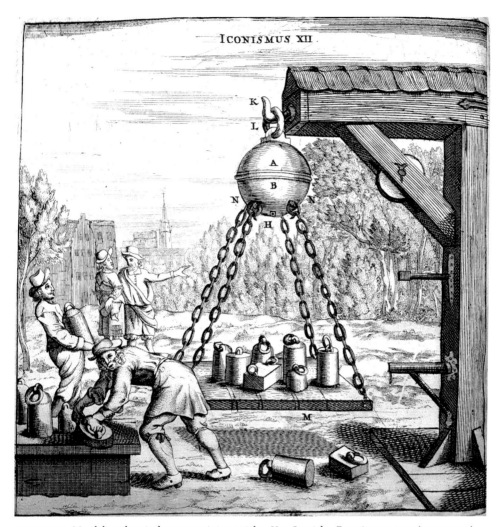

FIGURE 8.3 Magdeburg hemispheres sustaining weights. Von Guericke, *Experimenta nova (ut vocantur)*
Magdeburgica de vacuo spatio (Amsterdam, 1672), 106.
ETH-BIBLIOTHEK ZÜRICH

observing that in an inverted U-shaped container, whose arms are different in
size (see Figure 8.4, from Morley's notebook), the water reaches the same level
in both arms, even if one of them contains more water.[38] Of course, the way
to make sense of de Volder's words is to recognize that the pressure of water

---

38   MS., London, MS. Sloane 1292: De Volder (professor) and Morley (student), *Experimenta*,
      fols. 133ʳ–133ᵛ (Figure 8.4 is at fol. 133ʳ).

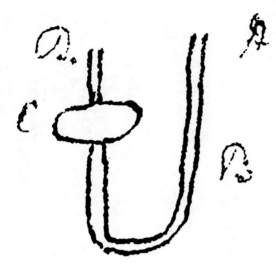

FIGURE 8.4
An inverted asymmetrical syphon used to demonstrate that the pressure of water depends only on its height. MS., London, MS. Sloane 1292: De Volder (professor) and Morley (student), *Experimenta philosophica naturalia* (1676–1677), fol. 133ʳ.
BRITISH LIBRARY

depends only on its height, not on its overall quantity and weight: namely, that the weight of a certain volume of water has to be differentiated from the pressure it exerts on the bottom of its container.

This principle—the idea that the pressure of water depends only on its height—is instantiated in the so-called 'hydrostatic paradox', which is presented by de Volder in the last experiment (28) reported by Morley. Namely, that the pressure exerted by the water in a cylindrical container terminating in a much wider circular base (an inverted-T-shape when viewed side on) is equal to the one exerted by the water filling a cylindrical, not T-shaped, container with a base and height equal to that of the inverted-T-shaped one. De Volder reports that the first systematic treatment of the paradox was by Stevin—who treats the principle as proposition 10 of his *Principles of Hydrostatics* (*De beghinselen des waterwichts*, 1586; later published as *De hydrostatices elementis*, 1605), and that Boyle, who considered it in his *Hydrostatical Paradoxes*, did not accept the experimental evidence given by Stevin therein.[39]

De Volder too attempted to provide an experimental proof of the paradox. It was described by Lufneu in 1685, reporting an experiment he witnessed around 1677–1678—so this is, in all probability, one following experiment 28 as reported by Morley. De Volder used an instrument of his own devising, then known subsequently as a 'cylindrum Volderi' (see Figure 8.5),[40] which was an improved

---

39   MS., London, MS. Sloane 1292: De Volder (professor) and Morley (student), *Experimenta*, fols. 140ʳ–141ʳ; Stevin, *Hypomnemata mathematica* (Leiden, 1605), 119–121 (I will henceforth refer to the Latin edition); Boyle, *Hydrostatical Paradoxes*, Paradox 6, scholium.

40   Lufneu, "Memoire communiqué," 385. In 1687 Lufneu published a defence of the paradox

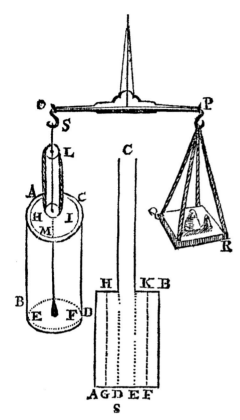

FIGURE 8.5
Instrument aimed at demonstrating the
hydrostatic paradox (De Volder's cylinder). Lufneu, "Memoire communiqué
[...] sur une expérience curieuse d'hydrostatique," *Nouvelles de la république des
lettres*, (April 1685), 385.
YALE UNIVERSITY LIBRARY

model of an instrument first theorized by Stevin in his *Preamble to the Practice
of Hydrostatics* (*Anvang der waterwichtdaet* or De *initiis praxis hydrostatices*;
appended to his *Principles of Hydrostatics*), and then built and used by Boyle.[41]
This instrument was nothing but a cylinder with a much wider circular base
(inverted-T-shape when viewed sideways); the base was moveable, i.e., it could
be lifted into the larger cylinder, and the top was linked by a chain to the arm
of a balance. This allowed the experimenter to measure the pressure exerted
by the water in the container—a pressure different from the bare weight of the
water contained in it. In turn, de Volder aimed at measuring the upwards pres-

---

against the criticism of a certain Moïse Pujolas (Fellow of the Royal Society from 1695),
providing more details on de Volder's demonstration: Lufneu, "Réponse." The 'cylindrum
Volderi' became one of the items sold by the Musschenbroek workshop at Leiden: see De
Clercq, "Exporting Scientific Instruments."

41   Stevin, *Hypomnemata mathematica*, 147; Boyle, *Hydrostatical Paradoxes*, Paradox 6, scho-
      lium.

sure exerted by the water from the lower part of the container on the internal surface HMI of the larger, lower section of the container. Therefore, he made this surface adjustable, and measured the upwards pressure by putting weights on it—as Edme Mariotte (ca. 1620–1684) did with a simpler instrument around the same time (first described in 1678).[42] According to Lufneu's account, this experiment—unlike Boyle's—was successful: de Volder, by measuring with the balance the pressure exerted by water, experimentally proved that the pressure on the bottom EF is the same as that exerted by the weight of water filling a cylindrical container with base EF and height EL.[43]

When one looks at the 1705 inventory of the *theatrum*, one can observe that it contains a broader collection, including instruments aimed at showing the rules of motion.[44] In 1676–1678, however, de Volder seems to have focused mostly on hydrostatical and pneumatical questions, certainly under the influence of Boyle. In fact, many of his experiments are nothing but repetitions of those described in Boyle's *New Experiments Physico-Mechanicall* and *Hydrostatical Paradoxes*. Other important sources are Stevin and Von Guericke, while Descartes's theory of cohesion is overtly criticized in experiment 2. De Volder's didactical experimental programme, thus far, was decidedly independent from his overt appreciation for, or association with, Cartesianism in the 1670s; not just because he relied on experimental philosophy sources more recent than Descartes (who did not develop a full-fledged experimental programme), but also because he overtly attacked a Cartesian theory in his lectures. In the explanation of the processes underlying water and air pressure, however, he reverted more to Descartes, as I clarify in the next sections.

4   De Volder's *physica theoretica*

In order to shed more light on de Volder's experimental lectures, we need to look at their background, that is to say, at the ways in which these fit with de Volder's *physica theoretica*. These are his "principles and doctrines" which, as he told the Curators of Leiden University, can be taught by the *physica experimentalis*. Such principles can be found, first of all, in his *On the Principles of Natural Things*. The physical principles defended in this series of disputa-

---

42   Du Hamel, *Philosophia vetus et nova* (Paris, 1678), III, 415–416.
43   Lufneu, "Memoire communiqué," 386–387; Boyle, *Hydrostatical Paradoxes*, Paradox 6, scholium.
44   Molhuysen, *Bronnen*, vol. 4, 104*.

tions are nothing but matter and motion, understood in Descartes's terms as extended substance and as the passage of a body from the proximity to one body through to the proximity to another one; to these, the ideas of *situs* (viz. the reciprocal disposition of parts), figure and size are added. For de Volder, these ideas have to be accepted as the first notions in physics since they: (i) are clear and distinct, (ii) are not the effect of any other natural cause, (iii) do not involve the idea of mind, and (iv) can explain every phenomenon.[45] Despite their evident Cartesian overtones, such principles are not traced back to any particular author. In fact, de Volder shows an appreciation for all those "dedicated to corpuscular philosophy, as the Englishmen call it," namely Pierre Gassendi (1592–1655), Francis Bacon (1561–1626), Descartes and Boyle, whose approach de Volder praises as being "mechanical," certainly under the influence of Boyle himself, as de Volder's categorization matches Boyle's idea of "the corpuscularian or mechanical philosophy."[46] At the same time, and just like in his *Experimenta* (experiment 2), he criticized Descartes's theory that rest is the cause of cohesion, explaining that this cannot be granted since a solid body can be easily moved.[47] While accepting Cartesian notions as the basis of physics, de Volder thus adopted a liberal approach towards other authors, including Boyle, as well as a 'mechanical' approach to the study of nature.

Such a 'mechanical' approach is particularly evident from his *On the Weight of the Air*. Each of the five disputations of this series has a different aim. In the first, de Volder gives some evidence for the weight or *gravitas* of the air, such as the *esperienza* of Evangelista Torricelli (1608–1647), or the case of the Magdeburg hemispheres—already addressed in his *Experimenta*.[48] Accordingly, de Volder criticizes some straw-man objections, representative of an Aristotelian standpoint, to the idea that air has a weight and it is not an absolutely light body (disputation 3) as well as the idea of the fear of a vacuum (disputation 4). Eventually, in disputation 5, he details his measurement of the weight of a separate volume of air—also described in his *Experimenta*—performed by

45   De Volder (praes.), *De rerum naturalium principiis* (Leiden, 1674–1676), disputation 14, theses 3–13.

46   "Quae principia dudum reiecta nostro demum saeculo in lucem revocarunt Gassendus, Verulamius, Cartesius, Boylaeus, et quantum est ingeniosorum hominum, qui corpusculari, ut Angli vocant, addicti sunt philosophiae," De Volder (praes.), *De rerum naturalium principiis*, disputation 3, thesis 5. See also disputation 10, thesis 8: "non possum non laudare mechanicam explicandi rationem, qua per figuram, magnitudinem, et motum rerum explicantur phaenomena." See Boyle, *The Excellency of Theology*, 51.

47   De Volder (praes.), *De rerum naturalium principiis*, disputation 10, thesis 11.

48   De Volder (praes.), *De aëris gravitate* (Leiden, 1676–1678), disputation 1, theses 1–2.

weighing, on a balance, two cohering hemispheres deprived of air by means of the pump.[49] The natural-philosophical backbone of the disputations, in turn, is presented in disputation 2, in the form of two laws which he draws from the first postulate of hydrostatics set out in Archimedes' *On Floating Bodies.* Namely:

> the first one is, in each fluid at rest, each surface of the fluid parallel to the horizon is pressed equally. The second one, if each surface of a fluid parallel to the horizon is unequally pressed, the part [which is] more pressed expels that one, which is less pressed.[50]

De Volder overtly uses these laws to explain two main cases: the 'forced' immersion of a body lighter than water into a basin full of water, and the cohesion of the Magdeburg hemispheres. In both cases, the water or air which is around such bodies exerts a downwards pressure which, after a sort of upwards turn, presses from below the lower part of the bodies, exactly as in a balance, where the heavier body pushes the lighter one upwards.[51]

This is the same model used by de Volder to explain the hydrostatic paradox, as reported by Lufneu: given Archimedes' postulate, in a condition of equilibrium each imaginary surface in a volume of water undergoes the same pressure, from above and below. Thus, in Figure 8.5, the internal surface HMI of the bigger cylinder undergoes, from below, the same pressure underwent by the water at the same level HMI in between the two cylinders. From above, in turn, the water surface HMI in between the two cylinders receives an equal pressure from the water above it (i.e., the portion of water contained in the smaller cylinder), while the internal surface HMI of the bigger cylinder undergoes either an equal pressure from the weights put on it, or, in the case that the cover (lighter than water) is fixed to the bigger cylinder by screws, this counteracts the pressure from the water below it.[52] At this point, there is equilibrium because there is no

---

49   Cf. MS., London, MS. Sloane 1292: De Volder (professor) and Morley (student), *Experimenta*, fols. 137^r–139^v.

50   "Prima sit, in fluido quovis stagnante unamquamque fluidi superficiem horizonti parallelam aequaliter premi. Altera vero, si superficies quaevis fluidi horizonti parallela inaequaliter prematur, partem magis pressam expellere eam, quae premitur minus," De Volder (praes.), *De aëris gravitate*, disputation 2, thesis 2. See Archimedes' postulate: "the fluid is of such a nature that of the parts of it which are at the same level and adjacent to one another that which is pressed the less is pushed away by that which is pressed the more," Dijksterhuis, *Archimedes*, 373.

51   De Volder (praes.), *De aëris gravitate*, disputation 2, theses 4–8.

52   Lufneu, "Memoire communiqué," 387–388. See also Lufneu, "Réponse," 242.

displacement of water, regardless of its potential causes: either the downwards pressure of the water contained in the smaller cylinder, or the weights put on the upper surface of the bigger cylinder, or the resistance to displacement of such a surface, when it is fixed to the container, while pressed upwards by the water in the bigger cylinder.

5   Why We Do Not Feel Pressure under Water?

This Archimedean model of floatation is presented by de Volder—in his *On the Weight of the Air*—together with his explanation of the lack of a sensation of pressure under water, which is based on Descartes's explanation. The discussion represents an ideal case-study of de Volder's approach to dealing with different natural-philosophical standpoints. Descartes's explanation is given in the context of his theory of weight, as presented in chapter 11 of his *The World*, and underlying *Principles* IV 20–27. For Descartes, weight is nothing but an effect of the tendency of each part of matter to pursue its movement according to straight lines. In the case of the matter rotating around the centre of the Earth, some parts have more force to pursue such a movement than others, thanks to their extreme speed: these are the particles of subtle or celestial matter, which recede from the centre of the Earth, pushing downwards any other kinds of body, namely visible bodies, or terrestrial matter. The weight of a body composed only of terrestrial matter is therefore equal to the force of 'downwards extrusion' exerted by a volume of subtle matter equal to the volume of such a terrestrial body. However, since there are no volumes of purely homogeneous kinds of matter, what effectively determines the weight of a body on Earth, for instance, a stone in mid-air, is (i) the excess of celestial matter present in a volume of air equal to such a body which would occupy its place in the case of a descent, with respect to the excess of terrestrial matter in the stone with respect to that contained in such a volume of air, plus (ii) the movement of the particles of the air which, being more subtle than those of the stone, behave as if they were parts of celestial matter. If two bodies, therefore, have the same quantities of subtle and terrestrial matter, and their parts of terrestrial matter have the same force, they are simply in equilibrium: it is the excess of one kind of matter over the other that is instrumental in determining the weight of a body in a medium.[53]

---

53   For a discussion of Descartes's theory of weight, see Schuster, *Descartes-Agonistes*, 479–495.

This can be rendered in hydrostatic terms. For Descartes, indeed, such a theory of weight explains why, as he put it in *The World*, "a man at the bottom of very deep water does not feel it pressing on his back any more than if he were swimming right on top."[54] Indeed, for Descartes such a man is not pressed by all the water above him, but only by that volume which, in the case of his descent, will fill its place, and with which there is an actual 'contest': in which the lighter part, provided with more force, extrudes the heavier downwards, thereby making a man under water feel a pressure.[55] Or, as he put to Marin Mersenne (1588–1648), a swimmer in the middle of a basin does not feel the pressure of the water above him because, if pressed downwards by this water, a volume of water equal to that of the swimmer would ascend in his place, and would prevent the descent of the water above him. Thus, would this swimmer feel only the pressure of a volume of water which, rising to his place, pushes him downwards. On the contrary, if this swimmer were to be like the stopper in a hole in a basin, with a space without water below, he would feel the pressure of the whole column of water, as in that case, if he would move downwards, the whole column of water will follow him, without other water rising at his place and counterbalancing the descending water.[56] According to Descartes' model there is no increasing, internal pressure in water in a condition of equilibrium, as all its sub-volumes counterbalance each other. In a condition of equilibrium, in other words, water is weightless on water. This is the tenet Descartes defends in *Principles* IV 26, overtly entitled "why bodies do not gravitate [when] in their natural places." He maintains that in a bucket "the lower drops of water, or of another liquid, are not pressed upon by the higher ones," i.e., any "drop of water [...] is not pressed upon by the others [...] situated above it," for the reason that

> if these were carried downward, other drops [...] [below it] would have to ascend into their place; and since these drops are equally heavy, they hold the former in equilibrium and prevent their descent.[57]

This is nothing more than is seen in an Archimedean model, where different volumes of water counterbalance each other: in Descartes's interpretation,

54   Descartes, *Le monde* XI, AT XI 77, *The World*, 50.

55   Descartes, *Le monde* XI, AT XI 77, *The World*, 49–50.

56   Descartes to Mersenne, 16 October 1639, AT II 587–588. This letter would have been available to de Volder in the edition of the correspondence by Claude Clerselier; Descartes, *Lettres*, vol. 2, 183–184.

57   Descartes, *Principia* IV 26, AT VIII-A 215–216, *Principles* 193.

FIGURE 8.6 Idealized representation of the transmission of pressure in water accord-
ing to Descartes. Descartes, *Principia philosophiae* (Amsterdam, 1644),
II 66, 63.
LIBRARY OF CONGRESS

such an equilibrium leads to a loss of weight of water in water. Or, as he puts it in
Principles II 56–57 (devoted to the explanation of the behaviour of a solid body
immersed in water, and equiponderant with it, as in Figure 8.6), "the particles
of fluids tend to move with equal force in all directions," and, being constantly
in movement (in order for them to constitute a fluid body), they move according to innumerable circular paths, which do not press in one direction more
than in another, and allowing an easy displacement of the solid body once their
movement is determined in one direction by a small, external force.[58] So far, for
Descartes the particles of water exert a pressure in all the directions, though,
the whole fluid is in equilibrium as none of the parts exerts a bigger pressure
(in any direction) than the other.

Descartes's explanation of the lack of a sensation of pressure under water
was criticized by Boyle in his *Hydrostatical Paradoxes*, a text based on the
same Archimedean theory of floatation adopted by de Volder.[59] He does so,
in Appendix 2 of his treatise, by attacking Descartes's idea that water does not
press on water.[60] He refuted this, as de Volder was to do in experiment 23, by
showing how the level of certain fluids, contained in tubes dipped in water,
varies in accordance with the depth they reach, i.e., in accordance with the
varying pressure of water (in Paradox 1). In a nutshell, for Boyle, Descartes's

58   Descartes, *Principia* II 56–57, AT VIII-A 71–72, *Principles* 63.
59   Boyle, *Hydrostatical Paradoxes*, 8–11.
60   Boyle, *Hydrostatical Paradoxes*, 229–236.

theory of floatation does not allow for an explanation of why a certain fluid is kept at equilibrium with water at different depths, i.e., it is inconsistent with Archimedes' model of floatation.

Hence, Boyle reverts to the explanation for the lack of a sensation of pressure adopted by Stevin, who, in proposition 3 of his *Preamble to the Practice of Hydrostatics*, claims that we do not feel the pressure of water on account of the uniformity with which it presses on all parts of the body (as de Volder claimed in experiments 1–2): so that no part is more dislocated (*luxatur*) by water than any other. Only in the case where a person is positioned above a hole in the water container (like the stopper in a hole in a basin) would such a person feel the pressure of the water, as such a pressure would not be uniform on him or her.[61] This is the same solution adopted by Von Guericke, and partially refined by Boyle by pointing also to the "texture" of the human body, which "is so strong, that, though water be allowed to weigh upon water, yet a diver ought not to be oppressed by it."[62]

As to de Volder, he moves in a limbo between Descartes and Boyle. As seen above, in experiment 23 he overtly criticizes, following Boyle, the idea that water does not press on water. Furthermore, in experiment 2, he adopts an explanation for the lack of a sensation of pressure under water, akin to those of Stevin and Boyle, where he claims that:

> the compression of the water [...] is not felt when we are in a river, or sea, because all the water compresses equally on all sides. But were there in the bottom of the sea a tubulus, hole, or cavity, which had not one drop of water in it, we should feel all the weight of the water then to press, and ponder, upon us, and with huge violence to force us into that hole, whereas were the hole full of water we should feel no such violence forcing us into it.[63]

A Cartesian explanation of the lack of a sensation of pressure under water is, in turn, given by de Volder in his *On the Weight of the Air* (disputation 3), in which he claims that the water ascending in the place occupied by a person who sinks in water counterbalances—as on two plates of a balance—the pressure of the water above that person, constituting a sort of "wall" (*obex*) by

---

61   Stevin, *Hypomnemata mathematica*, 148–149.
62   Boyle, *Hydrostatical Paradoxes*, 246; Von Guericke, *Experimenta nova*, book 3, chapter 30.
63   MS., London, MS. Sloane 1292: De Volder (professor) and Morley (student), *Experimenta*, fol. 81ʳ.

which the person is protected.[64] In any case, in his *On the Weight of the Air*, de Volder seems to smooth the edges of his Cartesian positions: while adopting a Cartesian theory of the lack of a sensation of pressure, he confines his overall theory of weight—based on Descartes's theory—to the corollaries of the disputations, and refrains from discussing the vexed question of whether elements gravitate to elements.[65] As to this, de Volder did not deal, in the text of disputation 1, with the thesis that they do so as a consequence of the idea that air has a weight, and thereby presses on itself, despite Morley in his *Experimenta*, nevertheless reporting this as having been printed.[66]

De Volder's Cartesian approach is more overtly declared in his commentaries on Descartes's *Principles of Philosophy*, which are extant to us in the form of handwritten *dictata* to students.[67] De Volder left two different commentaries on Descartes's *Principles of Philosophy*. One commentary has survived in two copies, one of which is partial and dated 1690 by the copyist, and is extant

---

64   De Volder (praes.), *De aëris gravitate*, disputation 3, theses 7–8.

65   De Volder (praes.), *De aëris gravitate*, disputation 3, corollary 11, and disputation 4, corollary 7.

66   In experiment 24 (held on 28 January 1677): "this experiment was [...] relating to the former ones, viz. the equiponderosity of material things with water, or rather as de Volder printed these days in some theses which he presided, *omnia corpora, quae innatant liquori, t[antu]m h[ab]ent ponderis, quantum liquor a corpore suo ex loco expulsus ponderat*. As also this experiment did relate, unto this thesis likewise then printed by him: *aquam, et aërem esse corpora gravia, cum r[ati]o, tum exper[ienti]a docet, ac proinde falsum est, elementa in suis locis non esse gravia*," MS., London, MS. Sloane 1292: De Volder (professor) and Morley (student), *Experimenta*, fols. 130ʳ–130ᵛ. The use of Latin and the fact that such a Latin text is underlined in the manuscript (rendered in italics in the transcription) indicates that it was taken *viva voce* from de Volder, i.e., it was a dictated part of the notebook. None of these Latin sentences can be found verbatim in the disputations. The first sentence, however, is derived from Archimedes' postulate, whose discussion dominates disputation 2 (held on 15 May 1677: usually, the text of disputations was printed beforehand: Wiesenfeldt, "Academic Writings"). In turn, the first part of the second sentence amounts to what de Volder maintained in thesis 2 of disputation 1 (held on 14 October 1676), while the second part, ("[...] *falsum est, elementa in suis locis non esse gravia*") is notably absent from the printed text. It might be that de Volder himself reported, during the lecture, that he had printed such a thesis, as entailed by his idea that air is heavy, or, alternatively, that he meant to have had printed only the first part of the sentence: in any case, he did overtly relate his (unprinted) claim to his printed text. Experiment 24 was aimed at demonstrating that mercury presses on water like air does. It was a variation of Torricelli's experiment, and performed by inserting a tube filled with mercury in a basin filled with water, after which the mercury descends in the water until its level in the tube reaches a condition of equilibrium with the weight of a volume of water equal to that of the portion of the tube immersed in water.

67   See Chapter 4, in this volume.

at The Hague. The other copy is undated, is complete and extant at Hamburg (hereafter: *Notulae*). The original redaction of this commentary, however, can be traced back to before or around 1673, because it contains a wrong quantification of centrifugal force.[68] In turn, another commentary—which has survived in a much greater number of undated copies, one of which is also extant at Hamburg (hereafter: *Dictata*)—provides a correct quantification, consistent with the exposition given by Huygens in his *The Pendulum Clock* (1673), so that it can be dated to during or after the same year.[69] Neither of the commentaries alters the essence of Descartes's explanation of the lack of the sensation of pressure under water as given in *Principles* IV 26, which is integrated in the *Dictata* with the account given in the aforementioned letter to Mersenne.[70] The text of de Volder's *Notulae* is particularly interesting as it expressly addresses Stevin and Boyle's solution:

> here it is possible to give reason of that issue, about which it is debated so vehemently, namely, why divers do not feel, in water, its weight. Stevin imagines a certain container, and in its bottom a man parallel to the horizon: what happens [when] water is poured in the container? He does not feel water at all. But if there is a hole in the bottom, by which the water can descend, and if [a man] covers it with [his] body, he then eventually feels the water completely. He [Stevin] (and Boyle follows him) believes that water actually presses, but because it presses uniformly, one does not feel it. Which if it were plausible, it will be plausible also this, namely that [someone] completely squeezed by a press does not feel pressure. In fact,

---

68   MS., Hamburg, Cod. Philos. 273: De Volder, *Notulae*, 156. The copy extant at the Koninklijke Bibliotheek at The Hague (Ms. 72 A 7) contains also parts of the other commentary.

69   MS., Hamburg, Cod. Philos. 274: De Volder, *Dictata*, fol. 66. Other copies are extant: MS., Leiden, BPL 2841; MS., London, MS., Sloane 1216; MS., Warsaw, BN Rps 3365 II; MS., Pretoria, MSD27. The Warsaw manuscript includes also a commentary on Descartes's *Meditations*. The manuscripts of de Volder's commentaries extant at Hamburg come from the private library of Zacharias Conrad von Uffenbach (1683–1734).

70   MS., Hamburg, Cod. Philos. 274: De Volder, *Dictata*, fol. 100: "[...] deduci potest istius vulgaris experimenti in quo enodando plurimi frustra insudarunt, cur nimirum fiat, quod urinatores aliisque homines sub aqua degentes nullum sentiant aquam supra incumbentis pondus: huius enim ratio ex iis, quae hoc paragrapho adferuntur, manifesta est. Etenim si homo constitutus in puncto ex. gr. 3, premeretur aqua 3, 4 deorsum ut occuparet locum 2, necessario aqua, quae in 2 est, eadem mole 1 est corpus hominis ascenderet versus 3, quae cum aequali impetu huic ascensui resistat, efficit etiam, ut homo in 2 nullo pacto deorsum premi queat, unde nec mirum est, eum nullum supra se sentire pondus, quippe cuius tota vis in ea consistit pressione."

if a body is solid, it does not feel pressure, or in the case that all the parts in man are full of water. However, since the [human] body is hollow, not filled with water but by air, it feels [the pressure] completely. Therefore there must be another explanation, which, anyway, Boyle wanted to criticize. However, it is the most true, and even if Boyle testifies that he is not satisfied by our author, he, however, does not propose anything which challenges [it] in any way.[71]

De Volder thus reverts to Descartes's solution, which is further explained by reducing it to an abstract, static model: de Volder claims that we do not feel pressure in water, in the same way as when we 'sustain' a weight on a balance on which equal weights are appended, we do not feel any pressure from it. Indeed, in both cases equal weights sustain themselves.[72] The same reductionist approach is to be found in his commentary on Descartes's explanation of the behaviour of a body in equilibrium with a fluid (*Principles* II 56–57), in which it is surrounded by particles pressing in all directions (see Figure 8.6), and forming infinite circles around it. According to de Volder's reading (in his *Notulae*), one can avoid the difficulties entailed by hypothesizing infinite circles of matter in motion—which amounts to state that the fluid is not in equilibrium, as the slightest bigger force exerted by one particle can re-direct the movement of

---

71 See the next note.

72 MS., Hamburg, Cod. Philos. 273: De Volder, *Notulae*, 235–236: "[...] posse hic rationem dari rei, De qua tam acriter disputatur, id est: cur urinatores non sentiant infra aquam eius gravitatem. Stevinus fingit vas quodpiam, et in eius fundo hominem horizonti parallelum, vasique aquam infundit, quod fit? Non sentit omnino aquam. Si vero in fundo sit orificium, per quod aqua descendere potest, idque si corpore tegat, omnino tunc demum sentiet aquam. Is, eumque secutus Boyle, putat aquam premere quidem: sed quia undique premit aequaliter, non sentire. Quod si verosimile sit, etiam et id verosimile erit, quod scilicet omnino a torculari pressus non sentiet pressionem. Verum quidem enim si corpus solidum esset, non sensurum pressionem, aut si omnia in homine essent aqua repleta. Cum vero cavum sit corpus, non aqua sed aere repletum, omnino sentiet. Alia ergo causa esse debet, quam licet Boyle voluerit impugnare. Verissima tamen est, et quamvis Boyle testatur sibi ab autore nostro non esse satisfacturum, nil tamen, quod aliquo modo urget, profert. Concipiamus itaque hominem in aquis mediis constitutum, non sentiet (si premeretur) aquam, nisi deorsum orientem, si vero ostendamus hanc pressionis vim, non ab omne sed alia aqua sustineri, causa certa est, concipiamus ergo cylindrum, aquam ipsi incumbentem, si hic descenderet, alius ipsi aequalis ascendere deberet, cum vero sint aequilibrio, non potest unus descendere, nec premere. Si vero concipiamus aperturam quam tegit omnino, sentiet pressionem quia tunc una aqua descendere potest non ascendente alia, sed aere, ut si duo habeamus bilances, et utrisque imponamus 100 ll, si quis manui sustineat alterutram, non sentient pressionem, quia ab altera parte est aequalis."

all the other particles, like when a small weight is added to a balance—just by considering the pure equilibrium of forces in a medium.[73]

6   Conclusion

As a conclusion, is worth reminding ourselves that for de Volder the explanation of the phenomena of cohesion of bodies in open air—like the Magdeburg hemispheres—and that of floatation (including the hydrostatic paradox) are based on the same rationale, namely Archimedes' model. So that, according to de Volder's *Experimenta*, both in the case of the hemispheres that sustain a weight dependent only on their diameter, and not on their volume (experiment 3), and in that of the hydrostatic equilibrium of the water contained in an asymmetrical, U-shaped container (experiment 25), shows that fluids press only by lines perpendicular to the horizon, upwards and downwards, and not by lateral or oblique lines. Of course, it is possible to make de Volder's statement acceptable by interpreting it as meaning that the pressure of fluids depends only on their height: so that it can be transmitted along oblique lines, but it increases or decreases only according to vertical lines. Experiment 3, at most, can prove that what matters in the cohesion of hemispheres is the pressure exerted perpendicularly with respect to their line of cohesion viz. their diameter, though this is not what de Volder states; moreover, it could prove that pressure depends only on the height of a fluid only if it were performed at different heights: but, again, this was not done by de Volder in the course of this experiment.

Ultimately, de Volder did not account for the phenomenon of the lateral pressure of fluids: a phenomenon which he could not ignore, it having been described and accounted for by Stevin himself, from whom de Volder borrowed, in his *Experimenta*, the explanation for the lack of the sensation of pressure under water.[74] Notably, according the inventory prepared by him in 1705, when he left his academic post, the Leiden *theatrum* was provided with a "glass

---

73   MS., Hamburg, Cod. Philos. 273: De Volder, *Notulae*, 117–118: "[...] dico demonstrationem nostram non uti talibus circulis, sed fluidi aequilibrio, quod si non facere, pars quae praevaleret, tolleret motum particularum, in contrarias partes motarum, unde haec aucta viribus reliquas secum aget, versus easdem partes, atque adeo non stagnans amplius, sed motum erit fluidum [...], si enim tot sunt particulae, quae impellunt versus orientem ac versus occidentem, necessario manebit eodem in loco. Si vero plures ab oriente, movebitur versus occidentem: quod in stagnante esse nequit. Res haec commodius explicari nequit quam exemplo bilancis.."

74   Stevin, *Hypomnemata mathematica*, 146–147.

with an opening in the middle to demonstrate the lateral pressure of the air,"[75] an instrument, however, most probably used after 1676–1678. Still, his strict reliance on an Archimedean model of floatation, in which only upwards and downwards pressures are taken into account, and his reductionist approach to the model of the lever both in hydrostatics and in pneumatics did not allow him to account for lateral pressure in 1670s. In other words, de Volder conceived the equilibrium of liquids as that of solids put on a scale (in which forces are exerted upwards and downwards). De Volder failed to capture a conceptual shift well described by Chalmers, namely the passage in the conceptualization of pressure, begun with Stevin and concluded with Newton. This passage went from the 'common' interpretation of pressure, understood as pressing or weighing—in principle, not differentiated from weight itself—to pressure understood in a 'technical' sense:

> a key difference between pressure in the common sense and pressure in the technical sense is that the former relates to forces on bounding surfaces between media whereas the latter refers to forces within the body of media. Another is that, from the technical point of view, pressure is a scalar not a vector. Directed forces, such as those that occur at the boundary of a liquid are determined by variations of pressure, the gradient of pressure in technical terms, rather than by pressure itself. The technical concept of pressure in fluids breaks from the directedness implicit in the verb 'to press' from which 'pressure' originally derived and which is presupposed in the concept of pressure in its common sense.[76]

As seen above, fluids, according to de Volder's printed texts, and following Descartes, do not ultimately press on themselves. In the *Experimenta*, following Boyle, de Volder argued for their internal pressure. Even in his interpretation of Descartes's *Principles* II 56–57, de Volder does not take into account the pressure exerted by the particles of fluids on each other.

De Volder's appropriation of Stevin's and Boyle's ideas—which informed his experimental teaching—was ultimately inconsistent with his strict reliance on an Archimedean model and with his Cartesianism, which did not provide him with the conceptual means to describe the behaviour of fluids. From 1698 onwards, Rohault's *Treatise on Physics*, which shows a reliance on Boyle's solution to the issue of the lack of the sensation of pressure under water,

---

75   Molhuysen, *Bronnen*, vol. 4, 104*: "Een glas met een openingh in het midden ad demonstrandam aëris pressionem lateralem."

76   Chalmers, *One Hundred Years of Pressure*, 6.

was apparently adopted as a textbook by de Volder but did not help him in overcoming his difficulties—although we have no direct insights on his use of Rohault's text.[77] Other factors, certainly, were behind his late dissatisfaction with Descartes's ideas: in particular, with his idea of material substance, which not allowing for any argument for its activity made it necessary to revert to an external mover, as discussed in his correspondence with Leibniz. It was difficulties such as these, combined also with his early teaching of Boyle's experiments at Leiden, that led ultimately to a divergence between his experimental and theoretical physics.

Acknowledgments

The research leading to this publication has received funding from the European Union's Horizon 2020 research and innovation programme under the Marie Skłodowska-Curie grant agreement No 892794 (*READESCARTES*), and, previously, from the Swiss National Science Foundation under the Spark grant number CRSK-1_190670 (*Testing a Multi-Disciplinary Approach to an Unexplored Body of Literature: The Case of Cartesian Dictations*). Special thanks go to the Forschungszentrum Gotha der Universität Erfurt.

---

77   Rohault, *Traité*, 76. See n. 19 above.

CHAPTER 9

Medicine and the Mind in the Teaching of Theodoor Craanen (1633–1688)

Davide Cellamare

1   Introduction

René Descartes was no university teacher. One aspect of his philosophy that made it difficult for Cartesianism to find its place within the early modern university was the fact that Cartesian philosophy seemed incomplete when compared to Aristotelian philosophy, which Descartes's new ideas aimed to replace.

The tradition of Aristotelian commentaries and textbooks that had dominated European universities, since their establishment around 1200, encompassed a great array of disciplines, including logic, physics, metaphysics, and ethics. By contrast, as Roger Ariew observes, Descartes had not manged to produce much more than a general metaphysics and a partial physics.[1] These two disciplines were of course very important in university teaching; Descartes himself, in the French Preface to his *Principles of Philosophy* (1647), had described them as respectively the roots and the trunk of the tree of philosophy. In the same context, Descartes had imagined the branches of the tree as representing the other disciplines that should develop out of the roots and the trunk, notably, mechanics, morals, and medicine.[2] However, he devoted comparatively less space to these disciplines. His failure to produce systematic accounts of ethics and medicine made it even more difficult for his new philosophy to replace the old one.

These well-known aspects of Cartesian philosophy notwithstanding, recent scholarship has highlighted Descartes's keen interest in medicine. Whilst he did not produce works in pathology and therapeutics, he nevertheless devoted much space to physiology. Justin E. Smith, for instance, has pointed out that Descartes himself considered the preservation of health as the principal end of his studies.[3] Tad Schmaltz, moreover, has drawn attention to the fact that it

---

1   Ariew, *Descartes and the First Cartesians*, IX.
2   Descartes, *Principes de la philosophie*, *Préface*, AT IX-B 14.
3   Smith, "Heat, Action, Perception," 106. More recently, the theme of medicine in Descartes's work has been explained in Cook, "Princess Elisabeth's Cautions."

was at the Utrecht faculty of medicine, and more precisely thanks to the physician Henricus Regius (1598–1679), that Descartes's philosophy was first introduced into Dutch universities. As Schmaltz has documented, Regius inspired a network of Dutch Cartesian physicians, which contributed to the diffusion of Descartes's ideas, alongside a group of Dutch Cartesian theologians including Abraham Heidanus (1597–1678) and Christoph Wittich (1625–1687), and philosophers such as Johannes Clauberg (1622–1665) and Johannes De Raey (1622–1702).[4] But what was this network of Cartesian physicians teaching? Did their teaching include the study of pathology and therapeutics that Descartes himself had not carried out? And if so, was Cartesian pathology limited to an application of Descartes's mechanical philosophy, or did teaching concerning illnesses contribute to the explanation of Descartes's ideas concerning the human body and human nature?

As a case study concerning Cartesian medicine in the United Provinces, this chapter addresses the medical and philosophical teaching given by the Cartesian physician, Theodoor Craanen, at the short-lived Academy of Nijmegen and at the Statencollege of Leiden.

Very little has hitherto been written about Craanen, whose work has not enjoyed the same scholarly fame as that of the above-mentioned Dutch figures. However, Theodoor Craanen was part of this constellation of Cartesian stars, some of whom, as we shall see, were his direct teachers and very influential on his work.

But Craanen also shone his own light. He influenced such a large number of students—at Nijmegen and Leiden—that Antoine M. Luyendijk-Elshout, in one study (possibly the only study) devoted to him, refers to Craanen's students as a fully-blown school, namely "the Mechanical Philosophical School of Theodoor Craanen."[5] This school, Luyendijk-Elshout explains, produced theories of the human body derived from the mechanical philosophy that originated from the revival of Greek atomism, and reflected recent influences, such as the works of William Harvey (1578–1657) and Walter Charleton (1619–1707), the iatrochemical school of Franciscus de le Boë Sylvius (1614–1672), and the physiological works of Santorio Santorio (1561–1636) and Alfonso Borelli

---

4   Schmaltz, *Early Modern Cartesianisms*, 228, 255–256.
5   Luyendijk-Elshout, "*Oeconomia animalis*," 295–307. Luyendijk-Elshout supplies an overview of the works of the students who were influenced by Craanen, such as Cornelis Bontekoe (1647–1685), Benjamin Van Broeckhuysen (1647–1686), Heydentrijk Overkamp (1651–1693), and Johannes Broen (1662–1703). Luyendijk-Elshout also outlines criticisms of some of the core concepts of Craanen's mechanical school, and their eventually being superseded by the medicines of Burchard de Volder (1643–1709) and Herman Boerhaave (1668–1738).

(1608–1679). All these influences may be found in the school of Craanen, who came to occupy the chair that was left vacant in Leiden after Sylvius's death. Whilst using the latter's chemical physiology to an extent, Craanen put forward a mechanistic concept of the human body, which was largely based on Cartesian philosophy.[6]

The vast majority of Craanen's ideas may be found in the student disputations over which he presided, as well as in posthumous works, which were published by his students on the basis of class notes. Here, I shall therefore look at Craanen's ideas as they developed in the classrooms of Nijmegen and Leiden.

Craanen's teaching bears witness to a complex interaction between his own ideas in medicine, Descartes's physiology, and the teachings he received in Utrecht from Regius. On the one hand, he based his medical teaching on the new mechanistic physiology of Descartes, as found in the ground-breaking *Passions of the Soul* (1649) and the *Treatise on Man* (1662), as well as on Regius's medicine; on the other, I shall show that Craanen did not limit himself to repeating Descartes's and Regius's ideas or to developing human pathology on the basis of Cartesian physiology alone. In fact, Craanen used medicine in order to find proofs of central ideas in Descartes's conception of the human being, which would, in Craanen's view, have otherwise remain unexplained. Importantly, Craanen thought that the study of an uncommon illness—namely 'catalepsy'—was the only way to prove Descartes's famous theory according to which the mind was joined to the human body through the pineal gland, whilst being itself an immaterial substance, really distinct from the body.[7]

Catalepsy today is sometimes synonymous of 'catatonia'—namely, a condition (or a symptom associated with other psychiatric disorders) in which patients cannot move normally, characterized by a prolonged rigidity of the

---

6   Luyendijk-Elshout, "*Oeconomia animalis*," 295–298.

7   In this article, I shall use the terms "soul" and "mind" as synonymous, when dealing with Craanen's work. In fact, in the documents that I consider, Craanen uses the terms "anima" and "mens" interchangeably. "Anima," according to his usage, has nothing to do with the Aristotelian principle of life, but corresponds to the Cartesian "mens." In Craanen's work, both "anima" and "mens" signify Descartes's *res cogitans*, namely an immaterial substance, devoid of any extension, whose task is only to think. Consider, for instance, the following passage in Craanen's *Tractatus*: "Cartesius, in secunda meditatione dicit animam esse conscientiam [...] ac proinde exclamat tandem animam esse meram et solam cogitationem. Hanc proinde cogitationem ab omnibus rebus corporeis separat." Craanen is probably here referring to Descartes, *Meditationes* II, AT VII 27: "Nihil nunc admitto nisi quod necessario sit verum; sum igitur praecise tantum res cogitans, id est mens, sive animus, sive intellectus, sive ratio."

muscles.[8] In Craanen's time, catalepsy was described as a state in which people were reportedly found to be suddenly immobile, as if frozen, while returning to their healthy state in an equally abrupt way.

Luyendijk-Elshout describes Craanen's teachings about human pathology as containing both "a serious system of physiology" and several "old wives' tales."[9] To some extent, Craanen's views concerning catalepsy are so odd as to warrant Luyendijk-Elshout's dismissive comparison with folkloric superstition. But as odd though they may be, Craanen's ideas recur in a fairly coherent manner in his teaching in Nijmegen and Leiden, and they play a central role in his system of (Cartesian) medicine and anthropology.[10]

Whilst Craanen's teaching was devoted to a variety of topics, his teaching about catalepsy is a constant feature, and becomes more and more precise over the course of his academic career. The key ideas disclosed in our study of Craanen's teaching concerning catalepsy and the mind-body relationship show him to be an influential, albeit forgotten, exponent of Cartesian philosophy in the United Provinces. An interrogation more generally of what Craanen taught to his students in Nijmegen and Leiden can make a significant contribution to our knowledge of the transformations and diffusion of Cartesian philosophy through its teaching in the classroom.

2   Theodoor Craanen and His Teaching in Nijmegen and Leiden

Let us first introduce Craanen as a man and personality, as well as his intellectual and institutional contexts. In so doing, we shall also introduce the documents which give us the evidentiary record of Craanen's teaching in Nijmegen and Leiden that form the basis of what we study in this chapter.

Theodoor Craanen was born in Cologne ca. 1634. He spent his career in the German and Dutch lands, studying first in Utrecht, between 1651 and 1655,

---

8   See Coffey, "Catatonia in Adults."
9   Luyendijk-Elshout, "*Oeconomia animalis*," 299, 301.
10   The expression 'Cartesian medicine' is as far from straightforward as the lone adjective 'Cartesian.' Here, 'Cartesian medicine' shall be taken as referring simply to Craanen's endeavour to develop his medical ideas in the framework of the mechanist physiology of Descartes. As we shall see, Craanen explicitly declares his intention of treating the human body and its pathologies on the basis of Descartes's physiology. Craanen is identified as "médicin cartésien" in Caps, *Médicins cartésiens*, 578, a book to which I refer the reader for an in-depth discussion of expressions such as 'Cartesian medic' and 'Cartesian medicine.' On Dutch Cartesian medicine and Craanen's belonging to it, see Schmaltz, *Early Modern Cartesianisms*, 228–283, especially 278.

under Descartes's early ally, Henricus Regius, whose work had an important impact on Craanen's teaching.[11] Craanen would later give classes on Regius's *Praxis medica* (1657), the notes from which were then to be published in his *Opera omnia* (1689), under the title *Annotationes in praxin Henrici Regii*.[12] As we shall see, Regius's work played a central role in Craanen's views concerning catalepsy and the mind.[13]

Between 1656 and 1661, Craanen studied medicine and philosophy in Duisburg, under the famous Cartesian, Johannes Clauberg, whose influence reached Craanen's next academic affiliation, namely the Academy of Nijmegen. This short-lived institution existed between 1655 and 1679, when it closed amidst a plague epidemic and financial struggles in the aftermath of the French occupation (1672–1674). During its short existence, the Academy of Nijmegen (1655–1679) styled itself as stronghold of Cartesian philosophy, by hiring outspoken Cartesians for the chairs of philosophy and theology. In 1656, the teaching of theology was given to Christoph Wittich, who had formerly worked shoulder to shoulder with Clauberg, first in Herborn (1651–1653) and then in Duisburg (1653–1656). In these German universities, Wittich started his life-long endeavour of harmonising Cartesian philosophy and Reformed theology; a programme that he would further develop in Nijmegen and Leiden.[14]

---

11 Craanen's date of birth seems unclear. Whilst literature often identifies 1620 as Craanen's year of birth, this seems very unlikely. This would mean, in fact, that he would have been quite old when starting his studies in Utrecht, at which time he would have been thirty-one years old. An alternative source, the Matriculation Catalogues of the University of Duisburg (accessed 16 March 2022), where Craanen was later to move in 1656, describe him as having briefly enrolled in Leiden, in 1655, at the age of twenty-one: "Theodorus Cranen, Coloniensis, aetatis annorum 21, phil. et medic. stud. Leida huc venit 6 Novembris." Based on this, I think it is more plausible that Craanen was born approximately in 1634.

12 Theodori Craanen, ... *Opera omnia, nunc demum conjunctim edita: tomus prior est Tractatus* physico-medicus De homine, *in quo status ejus tam naturalis, quam praernaturalis, quoad theoriam rationalem mechanice demonstratur: cum figuris aeneis & indicibus tam capitum, quam rerum & verborum locupletissimis. Tomus alter continens Observationes, quibus emendatur & illustratur Henrici Regii* Praxis medica, *medicationum exemplis demonstrata* (Antwerp, 1689).

13 In Utrecht, Craanen also studied under Johannes De Bruyn (1620–1675), as documented by Chapter 7, in this volume.

14 On the themes of Clauberg and his influence in Duisburg and Nijmegen, and the activities of Clauberg and Wittich in Duisburg, see Trevisani, *Descartes in Germania*, 25–34 and *passim*. A number of studies have been devoted to Wittich's attempt to harmonize Cartesian philosophy and Reformed theology; among others, see Del Prete, "Y-a-t-il une interprétation cartésienne de la Bible?"; Vermij, *The Calvinist Copernicans*, 146–148,

The teaching of philosophy and medicine in Nijmegen was first assigned to Gulliaume Soudan (fl.1651–1663), who taught Descartes's *Principles of Philosophy* (1644) and Clauberg's logic.[15] Starting 1558, the Board of the Curators asked Soudan and Wittich to periodically give classes in each other's subjects. In his capacity as professor of philosophy, Wittich taught a course, in 1664, during which he meticulously explained Descartes's *Meditations on First Philosophy* (1641) and the first part of Descartes's *Principles*. In this course, Wittich tried to show the compatibility of Descartes's metaphysics with the Calvinist faith, a subject on which I have written elsewhere.[16]

Craanen was called to teach philosophy and medicine in Nijmegen in 1661, after Soudan was expelled from the Academy for adultery. Wittich had initially sponsored his friend Clauberg as a successor to Soudan, but the Academy could only afford Craanen.[17]

We can form an idea of what Craanen's activity in Nijmegen entailed by looking at a doctoral thesis defended under him, by a certain John Teiler, in 1668—this, in fact, is the only doctoral thesis in philosophy ever defended during the short history of the Academy. The thesis was entitled *De anima philosophiae, sive rectae rationis ab errore discernendi ratione dissertatio*; a defence of the Cartesian method and metaphysics, which were then taught to the Nijmegen students.[18] More substantial information about Craanen's teaching in Nijmegen can be gathered from a series of at least eleven disputations over which Craanen presided, together with Wittich, between 21 November 1663 and 28 June 1665. These disputations belong to the main documents addressed in this chapter.

Craanen's Nijmegen disputations bear the overall title: *On Human Nature. On the Conjunction of the Soul with the Body* (*De natura hominis. De conjunc-*

256–271; Eberhardt, *Christoph Wittich*; Kato and Sakamoto, "Between Cartesianism and Orthodoxy."

15   In 1657 Wittich writes to Clauberg: "Nos hic vivimus, praesertim cum Dn. Soudan jam totus sit noster factus, institutis duobus Collegiis, altero in tuam Logicam, altero in Principia Cartesii, quod publice significavit se facturum affixo programmate." This transcription of the letter is found in Bots, "Témoignages sur l'ancienne université de Nimègue," 231. The original is preserved at Leiden University Library, BPL., 293B. Wittich refers to Clauberg, *Logica vetus et nova quadripartita, modum inveniendae ac tradendae veritatis in Genesi simul et analysi facile methodo exhibens* (Amsterdam 1654).

16   On Wittich's course, see Cellamare, "A Theologian Teaching Descartes." On the Academy of Nijmegen and the classes timetable, see van Meerkerk, "Een filosofische lessenserie," 58. van Meerkerk refers to Raadssignaten 13 jaunuari 1658, Gemeentearchief Nijmegen, Oud-Archief, inv. nr. 106, 11–12.

17   See Bots, "Témoignages sur l'ancienne université de Nimègue," 232–234.

18   On Teiler's thesis, see van Meerkerk, "The Right Use of Reason," 77–102.

tione animae cum corpore).[19] Seven of them—starting with disputation number IV and finishing with number XI—are extant in a collected volume, which is found at the University Library of Erlangen.[20] In the volume, the first three disputations and the sixth are unfortunately missing. Even so, it is evident that Craanen orchestrated these exercises, not just in their form, but also in their content, and that he conceived them as part of a single project.[21] The theses included in the Nijmegen disputations are numbered and contain references to other theses defended in previous and subsequent disputations in the series. This happens, especially in the beginning of each disputation, by means of expressions such as "having seen that ... it remains to be proven that ...," or "on the basis of what we have seen until now ..."

Craanen's disputations on human nature are interesting for at least two reasons. First, from the point of view of the teaching of Cartesian philosophy, they show that seventeenth-century students at Nijmegen were receiving a rather thorough instruction in Descartes's work. Next to Wittich's teaching of the *Meditations* and *Principles*, Craanen's disputations bear witness to the students' knowledge of Descartes's *Passions of the Soul*. Secondly, from the more specific perspective of Craanen's teaching, the Nijmegen disputations contain Craanen's first formulation of the following idea: the identity of the seat of the soul in the body can be proven *a posteriori*, only by observing the causes and symptoms of one specific illness, namely catalepsy. Craanen was to repeatedly revisit this idea throughout his career, which, from 1670, continued in Leiden. There, Craanen was reunited with his colleague Wittich, who took the chair of theology in Leiden, having left Nijmegen in 1671.

Given its steady decline, both Wittich and Craanen had good reasons for leaving the Academy of Nijmegen when they did. However, whilst Leiden offered greater prestige, it did not offer them the safe haven for Cartesian philosophy that they had found at their former university. The policy of Leiden University's Senate was shaped by the 1645 Utrecht decree which, in the wake of the

---

19   Craanen (praes.), *Exercitationes philosophicae De natura hominis. De conjunctione animae cum corpore* (Nijmegen 1663–1665). Although these texts are titled *exercitationes* (and not *disputationes*), I will refer to them as "disputations" because, in fact, just like in university disputations, these exercises address a disputed question and have one or more professors presiding over the "exercise" and a student defending his theses as a *respondens*.

20   As Edwin van Meerkerk, "Een filosofische lessenserie" observed, a copy of Disputation IX is also found separately at the Provincial Library of Arnhem. I numbered the pages of the extant disputations, from 1r (the title page of disputation IV) to 35r (the last page of Disputation XI).

21   On university disputations and their authorship, see Chapter 7, in this volume and Friedenthal, Marti and Seidel, eds., *Early Modern Disputations*, 8–9.

so-called "Utrecht Crisis," had forbidden above all the teaching of Cartesian philosophy. Leiden was also subject to another decree emerging from the volatile atmosphere surrounding the early diffusion of Cartesianism in the United Provinces. In a 1656 decree, the States of Holland and West Friesland (under whose jurisdiction Leiden fell) denounced the abuse of philosophy in theological matters, and more specifically identified Cartesian philosophy as responsible for an unacceptable crossing of the boundaries between philosophy and the interpretation of Scripture.[22]

The Statencollege of Leiden expected professors to teach logic and metaphysics according to the Aristotelian philosophy as found in the textbooks of the former Leiden professor Franco Burgersdijk (1590–1635).[23] Following his appointment as a teacher of both these disciplines, this injunction applied directly also to Craanen, who would nevertheless refuse to comply with it.

It is therefore perhaps unsurprising that, in 1673, the curators decided that it would be preferable for Craanen to stop teaching philosophy and appointed him instead to the chair of medicine, which was still vacant following the death one year previously of the well-known Franciscus de le Boë Sylvius.[24] Together

22   For the text of the decree, see *Register van Holland en Westvriesland*, 803–807; Knuttel, *Acta der particuliere synoden van Zuid-Holland*, 36–40. On the Utrecht Crisis and the following rulings, see Dijksterhuis, Serruier and Dibon, eds., *Descartes et le Cartésianisme hollandais*; Thijssen-Schoute, *Nederlands cartesianisme*; Cook, *The New Philosophy in the Low Countries*, 115–149; Verbeek, *La querelle d'Utrecht*; Verbeek, *Descartes and the Dutch*; Vermij, *The Calvinist Copernicans*.

23   In the resolutions of the curators of 1673, one reads: "Is mede eyndelyck verstaan dat den heer Professor Cranen volgens de Statuten van 't voors. Collegie geen andere Logicam ofte Metaphysicam in Collegio Ordinum sal mogen doceren ofte interpreteren als di van Franco Burgersdicius"; see Molhuysen, *Bronnen*, vol. 3, 271. The resolutions of the Statencollege of 1673 mark a change in the way in which Burgersdijk's works were perceived in Leiden. As Chapter 2, in this volume, shows, Burgerdijk's philosophy was repudiated at Leiden in 1641. The difference might of course be explained by the gap of more than thirty years between the two decisions (one taken at the University, the other taken at the Statencollege). The 1673 ruling does not name any textbook in particular, but it might be interpreted as referring to the following ones: *Franconis Burgersdicii institutionum logicarum libri duo, decreto illustriss. ac potentiss. DD. ordinum Hollandiæ et West-Frisiæ, in usum scholarum ejusdem provinciæ, ex Aristotelis præceptis nova methodo ac modo formati, atque editi* (ex officina Abrahami Commelini, 1645); *Franconis Burgersdici Institutionum metaphysicarum libri duo* (Apud Hieronymum de Vogel, 1640). On Burgersdijk, see Bos and Krop, *Franco Burgersdijk*; Gellera, "A 'Calvinist' theory of matter?" On the 1641 Leiden regulations, also see Chapter 3, in this volume.

24   Craanen was removed from the teaching of metaphysics, and from his role as "subregens" of the "Collegium Theologicum," amid disagreements with the then rector Frederick Spanheim. As emerges from the acts of the curators of 14 and 28 August, and 8 November 1673, Craanen presided over a disputation (which was defended by Abrahamus Weer-

with his chair in philosophy, Craanen also lost to the Aristotelian Gerard de Vries (1648–1705) his position as *subregens* of the "Nederduytsche Collegium Theologie."[25] de Vries, who was also allowed to give private classes, was to quit Leiden and move to Utrecht only one year after his appointment, citing what he depicted as an unfavourable atmosphere in Leiden. More specifically, he lamented a clear preference for Cartesian philosophy at Leiden and deemed Utrecht more suitable for his Aristotelianism. In the resolutions of the Leiden curators, we read that Craanen himself was one of the reasons for the departure of de Vries, whom we shall encounter again among the critics of Craanen's medicine.[26]

Craanen taught medicine in Leiden until 1686, after which he spent the last two years of his life as physician in ordinary to the Elector of Brandenburg (1620–1688). During his Leiden period, Craanen's teaching activities were not limited to Descartes's ideas alone but embraced wider influences. As Theo Verbeek explains in this volume, most Dutch Cartesian physicians were influenced by Sylvius, and Craanen was no exception in this regard.[27] Moreover, he devoted classes to Daniel Sennert's natural philosophy and medicine, as we know from one of the notebooks kept by his students entitled *Observationes excerptae ex praelectionibus publicis, privatisque collegiis Theodori Craanen, quibus emendator et illustrator v. Institutionum Liber Danielis Sennerti, De auxiliorum materia* (Leiden 1687). One year before the publication of this document, Craanen's *Examen in Institutionum in Dan. Sennert Librum quintum epitomes Institutionum tractantem de auxiliorum materia* had appeared as an appendix to the above-mentioned notebook on Regius's *Praxis medica*.

---

den), in which the following corollary was added: "Quod qualiscunque de Deo dubitatio dicatur neque pia neque honesta esse." Spanheim, however, alleged that this corollary was attributed in the disputation to scholastic philosophy, elsewhere maligned in the same disputation. According to Spanheim, the above-mentioned corollary was in fact part of a defence of Cartesian doubt. Craanen was summoned by the University Senate, but the curators considered his subsequent efforts to teach according to Aristotelian philosophy as somewhat forced, rather than deriving from his own free will ("niet met de veryste lust en ijver, nemaer als door dwang"). Craanen was thus deemed unfit to teach metaphysics according to Leiden standards and as better suited for the chair of medicine. The relevant acts of the curators are found in Molhuysen, *Bronnen*, vol. 3, 274–278.

25 Molhuysen, *Bronnen*, vol. 3, 278.

26 Molhuysen, *Bronnen*, vol. 3, 278: "ende nademael C. en B. van besijden waren beright dat den Professor Kraanen indirectelyk tot de oorsaek van het vertrek van D. de Vries hadde gecoopereert." On de Vries and his attitude towards Cartesianism, see Chapter 10, in this volume.

27 On Craanen's combination of Cartesian theories on subtle matter with Sylvius' iatrochemical principles, see Strazzoni, *Dutch Cartesianism*, 139.

In Leiden, Craanen also taught classes that were to be published in his *Opera omnia* (1689), under the title *Oeconomia animalis*. As Luyendijk-Elshout has pointed out, Craanen's *Oeconomia* bears witness to the encounter of Dutch Cartesianism with a larger tradition of physiology, which first received the title of "oeconomia animalis" by Walter Charleton and which devoted particular attention to the theory of bodily fluids, in the framework of a general theory of the human being. In the hands of Craanen, this encounter of traditions developed into a "mechanical school" of physiology.[28]

Many of the themes of the *Oeconomia* are also found in another and perhaps more complete expression of Craanen's Cartesian medicine, to which I shall devote a large portion of this chapter—namely, his *Tractatus physico-medicus de homine*.[29] This book, was published by Craanen's student Theodoor Schoon, in 1689. The views contained in it, however, can be traced back to 1679 or earlier, as is evident from a manuscript preserved in the Sloane Collection of the British Library, which is entitled *Dictata, ut videtur, in Theodori Craanen tractatum physico-medicum de homine* (hereafter: Ms. 1274 (Sloane)). This manuscript contains class notes taken by the Englishman Christopher Love Morley between 1677 and 1679, when he studied in Leiden.[30] It is difficult to say whether Ms. 1274 (Sloane) is one of the notebooks used by Schoon in the composition of the *Tractatus de homine*, which differs in wording from Ms. 1274 (Sloane). Yet, besides these differences and with some other smaller variations (mostly, the order in which some themes are treated), the two documents have the same general structure and present the same views. I shall devote specific attention to these documents in the fourth part of this chapter. For now, suffice it to say that one of the most important views that are put forward in Craanen's Leiden lectures concerns Descartes's understanding of the mind and the body, as well as Craanen's own idea that "catalepsy" offers the only proof of the seat of the connection between the two in the pineal gland. These themes, as we shall see, are the central pillars of the edifice of Craanen's

---

28   Luyendijk-Elshout, "*Oeconomia animalis*," 295–299 and *passim*.

29   The complete title of this book is: *Tractatus physico-medicus de homine, in quo ejus tam naturalis, quam praeternaturalis, quoad theoriam rationalem mechanice demonstratur* (Leiden 1689). This book was republished in 1722, in Naples, by the typographer Felice Mosca.

30   Information about Morley's biography can be found in the preface to a collection of notes of *dictata* he took in Leiden, which is entitled: *Collectanea Chymica Leydensia, id est, Maëtesiana, Margraviana, Le Mortiana. Scilicet trium in Academia Lugundo-Batava Facultatis Chimicae, qua publice, qua privatim professorum ... non solum ostenderunt, verum etiam suis verbis dictarunt ....* (Leiden 1689). On Morley's class notes, see Chapter 8, in this volume.

medicine. His Leiden teachings on medicine and the mind-body relationship may be viewed as an elaboration on the ideas first presented in his Nijmegen disputations, to which we will now turn.

3 Catalepsy in the Nijmegen Disputations on Human Nature

The overall title of the Nijmegen disputations—namely, *On Human Nature: On the Conjunction of the Soul with the Body*—points to a rather broad field of enquiry. However, most of the extant texts set out with a more specific purpose, namely to identify the seat of the soul in the human body. After having established that this seat is in the pineal gland, the disputations move on to describing the mechanisms through which the pineal gland, the nerves, the fibres, and the spirits in the human body make it possible for the soul to move the body and to be affected by the external and internal bodily stimuli.

Although Cartesian sources are not explicitly cited, it is clear that the students are asked to defend views that are found in Descartes's *Passions of the Soul*. Moreover, many of the corollaries added by the students at the end of the disputations bear witness to their knowledge of Descartes's *Principles* and *Meditations*. For instance, we find corollaries stating that the world is indefinitely extended, or that the cause of errors is the will, which—having a broader scope than that of the intellect—often extends itself to things that are not perceived clearly and distinctly by the intellect.[31] As we have already seen, at the time of the disputations, Descartes's *Principles* and *Meditations* were being taught to Nijmegen students by Wittich, who was present, together with Craanen, during the Nijmegen disputations, as we may gather from the frontispieces of the extant texts.[32]

In what follows, I shall focus on the first three of the extant texts on "the conjunction of the soul with the body," which conclude that the pineal gland is the point of conjunction between body and soul. Interestingly enough, two of these three texts are subtitled *et speciatim de catalepsi* ("in particular, concerning catalepsy").[33] But what does catalepsy have to do with Descartes's views

---

31 Craanen (praes.), *Exercitationes*, 9ᵛ: "Mundus est indefinite extensus"; "Causa errorum est voluntas; qui cum latius pateat quam intellectus, saepe se extendit ad ea quae intellectus non clare percipit." These corollaries echo Descartes's views on the indefinite extension of the world in *Principia* I 26–27 (AT VIII-A 14–15), as well as on the will as the source of error, in *Meditationes* IV (AT VII 56).

32 On Cartesian corollaries in students' disputations, see Chapter 7, in this volume.

33 The extant texts bearing the reference to catalepsy in their title are Disputations V and VII.

on the mind-body relationship and the pineal gland? What explains Craanen's recourse to the description of this illness in order to validate the role of the pineal gland as the seat of the soul? The answer to these questions, as we are about to see, appear to devolve on Craanen's conviction that the seat of the soul cannot be established either *a priori* or on the basis of Descartes's own arguments.

The first of the extant Nijmegen disputations, which was defended on 21 November 1663, is in fact the fourth in the series, meaning that, unfortunately, we cannot know exactly to what the defendant, Henricus Van Huyssen refers when he writes that the previous disputations have proven that the soul must be found within the region of the brain and not anywhere else in the body. In the absence of a documentary record of the first three disputations, Craanen's Leiden lectures may help to shed some light on this question, however, as we shall see later. At any rate, having established this conclusion, Van Huyssen writes that one might wonder whether the soul is joined to the whole brain, to multiple parts, or to one specific part of it. According to the disputation, joining the soul to multiple parts of the brain or to the whole of it would be less suitable to God's wisdom than a more simple designation of one single part of the brain as the seat of the soul. Considered alongside the fact that all of the filaments of the brain that are devoted to sense and arbitrary motion terminate in one single point of the concave surface of the brain, this position seemed that much more tenable.[34]

Although it recognises all these things as being very plausible, Van Huyssen's disputation argues that it is impossible to conclusively determine *a priori* at which point in the brain the aforementioned filaments converge. This knowledge must instead be obtained in another way, *a posteriori*. Moreover, the unlikely possibility that the mind is joined to the whole brain or to multiple parts of it cannot be entirely ruled out at this stage.[35] But what is the *a posteriori* way to which Van Huyssen refers?

Disputations VI is not extant, but we may be sure that this missing text was also devoted to catalepsy, given that Disputation VII is subtitled "et speciatim de catalepsi tertia," instead of *secunda*.

34   Craanen (praes.), *Exercitationes*, 2ʳ–3ᵛ.

35   Craanen (praes.), *Exercitationes*, 3ᵛ: "At quoniam supra memoratus iste omnium filamentorum cerebri, sensui et motui arbitrario dicatorum, situs hic nondum innotescere potest, alia et a posteriori nobis demonstrandi via impraesentiarum ineunda est, quae nos ad immediatam animae sedem perducat. Hanc autem determinare non adeo impromptu esse, diffiteri non possum, quoties considero aeque commode mentem omnibus vel aliquot cerebri partibus simul qua muni soli, quam versus reliquae spectent, alligatam intelligi posse."

One might think of Descartes's arguments in the *Passions of the Soul* as a plausible option in this regard. There, Descartes had in fact argued that the mind, whilst being joined to the whole body, exercised its functions more particularly in the pineal gland than in any other part of the body. In a very well-known passage in the *Passions*, Descartes dismissed the idea that the soul was in the heart and determined instead that the brain (to which sense organs are related) should be the part of the body where the soul most directly exercises its functions. He famously added that whilst all parts of the brain are double (just like all the organs of the external senses), we can only have one thought about a single object, at any given time. Therefore, there must be a part of the brain where the two images of an object coming from the two eyes (or any other twofold impression coming from the sense organs) come together in a single image or impression, which prompts sensations in the soul. The pineal gland is the only site at which this can happen, according to this theory, as it is the only non-double part of the brain.[36]

Van Huyssen's disputation offers an outline of Descartes's arguments, but does not regard these as conclusive. The objection could be raised, for example, that the fact that the parts of the brain and the body that serve sensation are double does not necessarily mean that the soul has to be found in one non-double part of the brain; nothing, in principle, can prevent God from joining the soul to the whole body and from producing a truthful perception of a single object in the soul (or, for that matter, an untruthful representation of multiple objects in the presence of only one external object) when the multiplicity of nerves present in the body are affected by a given object.[37] Finding Descartes's *Passions of the Soul* wanting in this regard, Van Huyssen looks for his *a posteriori* proof elsewhere:

> Nor do any of these opinions concerning the immediate seat of the soul lack some probability. In order for the true one among these [opinions] to be chosen, we shall opportunely direct our mind to *catatonia* or *catalepsy*, a sickness that has tormented the genius of philosophers and physicians no little. For this reason, its consideration will be very much recommended. It will first be opportune to provide a brief account of it, so that, afterwards, in enquiring into the causes of the particular phenomena [related to the sickness], we are, unsuspecting, as if whilst doing something else, reconducted to the immediate seat of the soul.[38]

---

36   Descartes, *Passions de l'âme* I 30–33, AT XI 351–354, CMS I 340–341.
37   Craanen (praes.), *Exercitationes*, 4ᵛ–4ʳ.
38   Craanen (praes.), *Exercitationes*, 4ʳ: "Neque cuilibet harum opinionum de animae imme-

This, then, is the abrupt manner in which we are told that the description of catalepsy and of its symptoms will lead us to determining the part of the brain to which God joined the soul, an argument that, to the best of my knowledge, is not found in Descartes's works.

Van Huyssen's discussion breaks off at this point, thus leaving to the next disputation the task of describing the symptoms and the causes of catalepsy. Disputation v of the Nijmegen series was defended by Gualterus Van Boshuysen, on 28 November 1663 and is entirely devoted to describing the symptoms and the causes of catalepsy.[39] For what concerns the causes, Van Boshuysen expatiates at length on disproving two opinions, according to which the abrupt state of immobility in which those affected by catalepsy are plummeted might depend on the freezing or coagulation of the animal spirits. These theories are ascribed respectively to Galen and Daniel Sennert.[40] For the present discussion, it suffices to know that, according to Van Boshuysen, none of these views can be true. Both freezing and coagulation would make the bodily spirits take up less volume in the body, hence making the limbs flaccid. This flaccidity is not observed among the symptoms of catalepsy.[41]

At the time of Van Boshuysen's disputation, the explanations of catalepsy in terms of freezing and coagulation had already been criticized by Craanen's teacher, Henricus Regius. In the *Praxis medica*, Regius had argued that the freezing or coagulation of the animal spirits would immediately imply the cessation of breathing, and ultimately death.[42] Whilst Craanen did not comment

---

diate sede suam deesse probabilitatem. Ex hisce ut vera eligatur, opportune mentem convertemus in Catochen seu Catalepsin, morbum qui philosophorum et medicorum genium non parum torsit. Quo igitur nomine quoque ejus contemplatio erit commendatissima. Illius autem succintam historiam primo referre conveniet, ut dein in causas singulorum phoenomenorum inquirendo, ad immediatam simul animae sedem, quasi aliud agendo, imprudentes devolvamur."

39   This disputation, together with the next one in the series are the above-mentioned texts that are subtitled "et speciatim de catalepsi."

40   No precise reference to any works by Galen and Sennert is made in the text. In the case of Sennert, the relevant discussion is found in *Pratica medicina* (Wittenberg, 1636), 584–589. Sennert and other authors, such as Hyeronimus Capivaccius and Johannes Schenck (see below, n. 46), refer to two loci in Galen's work, namely: the second commentary of *In Hippocratis prorrhetici librum primum commentariorum libri tres* and *In Hippocratis Aphorismos commentaria*, section II, aphorism 3.

41   Craanen (praes.), *Exercitationes*, 7ʳ–8ᵛ and *passim*.

42   Regius, *Praxis medica*, 309: "Alii volunt hunc affectum a vapore spiritus in cerebro et toto corpore congelante vel coagulante originem ducere. Sed si haec vera eius esset causa, respiratio et pulsus, quae spirituum ope peraguntur, hic non possent esse integra. Nam coagulatio illa magna, ut spiritus totius corporis, ita etiam respirationi et pulsui inservientes congelaret, vel coagularet, iique condensati et humoris vel pruinae specie in ventriculus

on the *Praxis medica* until a later stage of his career, he was probably familiar with Regius's theories, which were published in 1657 and first defended in
a disputation presided over by Regius in 1645.[43] As we shall see, the affinities
between Regius's work and the Nijmegen disputations are not limited to the
criticism of past theories of catalepsy. Together, in fact, they amount to a very
thoroughgoing explanation of this illness. In order to understand this explanation of catalepsy, let us revisit Van Boshuysen's discussion.

Contrary to what the freezing or coagulation of the spirits would determine, neither flaccidity of the limbs nor the absence of breathing are observed
during catalepsy. On the contrary, rigidity of limbs is one of the main features
of the illness. In fact, as Van Boshuysen's disputation states, those affected by
catalepsy remain immobile, just like statues, in the last position in which they
were before the sickness manifested itself; their pulse and breathing remaining
regular. This means that the sickness does not affect the blood, which keeps
normally fermenting in the heart; nor does it affect the animal spirits, of which
only a regular inflow can ensure the necessary movements of the diaphragm
for the continuation of breathing.[44]

But, if the causes of catalepsy are not to be found in the freezing or coagulation of the animal spirits, to what are we to attribute the onset of this illness?
Whilst Van Boshuysen criticizes Galen and Sennert, he does not push his own
discussion much further than a general symptomatology of the disease. One of
the corollaries he attaches to his disputations, however, gives us a hint about the
general framework within which we may expect to find the causes of catalepsy,
containing as it does the following remark: "Peripatetic philosophy is useless in
medicine" ("Philosophia Peripatetica est inutilis in medicina").[45] It is through
Cartesian philosophy that the causes of catalepsy will be explored in the rest
of the Nijmegen disputations.

In fact, the relationship between Cartesian philosophy and Craanen's explanation of catalepsy is double. On the one hand, the physiology of Descartes

---

cerebri decidentes, non possent cerebri molem diutius sustinere, nec spirituum sufficientem copiam musculis sustinere, nec spirituum sufficientem copiam musculis thora-
cis suppeditare; unde primum apoplexia aegrum prostereret, et deinde, respirationis et
pulsus usu ablato, mors necessario ipsum invaderet."

43 The text on catalepsy found in Regius's *Praxis medica* (Utrecht, 1657) may also be found in
the 1645 Utrecht disputation: *Medicatio viri catalepsi laborantis*, defended, under Regius,
by Cornelius Godefridus Elbarchs. In the intervening period, Regius's views concerning
catalepsy may be found in a disputation *pro gradu*, which was defended by his student,
Petrus a Wouw, in Utrecht, in 1653: *Disputatio medica inauguralis de catalepsi*. I thank Erik-
Jan Bos for pointing out to me the existence of these two disputations.

44 Craanen (praes.), *Exercitationes*, 6^r–7^v.

45 Craanen (praes.), *Exercitationes*, 9^v.

proves *utilis* in the medicine used to treat catalepsy. On the other, the determination of the causes of catalepsy will lead to a proof of what, according to Craanen, Descartes's arguments had only insufficiently shown: namely, that the pineal gland is the seat of the soul. This is the subject of Disputation VII, which was defended by Paulus Grondt, in February 1654.

3.1   *Catalepsy and the Seat of the Soul*

As I have already pointed out, Disputation VI is not extant; therefore, we are missing some of the steps between the discussions of Van Boshuysen and Grondt. However, Craanen's Leiden lectures, discussed in the section 4 of this chapter, will help us fill this gap.

The disputation defended by Paulus Grondt addresses the following question: since we have proven that, during the symptoms of catalepsy, there are enough animal spirits left in the limbs and that the spirits cannot undergo freezing or coagulation, how does it happen that cataleptic patients suddenly remain immobile in the last position in which they were before the onset of the sickness? In his reply, Grondt starts off his discussion by explaining that those affected by catalepsy remain conscious during the symptoms, as allegedly confirmed by their own testimony afterwards. People affected by catalepsy remember having been aware of their surroundings and conscious of their own desire to move and react to the environment during catalepsy. They also remember, however, their complete inability to do so.[46]

Craanen put particular emphasis on the non-interruption of perception and will reported by those afflicted with catalepsy. Besides pointing to the independence of the mind from the body, the fact that patients seemed to retain their ability to sense and will whilst in the grip of catalepsy, makes Craanen and with him Paulus Grondt draw the following conclusion: The immobility observed during catalepsy does not depend on the mental abilities of perceiving and willing (which are allegedly intact), but on the mind's incapability to determine the animal spirits in such a way as to react in the desired way. This incapability must depend on the wrong disposition of one or more parts of the brain, which are used by the soul in directing the animal spirits in order to move the body. Once this or these parts are found, we will have determined the cause of catalepsy and thereby—Craanen claims—also the seat of the soul.[47]

---

46   Grondt does not use first-hand reports, but refers to other sources, such as Hyeronimus Capivaccius, *Medicina practica* (Venice 1594), 38 and Johannes Schenck, *Observationes medicae de capite humano* (Basel 1584), CIV–CVII.

47   Craanen (praes.), *Exercitationes*, 11ᵛ: "Relinquitur ideoque, ut persuasum habemus, mentis

The rest of Grondt's disputation aims to outline the physiological and ana-tomical conditions that one or more parts of the brain need to satisfy, in order to be identified as responsible for the direction of the animal spirits. This out-line contains hardly any observation-based material and follows what Craanen and his students had already adopted from Descartes.

It may be useful here to enumerate some of main features identified by Grondt's long discussion. The immediate seat of the soul must be able to determine the motion of animal spirits (which gush out of the choroid plexus) in all possible directions, in such a way as to make the body do what the mind wants it to do (e.g., to walk, or to move one's mouth in order to speak). The bodily seat of the soul needs therefore to be extremely mobile and to be placed at the encounter of the two ventricles of the brain, whence the spirits gush out.[48] Whilst mobile, the relevant part of the brain must be kept in balance by two little strings and its shape must be such as to enable it to intercept all the relevant animal spirits. Grondt thinks that this could only be performed by a spherical item, or better still a gland-like part, which is slightly oblong. This type of shape is called a "spheroid" in geometry or "pineal" in medicine and is fit to intercept and determine the animal spirits from the brain to the limbs. Even-

---

potestatem determinandi motum spirituum conditionatam tantum esse; eamque depen-dere a dispositione alicujus partis, unius vel plurium, intra cerebrum, qua aut quibus mediantibus, tanquam sede immediate, spirituum determinationi praesit."

48   Grondt's disputation carefully points out that neither the seat of the soul nor the soul itself can set into motion the animal spirits, but only have the task of determining the direction of the constantly moving animal spirits. Otherwise, the seat of the soul would introduce new motion into the universe, which would violate Descartes's fundamental physical law, according to which the quantity of motion in the universe is constant; Craanen (praes.), *Exercitationes*, 11ʳ. Grondt's point reflects an interpretation of Cartesian physics that is found in the works of Craanen's teachers, Clauberg and Regius. McLaugh-lin, "Descartes on Mind-Body Interaction," 165 and *passim* pointed out that Descartes never explicitly states that the mind does not produce new movement but only determ-ines the already existing movement of the spirits. McLaughlin ascribes this interpretation of Cartesian philosophy to Clerselier, in a letter to de La Forge of 4 December 1660, as well as to Clauberg's *Corporis et animae in homine conjunctio plenius descripta*, XXVI, 4 (in Clauberg's *Physica* [1664]). McLaughlin calls the above-mentioned interpretation a "Leibnizian reading," because it was clearly formulated in Leibniz's *Monadology* § 80 (Leibniz, *Die philosophischen Schriften*, vol. 6, 620–621). More recently, Bos, "Descartes and Regius on the Pineal Gland," 105–108 has traced the above-mentioned reading even further back in history and ultimately to Descartes himself. Bos found said reading in Regius's *Physiologia* (Utrecht, 1641) and observes how Regius's relied on the unpublished manuscript of *The World* that he received from Descartes himself. It seems very plausible to me that Craanen and his student Grondt echoed the views that they found in Regius and Clauberg.

tually, the so-called pineal gland, which satisfies all of the above-mentioned requirements, is identified as the part of the brain that is responsible for the direction of the animal spirits, hence, as the immediate seat of the soul in the body.

To sum up, when taken together, these three disputations outlined above present roughly the following argument: catalepsy cannot be due to a temporary lack of spirits or to their freezing or coagulation in the limbs, as the theories then available had proposed. The problem instead occurs when the part of the body that is responsible for determining the motion of new spirits into the rest of the body fails to do so. Having ascertained that this part is the pineal gland, Grondt explains that catalepsy takes place when this gland is hindered from determining the motion of the spirits, in such a way that the body cannot move, even when the soul is willing such motion to occur. But this very impediment is what affords us a crucial insight: the fact that the soul is impeded from determining the movements of the body when the pineal gland does not work properly reveals this gland to be the immediate instrument of the soul in the body.

Now, linking a discussion concerning catalepsy to Cartesian theories concerning the pineal gland may seem strange and may legitimately be regarded as questionable, to say the least. Moreover, Craanen's point that catalepsy must have something to do with the part of the brain that the soul uses to determine the spirits sounds quite arbitrary (if not like a begging of the question). These oddities may be explained, however, by considering their source (never explicitly mentioned in the disputations), namely Regius.

It was from Regius, in Utrecht, that Craanen must have heard the idea that catalepsy should be explained as a hindrance of the part of the brain that determines the motion of the animal spirits (and that this part was the pineal gland). We remember that Regius expressed his views concerning catalepsy, first in two disputations (of 1545 and 1653), and subsequently in his *Praxis medica*. There, Regius cites the case of an elderly man, melancholic by temperament (*temperamento melancholicus*), who experienced the symptoms of catalepsy whilst studying. Regius explains the symptoms as follows:

> The disease by which the sick person is affected is a melancholic distemper affecting the whole brain in a lighter way, but chiefly and more violently the common sense or conarion; which removes all sense and motion—with the exception of breathing and the pulse—and keeps the patient in the state in which he was before being seized by this condition. It is called catalepsy or catochus by the physicians. The original cause of this disease is the continuous study of letters, to which our eld-

erly [man] devoted himself for entire days. For this reason, then, assisted by the old age, melancholic humours were brought forth, which are the antecedent cause of this condition.[49]

This explanation is very likely to be the source of the discussion found in Craanen's Nijmegen disputations. However, one should not fail to notice that Craanen's ideas differ from those of Regius in at least two important ways.

First, according to Regius, the melancholic humours that impede the movements of the pineal gland during catalepsy only leave the pulse and breathing intact, as they remove not just motion but also sensation. Instead, in the Nijmegen disputations, we are told about patients who remember and report that they were completely aware of their surroundings during their cataleptic episodes.[50]

Secondly, Regius limits himself to explaining catalepsy as a disturbance of the pineal gland's activity. His discussion of catalepsy is purely medical and Regius does not use it to infer conclusions about the seat of the soul. This last inference instead constitutes the further step taken by the Nijmegen disputations and it amounts to an important element of novelty contained in Craanen's teaching: the study of catalepsy has more than just medical importance, as it can be used to prove what Craanen thought Descartes had correctly seen but insufficiently demonstrated: the location of the seat of the soul in the pineal gland.

---

49 Regius, *Praxis medica*, 308: "Morbus, quo affligitur aeger, est intemperies melancholica, cerebrum totum leviter, sed sensorium commune sive conarion, praecipue graviterque occupans; quae, salva respiratione et pulsu, omnem sensum et motum tollens, aegrumque in eo statu, in quo erat, cum ab hoc affectu primum corriperetur servans, *Catalepsis* sive *Cathocus* a medicis appellatur. [...] *Causa procatarctica* hujus morbi est perpetuum literarum studium, cui noster senex, totos dies inter libros sedendo, operam dedit. Hinc enim, adjuvante senili aetate, melancholici humores, qui *antecedens* hujus affectus sunt *causa*, fuerunt geniti."

50 As is well known, Regius's views on mind, body, and their union differed from those of Descartes in that they retained Descartes's physiology but not his metaphysics. Regius identified sensation as the sole source of thinking (hence rejecting innate ideas) and, in his 1647 Corollaria to the Utrecht disputation *Medicatio viri cachexia leucophlegmatica affecti*, went so far as to purport that the soul might be a mode of the body. It would be interesting to explore the ways in which Regius's ideas might have influenced his views concerning catalepsy and sensation. This would be beyond the scope of this chapter, however. Concerning Regius's epistemology and philosophy of mind, see Verbeek, ed., *Descartes et Regius*; Clarke, "The Physics and Metaphysics of the Mind"; Bellis, "Empiricism without Metaphysics"; Kolesnik-Antoine, *Le rôle des expériences*; Bos, "Henricus Regius"; Bos, "Descartes and Regius on the Pineal Gland."

Of course, whilst introducing these interesting ideas, the Nijmegen argument can appear incomplete and unsatisfactory. This might in part be due to the fact that some of the disputations in the series (especially Disputations III and VI) are missing. The story we gather from Craanen's Nijmegen students, moreover, can also give rise to several problems. First, whilst Henricus Van Huyssen and his fellow students might be justified in presenting their argument as *a posteriori*, they could not present their discussion as one in which observation plays any significant role. After having defined catalepsy as an impediment of one or more parts of the brain in sending the animal spirits out to the limbs, the argument moves on to deductively determine which this or these parts are, entirely on the basis of pre-stipulated conditions (such as the mobility and shape of the part, etc.) that need to be satisfied in order for that part of the brain to be deemed to have the potential to determine the motion of the spirits in the body. This way of proceeding makes the argument not only lack an observational basis, but also somewhat circular, given that some of the anatomical conditions stipulated by Craanen, as well as the identification of the item to which they correspond (the pineal gland) were already found in Descartes's work.[51]

Secondly, given the inconclusive nature of what we can read in the disputations, we may still wonder why the part of the body that determines the motion of the animal spirits (and hence the immediate seat of the soul) must necessarily be found in the brain.

Thirdly, and more importantly, whilst the Nijmegen disputations are titled "on the conjunction of the mind with the body," they leave important aspects of this theme unanswered. In fact, granting that the pineal gland is identified as the seat of the soul, one might still wonder about one crucial point: how exactly do the conjunction and the interaction between soul and body (in the pineal gland) take place? Fortunately, Craanen answered these questions in his Leiden lectures. Herein he also presented catalepsy in the framework of his broader study of human nature. It is to these lectures that we next turn our attention.

---

51   I have been unable to find a specific *locus* in the work of Descartes, in which the anatomical configuration of the pineal gland is stated in the same clear way as occurs in Craanen's treatment of it in the disputations. However, in at least two letters, Descartes describes the pineal gland more clearly than in the descriptions contained in the *Passions of the Soul* or the *Treatise on Man*; see Descartes to Meyssonnier, 29 January 1640, AT III 18; Descartes to Mersenne, 30 July 1640, AT III 120. I am grateful to Erik-Jan Bos for highlighting to me the availability of these letters since 1659. On the anatomy of the pineal gland in Descartes's work, see Shapiro, "Descartes's Pineal Gland."

4   Craanen's Lectures on Human Nature in Leiden

Information concerning Craanen's teaching in Leiden may be found in the above-mentioned *Tractatus physico-medicus de homine* and Ms. 1274 (Sloane). In these two documents, Craanen's views about catalepsy are presented in the broader framework of a complete theory of human nature and the sicknesses to which it can fall prey. Because Ms. 1274 (Sloane) is less complete than the published version of Craanen's classes, I shall mainly consider the *Tractatus*. However, I will make references to Ms. 1274 (Sloane) whenever pertinent. The study of these documents will help us resolve some of the problems formulated at the end of section 3. Moreover, a closer look at Craanen's Leiden classes will allow us to better appreciate the centrality of this strange theme of catalepsy in the medical teaching that Craanen's students received. In so doing, we will gain a better understanding of the ways in which Cartesianism intersected with Leiden University medical teaching in the second half of the seventeenth century.

A complete account of Craanen's Leiden courses is beyond the scope of this article. However, I shall at least outline their structure and some of the relevant themes they present, in order to focus on Craanen's ideas about medicine and the mind-body conjunction.

The Leiden curators, as we have seen, removed Craanen from the chair of metaphysics in 1673, as he did not comply with the requirements for the teaching of philosophy in the Aristotelian way. The curators thought that moving Craanen to the chair of medicine would be less damaging, but to judge from his medical classes on human nature, Craanen must have cared very little about the curators' worries. He opens his classes by stating that a proper understanding of human nature should include both the anatomical explanation of the body and philosophical knowledge of the soul.[52] But the requisite philosophy for the study of human nature, Craanen claims, is not the verbal scholastic philosophy of the Peripatetics, but the real and modern philosophy, which derives from the right use of reason and which agrees with Cartesian philosophy.[53]

In order to properly understand human nature, Craanen sets out to follow what he considers to be the method used by Descartes in the *Treatise on*

---

52   The idea that the study of human nature should include an anatomical description of the body, next to the study of the soul, was very common in the century preceding Craanen's work. See De Angelis, *Anthropologien*, 198-202 and *passim.*; Cellamare, "History of a Productive Failure"; Cellamare, "Confessional Science"; Cellamare, "Soul."

53   Craanen, *Tractatus*, 1.

Man.[54] The *Tractatus* studies the human body, the human soul, and their conjunction. But in order to understand the conjunction between mind and body, these two parts of human nature must first be considered as two different things.[55]

Because the body and the soul differ ontologically from each other, Craanen's treatise sets out to consider first the body as such (viz. not insofar as it is united with a soul), both in its state of health and in its state of illness. In this case too, Craanen explains, his teachings will follow Descartes's understanding of the body as a machine. It is not by virtue of its conjunction with the soul that the human body lives. Instead, just like the bodies of plants and brute animals, it lives through the workings of purely mechanical organs and spirits that are fit to perform the functions of life. These bodily organs and spirits are just like the gearwheels and weights of a clock. But, just like the clock will not tell the right hour if its parts are badly arranged, so too the body whose organs are badly disposed will be subject to illness.[56]

The first function of Craanen's mechanical body is the *coctio*, or digestion, of food, which happens through fermentation. In the stomach, this fermentation turns food primarily into the nutrients for the body. In the heart, fermentation turns food into blood, which is constituted by diversely-shaped particles for distribution to the appropriate parts of the body, through pores of varying geometrical shapes.[57] This fitting of the blood's corpuscles into pores of the corresponding shape—which, in the *Tractatus*, is explained also by means of illustrations added by the editor Theodore Schoon—is identified by Craanen as health, or as the good 'clockwork' functioning of the body. Illness occurs as a result of a geometrical mismatch between pores and particles.[58] By starting

---

54   Craanen, *Tractatus*, 3: "In hoc opera conabimur methodum eandem tenere quam secutus fuit philosophus noster Cartesius in suo tractatu de Homine."

55   Craanen, *Tractatus*, 2. This, in fact, is what Descartes sets out to do, in the opening lines of the *Treatise on Man* (AT XI 119, CSMK I 99).

56   Craanen, *Tractatus*, 2, 5–6; MS., London, MS. Sloane 1274: Craanen (professor) and Morley (student), *Dictata, ut videtur, in Theodori Craanen tractatum physico-medicum de homine* (1677–1679), 1.

57   On Craanen's use (and combination) of Sylvius's iatrochemistry and Cartesian elements (notably that of "subtle matter") in the account of digestion that is found in the *Tractatus*, see Strazzoni, *Dutch Cartesianism*, 138–139.

58   A study of the illustrations included in the *Tractatus* goes beyond the scope of this article. Some short remarks about Schoon's engravings may be found in Luyendijk-Elshout, "*Oeconomia animalis*," 300. There, some of the illustrations of the *Tractatus* are recognized as copies of illustrations by anatomists such as Richard Lower, Thomas Willis, and Theodoor Kerckring, often adapted to Cartesian theories in the style of Schuyl, de La Forge, and Van Gutschoven.

from his corpuscular theories on digestion and fermentation, Craanen sets out first to consider the healthy human body: it is only by knowing the body in its state of health that we may properly understand its illnesses.

Most of what follows in the *Tractatus* (as well as in Ms. 1274 (Sloane)) is an explanation of the clock-like workings of the human body, its functioning, and its malfunctioning. After digestion and the parts involved therein, Craanen addresses a large number of subjects related to blood circulation, breathing, the workings of the brain, the nerves, the animal spirits, and the organs of sensation. Craanen concludes his explanation of the human body by addressing generation; Ms. 1274 follows the same structure.

For virtually each of the above-mentioned topics in Craanen's Leiden lectures, we are given detailed information about the relevant pathology (from fever and diarrheal, through to apoplexy, frenzy, and much more). Most interesting, for the present discussion, is the fact that catalepsy is the one illness that Craanen does not treat in the parts of his lectures devoted to the body, but deals with it instead in those parts that consider the soul and its conjunction with the body.

In Ms. 1274, Craanen discusses catalepsy in the chapter "De mente humana et primo ubi ea sit," a sudden shift in the manuscript after its explanation of several illnesses of the body. In the case of the *Tractatus*, catalepsy is discussed in the chapter "De anima ejusque passionibus," to which the text transitions in a more gradual way than in the equivalent shift in Ms. 1274. In fact, having concluded its discussion of the body, the *Tractatus* turns to the topic of sensation, which is considered to belong partly to the body and partly to the soul.

Providing a full outline of Craanen's account of sensation would represent too great a digression from our discussion. However, there is one aspect of his distinction among three levels of sensation that will lead us right to the crux of his ideas concerning the mind-body relation and the seat of the mind in the body. According to Craanen, the first level of sensation is wholly corporeal and amounts to the local motion impressed by the external object on the sensory organ. The agitation of the particles (*particularum agitatio*) impressed by the object on the organ is transferred to the brain through the same nerves that serve locomotion.[59] This transmission of motion is received by the spirits that are present in the ventricles of the brain, which spirits depict the idea of the object to the pineal gland (*in superficie glandulae dipingitur idea*). This transmission of motion is followed by the second level of sensation, namely

---

59   In *Tractatus*, 516, Craanen argues that the same nerves serve both sensation and locomotion.

the mind's perception of the object. This second level, as well as the third level of sensation, through which the mind formulates judgments about the object, are entirely incorporeal.[60]

Craanen's distinction between the corporeal and incorporeal levels of sensation is not underpinned by extensive arguments in the text, but seems simply, to a considerable extent, to recapitulate Descartes's ideas.[61] But besides lacking any original argumentation, Craanen's story also raises two fundamental questions. First, if the first level of sensation is corporeal and constituted by local motion, how does it cause the two incorporeal levels of sensation? And how can these two latter levels cause the body to move? Secondly, where is the mind that represents and judges ideas and sets the body in motion?

4.1 *Catalepsy and the Mind-Body Connection in Craanen's Leiden Courses*

In the *Tractatus*, Craanen replies to our questions concerning the seat of the soul in the body at the end of the section devoted to "The soul and its passions," and in the subsequent section, "On catalepsy." The main views that are expressed in these sections do not differ too much from those that we found in the Nijmegen disputations. However, the discussion in the *Tractatus* provides us with the answers to a number of fundamental questions that the extant Nijmegen disputations leave unanswered. Notably, Craanen's *Tractatus* explains the nature of the conjunction between body and soul in a framework that we may consider occasionalist, at least as far as the mind-body relationship is concerned.[62] Of course, as recent scholarship has documented, the category of 'occasionalism' refers to a rich and nuanced array of post-Cartesian theories.[63] In what follows, however, it will become clear that Craanen denied any immediate causation between mental and bodily states, leaving to God the task of causing the correspondence between them.

---

60   Craanen, *Tractatus*, 517.

61   Craanen's distinction between three levels of sensation reflects several *loci* in the works of Descartes. As far as I can see, the formulation of said distinction in Descartes's work that most resembles Craanen's is found in *Meditationes, Responsiones* VI, AT VII 436, CSMK II 294–295.

62   The term "occasion" (in its ablative, *occasione*) is used, albeit in passing, in Craanen (praes.), *Exercitationes*, 4ʳ, to characterize the ways in which the soul may have specific perceptions when specific motions of the nerves take place.

63   An extensive discussion of post-Cartesian occasionalist theories (ranging from the mind-body relationship to intra-substantial [mind-mind, body-body] causation), as well as references to the existing scholarly debate may be found in Schmaltz, *Early Modern Cartesianisms*, 204–227; Renz and Van Ruler, "Okkasionalismus."

Before he turns his attention to the soul, its passions, and its seat, Craanen
concludes his section on the different levels of sensation by addressing the connection between the soul and the body. Interestingly enough, we are told that,
properly speaking, there is no interaction between the agitation of the particles
and the nerves, on the one hand, and the ideas that are represented in the mind,
on the other:

> There is no consequence between the bodily movements and the ideas
> in the mind and the attention of the latter. Therefore, God could connect
> those things that cannot be connected, and which oppose each other in
> all respects. This which must therefore be a very firm argument for the
> existence of God.[64]

According to Craanen, the movements of the nerves and the incorporeal ideas
in the mind are ontologically distinct aspects of sensation and cannot therefore interact with each other. As a consequence of this ontological distinction,
there only remains one agent that can bring about the relation between mind
and body: God.[65] But what exactly does Craanen mean when he says that mind
and body are connected with each other by God?

A clearer formulation of this idea may be found at end of those parts of his
course that are devoted to the soul and its passions. Just as announced in the
beginning of the *Tractatus*, after having given an account of the body as such,
Craanen's classes turn to the soul considered independently of the body. Only
after this does the *Tractatus* address the conjunction between the two parts of
human nature. I shall not deal with Craanen's discussion of the soul and its
passions, which to a large extent restates the ideas that Descartes expressed in
The Passions of the Soul and in the *Meditations*; notably, that the mind is not

---

64   Craanen, *Tractatus*, 523: "Non datur consequentia inter motus corporeos et ideas mentis,
      ejusque attentionem. Deus igitur potuit connectere ea, quae connecti nequeunt, quae sibi
      in omnibus contrariantur, quod igitur debet esse solidissimum pro existentia Dei argu-
      mentum."

65   Incidentally, Craanen thinks that the fact that the correspondence between mental and
      bodily states can only be brought into existence by God amounts to a proof of God's exist-
      ence. As Craanen explains, a few lines later in the text, this argument also neutralizes the
      accusations of atheism that are often put forward against "those innovators, whom they
      call Cartesians." Craanen, *Tractatus*, 523: "Ex his et similibus possumus colligere quam
      injuriose Neotericos quos Cartesianos vocant, pro atheis habeant, cum multa solidiora
      et ex rationis fonte emanantia argumenta, pro Dei existentia copiosora proferunt, quam
      aliquis veterum unquam praestitit."

extended and that the mind is pure thought (*meram et solam cogitationem*).[66] I shall instead focus on Craanen's ideas concerning the mind-body relation and show the importance of Craanen's treatment of catalepsy in the framework of his conception of human nature.

One of the first questions to ask is how are we to interpret Craanen's above-mentioned statements concerning the role of God in joining the soul with the body? In the Nijmegen disputations, we are told that the mind is united to the body through the pineal gland. Did Craanen's Leiden lecture shift away from this conception? Or does the pineal gland still play a role in the conjunction that God produces between mind and body? If Craanen denies any direct causality between mind and body, does it even make sense to try to identify what the seat of the soul in the body is?

In the Leiden classes, Craanen does not seem to have changed his mind vis-à-vis the Nijmegen positions. If anything, his earlier ideas receive a more precise explanation in the *Tractatus*. Here, for instance, we are told that the mind cannot be in the whole body because we observe that several parts of the body can be removed whilst the mind remains intact. On this basis, Craanen concludes, the seat of the mind must be limited to the brain (something that we cannot gather from the extant Nijmegen disputations). Moreover, Craanen argues, that since several parts of the brain may also be removed without this damaging the mind, the seat of the soul must be in one specific part of the brain. At this point, Craanen sets out to prove that the soul is joined to one single part of the brain, namely the so-called conarion or pineal gland.[67] Just as in the Nijmegen disputations, Craanen argues for Descartes's idea that the pineal gland is the immediate seat of the soul in the body, relying solely on the evidence derived from the study of catalepsy.

In addition to this, however, Craanen taught his Leiden students that the concepts of 'conjunction' and 'seat of the soul' should be understood in a very specific way, introducing his discussion of catalepsy by means of the following disclaimer.

---

66   Craanen, *Tractatus*, 524.

67   Craanen, *Tractatus*, 536: "Dicemus nos in sequentibus, proprie uni parti animam esse junctam, quae particula toti juncta, at non immediate, anima omnibus membris est juncta, quia plurima membra tolli possunt, remanente tamen illaesa anima, et cogitationibus integris produrantibus; sic simulac homo vel animal detruncatur, statim cessant omnes functiones animales et motus arbitrarii, nulla sensatio superest, quod signum est in capite esse immediatam conjunctionem. [...] Non cerebri toti, quia experimur notabilem portionem cerebri tolli posse, functionibus animalibus illaesis manentibus, nec anima quiquam detrimenti inde patitur, sanantur sine aliquo incommode remanente, quod sane fieri non posse si anima esse juncta toti cerebro. Probabimus itaque in sequen-

Properly speaking, one cannot say that the conarion, or any other bodily part, is the seat of the soul. Terms such as 'seat,' 'conjunction,' 'connection,' and so on, only apply to corporeal things; in the context of Craanen's classes, they are admittedly used as a concession to the common way of speaking. Craanen refers to de La Forge's *Tractatus de mente humana*, in which three modes of conjunction are distinguished. Local conjunction can occur between bodies when they come into contact with each other. Conjunction can also happen between minds, through love. But the conjunction between mind and body can only happen through the will (*per voluntatem*), when the act of willing proceeds outside the mind.[68]

According to Craanen, however, this act of will must be qualified as a general decree through which God makes the body move, whenever the mind wills it. Just like we were told, at the end of Craanen's discussion of sensation, that no causality is involved between the corporeal and incorporeal aspects of sensation, here too we read that, strictly speaking, there is no causality between the will and bodily movements. We only know that certain actions of the body happen *at the same time* when the mind wills them.[69] Next to this is the other decree of God, through which when we are affected by external objects, the movement of the filaments of our nerves and of the animal spirits is communicated to the brain, and whenever this happens, corresponding thoughts occur—without any causal connection—in our soul. Our inquiry into the conjunction between mind and body ought therefore to be interpreted as a question concerning which part of the body God uses to produce the correspondence between the movements of the body and the thoughts of the soul.[70]

---

tibus, animam esse junctam exiguae cerebri particulae, in ejus meditullio sitae, quod conarium seu piniglandula vocant." In MS., London, MS. Sloane 1274: Craanen (professor) and Morley (student), *Dictata*, fol. 249, Craanen more specifically argues that parts of the brain can be removed without any damage to senses and movement: "Respondeo, animam non unita esse toti cerebro, qua maxima moles cerebri non inservit motui nec sensui, sed sit aliqua eius pars, in qua anima immediate remanet."

68   Craanen, *Tractatus*, 538. The relevant discussion is found in de La Forge, *Tractatus de mente humana* (Amsterdam, 1669), 96.

69   Craanen, *Tractatus*, 538–539: "Totum hocce negotium consistere in privilegio quod generali decreto fecit Deus, obsequendi voluntatibus mentis circa motus sui ergastuli, quando rite fiunt. [...] Nec putandum mentem revera facere ex se hunc vel illum membri alicujus motum ..., sed scimus hos motus tantum fieri eo ipso momento quo mens de iis cogitat." Besides in de La Forge, Craanen might have read similar ideas in the *Corporis et animae in homine conjunctio*, written by his teacher, Clauberg. However, Craanen never mentions Clauberg, and the latter's form of occasionalism, moreover, seems much less straightforward than that defended by later Cartesians, as has been shown by Schmaltz, *Early Modern Cartesianisms*, 177–181.

70   Craanen, *Tractatus*, 542: "nunc quaerendum restat quanam in parte cerebri anima resi-

Predictably, Craanen's discussion moves on to establish the identity of this part as the pineal gland. Having already explained that the soul cannot be in the whole body or in the whole brain, Craanen rules out the possibility that God uses the animal spirits as the means through which the correspondence between body and soul is orchestrated. Because the animal spirits are the main vehicle through which the mind issues movements into—and receives movements from—the body, it would not be unreasonable to think that God uses the animal spirits to operate the conjunction between mind and body. This hypothesis can only be excluded by looking at the symptoms and causes of catalepsy:

> But there is a single sickness in the nature of things that teaches that God did not preserve [things] in this way, nor did he use the spirits immediately, which [sickness] is called catalepsy.[71]

The *Tractatus* thus continues with a section *De catalepsi* (Ms. 1274 treats catalepsy in the section devoted to the soul and its seat), which sees the return of the by now familiar theories borrowed from Regius's *Praxis medica*.

Catalepsy occurs when the part of the brain that is responsible for sending out the determinations (*determinationes*) of the motion of the animal spirits to the limbs is rendered immobile. In Ms. 1274, just as in the Nijmegen disputations, we find again more or less credible reports of patients who claim to remember everything that was happening around them whilst they

---

deat, hoc est, quanam parte Deus utitur ad nostras voluntates exequendas et imperia mentis nostrae per corpus distribuenda; aut secundo, qua mediante facti Deus oriri in anima nostra speciales cogitationes ex certis actionibus." Craanen does not elaborate on whether the decree through which God coordinates mind and body applies to the single actions and passions of this or that individual or whether we have to interpret God's decree as metaphysical law, according to which certain states of the body correspond to certain states of the soul and vice versa. The second scenario seems to be suggested by the fact that Craanen qualifies God's decree as "general." Moreover, Craanen also makes sure to point out that God's general decree does not eliminate our free choice nor is God the cause of our sinful actions. On the one hand, no action can be performed that is not assisted by God ("Deus haec omnia exequatur"); on the other, all actions proceed from God insofar as they are good. But the mind is free to follow its bad intentions ("omnis actio procedit a Deo, quatenus bona est, sed mentis intentio est pessima, quod velit injuriam fieri suo proximo"); see *Tractatus*, 540. Just to what extent Craanen is successful or devotes enough attention to squaring his type of occasionalism with free will and the occurrence of evil actions in human affairs is a point that lends itself to further enquiry.

71   Craanen, *Tractatus*, 543: "at unicus est morbus in rerum natura qui docet hunc modum Deum non servasse, nec enim uti spiritibus immediate, qui dicitur catalepsis."

were immobilized by catalepsy.[72] We remember that Craanen's views in Nijmegen differ from those of Regius, insofar as they are not limited to a medical explanation of catalepsy, but they are used to determine the seat of the soul in the pineal gland. Craanen's Leiden classes differ even further from Regius's teaching than the Nijmegen disputations. Craanen now also incorporates Regius's explanation of catalepsy into the type of occasionalist understanding of the mind-body relation that I described above. Within this scheme, Regius's explanation of catalepsy becomes the only argument capable of demonstrating that God does not use the animal spirits as the seat of the soul, and that he coordinates the body and the soul through the pineal gland.

Certainly, given the absence of any direct causality between the mind and body, one may legitimately wonder why God would ever need the pineal gland to produce the correspondence between mental and bodily states. In fact, Craanen too writes that, because mind and body are mutually coordinated through a divine decree, God could have chosen to orchestrate the commerce between the two human parts in whatever way he wished. Were it not for the observation of the symptoms of catalepsy, we would be left wondering about this point. But the case of catalepsy points to the existence of a particular correspondence between whatever happens in the pineal gland (including its illnesses) and that which happens in the mind.[73]

As indicated previously, people who are affected by catalepsy stay in the exact same position in which they were before the onset of the sickness. This shows that there are sufficient spirits present in the limbs during catalepsy, else the body could not retain the position in which it was. If the soul were in the animal spirts, why, Craanen asks, given that these spirits are abundantly present in the limbs during catalepsy, would the soul be prevented from moving the limbs in which it is present?[74] Having ruled out that catalepsy corresponds to a lack of spirits in the limbs and that the soul is in the spirits, Craanen establishes the cause of catalepsy as the lack of new determinations of the motion

---

72 MS., London, MS. Sloane 1274: Craanen (professor) and Morley (student), *Dictata*, fol. 254: "Hi qui Catalepsi laborant, ut isti postea retulerunt." In the *Tractatus*, 544, we find other reports of persons who were allegedly conscious during episodes of catalepsy.

73 Craanen, *Tractatus*, 545–546: "Deum non opus abuisse conario, eumque potuisse illa efficere sine conario in corpore nostro, de hoc non facile determinare licet, nec potentiam Dei limitare videremur et sic potuisse imperia nostrae mentis non tantum distribuere per corpus, quod nunc nostrum vocamus, sed etiam per quolibet aliud, quod tamen nemo crediderit unquam. [...] Deum voluisse uti particula quadam tanquam conario ad imperia mentis exequenda, hoc catalepsi nos docet, in quo desunt solae determinations spirituum, quibus Deus utitur ad hoc vel illud membrum movendum."

74 Craanen, *Tractatus*, 544–545.

of the animal spirits in the limbs. The failure in sending out new animal spirits is in turn explained—just like it was by Regius—as the obstruction (caused by a mucous substance) of the part of the body which is responsible for determining the motion of the spirits into the limbs: i.e., the pineal gland.[75]

Whilst Regius discusses catalepsy among the other illnesses that are found in his *Praxis medica*, Craanen considers it as no ordinary illness. In the Leiden classes in which he deals with catalepsy, we are shown that just as God is the ontological item that recomposes the fracture between mind and soul—which, according to Craanen, Descartes correctly discovered—, so catalepsy is our only epistemological access to the way in which God harmonically orchestrates the correspondence between the *res cogitans* and the *res extensa*.

5   Conclusion

In the beginning of his Leiden classes, Craanen sets out to first offer an account of the healthy body, arguing that looking at a fully formed and healthy body is necessary to understand the illnesses to which a body may become susceptible. However, the contrary path applies to Craanen's explanation of the conjunction between the body and the mind. In this case, information concerning how the two parts of the human are arranged by God may only be disclosed through the study of one illness of the body, namely catalepsy.

Medical and philosophical reflections concerning this illness run through the entirety of Craanen's career: from his Nijmegen disputations, through to his mature teachings in Leiden. There, Craanen was removed from the teaching of philosophy to the chair of medicine, when the curators deemed his Cartesianism incompatible with an acceptable way of teaching philosophy. However, it was exactly through the teaching of medicine that Craanen, on the one hand, introduced his students to Descartes's ideas concerning the mind and the body and, on the other, sought to improve the parts of Descartes's doctrine that he deemed incomplete and insufficient. Craanen did this by heavily relying on Henricus Regius's medicine. From him, Craanen borrowed the idea that catalepsy is ultimately an illness affecting the pineal gland. With this idea Craanen

---

75   The idea that catalepsy (at least in the way in which Regius and Craanen explained it) is the only possible proof that the seat of the soul is in the pineal gland and that this gland should be mobile and determine the motion of the animal spirits is also clearly stated in MS., London, MS. Sloane 1274: Craanen (professor) and Morley (student), *Dictata*, fol. 252: "Glandulam pinealem esse sensorium commune illudque debere esse mobile liquet ex contemplatione morbi Catalepseos."

thought it possible to prove, more conclusively than Descartes, what he had first read in the latter's books, namely that the pineal gland is the seat of the soul. According to Craanen, were it not for the evidence provided by the pathology concerning catalepsy, Descartes's theories would still lack a conclusive proof.

All of this was taught by Craanen at a time when Descartes's views on the pineal gland were already in decline, owing largely to their lack of empirical evidence.[76] Craanen's medicine itself was criticized for the lack of an observational basis by Burchard de Volder. In his *Oratio de rationis viribus* (1698), de Volder attacked the type of philosophical speculation that was found among Craanen's pupils, and criticized what he regarded as an excessive reliance on deductive arguments in Craanen's *Oeconomia animalis* and in the *Tractatus de homine*.[77]

Some years earlier, Craanen's theory of catalepsy was more specifically targeted in Utrecht, in a *Disputatio medica de catalepsi* (1692), presided over by none other than Craanen's old foe, Gerard de Vries.

In this disputation, Craanen's views are criticized by the defendant, Adolph Schröder. He argues that the explanation of catalepsy given in Craanen's *Tractatus de homine* could make sense if the pineal gland was in fact the seat of the soul, but that such is not the case. The *Disputatio* instead defends the view that the seat of the soul, or common sense, is in fact in the spinal marrow, where all the nerves originate and the spirits are sent to our bodily machine (*machina nostra corporea*); catalepsy should therefore be explained instead as an illness during which the spinal marrow is incapable of emitting and receiving the animal spirits.

Whilst Schröder attacks the credibility of the views of Descartes and Craanen on the pineal gland, the main substance of Craanen's (and Regius's) explanation of catalepsy is retained, albeit with the small difference that the item which is affected by catalepsy is not the pineal gland, but the spinal marrow.[78]

---

76   A well-known contemporary criticism of Descartes's mistaken anatomical assumptions concerning the pineal gland is found in Niels Steensen, *Discours de Monsieur Stenon sur l'anatomie du cerveau* (Paris, 1669).

77   See Luyendijk-Elshout, "*Oeconomia animalis*," 304; Strazzoni, *Dutch Cartesianism*, 138–139. Leibniz seemed to be of a different opinion, as pointed out by Caps, *Médicins cartésiens*, 578–579. Leibniz (in a letter to Christian Huygens, of 2 March 1691) considers Craanen as the only exception to the excessively speculative tendency of the Cartesians.

78   De Vries (praes.) and Schröder (resp.), *Disputatio medica de catalepsi* (Utrecht, 1692), 10, 13: "Mentis sedem in qua sentire, percipere, meminisse, imaginari suasque functiones exercere comperitur, seu sensorium commune pro primario ratione causae radicalis affectus hujus domicilio habere, hac adducimur ratione, quod videamus mentis cum corpore

Despite its lack of an empirical basis and its reliance on what Luyendijk-Elshout called "old wives' tales," Craanen's theory about catalepsy and the pineal gland formed a very central part of the teaching of medicine and philosophy that Nijmegen and Leiden students received in the second half of the seventeenth century; this teaching, as we have seen, reached beyond the walls of these two universities and was still being discussed at the end of the seventeenth century. In his medical classes, Craanen introduced his students to central notions in Descartes's physiology and in Regius's medicine; but he also taught to a new generation of physicians that the study of pathology was indispensable to a proper understanding of Cartesian philosophy and to correct its shortcomings.

Acknowledgments

This work was supported by the NWO (Nederlandse Organisatie voor Wetenschappelijk Onderzoek) and the FWO (Fonds Wetenschappelijk Onderzoek) under grant number 326-69-001 (Samenwerking Vlaanderen-Nederland). The last phases of the realization of this volume were supported by an FWO "Senior Postdoc" grant (grant number 12ZW521N).

I would like to thank Erik-Jan Bos, Christoph Lüthy, Mattia Mantovani, Carla Rita Palmerino, as well as the two anonymous referees for their useful comments on the various drafts of this chapter. I would like to thank Christoph Lüthy and Jan Papy for their support during the realization of this volume and during the research project of which this volume is part.

commercium pro majori vel minori mali hujus gradu, plus minusve esse interceptum, hincque sensus, motusque voluntarios, modo omnes modo quosdam tantum cessare. Sensorium autem commune merito ibi collocamus, ubi mens firmissimam suam cum corpore unonem demonstrat. Hoc autem ibi fieri, ubi principium ac origo nervorum omnium ... Nervos autem hisce actionibus dicatos omnes, in cerebro et imprimis ex medulla spinali oblongata originem habere satis jam per experimenta in corporum dissectionibus facta, oculatissimi seculi nostri anatomici comprobarunt ... Renatus des Cartes una cum asseclis suis, uti sedem hujus mali in glandula cerebri pineali dicta collocant ... Quam sententiam acriter defendant Brockhuysen in Oecon. Anim. 102 p. 599 et Craanen in Tr. De Hom. XCI p. 545. Sed collabente hoc fundamento, glandulam hanc nostri affectus sedem esse, collabitur simul quiquid illo superstruitur."

CHAPTER 10

Cartesius Triumphatus: Gerard de Vries and Opposing Descartes at the University of Utrecht

Daniel Garber

The University of Utrecht holds a special place in the history of the reception of Descartes's philosophy. In the early 1640s, Descartes became involved in an explosive controversy over his views. Initially sparked by a series of disputations advanced by one of his disciples, Henricus Regius (1598–1679), it eventually brought in the theologian Gisbertus Voetius (1589–1676), Martin Schoock (1614–1669), and involved a nasty exchange of pamphlets from both sides, a condemnation by the academic senate of the University, and a legal action against Descartes initiated by the Utrecht city council.[1]

In this chapter, I will not deal with this important episode. Instead, I would like to return to the University of Utrecht about thirty or forty years later and examine the ways in which Descartes was treated by a later generation of philosophers in the university. In particular, I would like to look at how Descartes figures in the work of one of its professors of philosophy in the later part of the seventeenth century, Gerard (or Gerardus) de Vries (1648–1705).[2] De Vries was a student of Voetius, and unsurprisingly, was definitely not a Cartesian. But, at the same time, he was absolutely obsessed with Descartes and Cartesianism. A prolific writer and a popular teacher, an examination of de Vries's lectures and disputations can give us a view of how students at Utrecht in the 1670s and 1680s were taught about Descartes.

1 Gerard de Vries

Gerard de Vries was born in 1648 and died in 1705.[3] He entered the University of Utrecht in 1667, where he studied under Gisbertus Voetius, the figure who made

---

1 For excellent accounts of the disputes in Utrecht, see Verbeek, *La querelle* and Verbeek, *Descartes and the Dutch*, ch. 2.
2 On de Vries, see also Chapter 9, in this volume.
3 This biographical sketch is based on the accounts of his life in Van Bunge, "Vries"; Fournier, "Vries"; Schuurman, "Continuity and Change"; Israel, *Radical Enlightenment*, 479–480.

© DANIEL GARBER, 2023 | DOI:10.1163/9789004524897_012

Descartes's life miserable earlier in the 1640s. In 1671 de Vries was allowed to teach private courses in the university, but the French occupation of Utrecht in 1672 forced his students to leave for the University of Leiden, and de Vries followed them. However, he did not like the atmosphere at Leiden, which was strongly favourable to Cartesianism; de Vries complained bitterly about the "persecution, affronts and insults of those devoted to the Cartesian philosophy who, with great impudence endeavour to subdue and destroy the old peripatetic philosophy."[4] And so it was no surprise that in the following year, 1673, de Vries returned to Utrecht, where he spent the rest of his career until his death in 1705.

There is a wonderful account of the philosophical atmosphere in Utrecht in this period in the diary of Edmond Calamy (1671–1731), a student who arrived in Utrecht from England in 1688, and later returned to become a rather prominent non-conformist minister. Calamy reports that there were just two main professors in philosophy at the time, de Vries and Jan Luyts (1655–1721), who had been a student of de Vries.[5] Calamy described de Vries as "the chief philosophy professor."[6] He goes on to report that:

> The main differences then in the University were about the old philosophy and the new, and between the Cocceians and the Voetians. The old philosophy was chiefly adhered to by de Vries, who was a great enemy to the distinguishing principles of Descartes, and particularly his innate ideas ..."[7]

De Vries is reported to have been "very civil to the English, and free in conversing with them."[8] Calamy seems to have enjoyed very much his contact with the old professor. He reports that "in all his lectures, whether public or private, [he] was used to intermix a variety of historical passages that were entertaining."[9] Particularly noteworthy to Calamy were stories that de Vries told about his own professor, Voetius, and about Anna Maria van Schurman (1607–1678),

---

4   Molhuysen, *Bronnen*, vol. 3, 291, quoted in Israel, *Radical Enlightenment*, 479. On Cartesianism at Leiden, see Verbeek, *Descartes and the Dutch*, ch. 3, and Ruestow, *Physics*, chs. 3–5.

5   See Calamy, *An Historical Account*, vol. 1, 143.

6   Calamy, *An Historical Account*, vol. 1, 158.

7   Calamy, *An Historical Account*, vol. 1, 157.

8   Calamy, *An Historical Account*, vol. 1, 158.

9   Calamy, *An Historical Account*, vol. 1, 160.

the woman very learned in philosophy, languages, and theology, who had also known Descartes during the period of his difficulties at Utrecht.[10]

Another very sympathetic contemporary portrait of de Vries can be found in a letter that the English dissenting theologian John Shute (later John Barrington, 1st Viscount Barrington) (1678–1734) wrote to his friend John Locke (1632–1704) on 27 June 1701, recalling his teacher at Utrecht, where he studied from 1694 to 1698:

> Mr de Vries has a great deal of that chearfullness, that willingness to oblige, that easiness and compliance, that affability, that candor, that communicativeness which as it makes you, so it renders him too a mighty ornament to Learning; freeing it from the Objections of Sullenness and Pedantry; and rendring it agreable as well as usefull: And tho' I must own Sir, that these mighty qualifications want no other argument to sett the Person, that possesses 'em as high as may be in our Esteem, yett I cannot forbear to say, that the particular share Mr de Vries gives me in his affections and the mighty share he has in my best esteem adds this to their vallue in you Sir that they represent Mr de Vries to me, who I thought, before I had the honour to see you, the best Man and the best Thinker of this Age.[11]

2 Disputing at the University of Utrecht

Calamy and Shute were not alone in being charmed by de Vries. De Vries seems to have been a very popular teacher. He left behind records of an enormous number of disputations over which he presided, attesting to his popularity among students, and not only the English students. While he published some texts that are largely independent of academic disputations,[12] a great deal of de Vries's published output seems to be the record of disputations where he presided over the academic exercise of a student. The typical form of these essays associated with disputations can be illustrated by a text that I will examine in more detail below, *Dispuatio philosophica ... An philosophia Cartesiana*

---

10 On Voetius, see Calamy, *An Historical Account*, vol. 1, 161–162; on van Schurman, vol. 1, 163–165.

11 John Shute to Locke, 27 June 1701, in Locke, *The Correspondence*, vol. 7, 354. See the discussion of Shute in Schuurman, "Continuity and Change," 293.

12 See, e.g., de Vries, *Narrator confutatus* (Utrecht, 1679), *Logica compendiosa* (Utrecht, 1684), *De Renati Cartesii* Meditationibus (Utrecht, 1691).

sit Aristotelicae, prout hodie docetur, praeferenda? ["Philosophical disputation ... Whether the Cartesian philosophy is to be preferred to the Aristotelian, as it is taught today"] (1676).

The title page announces the text as an academic disputation (*"Disputatio philosophica"*), listing the title of the disputation, the president, de Vries, the name of the student and where he is from, and the date and time of the disputation. After a page of dedications to various people whom the student would like to thank, there follows a series of fourteen theses on the subject of whether the Cartesian philosophy should be preferred to the Aristotelian philosophy, as currently taught, written by de Vries, the president (*"sub praesidio"*) of the disputation. At the end, at the bottom of the last page are is a series of twelve questions or statements, designated as *"corrolaria,"* that is "corrolaries":

1.   No complete substance perishes naturally.
2.   Whether there are small vacua? For discussion.
3.   No one can properly be said to have done something unwillingly.
[...]
12.   A line is not a shape ... etc.[13]

While some of the questions may relate to the contents of the long text that preceded them, about Aristotelianism vs. Cartesianism, most seem quite unrelated to that text; indeed, most are quite unrelated to one another, a mish-mash of metaphysical, physical, mathematical and theological questions.

In the actual ceremony, one supposes that the main text was actually delivered as an oration by the president of the session, de Vries. This is confirmed by the testimony of John Locke, who attended one such ceremony under the presidency of de Vries when he was in Utrecht in April and May 1685. Locke recorded his impressions in his notebook.[14] Locke reports that at eleven o'clock on the morning of Thursday, 10 May 1685,

> [the candidate] with the professors of the University came into the great Church. [...] The professor De Vryes the promoter tooke his place in an high deske & the candidate in a lower under him & the professors their seats on both sides in the quire of the Church the place for this solemnity.

---

13   De Vries (praes.) and Broedelet (resp.), *Disputatio philosophica, qua disquiritur, An philosophia Cartesiana sit Aristotelicae, prout hodie docetur, praeferenda* (Utrecht, 1676), *corollaria*: "1. Nulla fubstantia complet interit naturaliter. 2. An dentur vacuula? Quodl[ibit]. 3. Nemo proprià dicendus est aliquid invitus fecisse. [...] 12. Linea non est figura."

14   The text is from Ms. Locke f.08, 269–271. I would like to thank Paul Schuurman for sharing his transcription of the text with me. He reports some other details of the text in Schuurman, "Continuity and Change," 292–293.

The ceremony began with some music, which "was followed by a speech of Mr de Vries." After his oration, de Vries called the candidate to the "high deske," pronounced him "*Artium Magistrum et Philosophiae Doctorem*," and gave him the symbols of his new status. More music followed, as did a speech by the candidate. The whole party marched out of the church to more music and made their way to a feast waiting for them at the house of the father of the candidate. The main point of the exercise was to bestow an academic degree on a student, in this case both a Master of Arts and Doctor of Philosophy. But at the heart of the ceremony was a speech from the president, de Vries, and one by the candidate, whose name Locke didn't record.[15] Presumably the president's speech was what corresponded to the longer text in the pamphlets de Vries published in connection with such events, but it is unclear what the candidate spoke about.

This is the general form of the individual disputations, published as individual pamphlets. But on a few occasions, de Vries published a longer text, apparently delivered over a series of disputations. And so, the *Introductio historica ad Cartesii philosophiam*, to which we will later return, is a series of seven such pamphlets published between 1683 and 1686, each corresponding to a single disputation. In this series of individual disputations, the theses are numbered successively, from one to fifty-one, and meant to form a single coherent text, while each individual pamphlet has its own series of questions for the individual candidates. In the case of the *Exercitationes rationales de Deo, divinisque perfectionibus ...* (1685), we have a series of "exercises"—in this case, on God and divine perfections, among the other topics—texts that were most likely delivered first as public lectures at such disputations. At the beginning of the book, there is a list of the fifty-nine students who participated in the ceremonies in which the texts were given publicly. (Included among them is Johannes [Jan] Luyts, later to become a professor himself at Utrecht, as Calamy attests. Luyts also contributed one of the dedicatory poems to the volume.) Ten years later, in 1695, the book was reprinted in an expanded edition with a slightly different title, but still with the listing of students who had participated in the relevant disputations.[16] The case of another of de Vries's books, *De natura*

---

15   Paul Schuurman has identified the student as Cornelius Beerninck; Schuurman, "Continuity and Change," 293.

16   The title of the 1685 version is: *Exercitationes rationales de Deo, divinisque perfectionibus. Accedunt ejusdem dissertationes de infinito; nullibitate spirituum;* [...] *In quibus passim quae de hisce philosophatur Cartesius cum rectae rationis dictamine conferuntur.* The title of the 1695 edition is: *Exercitationes rationales de Deo, divinisq[ue] perfectionibus. Nec non philosophemata miscellanea. Edito nova, ad quam, praeter alia, accedit diatriba singularis*

Dei et humanae mentis determinationes pneumatologicae (1687), is similar; it too seems to have been composed of orations originally given at academic disputations.

3   De Vries: *Philosophus liber*

De Vries's texts generally show an overwhelming preoccupation with Descartes and Cartesian philosophy. Over and over again in his writings and lectures, de Vries cites Descartes and the Cartesians to show the inadequacy of their views. Indeed, among his contemporaries, de Vries was known as the *malleus cartesianorum*, "the hammer of the Cartesians."[17] To the title of the 1685 collection of disputations, *Exercitationes rationales ...*, de Vries added: *In quibus passim quae de hisce philosophatur* CARTESIUS *cum rectae rationis dictamine conferuntur* ("In which throughout what DESCARTES philosophized about these things is compared with the dictate of right reason"). This is, indeed, what the volume is like: Descartes's views and those of the Cartesians come up again and again to be refuted. And this, indeed, is what much of his published writing is like as well. It is not surprising that he is often credited as the Latin translator of Father Gabriel Daniel's popular anti-Cartesian fantasy novel, *Voiage du monde de Descartes* (1690).[18]

We will, of course, have to say something about the specific criticisms that de Vries offered. But before turning to them, let me begin with some general comments on de Vries's own philosophical orientation, his general view of the intellectual world as it stood in his day, and what he saw as his own position.

Later we will discuss de Vries's *Introductio historica ad Cartesii philosophiam* (1683–1686) in more detail, but in the course of his discussion there, he offers

---

gemina altera, de cogitatione ipsa mente; altera, de ideis rerum innatis. The 1695 version has everything that the 1685 version has, though in somewhat revised form in places, but it also contains additional material, most notably a diatribe against innate ideas, given at the end with its own title page and separate pagination. Both contain the names of the students from whose defences the texts were extracted. One of the interesting additions to the 1695 version is an *Exercitatio* [...] *de officio philosophi circa revelata* (*Exercitationes*, 473–506) on Descartes's views on revelation derived from his *Principia philosophiae* I 76. It also survives in part as an independent pamphlet, *Exercitationis* [...] *de officio philosophi circa revelata* [...] *pars prima* (1687), wherein only *pars prima* seems to have survived. I suspect that most, if not all of the *Exercitationes* in both versions may have been published as independent pamphlets, even if they no longer survive in that form.

17   Fournier, "Vries, Gerardus de."
18   Van Bunge, "Vries, Gerard de," 1053.

an interesting account of his own understanding of the history of philosophy in his century. After bringing up Ramism, de Vries notes:

> Furthermore, concerning the strengths and weaknesses of this Ramist philosophy, this is not the place to discuss it, nor is it my intention. But I clearly didn't want to pass over a noble sect of philosophers in silence, since I was trying to sketch out the shape of philosophy such as it has presented itself in recent times in the European schools. Concerning the Patrizis, the Gorlaeuses, the Keplers, the Galileos, the Bassons, the Telesios, the Campanellas, the Boots, the Bacons, the Gilberts, the Berigards, and numerous other *novatores*, I won't add even a word, since this whole bunch never came together into a unity so as to make up a proper sect. For now, it is sufficient to have known from what I have said that for about the last fifty years, the divided kingdom of the philosophical world has contained those three sects that we have just surveyed: the Scholastics, the more liberal Peripatetics, and the Ramists. The former, having already been taught to philosophize in a more restricted way to some extent, has built its nest in the monasteries and schools associated with Rome. The second, which deserves to be called Eclectic Aristotelianism, has made a place for itself in the Protestant Academies. The last has thoroughly penetrated the gymnasia, especially in Germany with such success that it could have fought quickly with the others for dominance, had not Cartesianism, raising its head unexpectedly, suddenly perturbed matters in an altogether remarkable way.[19]

De Vries is interested here in the state of philosophy before Descartes entered. It is interesting here whom he includes, and whom he excludes. Those he

---

19   "Caeterum, de ipsius Philosophiae Ramaeae [sic] seu praestantia seu vitiis pronunciare neque loci hujus est, neque instituti. Tantum nobilem Philosophorum Sectam silentio plane praeterire non debebam, quando, qualis nuperrimè fuerit in Europaeis scholis rei Philosophicae facies, delineandam susceperam. De Patriciis, Gorlaeis, Keppleris, Galilaeis, Bassonibus, Telesiis, Campanellis, Bootiis, Verulamiis, Gilbertis, Berigardis, Novatoribus aliis innumeris, ne verbum addo; quum tot eorum nunquam coierint in unum, ut justam familiam facere potuerint. Nunc sufficiet intellexisse ex dictis, abhinc quinquaginta circiter annos orbis Philosophici divisum regnum tenuisse tres illas, quas recensuimus Sectas, Scholasticam, liberiorem illam Peripateticam, & Ramaeam. Illa, modestius nonnihil jam philosophari edocta, in Romanensium Monasteriis atque Scholis adhuc nidulabatur. Ista, quam jure merito Aristotelico-Eclecticam dixeris, in Protestantium Academiis sedem fixerat. Haec plurima, Germaniae imprimis, Gymnasia eo successu jam perreptare caeperat, ut cum caeteris de imperio contendere brevi potuisset, nisi subitò res pariter

excludes are the *novatores*: "the Patrizis, the Gorlaeuses, the Keplers, the Galileos," etc. This group made quite a splash earlier in the seventeenth century: they were the new philosophers, those who challenged Aristotelian orthodoxy, and set off on their own paths. When Descartes first came on the scene, he was numbered among the *novatores*.[20] But, as de Vries correctly notes, "this whole bunch never came together into a unity so as to make up a proper sect." The *novatores* had little in common with one another except for the fact that they all opposed Aristotle and Aristotelianism. But other than that, it was difficult to find common themes. For that reason, de Vries doesn't take them seriously: they are not a sect. But the sects that existed before Descartes were three: the old scholastic Aristotelians, which has its seat in Catholic institutions; what he calls the Eclectic Aristotelians, the kind of Aristotelianism taught in the Protestant universities; and the Ramists, popular not in the universities but in the German gymnasia. And then comes Descartes.

De Vries clearly identified with the Protestant Aristotelians, those to whom he here refers as the Eclectic Aristotelians. This is also reflected in a text he wrote somewhat earlier, *An philosophia Cartesiana sit Aristotelicae, prout hodie docetur, praeferenda?* (1676). The main theme of this disputation is the comparison between the Cartesian philosophy and the Aristotelian. But before entering into the comparison between the two, in Thesis II he addresses the question as to what he means by the Aristotelian philosophy in this context. The Cartesian philosophy, of course, considers its origin in "the noble and ingenious René Descartes." But Aristotelianism is more complicated. It is obviously named after Aristotle the Stagirite, teacher of Alexander the Great. But, de Vries writes, that by the Aristotelian philosophy he means not the philosophy which derives directly from the words of Aristotle himself, but

> [...] that which, principally grounded in an Aristotelian foundation, is everywhere widely accepted among the Reformed in their schools and academies, and which distinguished and blessed men, conquerors of the enemies of the Christian name, have embraced, and which, having changed certain things in need of alteration, many very learned men have defended.[21]

---

omnium mirum in modum perturbasset de improviso caput extollens CARTESIANISMUS ..." De Vries, *Introductio historica*, thesis XIV.

20 On the *novatores* and their significance for the period, see Garber, "Novatores."

21 "... per Philosophiam Aristotelicam non intelligo illam, quae Aristotelis verba promiscuè pro legibus habenda esse ducit ... sed eam, quae Aristotelicis potissimum innixa fundamentis, inter Reformatos in Scholis & Academiis ubique ferè recepta fuit, quamque

And so, the question addressed in the disputation becomes:

> Whether the Cartesian Philosophy, following Descartes both in its method and in its content should be preferred to the Aristotelian philosophy such as, and to the extent that, it is still taught in the Schools and Academies of the Reformed.[22]

Among other things, de Vries objected to the way in which Catholic scholasticism mixed Aristotle and divinity. In the preface to the reader in the *Exercitationes rationales de Deo, divinisque perfectionibus* (1685, 1695), de Vries emphasizes that he will carefully distinguish between Philosophy, based on reason, and Theology, based on faith. It is because this distinction was ignored that

> [...] the unhappy confusion of the one with the other formerly gave rise to that chimerical monstrosity of Scholastic Philosophy, or, indeed, what I might call Scholastic Theology, in which the domain of Aristotle was interleaved with the Divine Page.[23]

De Vries often emphasizes that though the ancients are his guide, his commitment is to follow truth wherever it goes. *An philosophia Cartesiana sit Aristotelicae, prout hodie docetur, praeferenda?* begins Thesis I with a quotation from Seneca:

> What then? Shall I not follow in the footsteps of my predecessors? I shall indeed use the old road, but if I find one that makes a shorter cut and is smoother to travel, I shall open the new road. Men who have made these

---

viri amplissimi & felicissimi hostium Christiani nominis expugnatores, amplexi sunt, & adhuc, mutatis, quae mutanda erant, quibusdam, defendunt multi viri doctissimi." De Vries (praes.) and Broedelet (resp.), *An philosophia Cartesiana*, thesis II.

22 "An Philosophia Cartesiana ..., tam in methodo quam rebus ipsis Cartesium sequens, sit Philosophia Aristotelica, qualis & quatenus in Scholis & Academiis Reformatorum adhuc docetur, praeferenda?" De Vries (praes.) and Broedelet (resp.), *An philosophia Cartesiana*, thesis II.

23 "Peperit olim infelix alterius cum altera confusio chimaeram illam monstrificam Philosophiae dicam, an Theologiae Scholasticae; in qua cum Divina Pagina divisum habebat imperium Aristoteles." This is taken from the (unpaginated) preface to the reader from the 1695 edition. In the 1685 edition, the last clause reads: "... in qua cum Aristotele divisum habebat imperium Divina Pagina." While this represents a genuine change—though relatively small—for the most part, the 1695 preface repeats the 1685 preface, with stylistic revisions in the Latin, but adds some additional material relating the additional content of the new edition. In general, I will quote from the edition of 1695.

discoveries before us are not our masters, but our guides. Truth lies open for all; it has not yet been monopolized. And there is plenty of it left even for posterity to discover.[24]

De Vries adds:

And so, we must allow the holy initiates of Philosophy that golden freedom, constraining man to this one thing, that emancipated from the opinions of his teachers, he is a slave only to nature and reason, by which man stands above the other animate creatures, but that, however, we take as guides those who came before us, who have striven to establish a testimony of their activity through their writings, and standing on what they left us, we should strive with assiduous effort in entering into the more secret parts of nature.[25]

In the preface to his *Exercitationes rationales*, he says something similar:

In all matters I have made use of the liberty of philosophizing, so that I have sought to confirm by reason what others say truly, not because it is said by others, but because it is true, and to refute by reason what others have said falsely, not because it was said by others, but because it is false.[26]

Throughout, de Vries is especially eager to contrast his attitude toward philosophy and toward truth with that of the Cartesians. First of all, he accuses Descartes and his followers of irrationally dismissing the ancient philosophers, "petulantly insulting" them, "as if they were stupid, dolts, soft in the head …"[27] And just as the more conservative Aristotelian philosophers were regularly

---

24   Seneca, *Ad Lucilium epistulae morales*, vol. 1, 240–241; quoted in de Vries (praes.) and Broedelet (resp.), *An philosophia Cartesiana*, thesis I.

25   "Ita lc. coacedenda est sacris Philosophiae initiatis aurea illa libertas, ad hoc unum hominem constringens, ut soli naturae rationique, quâ homo caeteris animantibus praestat, nulli praeceptorum opinioni emancipatus serviat; ut priores tamen, qui industriae suae testimonia monumentis literarum consignare aggressi sunt, tanquam duces suspiciamus, eorumque vestigiis insistentes assiduo conatu in secretioribus naturae intrandis, desudemus." De Vries (praes.) and Broedelet (resp.), *An philosophia Cartesiana*, thesis I.

26   "… ea per omnia usus [sum] philosophandi libertate, ut quae ab aliis vere dicta, non quia ab aliis dicta, sed quia vera, ratione confirmare; quae ab aliis falso dicta, non quia ab aliis dicta, sed quia falsa, ratione infirmare laboraverim." *Exercitationes rationales* (Utrecht, 1695), praefamen, fol.**r (the insertion in brackets is from the 1685 text).

27   "… animus lasciviâ plenus veteribus Philosophis petulanter insultet, quasi illi bardi, blenni, fungi (utor verbis Planti) fuerint …" De Vries (praes.) and Broedelet (resp.), *An philo-*

accused of mindlessly bowing to the authority of Aristotle (which, of course, de Vries denies of himself), he accuses the Cartesians of doing the same thing with respect to their master. In the preface to the *Exercitationes rationales*, de Vries complains that the "authority of the Cartesian name" is so great that any-one who doesn't hide anything "inconsistent with the meditations of the Noble Gentleman [i.e., Descartes]" is threatening their reputation. He continues:

> In short, it is just as in olden times, when no one paid attention to a philo-sopher unless all of his wisdom was set out in accordance with the views of Aristotle. Thus, though the person is changed, the story is really still the same.[28]

That said, de Vries claims not to reject Descartes's philosophy altogether. Quot-ing what Descartes wrote in a letter about Aristotle, de Vries writes:

> Therefore, what Descartes said elsewhere about himself, I might say frank-ly about myself: *How fortunate that man* (he refers to Aristotle; I refer to him) *was: whatever he wrote, whether he gave it much thought or not, is regarded by most people today as having oracular authority. So, there is nothing more I could wish for than, without departing from the truth, to be able to follow in his footsteps in all things.* Certainly, I have nowhere tried to depart from what pleases the Gentleman [i.e., Descartes], unless it seemed to me to approach closer to the truth.[29]

Descartes's point is that he happily follows Aristotle, when Aristotle is correct; de Vries's is that he happily follows Descartes under the same circumstances. It's just that it doesn't happen very often.

---

    sophia Cartesiana, thesis I. According to Lewis and Short, "fungus," literally "mushroom," was used as an insult of the order of "stupid" or "dolt."

28   "Tanta scilicet hisce diebus est Cartesiani nominis auctoritas, ut famae prodigus videri queat, suisque rebus consuluisse pessime, qui non mutam secum excolere Philosophiam satius habet, quam mutire quicquam Nobilis Viri meditationibus non consonum. Pro-rsus ut anteacta aetate nemo salutatus erat Philosophus, cujus non omnis ad Aristotelis sensum composita erat sapientia. Sic, mutatis personis, re vera eadem agitur fabula." De Vries, *Exercitationes rationales* (Utrecht, 1695), *praefamen*, n.p..

29   "Igitur quod de se alibi Cartesius, de me ipso dicam ingenue: *Cum ille homo* (Aristotelem is innuit; Ipsum ego) tam felix *extiterit, ut, quaecunque olim sive cogitans sive incogit-ans scriptitavit, hodie a plerisque pro oraculis habeantur, nihil magis optarem, quam ut a veritate non recedendo ejus vestigia in omnibus sequi possem.* Certe a Viri placitis nus-piam abscedere intendi, nisi quo propius mihi viderer accedere ad veritatem ..." De Vries,

As de Vries represents himself and his Cartesian opponents, it is he, the traditional philosopher who is open-minded and the seeker after truth, while it is Descartes who rejects any authority but his own, and the Cartesians who reject any authority but Descartes. It is worth pointing out that de Vries's openmindedness was not completely a matter of show. De Vries published a small book, *De R. Cartesii* Meditationibus *a Petro Gassendo impugnatis dissertatiuncula historico-philosophica* (1691) which included a summary and discussion of the arguments in Pierre Gassendi's *Disquisitio metaphysica* (1644), as well as many other (unflattering) discussions of Descartes and related topics. De Vries may also have had a role in the publication of an edition of Gassendi's *Disquisitio* that same year in Utrecht. I suspect, though, that his interest in Gassendi and his *Disquisitio* may have been less because of Gassendi's atomism or the apparently materialistic overtones of this text, and more to do with the fact that it was a strong critique of Descartes's philosophy.[30] De Vries also makes mention of a number of other non-Cartesian moderns, including Henry More (1614–1687), Pierre-Daniel Huet (1630–1721), and Pierre Poiret (1646–1719). What all of these have in common, of course, is that they explicitly attack Descartes.[31] De Vries was evidently very open-minded and eclectic insofar as he was quite happy to take his criticisms of Descartes not only from Aristotelian sources, but from the moderns as well. One should add, though, that there are interesting correspondences between de Vries's empiricism and his rejection of innate ideas and similar positions in Locke.[32]

       Exercitationes rationales (Utrecht, 1695), *praefamen*, n.p.. The quotation from Descartes is from Descartes to Plempius, 15 February 1638, AT I 522, CSMK III 79–80.

30   On de Vries and Gassendi, see Verbeek, "Gassendi et les Pays-Bas," 263–267.

31   There is a brief quotation from More's *Enchiridion metaphysicum* (London, 1671) in de Vries's *Exercitationes rationales* (1685), 317, though its source isn't directly cited. But in *Exercitationes rationales* (1695), 319–320 it is cited directly and quoted at some length. In both cases, de Vries uses it to refute the Cartesian claim that, properly speaking, the soul is nowhere. Both Huet, *Censura* (Paris, 1694) and Poiret, *Cogitationes rationales* (Amsterdam, 1685) are cited and quoted in the expanded letter to the reader at the beginning of the 1695 *Exercitationes rationales* and used to support de Vries's anti-Cartesianism.

32   De Vries's former student Shute also comments on the resemblance between de Vries's thought and that of Locke: "I'm persuaded you'll find a great deal of Satisfaction, in seing an effect which Truth allways has upon the minds of Men, that search for it with the same desires and the same application, in making 'em think alike in different Countrys and distant times. For you'll find Mr de Vries in many things, and some of 'em principall ones too, to entertain the same opinions, which you have advanc'd with so much strength and so much success" (Locke, *The Correspondence*, vol. 7, 353). It's not clear that Locke himself saw the connection, though. Schuurman, "Continuity and Change," 293 sees himself as exploring the connection between Locke and de Vries that Shute's letter suggests.

But even so, I think that we can see more than a little rhetoric in de Vries's comments about his Cartesian opponents, and about his own attitude toward authority and truth. It was a standard trope of the new philosophers to present themselves as impartial seekers after truth and to berate the dependence of their opponents on Aristotle, claiming that they were so blinded by Aristotle's authority that they wouldn't even consider alternative positions. And it is as much a standard trope for the Aristotelian side to retort with a similar criticism, claiming that it was *they* who sought truth, that the new philosophers were mindlessly hostile to tradition, and were interested only in fame and personal advancement. Or, as de Vries suggests, they are as bound to the authority of *their* masters as they accuse the Aristotelians of being toward the Stagirite. Both sides are equally correct—and equally wrong. Though there is no doubt that de Vries genuinely believed that he was open-minded and committed to reason and the truth, he was largely Aristotelian.

4   De Vries: *Malleus cartesianorum*

De Vries's commitment to Aristotle and Aristotelian philosophy is evident in his disputation, *An philosophia Cartesiana sit Aristotelicae, prout hodie docetur, praeferenda?* (1676), where he compares Descartes's philosophy directly with that of Aristotle, and where we see some of the specific criticisms that he has of Descartes's philosophy. The main text consists of a total of thirteen theses. The first two theses are introductory. In the first, from which we have already seen some excerpts, de Vries raises the question that he will discuss in the rest of the text. The Cartesians have become very prominent; they claim that theirs is the right way of doing philosophy, and shun the way of the Aristotelians with disgust. De Vries asks, "Is the Cartesian philosophy so many leagues more excellent than the Aristotelian?"[33]

   This is the challenge that de Vries attempts to answer. The main body of the examination takes up a number of specific issues relating to the comparison between Cartesian and Aristotelian philosophy. De Vries begins with the question of doubt. The Cartesian philosophy begins with universal doubt, "a word on everyone's lips" (Thesis III). In his *Metaphysics*, de Vries tells us:

However, while Locke may have been aware of de Vries, there is no reason to think that de Vries was aware of Locke, so far as I can tell.

33   "Sitne Cartesiana Philosophia tot parasangis Aristotelica praestantior." De Vries (praes.) and Broedelet (resp.), *An philosophia Cartesiana*, thesis I.

Aristotle lays down the firm foundations of the sciences, while he teaches that the sciences should take their origin from doubt, and then shows us the proper way of doubting.[34]

But what about Descartes? What does the *vir eruditus* recommend? Seeing that many things that we earlier thought to be true, turn out to be false, he thinks we should reject the true along with the false! De Vries comments:

O prudent philosopher! He [i.e., Descartes] is similar to a physician who, to cure a body affected with bad blood, decides to drain all of the blood completely, down to the very last drop![35]

Where, then, does Descartes leave us? In eliminating all of our previous beliefs, including those that are *per se notis*, we are deprived of the possibility of any demonstrations at all, de Vries claims (Thesis IV). And so, he concludes, "through this method, the foundation of all science is undermined."[36] And in this respect, he claims, the Aristotelian philosophy is much to be preferred to the Cartesian.

De Vries next turns to the question of the proof for the existence of God. He begins by setting out in elaborate detail and in geometrical form (with definitions, hypotheses) and axioms what he represents as "the vulgar and commonplace demonstration" for the existence of God, an argument that he attributes to the Aristotelian Philosophy (Theses V–VI). The argument in question is a straightforward first-cause argument, based on the axioms that there can be neither a progression of causes to infinity, nor can there be a circle in a series of causes (e.g., A causes B, B causes C, and C causes A). From this it follows, de Vries claims, that God exists.

De Vries then turns to the arguments for the existence of God in Descartes. He considers in succession the three arguments for the existence of God that Descartes gives in the geometrical presentation of selected arguments from the

---

34   "Aristotelis ... firma jactat fundamenta scientiarum, dum eas à dubitatione originem capere debere monet, et benè dubitandi viam praemonstrat." De Vries (praes.) and Broedelet (resp.), *An philosophia Cartesiana*, thesis III. De Vries refers here to "1.2. *Metaphysics*," probably a reference to *Metaphysica* A 2, 982ᵇ12 ff., where Aristotle suggests that philosophy begins in wonder.

35   "Prudentum Philosophum! Medico similem, qui, ut corpus maligno sanguine levet, omnem promiscuè sanguinem ad minimam usque guttam exhaurire instituit." De Vries (praes.) and Broedelet (resp.), *An philosophia Cartesiana*, thesis III.

36   "Apparet itaque per hanc methodum subrui fundamentum omnis scientiae." De Vries (praes.) and Broedelet (resp.), *An philosophia Cartesiana*, thesis IV.

Meditationes in the appendix to the *Secundae Responsiones*.[37] Considered first is the ontological argument, from the claim that necessary existence is contained in the concept of God (Thesis VII). To this, de Vries offers two responses. The first is that even though something may be part of the concept of a thing, that doesn't prove that the thing itself actually has that property. In just the same way, it doesn't follow that we can affirm that whatever is found in a picture is contained in the thing pictured, unless beforehand one has established that the thing pictured is altogether similar to the thing itself. As a consequence, de Vries claims that the proposition "is not absolutely true." The second objection is that even if the concept of a thing contains infinite perfections, it doesn't actually have them unless that thing actually exists.

The second proof that de Vries considers is the proof that proceeds from the idea one has of God: I have an idea of God as infinitely perfect, an idea that I could only have if it were caused by a being that was infinitely perfect. To this de Vries objects, first, that it isn't universally true that an idea cannot represent more perfections than the cause of the idea actually has. And secondly, he denies that we can positively conceive of God's infinite perfections (Thesis VIII).

The third argument de Vries considers is from the fact that I cannot conserve myself in existence, so there must be something else that does. Since I have ideas of the many perfections I lack, those ideas must be in the cause that conserves me. But if the cause that conserves me has sufficient power to conserve me, it has sufficient power to give itself all of those perfections, and since it strives for everything that contributes to its goodness, the conserving cause will actually have all of those perfections. Therefore, that conserving cause must be God. To this de Vries responds that that which conserves something is not necessarily endowed with the faculty for giving itself all the perfections. Food, for example, sustains us but is not perfect to the highest degree (Thesis VIII).

In summary, de Vries concludes, the Aristotelian philosophy is more suited to piety than is the Cartesian philosophy. Whereas the Aristotelian philosophy firmly proves the existence of God, the Cartesian philosophy not only fails to prove God's existence with adequate rigor, but even begins by calling the existence of God into doubt. And whoever does not know that God exists, also cannot respect his law (Thesis IX). Furthermore, de Vries claims, the Cartesians don't respect the Holy Scriptures. (Note here how we have moved from Descartes to the Cartesians; later we shall see important ways in which de Vries distinguishes Descartes from his followers.) (Thesis XI). And insofar as the

---

37   See Descartes, *Meditationes, Responsiones* II, AT VII 166–169.

Cartesian philosophy takes clear and distinct perceptions as the norm of truth, they would seem to exclude propositions such as the fact that God has known everything from eternity, yet doesn't interfere with creaturely freedom, which, one presumes, can only be known through revelation of some sort (Thesis XII).

De Vries ends the lecture with a series of objections that the Cartesians could make to the Aristotelian philosophy, with the answers that the Aristotelian could give (Thesis XIII). De Vries's conclusion: "the Cartesian philosophy is not to be preferred to the Aristotelian, as it is taught today, but the contrary."[38]

In this lecture, de Vries focused on a few issues in the Cartesian philosophy that he opposed: the universal doubt, the proof of the existence of God, and the way that the Aristotelian philosophy taught in the schools is better suited to piety than is the philosophy of Descartes. These anti-Cartesian themes are repeated, elaborated, and supplemented by other attacks on Descartes in de Vries's other writings. In the *Exercitationes rationales* of 1685, where many lectures that originally appeared connected with disputations are collected, he begins with twenty-six *exercitationes* on the subject of God, in which the philosophy of Descartes is constantly being "compared with the dictate of right reason," as the subtitle of the book informs the reader. This is followed by six further *exercitationes* and *dissertationes* on a variety of subjects in which Descartes's philosophy is constantly at issue, including infinity, whether spirits exist in places, the automata of human and animal bodies, whether God can create contradictories, the use of the senses in philosophizing, whether the human mind is thought alone, and the activity of animals. In the 1695 expanded edition of this collection, these earlier lectures are augmented by an extensive attack on innate ideas, an issue that, along with his defence of the senses, links de Vries with the philosophy of John Locke, whose *Essay Concerning Human Understanding* had come out a few years earlier in 1690, as mentioned above.

5   De Vries: The Historical Descartes and the Cartesians

So far, de Vries is not so unusual among conservative Dutch professors. He strongly favours the Aristotelian philosophy in some form or another over Descartes and the Cartesians, while at the same time he thinks of himself as open-minded and committed to following the truth. But even though he is an

---

38   "Cartesianam Philosophiam Aristotelica, prout hodie docetur, non esse praeferendam, sed contra. q.e.d." De Vries (praes.) and Broedelet (resp.), *An philosophia Cartesiana*, thesis XIII.

enthusiastic critic of Descartes, it is difficult to find much that is original and penetrating in his critiques. While he was known as the *malleus cartesianorum*, the "hammer of the Cartesians," one might suppose that it was because of the bluntness of his attacks rather than their incisiveness.

But now I would like to turn to a curious writing that doesn't seem to fit neatly into de Vries's anti-Cartesian diatribe: the *Introductio historica ad Cartesii philosophiam*, the text that initially drew my interest in de Vries. As I noted earlier, the text is actually a series of seven disputations, each published in a separate pamphlet, starting in 1683 and finishing in 1686. It is clear that these short publications are intended as a series. Each pamphlet has the same title, distinguished only by the name of the degree candidate and the part number (beginning with *Introductionis historicae ad Cartesii philosophiam pars prima* and going to *pars septima*). The theses are numbered consecutively; that is, in *pars prima* are found theses one to four, in *pars secunda*, theses five to fourteen, etc. There is a total of fifty-one theses, preceded by a preface at the beginning, and ending with a *mantissa praesidis*, an "addition from the president," that is, from de Vries, following Thesis LI. Each pamphlet has its own set of philosophical questions, following the theses, presumably to be addressed by the candidate of the day. The seven pamphlets together come to one hundred and four pages.

What did de Vries mean by his title, an historical introduction to the philosophy of Descartes? The preface is illuminating in this respect. De Vries begins by noting that

> [...] it used to be that one needed to attend to the commonplace, that *it is not who has said it, but what was said*, when advancing an opinion which was treated by others as a truth.[39]

But while de Vries agrees with this, at the same time, he thinks it is important to understand the person whose views are being examined. For this reason, de Vries thinks that, before judging new ideas, it is helpful to begin by

> [...] gathering together certain things, about the intellectual temperament [*genius*] of the author in whose brain they were conceived, about the circumstances in which they were born, about the ways in which they

---

39　"Tritum jam olim, in expedenda opinionum quae ab aliis traduntur veritate, Non quis dixerit, sed quid dictum sit, esse attendendum." De Vries, *Introductio historica, praefamen*, n.p..

were developed, and finally about the fortune that they experienced in the variety of judgments of the learned.[40]

The bulk of the text that de Vries presented, unlike much of his other work, is actually about Descartes the person, and much less about the doctrines: it is a genuine *introductio historica*.

The historical account of Descartes begins in Thesis XV, in part 3, that is, the third pamphlet. De Vries begins by distinguishing between Descartes and his followers, the Cartesians: it is Descartes himself who is largely at issue (Thesis XV). After discussing some of the influences on Descartes, and whether he should be called "Descartes" or "Cartesius," he turns to a description of the person (a *prosopographia*) (Thesis XIX), including his birth and early youth (Thesis XX), his study of medicine (Thesis XXI), his life-habits (Thesis XXII), including his habit of staying in bed long into the morning (Thesis XXII), and so on. He notes that Descartes was a very talented mathematician (Thesis XXV), and that he preferred his own thought to reading books (Thesis XXIV). There is considerable discussion of Descartes on religion, his Catholicism and his relation to Protestantism. Part 5 ends with an account of Descartes's birth and natal astrology. Parts 6 and 7 of the *Introductio historica* follow pretty closely the history of himself that Descartes presents in the *Discours de la méthode*, going from his early education at La Flèche, through his military education, and ends with the beginning of the philosophical project in universal doubt (Theses XXXVI–LI).

All of this biographical material is documented carefully, drawn from published letters of Descartes and his philosophical writings, from published biographical sketches. (Lipstorp and Clerselier are a particularly important sources, but others are consulted as well.) In general, de Vries's account of Descartes is not unflattering. He does complain about his arrogance (Thesis XXVIII) and his tendency toward anger in intellectual disagreements (Thesis XXIX). But there are many ways in which de Vries is also very complimentary. However, this curious focus on biography and the man Descartes does raise a question: how does this text fit into de Vries's larger programme? The point at which the project begins and the point where it ends can give us some hints.

---

40   "Nimirum haud rarò evinire solet, ut, ad altius introspicienda dogmatum mysteria, conducat plurimum, perspecta habere in antecessum quaedam quasi praecognita, & de genio auctoris in cujus cerebello ea concepta, & de occasionibus quibus nata, & de modis quibus educata, & denique de fortuna quam ex vario eruditorum judicio experta fuerunt." De Vries, *Introductio historica, praefamen*, n.p..

Let's begin with the ending. After the last thesis, de Vries presents a short addition to the theses, what he calls a "*mantissa*." The text begins:

> We see that Descartes, preparing himself for his doubt, anxiously has completely exempted matters of faith from the examination of reason. Whatever else he does, this is good![41]

To this, de Vries contrasts those who insist on applying reason to questions of faith:

> It wasn't so for these ones [i.e., the Cartesians] for whom: *the divinity of scripture, on which all of its authority rests, can be based only on reason,* especially *if there is no argument why we can be deceived in this reason any less than in others.*[42]

This, de Vries claims, is the position of the Cartesians,[43] which to Christian ears he finds horrifying: If faith is based on reason, then faith cannot extend any further than reason. On the other hand, de Vries argues that Descartes himself (unlike his followers) respects the distinction between reason and faith, between philosophy and theology, basing his discussion on a passage from the *Second Replies* together with some letters.[44] The Cartesians, de Vries claims, have forgotten the position of their Descartes himself, which leads them toward Socinianism and atheism. De Vries ends the *mantissa* as follows:

---

41   "Vidimus Cartesium, ad Dubitationem suam semet ipsum accingentem, sollicitè admodum à Rationis examine exceptisse res Fidei. De caetero quicquid sit, hoc benè!." De Vries, *Introductio historica, mantissa.*

42   "Non saltem sic isti, quibus *Scripturae Divinitas, quâ auctoritas ejus omnis nititur, non aliundè quam ex ratione adstrui potest.* Imprimis, *si nullam sit argumentum, cur in hac ratione minus quam in aliis falli possimus.*" De Vries, *Introductio historica, mantissa.* The quotation translated in italics is from Gisbertus Wesselus Duker, *Disputatio philosophica inauguralis de recta rationcinatione* (Franeker, 1686), thesis VIII. The *Disputatio* was originally given in Franeker and is partially reprinted in the appendix to Huber, *De concursu rationis* (Franeker, 1687), 2. (Note that the appendix is paginated separately.) Duker's original 1686 *Disputatio* is very rare, and I have not been able to see a copy. Duker's assertion caused a massive uproar in Franeker and elsewhere; de Vries's response is one of many. On the celebrated incident, see Goudriaan, "Ulrik Huber" and Bordoli, *Dio, ragione, verità.*

43   See earlier in the text, thesis XXVII, for an account of how Cartesianism came to spread so widely.

44   See AT VII 147–148. The letters from which de Vries quotes are Descartes to Hogelande (?), August 1638, AT II 348 and Descartes to Hyperaspistes, August 1641, AT III 425–426.

How much more prudently in other matters the pious diligence of our civil leaders has recently been vigilant to suppress seeds with the same qualities planted by the same hand, in grass: without such care they [i.e., the seeds] would more and more certainly grow little by little into thorny weeds (SPINOSA *lolia*), which in the end would suffocate the entire planting of Christian faith and of a better mind.[45]

I suspect here that, by seeds of the same qualities, de Vries meant those that lead to heresy, and by the same hand, that of the Cartesians. The word for "thorny," "*spinosa*," is in small caps in the text: there is no doubt in my mind that this is a reference to Benedictus. The point is clear: de Vries's target is not Descartes, of whom he approves, at least in this respect, but the Cartesians, whose views are leading us directly to Spinozism and to the undermining of Christianity. His claim is that their doctrine is not only dangerous, but that it is also unfaithful to the doctrine of their master, Descartes.[46]

This is how the text ends. But it is also important to return to the beginning of the text, and call attention to the larger context of the discussion of Descartes and his history. At the end of the preface, de Vries writes: "[...] the reason for philosophizing in our time is not understood correctly unless it is considered along with those things that there were in a previous age."[47] And so, before entering into his account of Descartes in Thesis xv, de Vries begins with a history of philosophy as it existed before Descartes entered the scene.

De Vries is particularly interested in the development of scholasticism and the penetration of Aristotle and Aristotelianism into philosophy. His account begins with the combining of Aristotle—the most durable among the ancients —with Christianity, beginning with Albertus Magnus and St. Thomas Aquinas, and from them into the medieval sects of Christian Aristotelian philosophy. De Vries continues the story with the introduction of further Greek texts into Italy with Valla, Pico, and others (Theses I–V). De Vries then notes the importance of the Reformation, with Luther and Melanchthon, the initial rejection

---

45   "Quanto prudentius alibì nuperrimè vigilavit, ut in herba oppressa fuerint ejusdem indolis, sata ab eadem manu, semina, pia Procerum diligentia: absque qua cura certò certius excrescent ista paulatim in SPINOSA lolia, universam Fidei Christianae ac melioris mentis sementem tandem suffocatura." De Vries, *Introductio historica, mantissa*.

46   The theme is discussed further in de Vries, *Exercitatio ad Cartesii* Princip. Philos. *Part. 1 Art. LXXVI. de officio philosophi circa revelata*, in the *Exercitationes rationales* (Utrecht, 1695), 473–506. As noted above, this was originally published as a series of pamphlets starting in 1687, of which only one survives.

47   "Atquì non rectius nostri temporis philosophandi ratio intelligitur, quam si cum ea quae anteactae aetatis fuit conseratur." De Vries, *Introductio historica, praefamen*.

of Aristotelianism but the eventual emergence of a Protestant scholasticism, and the introduction of Ramus and Ramism (Theses VI–XIII). In Thesis XIV, de Vries ends with the very interesting summary of where philosophy was at the moment when Descartes entered, which was discussed above. As I noted then, de Vries sees that there were three main sects of philosophy before Descartes's arrival on the scene: the scholastic Aristotelians, the eclectic Aristotelians, and the Ramists. The attention paid by de Vries to Descartes and his philosophy in the series of disputations suggests strongly that he thinks that with Descartes, we have a fourth important sect.

The overall picture that emerges is that, even if he is wrong in some— indeed many—respects, Descartes presents a genuine alternative to the accepted philosophies, one that has to be taken seriously. In this regard, Descartes himself is not to be compared with the Cartesians, who are, perhaps, theologically dangerous in a way in which Descartes himself is not. In the course of the *Introductio*, de Vries talks of Descartes with some genuine admiration, using what Ciermans had written to Descartes in a letter of March 1638 to summarize his own view:

> In a word, I agree with those things which were said to Descartes: Although *in many things, I think that one may want a little more of the truth*, however *things are such that nothing that doesn't come from the pleasure of invention is worthy of special commendation*.[48]

Unlike his followers, the Cartesians, though Descartes may be wrong in many respects, he is a worthy opponent, a philosopher who presents a serious alternative to the dominant philosophies that had preceded him. Even if his philosophy should ultimately be rejected, it has to be taken seriously.

6   Conclusion

Where then does this leave us? What does this tell us about Descartes in the classroom? Even though de Vries was an opponent of Descartes and Cartesianism, Descartes was very present in de Vries's classroom: an enormous portion

---

48   "Verbo, subscribo illi, qui sic Cartesio: Quanquam *in multis plus aliquid veritatis desiderari posse puto, talia tamen sunt, ut nulla non ab inventionis amoenitate commendationem mereantur singularem*." De Vries, *Introductio historica*, thesis XXVI. The quotation is from Ciermans to Descartes, March 1638, AT II 56. De Vries rearranged the clauses somewhat, but the meaning is not significantly altered.

of his output was directed squarely at a close examination—and refutation—of the Frenchman's views. The students who attended these disputations and, I suspect, his classes could hardly have avoided learning what Descartes had to say on a wide variety of topics. But they also learned that what Descartes said was generally wrong. De Vries's students had to learn their Descartes well in order to succeed with their master: but they had to learn his philosophy in order to refute him.

One wonders, though, to what extent de Vries's strategy may have failed: in giving his students exposure to Descartes's ideas, might he have been introducing them to doctrines that they may have ultimately found more attractive than the Aristotelian doctrines that he himself favoured? One strategy for introducing forbidden ideas is to present them under the guise of a refutation. In such a presentation, one can offer a long and detailed (and attractive) account of the ideas that are represented as being worthy of rejection, followed by a rather less extensive and less convincing account of why they should be rejected.[49] I don't mean to suggest that this is what de Vries was doing: there is no reason to doubt that he was genuinely opposed to the new philosophy of Descartes and his followers. But, at the same time, it is not out of the question that his attempt backfired, at least with some of his students, and that his obsession with Descartes may have had the unintended effect of planting the Cartesian seed in their tender minds.

Acknowledgments

I would like to thank Paul Schuurman and Theo Verbeek for their invaluable help as I worked on this chapter. I would also like to thank the two anonymous readers for their very helpful comments. Unfortunately, I was not able to address all of their questions in the final version of the chapter, but they have stimulated me to think further about the issues raised here.

---

49   The classic statement of this view is in Strauss, *Persecution and the Art of Writing*, 24–25.

CHAPTER 11

Debating Cartesian Philosophy on Both Sides of the Channel: Johannes Schuler's (1619–1674) Plea for *libertas philosophandi*

Igor Agostini

> Descartes, ce mortel dont on eut fait un dieu chez les payens.
>
> LA FONTAINE, *Discours à Madame de La Sablière*

∴

1   *Status quaestionis*

Johannes Schuler (1619–1674) is the author of two polemic works on Descartes's *Principles of Philosophy*: the *Examinis philosophiae Renati Des-cartes specimen, sive Brevis et perspicua* Principiorum philosophiae *cartesianae refutatio* (hereafter, *Examen*) and the *Exercitationes ad* Principiorum *Renati Des-cartes primam partem* (hereafter, *Exercitationes*), published respectively in Amsterdam in 1666 and in Utrecht in 1667. Very little is known of Schuler himself: the two main studies are the chapter devoted to him by Sassen in his volume on the School of Breda, and a recent entry by Henri Krop for the *Dictionary of Seventeenth and Eighteenth-Century Dutch Philosophers*.[1]

Schuler was born in Bergen op Zoom in 1619 and was probably educated in Leiden. He became a preacher in Zundert in 1643, and from 1656 onwards, he ministered in Breda where he was governor of the Latin School. In 1662, he graduated at Leiden with Jacob Golius (1596–1667) as a supervisor. The following year, Schuler was appointed Professor of Philosophy at the *Illustre School te Breda* (*Collegium Auriacum*), also called the Orange College of Breda. His main work is *Philosophia nova methodo explicata* (hereafter, *Philosophia*), published in The Hague in 1663, three years before his *Examen*.[2]

---

1   Sassen, *Het wijsgerig onderwijs*, 94–101; Krop, "Schulerus," 899–902.
2   Schuler, *Philosophia nova methodo explicata* (The Hague, 1663).

Neither Sassen nor Krop devote much attention to Schuler's *Examen* and *Exercitationes*. This is not surprising: Schuler's writings on Descartes are not well known even in Cartesian scholarship. He was nevertheless a celebrity in his time. The Orange College of Breda was a renowned institution, and Schuler was a very esteemed professor there. Moreover, as we will see, his *Philosophia* turned out to be a particularly ambitious text and, at the end of the seventeenth century, Schuler's writings on Descartes were much better known. As Boullier points out in his *Histoire de la philosophie cartésienne*,[3] they had been mentioned as a very authoritative source in 1706 in the *Vindicatio philosophiae cartesianae adversus* Censuram *Petri Danielis Huetii* by Andreas Petermann.[4] In claiming that Huet's arguments were not new, Petermann wrote that they could already be found in Gassendi, More, Parker and Schuler, and that they had already been confuted by Descartes himself, by Antoine Le Grand, and by Fabrice Le Bassecour,[5] one of Schuler's colleagues in Breda, who wrote a defence of Descartes in response to Schuler's refutations.[6] Finally, it is worth noting that both the *Examen* and the *Exercitationes* were republished in Cambridge: the *Exercitationes* in 1682 (and again in 1686), and the *Examen* in 1685.

According to existing studies on the history of the universities in England and to histories of English philosophy, Schuler's works on Descartes constitute an unmistakable proof of the presence of Cartesianism in Cambridge.[7] And I think that it should be added that the English editions of Schuler's two works on Descartes can be taken as an example of what Rosalie Colie called the "intellectual contact between England and Holland in the seventeenth century."[8] Where Schuler is concerned, however, the nature of the intellectual contact remains to be explained in more detail. Scholars have seen the re-edition in Cambridge of his two books as a reaction against the spread of Cartesianism and its penetration in the English universities. In 1668, a decree by the Jesus College Cambridge vice-chancellor Edmund Boldero established that only Aristotle's philosophy should be the basis for disputations; yet despite his injuctions against disputations on Cartesian philosophy, Cartesianism gradually infiltrated the *curriculum*.[9] A decisive role in the English dissemination of

---

3   Boullier, *Histoire de la philosophie cartésienne*, 404.
4   Armogathe, "Early German Reactions," 302.
5   Petermann, *Philosophiae cartesianae* (Leipzig, 1706), 3. Huet's *Censura* was first published in 1690.
6   Bassecour, *Defensio cartesiana* (Amsterdam, 1671).
7   Cf. Kearney, *Scholars and Gentlemen*, 151; Gascoigne, *Cambridge in the Age of the Enlightenment*, 55; Henry, "The Reception of Cartesianism," 138; Hutton, *British Philosophy*, 67.
8   Colie, *Light and Enlightenment*, ix.
9   See Hutton, *British Philosophy*, 67, 43.

Descartes's ideas was played by Antoine Le Grand (1627/8–1699), a French Franciscan from Douai, sent by the Order to England in 1656. Le Grand is certainly Descartes's most important supporter in late seventeenth-century England.[10] In London in 1671, he published a *Philosophia veterum e mente Renati Descartes, more scholastico breviter digesta*, and in 1672, an expanded version of this work appeared. This constituted a system of Descartes's philosophy for academic usage (*ad usum juventutis academicae*), and was twice reprinted (in 1675 and 1678) and later translated into English (1694).

It has been argued that the *Examen* and the *Exercitationes* were reprinted in England primarily to counteract the influence of Le Grand.[11] This is undoubted, but I think that a deeper examination of Schuler's works can allow for the emergence of a more intriguing picture. In order to see how, a crucial requirement is to first exactly define Schuler's position in relation to Cartesianism. The few historians who have paid attention to this question have given different anwers.

In his *Histoire de la philosophie cartésienne*, in the pages devoted to the School of Breda, Boullier wrote:

> [...] Founded by the Prince of Orange in 1646, the prestigious school of Breda had been Cartesian since its birth. One of Descartes's friends, M. Pollot, was appointed to the Chair of Philosophy and Mathematics. Philosophy was also taught there by Pastor Jean Schuler. Jean Schuler is enlivened by the Cartesian spirit, he reproduces most of Descartes's doctrines, but he avoids mentioning him, probably because he was afraid of the decrees of the synods.[12]

Among Schuler's works, Boullier only mentioned *Philosophia*; and his only argument for classifying Schuler as a Cartesian thinker was that the School of Breda was a Cartesian school. This last point was in part corrected in the famous study on the Cartesian Reformed Scholastic by Josef Bohatec (1876–1954),

---

10   Le Grand is also the author of "one the last known utopias of the seventeenth century"— *Scydromedia* (London, 1699)—inspired by Thomas More's *Utopia*, but characterized by the idea that Descartes's ideas could be applied to a model society: see Patrick, "*Scydromedia*," 273–278.

11   See Hutton, *British Philosophy*, 67.

12   Boullier, *Histoire de la philosophie cartésienne*, vol. 1, 256: "Fondée par le prince d'Orange, en 1646, l'École de Bréda fut cartésienne dès l'origine. C'est un ami de Descartes, M. Pollot, qui tout d'abord fut nommé à la chaire de philosophie et de mathématiques. La philosophie y fut aussi enseignée par le pasteur Jean Schuler qui a reproduit la plupart des doctrines de Descartes, mais en évitant de le nommer, sans doute par crainte des décrets de quelque synode."

showing that the School of Breda was dominated by both Cartesian scholasticism *and* Eclectic philosophy, understood (I will discuss this point in more detail below) as that kind of philosophy that rejects any set of rules grounded on the principle of authority and follows only the rule of reason. However, Bohatec completely overturned Bouillier's account on Schuler: appealing to the authority of Daniel Georg Morhof, who, in his *Polyhistor, literarius philosophicus et practicus*, had presented Schuler as a "Semi-Scholasticus Cartesianus."[13] Bohatec went further and argued that Schuler was the "only exception" to Breda's Cartesian and Eclectic framework. Indeed, he did not hesitate to label him as an anti-Cartesian ("der Anticartesianer Joannes Schulerus").[14] This image of Schuler as an anti-Cartesian thinker is still the dominant view,[15] and it is this image, arguably, which supports the idea of the publication of Schuler's pamphlets in Cambridge as an anti-Cartesian move.

However, Bohatec did not do enough justice to Morhof, who seemed rather to suggest Schuler's affinity to the Eclectic philosophy, the presence of which in Breda was correctly highlighted by Bohatec himself, as we have seen. It is only Ferdinand Sassen, in the chapter which he entirely devoted to Schuler in his fundamental book on philosophical education at the School of Breda,[16] who took the decisive step towards a more appropriate characterization of Schuler's thinking. According to Sassen, Morhof's account is fundamentally correct: between Aristotle and Descartes, Schuler goes his own way and follows the Scholastics wherever they are consistent with Descartes. His *Philosophia* is actually addressed to those who are Aristotelians and wish to delve deeper into Cartesian philosophy.[17]

Sassen shed new light on Schuler. Though overlooking the *Examinatio*, which he did not mention at all, Sassen nevertheless paid attention to the later *Exercitationes* that scholars until that moment had neglected. According to Sassen, Schuler's later criticism of Descartes should be explained as a reaction against the growing influence of Cartesianism in the Netherlands, especially in theology. These writings show how Schuler became increasingly outspoken against Descartes over time,[18] although one can describe his think-

---

13   Morhof, *Polyhistor* (Lübeck, 1714), l. i, ch. 14, 101: "An hunc Scholasticis accenseam, ignoro. Seipsum inter Scholasticos reponit, ac fuit Anti-Peripateticus, et quidem Cartesianus: Quoniam vero more tractandi singulari usus est, et ubique Scholasticos, in quibus illi a Cartesio non dissentiunt, sequitur, quasi Semi-Scholasticus nobis sit Cartesianus."

14   Bohatec, *Die cartesianische Scholastik*, 51.

15   Cf. Borghero, "Osservazioni conclusive," 259.

16   Sassen, *Het wijsgerig onderwijs*, 94–101.

17   Sassen, *Het wijsgerig onderwijs*, 98–99.

18   Sassen, *Het wijsgerig onderwijs*, 99.

ing as anti-Cartesian only in a limited sense.[19] Sassen's account is essentially endorsed by Krop, who agrees with characterizing Schuler's philosophy as constituted mostly by Eclecticism. Krop does not present Schuler as a Cartesian thinker, but points out the ways in which Descartes influenced Schuler. He insists on the fact that Schuler's criticism of Aristotle outweighed his criticism of Descartes. In sum, Schuler is "an eclectic philosopher, who is open to Descartes and the new philosophies, beside Cartesianism."[20] By contrast with Sassen, moreover, Krop does not consider Schuler's opposition to Descartes as owing to a diachronic development of his own thought.

I think that Krop's account is fundamentally correct, and that he catches the very point missed by Bohatec, who rightly argued, *pace* Boullier, that an Eclectic philosophy had taken hold in Breda, but presented Schuler as an exception not only in respect of the Cartesian philosophy, but also of the Eclectism. All this, however, requires a more detailed analysis. In what follows, I dismiss the account of Schuler's eclectism given by Sassen which seems to me still too tied to the label 'semi-Cartesian', that Krop too does not emend. I will maintain that in Schuler the Cartesian doctrines, though adopted, merge into a broader philosophical syntesis in which they not only coexist with other doctrines, also scholastics, but are also adapted to them and, therefore, modified in their genuine meaning. Once we dismiss the unconfortable label of 'semi-Cartesianism', it is easier to present Schuler as an eclectic philosopher, to the further condition that the term 'eclectism' is taken not with the pejorative sense that it has acquired nowadays, but with the meaning that it had in the seventeenth and eighteenth centuries, where it designated a philosophy claiming to be independent from any dominant influence, and aspiring, precisely, to the *libertas philosophandi*. I will also argue that Henry More should be included among Schuler's sources, given that his doctrine of the spiritual extension is clearly endorsed by Schuler in his *Examen*, where he also adopts, against Descartes, both More's theses and argument. On this basis, I will advance the hypothesis that the English edition of Schuler's writings was not merely intended to criticize Descartes's philosophy, nor intended only to oppose the rise of Le Grand. Rather, the intention was to give proof, in Cambridge, of the measurable superiority of the doctrine of spiritual extension with respect to Cartesian philosophy.

19   Sassen, *Het wijsgerig onderwijs*, 100–101.
20   Krop, "Schulerus," 901.

2   Which Kind of Eclecticism?

What must in any case be considered certain is that, despite his criticisms of Descartes, Schuler was not an anti-Cartesian philosopher. This was also obvious to his contemporaries. Johannes Tepel, in his *Historia Philosophiae Cartesianae*, published in Nuremberg in 1674, defers to Schuler's authority in arguing that, in spite of its mistakes, Cartesian philosophy had opened the way to the freedom to philosophise. Tepel cites Schuler's observation in the Preface of his *Examen*, that Descartes was right in rejecting, in the name of the freedom to philosophise, scholastic philosophy and its obscure distinctions, and in trying to replace it with a clear and distinct philosophy.[21] Though Schuler elsewhere disagrees with Descartes, it is clear that he does not struggle against him in everything.[22]

Now, in claiming that Schuler did not oppose Descartes *in omnibus*, Tepel is not just referring to the freedom to philosophise. What he is suggesting is that Schuler's positions on philosophy, in certain respects, agree with those of Descartes; rather than merely presenting Descartes as a hero of the freedom to philosophise, Schuler actually endorses some of his doctrines, too.

That Tepel is correct in this respect is clear from the very beginning of Schuler's *Philosophia*. When, in the first *exercitatio* (*De philosophia, ejus subiecto*) of the first book *Metaphysica*, Schuler claims that the object of metaphysics is the human mind as distinct from the body,[23] he is clearly echoing the subtitle of the 1642 Amsterdam edition of *Meditations on First Philosophy*, "in which are demonstrated the existence of God and the real distinction between the soul and the body," though Descartes's concept of metaphysics is much more complex.[24] Schuler also echoes Descartes by: a) identifying the soul with the mind,

---

21   Tepelius, *Historia philosophiae cartesianae* (Nuremberg, 1674), 51–52: "Ceterum ne sola mala ex Philosophia Cartesiana provenisse putetur, id boni dissimulandum non esse reor, quod ope Methodi Cartesianae via *ad libertatem philosophandi* strata sit. Equidem bene fecit CARTESIUS, inquit JOHANNES SCHULERUS Philosophus Professor in *Praef. ad Lector. Exam. Philos. Renati Des Cartes*, laudoque hominis caetera institutum, quod in scholis trivialibus vulgo receptam Philosophiam, quibus variis ineptissimisque distinctiunculis a Scholasticis in eam introductis, innumerisque erroribus oppressa, misereque dilaniata, et quasi sepulta iacebat veritas, tandem reformare quasi ab ovo aggressus sit; quodque pro jure libertatis philosophicae, occultis quibusdam Peripateticorum, atque inexplicabilibus principiis rejectis, alia quaedam magis clara atque perspicua substituerit."

22   Tepelius, *Historia philosophiae cartesianae*, 52: "Haec SCHULERUS, quamvis alias parum CARTESIO faveat, et tantum non contra illum in omnibus pugnet."

23   Schuler, *Philosophia nova methodo*, vol. 1, lib. I (*Metaphysyica*), ex. 1, 1.

24   See Descartes's famous letters to Mersenne, 11 November 1640, AT III 239: "Je n'y traite pas seulement de Dieu et de l'Ame, mais en général de toutes les premières choses qu'on peut connaître en philosophant par ordre."

that is, with the principle of thought; b) claiming that this is the most evident principle insofar as it is not subject to doubt; c) inferring the existence of a *res cogitans* as the subject of thought; d) defining the mind as a thing (*res*) *cogitans, dubitans, judicans, intelligens*; e) affirming that its essence can be known from its properties, in particular from thought (*cogitatio*); and f) adopting the principle that the more properties of a substance we know, the more perfectly we understand its nature.[25] The second *exercitatio* is also influenced by Descartess, because Schuler adopts the Cartesian principle of evidence: whatever we evidently perceive as true is actually true.[26]

However, it is also clear that Schuler aims to adapt the Cartesian principles to a more traditional ontology, grounded in the notion of 'being'. From the third *exercitatio*, indeed, Schuler's metaphysics evolves into focusing on the traditional transcendental and predicamental property of the 'being' (*aliquid, ens et essentia, conceptus entis, substantia, existentia, duratio, ubi*, etc.). This results in an ontology in which Cartesian doctrines are inserted into a more traditional philosophical framework, in particular, as I will show in more detail below, with the soul becoming the object of metaphysics insofar as it is *ens cogitans*.

This sort of contamination also takes place in rational theology. It is certain, as both Sassen and Krop observed,[27] that Schuler adopts the *Third Meditation*'s proof of God's existence from the idea of God, which he presents as the first of his own three demonstrations of God's existence.[28]

In this case, Schuler not only adopts Descartes's argument, but also shows no hesitation in appealing to his authority. Indeed, although Schuler does not mention Descartes in the *De philosophia* book on natural theology, when building up to the formulation in *Metaphysica* of the *a posteriori* proof of God's existence he had mentioned Descartes repeatedly.[29]

---

25 Schuler, *Philosophia*, vol. 1, lib. I (*Metaphysica*), ex. 1, 1–2.

26 Schuler, *Philosophia*, vol. 1, lib. I (*Metaphysica*), ex. 2, 5.

27 Krop, "Schulerus," 900: "Schuler adopts Descartes's proof from the *Meditations*, observing that finite man is unable to cause our idea of an infinite God."

28 Schuler, *Philosophia*, vol. 1, lib. II (*De Deo sive theologia naturalis*), ex. 3, 123: "Argumentum secundum pro Existentia Dei desumi potest ab Idea Dei in nobis existente. Ubi notandum *Ideam* nihil aliud nobis esse, quam formam illam, quae per intellectionem, seu cogitationem menti imprimitur, ita ut Idea se fere habeat ad suum objectum, sicut Imago in speculo, ad suum exemplar [...] Cum ergo Idea Dei in mente mea sit, concipio enim Ens simplicissimum et perfectissimum. Ergo, etiam extra mentem meam Deus realiter et formaliter existit."

29 Schuler, *Philosophia*, vol. 1, lib. I (*Metaphysica*), ex. 21, 73: "Et ex hoc argumentum etiam existentiam Dei evidenter probari posse docet Cartesius in Medit. De Phil. Med. 3. quod scilicet idea illius in mente nostra reperiatur."

Moreover, as previously noted by Sassen, Schuler also endorses Descartes's *a priori* proof.[30] More particularly, we can ascertain that the first proof of his *Philosophia* strictly follows the formulation of the *a priori* proof of Descartes's *Rationes more geometrico dispositae*, as a textual comparison clearly shows.[31] However, Schuler also gives a third proof of God's existence, based on the succession of efficient causes, which is equivalent in fact to a reproposal of Aquinas's second way on efficient causality.[32] Moreover, if Schuler's proofs of God's existence clearly reveal Descartes's influence, the treatise on God's attributes follows a much more traditional path. This is particularly evident with regard to the concept of God's independence, which Schuler formulated in a traditional negative sense,[33] without reference to Descartes's controversial notion of the *sui causa* (which had been at the centre of heated discussion in the Netherlands).[34]

In its totality, Schuler's rational theology seems to be an eclectic patchwork, with one part that was clearly influenced by Descartes (God's existence),[35] and another part of much more traditional provenance (God's attributes). This attitude is also clearly evident in book VII, *Physica*, where on the one hand Schuler resorts to Descartes's doctrines of the vortices,[36] but on the other hand

---

30   Sassen, *Het wijsgerig onderwijs*, 96.

31   Schuler, *Philosophia*, vol. 1, lib. II (*De Deo sive theologia naturalis*), ex. 3, 122: "ego quidem hanc veritatem, esse Deum adeo perspicuam esse existimo, ut nihil clarius in rerum natura cognosci queat. Ita ut, si quis diligenter attendere velit, quid sit Deus, seu, quid nomine Dei intelligitur, ex eo solum, absque ulteriori discursu, evidenter cogniturus sit, Deum existere; eritque hoc ipso non minus notum, quam duo et tria esse quinque, aut totum esse majus sua parte." Cf. Descartes, *Meditationes, Responsiones* II, *Rationes more gemetrico dispositae*, AT VII 163–164: "Ex hoc enim solo, et absque ullo discursu, cognoscent Deum existere; eritque ipsis non minus per se notum, quam numerum binarium esse parem, vel ternarium imparem, et similia."

32   Schuler, *Philosophia*, vol. 1, lib. II (*De Deo sive theologia naturalis*), ex. 3, 125.

33   Schuler, *Philosophia*, vol. 1, lib. II (*De Deo sive theologia naturalis*), ex. 7, 137: "Est ergo Independentia Dei nihil aliud quam negatio causae, qua Essentia Dei concipitur, ut a nulla causa sive in esse, sive in fieri, sive in operari depedendens."

34   See, among others, the criticism by Revius in Goudriaan, ed. and transl., *Jacobus Revius*, 56 ff.

35   A third proof of God's existence, much more traditional, is from the order of efficient causes: cf. *Philosophia*, vol. 1, lib. II (*De Deo sive theologia naturalis*), ex. 3, 125.

36   Schuler, *Philosophia*, vol. 2, lib. VII (*Physica*), ex. 17, appendix 2, 113: "In hoc Schemate Universi, quasi immensi, varii cernuntur circuli, quos Vortices appellat Cartesius." Schuler also refers to Descartes's doctrines on the movement of the moon; *Philosophia*, vol. 2, lib. VII (*Physica*), ex. 32, 117: "Hanc sententiam amplexus fuisse videtur Cartesius, eamque explicat in sua *Philosophiae Part*. IV. §. 49." On the teaching of Descartes' theory of the tides in the Southern Low Countries see Chapter 16, in this volume.

openly rejects Descartes's conception of light, when (perhaps misunderstanding the crucial Cartesian identification between light and instantaneous pressure) he claims that it is not understandable how light can involve movement.[37] Descartes is also discussed in book IV, *Psychologia*, where Schuler strongly endorses Descartes's identification of the pineal gland as the physical locus for these functions.[38]

It is therefore clear that in his *Philosophia* Schuler is far from being an anti-Cartesian philosopher: not only does he unhestiatingly endorse some Cartesian theses, but he also sometimes appeals to Descartes's authority. Nevertheless, it would also be misleading to present him as a semi-Cartesian thinker, not only because he often rejects ideas by Descartes, but also because, when he adopts, explicitly or implicitly, Cartesian theses, these are incorporated into a broader philosophical system—a very eclectic one—in which they coexist with more traditional doctrines.[39]

Moreover, once integrated, the Cartesian theses are adapted to this eclectic system. It has been argued, as we have seen, that Schuler endorses the scholastic doctrines insofar as they are in conformity with those of Descartes. In the light of the analysis of his *Philosophia* outlined above, however, I think that it is rather the contrary that should be argued: far from adopting traditional doctrines inasmuch as they are consistent with Descartes, it seems to me that, by endorsing Cartesian doctrines, Schuler adapts them to the traditional doctrines. This is, technically, what was called an *accomodatio*, in which the original meaning of Descartes's doctrine is radically modified, if not entirely lost. A few clear examples will substantiate this point.

First, Descartes's *res cogitans* is actually transformed into an *ens cogitans*: "res, aut Ens cogitans, dubitans, judicans, intelligens." This change attests not only the substitution of a transcendental property (*ens*) to another (*res*), but is a clear dismissal of thought as the essence of the *res cogitans*. While asserting with Descartes that it is possible to have access to the essence of the soul through its actions, Schuler is far from claiming that one can adequately grasp the soul's essence: this essence can be known *si non adaequate, saltem inadaequate*. In the same way, one knows that fire is a being having the capa-

---

37   Schuler, *Philosophia*, vol. 2, lib. VII (*Physica*), ex. 26, 176: "Cartesius lumen nihil aliud esse voluit, quam motum quondam particularum minimarum, seu globulorum."

38   Schuler, *Philosophia*, vol. 1, lib. IV (*Psychologia*), q. 7, 251–252: "Renatus Cartesius in cerebro glandulam dari docet, in qua Anima suas functiones specialius quam aliis partibus exerceat [...] Et revera nulla alia cerebri pars huic fini magis apta videtur."

39   For the mixed attitude towards Descartes, especially in the Ducth context, see the Chapters 2 and 10, in this volume.

city of heating on the basis of its act of heating, and that the soul is a being having the capacity of thinking from its act of thinking.[40] The *res cogitans* is no longer—as it was for Descartes—a thing whose essence consists in the thought, but is a thing whose main act is thinking. This point will be re-established in the *Exercitationes*, where Schuler writes that it is not true that the thought constitutes the essence of the mind: "As concerns Thought, what Descartes adds, namely that it constitutes the nature of the thinking substance, seems to me no less absurd, not to say worst."[41]

Also, while Schuler endorses the principle of evidence, he does so not in a genuinely Cartesian sense. While, on the one hand, he says that whatever one evidently perceives as true is actually true, he at the same time affirms that even the existence of God depends on the evidence (*ne quidem certus esse possem de existentia Dei*), which clearly contrasts with Descartes's repeated claim that the criterion of truth is grounded on God's existence and veracity. Moreover, in his *Rational Theology—De Deo sive theologia naturalis*, after formulating the *a priori* proof of God's existence according to the *Rationes more geometrico dispositae*, Schuler reformulates the argument by the traditional terminology of the *ens quo maius cogitari nequit*.[42]

Descartes's proof is here connected by Schuler with Anselm of Canterbury's famous *Prosologion* argument, despite Descartes's own insistence in the *First Replies* that his demonstration differs from it.[43] Later, and even more explicitly in the *Exercitationes*, Schuler will point out that his *a priori* proof of God's existence is neither new nor invented by Descartes.[44]

These divergences from Descartes take us closer to the account of Schuler's Eclectism given by Sassen and, partly, by Krop, which seems to me still too tied to the 'semi-Cartesian' category employed by Morhof. If Cartesian doctrines are adopted, this happens in a composite framework in which Descartes's influ-

---

40   Schuler, *Philosophia*, vol. 1, lib. I (*Metaphysica*), ex. 1 2.

41   Schuler, *Exercitationes*, q. 50, 66: "De cogitatione, quod deinceps docet Cartesius, eam substantiae cogitantis naturam constituere, non minus absurdum, ne quid gravies dicam, mihi videtur [...] Cogitatio enim cogitantis est, non vero natura aut essentia ejus rei, quae cogitat." See Landucci, *La mente in Cartesio*, 144.

42   Schuler, *Philosophia*, vol. 1, lib. II (*De Deo sive theologia naturalis*), ex. 3, 22: "Breviter autem hoc loco argumentum pro Existentiæ Dei sic proponi: *Qui perfectius Ens non est, nec cogitari potest, illud etiam actu existit: Atqui perfectius Ens Deo non est, nec cogitari potest. Ergo, Deus etiam actu existit.*"

43   Descartes, *Meditationes, Responsiones* I, AT VII 115–116.

44   Schuler, *Exercitationes*, q. 17, 24: "Constat ergo hoc argumenti genus ad existentiam Dei demonstrandam non esse novum, aut recens a Cartesio excogitatum, sed antiquum et vulgare, quamvis interpolatum, aliisve terminis propositum nunc videatur."

ence is only one of the many voices heard, and in which they receive a new meaning for the very fact of being accommodated to this eclectic system. In my opinion, it would therefore be more accurate to define Schuler's philosophy as eclectic *tout court*, provided, however, that the term 'eclectism' shall be associated not with the sense that it has acquired nowadays, but more with the meaning that it had in the seventeenth and eighteenth centuries, which is well expressed in the description of the 'eclectic philosopher' given by Kant: "an independent thinker (*selbstdenker*) who acknowledged no allegiance to any school."[45] It is in this sense that Schuler is an eclectic philosoper; he is not interested in endorsing this or that philosophy. His main interest is to discuss both traditional and new philosophies, in order to adapt them to his own philosophy, in a sort of *accomodatio* of their doctrines to his own.

In order to grasp this point, it is crucial to note that 'eclectism', far from involving the negative evaluation that one is nowadays accostumed to associate with it, was at this time considered both a model by historians and an ideal in philosophy. In the most important history of philosophy of the eighteenth century, Johann Jakob Brucker's *Historia critica philosophiae*, 'eclectism'—which Brucker sharply distinguishes from that diseased reconciliation of utterly discrepant doctrines and sentences which is 'syncretism'[46]—is the philosophy that rejects any kinds of rules grounded on the principle of authority and that follows only the rule of reason.[47] According to Brucker, the rebirth of 'eclectic philosophy' coincides with the birth of modern thought: the heroes of eclectism are the philosophers whom we nowadays situate among the great classical philosophers of early modern philosophy: Bruno, Francis Bacon, Campanella, Hobbes, Descartes, Leibniz, and Thomasius.

All this should be kept in mind when approaching seventeenth-century eclectic philosophies such as Schuler's, all the more since Schuler aimed to present a very original philosophical project, characterized, as I have shown just now, as the first attempt to give a philosophical system grounded in the liberty of freedom.

---

45   Kant, *Gesammelte Schriften*, vol. 9, 31.

46   Brucker, *Historia critica*, vol. 4.1, period. II, pars I, lib. 3, 750: "Eclectica philosophandi methodus, dudum viris prudentibus commendata, et a maximi ingenii philosophis culta cum syncretismo confundi non debet."

47   Brucker, *Historia critica*, vol. 4.2, period. III, pars II, lib. 1, 4: "Ille solus nobis eclecticus philosophus est, qui procul ire iusso omni acutoritatis, venerationis, antiquitatis, sectae, similique praeiudicio ad unam rationis connatae regulam respicit." See Donini, "History of the Concept of Eclecticism," 199.

3   The Core of the Project: The System of the *libertas philosophandi*

The frontispiece of Schuler's *Philosophia* features depictions of seven virgins and four philosophers (Figure 11.1). As the *Imago titularis* explains, the seven virgins represent the *libertas philosophandi*: they are *Meditatio, Diligentia, Lectio, Docilitas, Iudicium, Iustitia, Experientia*. Their antitheses are the seven forms of slavery (*mancipia*): *Error, Segnities, Pertinacia, Arrogantia, Invidia, Praejudicium, Servitus*. Four philosophers are depicted, too: Plato, Aristotle and Democritus alongside Descartes, "Cartesius, aliique Libertatis Philosophicae amantissimi" whose aim it is—like that of the founder of Atomism, Leucippus —for the introduction of the *libertas philosophandi* into the schools. Under the title, the layout of the town of Breda is represented "quae similiter hanc libertatem unice spirat, et sperat."

All these references must be related to a historical context. It is known that in the Netherlands the *Judicium* of the Academic Senate of Utrecht University, under the influence of Gjisbert Voetius, established, as Descartes recalls in the *Epistola ad Dinet*, that all persons in charge of the teaching of philosophy should be sure to dissent in some individual opinions, "with moderate freedom being used here, following the example of other famous Academies so that the foundations of the ancient and traditional philosophy do not fall, and strive more and more in order that the tranquility of the Academy is rigorously preserved throughout."[48] However, after the *Leges* and *Statuta* of 5 September 1643 confirmed in Utrecht the *Judicium* of 1642, the Prince of Nassau decided to ask all of the Dutch universities to express their advice on Cartesian Philosophy. His question elicited a range of very different answers.

As observed by Paul Dibon, this shows the fundamental features of the *Nederland's Beschaving*, that is, the particularism and the pluralism famously described by the great Dutch historian Johan Huizinga. On 9 October of the same year, 1642, the Academic Senate of the *Illustre School te Breda*—strongly inspired by the School founder, Frederick Henri, Prince of Orange and his claim to the *libertas philosophandi*—declared that Cartesian philosophy was neither prescribed nor proscribed in Breda.[49] The text also clarifies their position on

---

48   Descartes, *Epistola ad P. Dinet*, AT VII 593: "modica libertate dissentiendi in singularibus nonnullis opinionibus, ad aliarum celebrium Academiarum exemplum, hic usitata: ita ut veteris et receptæ philosophiæ fundamenta non labefactent, & in eo etiam atque etiam laborent, ut Academiæ tranquillitas in omnibus sarta tecta conservetur."

49   "Philosophiam cartesianam in Illustri hac Schola Celsissimi fundatoris autoritate neque prohibitam neque receptam esse," quoted in Dibon, "Scepticisme et orthodoxie," 55–81, 65. On Breda's answer to Nassau's request, see Sassen, *Het wijsgerig onderwijs*, 92–93.

FIGURE 11.1 Frontispiece of Schuler's *Philosophia*.

Aristotle's authority. Aristotle's philosophy was recommended to be taught in schools, provided, however, that this happened with the same freedom that Aristotle himself gave his disciples. Aristotle achieved much, but he did not finish everything (*multa egit, sed non omnia tamen peregit*); much remains to be done, and this must not be precluded to posterity: many things still lie hidden, waiting to be discovered in the inner shrine of nature.[50]

It is therefore clear that in Breda, in the 1640s, the *libertas philosophandi* was not only tolerated, but also encouraged. This tendency was even strenghtened in 1656 when, because of the struggle between Cartesians and anti-Cartesians, the States of Holland promulgated a decree proclaiming that philosophy and theology each had their proper sphere, and should be kept separate, and that where overlap occurred and conflicts between theological and philosophical truth arose, philosophy professors must defer to the theologians and refrain from interpreting Scripture contentiously according to their principles. In this way, while formally condemning the Cartesian abuse of philosophising in theology, the decree, also due to pressure exerted by Johan de Witt (1625–1672), Pensionary of Holland (1653–1672), actually left intact the freedom to philosophise about everything that does not impinge directly on the interpretation of Scripture and central issues of theology.[51] At this moment, as Jonathan Israel pointed out, the United Provinces became the main source of a "powerful intellectual current, basically a modified, academic Cartesianism," which spread quickly across Germany and the rest of Northern Europe.[52]

These facts help to understand Schuler's insistence on the *libertas philosophandi* which, as Tepel remarked, also included a meaningful openness towards Descartes. Schuler's *Philosophia*, in fact, begins with this leitmotiv. Schuler presents his philosophy programmatically as the first attempt at a systematic philosophy inspired by the method (*methodus*) of the *libertas philosophandi*. On the one hand, it is necessary to throw off the yoke of Aristotle;

---

50   Dibon, "Scepticisme," 65. It is possible that the professor of philosophy Henricus Bornius (1617–1675) was one of the authors of this declaration of the Academic Senate of the *Illustre School te Breda*, because this presents many similarities, even textual ones, with the speech he gave in Leiden on 13 September 1653 on the topic *De vera philosophiae libertate* for the official inauguration of the special Chair of Philosophy.

51   Israel, *Dutch Republic*, 892–895. It is by claiming to this freedom to philosophise within the limits of philosophy and to its legitimation by the States of Holland's decree, that, later, after the anonymous publication of the *Tractatus theologico-politicus*, the Cartesians will sharply distinguish their *libertas philosophandi* from Spinoza's, that they condemned as overstepping these limits and undermining the peace of the Republic. See Israel, "The Early Dutch and German Reception," 81–82.

52   Israel, *Radical Enlightenment*, 29.

on the other, not one of the *recentiores* who has abandoned Aristotle has so far been able to explain all the parts of philosophy by the method of the *libertas philosophandi*. Now, Schuler presents his *Philosophia* as a systematic reconstruction of all parts of philosophy in accordance with the method of the *libertas philosophandi*; a systematic reconstruction, in seven volumes, that Schuler claims has never been attempted by anyone before him.[53] One can now appreciate the scope of Schuler's philosophical programme. If it is false to label him as an anti-Cartesian thinker, neither is it adequate to present him as a semi-Cartesian thinker. Schuler is really an eclectic thinker, and not in a retroactive, negative, sense of the term: he pursues an independent philosophical project, a very ambitious one indeed, consisting in the attempt to present a new philosophical system, inspired by the method of the *libertas philosophandi*. And this is not the whole story. It is necessary to better understand the position which he adopts in his later Cartesian pamphlets, which certainly show a more critical attitude towards Descartes, though this attitude should not be interpreted as a retraction of an alleged earlier semi-Cartesianism, if only because, as we have seen, Schuler had never been a Cartesian.

4   The *libertas philosophandi* in the *Examen*

As it was in the *Philosophia*, the *libertas philosophandi* is also the leitmotif of the *Praefatio ad lectorem* with which Schuler opens his *Examen*.

Evoking the *Philosophia*, Schuler says that he has always been a lover of philosophical freedom as all his previous philosophical inquiries attest. And this freedom is such that he does not feel bound (*quarere mancipia*) to the opinions of others, or even to his own. He hates and has always hated the kind of men who believe in what they say as if they were oracles, and think that they are not subject to errors. And there are philosophers of a type who follow Aristotle so slavishly that they think that the philosophy of the book of nature and truth dwells only within him and do not even dare to enter the sanctuaries of nature themselves. But they are false disciples. As Aristotle himself said, *Amicus Plato, sed magis veritas*; and he, moreover, harshly criticized his predecessors, Plato included. In this way, Schuler insists that Aristotle's self-styled followers are mere *philosophastri* who contradict the very freedom to philosophise reclaimed by their own master. In the seventeenth century, however, a new light had finally shone, with the genuine way of philosophising (*genuina*

---

53   Schuler, *Philosophia, Praefatio ad lectorem*, n.p..

ista philosophandi libertas) illuminating with its rays this very dense darkness.[54] First among all these philosophers was René Descartes, a man of acute intelligence (*acutissimi ingenii philosophus*), who taught the mind to suspend assent to things of which the truth is not certain and to be free of all prejudices and servitude. In this way, he restored the priceless *libertas philosophandi*.[55]

Schuler acknowledges that he approves very much of Descartes's purpose. Descartes, indeed, abandoned the distinctions and sophistry of scholasticism, replacing them with his mathematical and mechanical principles; these appealed to many ingenious minds, who, disgusted by Aristotle, have since all become Cartesians (*universo Aristotele relicto toti cartesiani facti sint*). Unfortunately, by so doing, they jumped out of the frying pan and into the fire (*e fossa, quod ajunt, in foveam inciderunt*). Why abandon Aristotle in order to follow Descartes, and likewise never depart at all from his opinions? "Descartes," writes Schuler, "was a man, and, as a consequence, the error is not unknown to him [...] and he is no less subject to errors than Aristotle and the others."[56] Aristotle was a highly-esteemed and authoritative man, and Descartes was a great defender of the truth, but he never claimed to be beyond error. Descartes must therefore be followed not as *dominus*, but as *dux*, like any other man, and like Aristotle, who sincerely strove for truth. For Aristotle, indeed, the truth was to be preferred to all friends. Like Descartes, Schuler claims that the truth is the only gateway to philosophy.[57]

For this very reason, Schuler wants to make the students of Breda aware of the great errors (*gravissimos* [...] *errores*) which he has discovered in Descartes's philosophy. Just as enormous mistakes have derived from some of the

---

54   Schuler, *Examen, Praefatio*, n.p..

55   Schuler, *Examen, Praefatio*, n.p.: "Huic autem rei inter cateros, quod nominare nihil refert, imprimis operam navavit Vir nobilissimus Renatus Des-cartes, acutissimi ingenii Philosophus, qui tantisper moderari ac cohibere nos mentis assensum ab iis in rebus Philosophicis voluit, de quorum veritate non certo constat, et mentem nostram ab omnibus praejudiciis omnino liberandam esse, turpissimumque servitutis jugum esse deponendum diligenter exhortatus est; quo ita tandem postliminio inaestimabilem illam philosophandi libertatem restauraret."

56   Schuler, *Examen, Praefatio*, n.p.: "At vero et Cartesius homo fuit, neque proinde et errare ab eo alienum [...] adeoque non minus quam Aristotelem aliosque erroribus obnoxium."

57   Schuler, *Examen, Praefatio*, n.p.: "Fuerit profecto Aristoteles vir omni laude et authoritate potens, fuerit et Cartesius libertatis in Philosophando acerrimus vindex et assertor, numquam tamen id confeceris, ut extra errandum sphaeram constitutus sit. Sequamur ergo nos non ut Dominos, sed ut Duces et Aristotelem et Cartesium, omnesque alios, quotquot sunt veritatis vere studiosi, sed usque ad aras. Ut enim Aristoteles veritatem omnibus amicis praeferendam, ita et nos cum Cartesio veritatem unicam in Philosophiae solio omnibus postabitis collocandam censemus."

false assumptions of ancient philosophy, so mistakes no less serious—and very dangerous for young students—have come from some false Cartesian principles. If the first principles of natural things are ignored, the whole understanding of these things is compromised, because small errors in the beginning become magnified at the end. According to Schuler, Descartes's primary error is the identification of the body with extension, because from this false hypothesis many errors are immediately derived. Since extension, for Descartes, means not to be contained in an indivisible point but to be diffused outside of it, he then concludes that no sensible quality belongs to the essence of the body. The problematic implication of this is that the body can exist without sensible qualities; so that, finally, the body vanishes into nothingness. As neither the place nor the space is a point, his questionable argument is that extension is identical to body and that, therefore, the void is impossible, even for the absolute power of God, since it is contradictory that something that is neither body nor matter is extended. From this, he also concludes that there are no indivisible atoms, and that the world is infinite, or rather indefinite, since any conceivable limit must be imagined as something extended, which necessarily entails the inclusion of body and being in its concept. Finally, he also concludes that God cannot create other worlds and other matter. These errors are committed at the very beginning of philosophy (*in ipso Philosophiae limine*) and are so serious that it is necessary to deal with them immediately and in the strongest possible way.[58]

Of course, Descartes was right in rejecting the errors of Scholastic philosophy in the name of the *libertas philosophandi*, and to replace them with some clearer and more distinct principles. But Cartesians were wrong in clinging so much to their mentor and trusting him as an oracle.[59] *Amicus Aristoteles, Amicus Descartes* but *usque ad aras*, because the truth cannot be found in the narrow human brain, but only in the great book of nature. One should not hesitate to move away from Descartes when the truth of his opinion is unclear, but, by the same measure, one could embrace his opinions when they are true.[60]

It is clear that Schuler does not present himself in the *Examen* as an anti-Cartesian philosopher—and indeed, he is not—but as a philosopher inspired

---

58   Schuler, *Examen, Praefatio*, n.p..
59   Schuler, *Examen, Praefatio*, n.p.: "Male faciunt Cartesiani, quod usque adeo se suo ipsi astrinxerint Praeceptori, ut vel quaevis ipsius dicta oracula putent."
60   Schuler, *Examen, Praefatio*, n.p.: "Hoc solum peto, ut libere mihi recedere liceat a Cartesio, ubi ipsius dicta veritati haud satis consentanea fuisse percepero, quemadmodum et contra me eadem amplexurum sancte recipio, quotiescunque veritatem iis contineri intellexero."

by the *libertas philosophandi* who criticizes Descartes in the name of this same freedom to philosophise. As Schuler said it himself, he distances himself from Descartes on the points in which Descartes has distanced himself from the truth. This picture is confirmed by Schuler's further discussion of the passage related to the Cartesian identification of matter and extension.

5   The *falsa hypothesis* (Identity between Matter and Extension) and
         Its Wrong Consequences

The *Examen* consists of 29 chapters, which cover the most important questions of the first part and some questions of the second part of the *Principles of Philosophy*. The fundamental point of Schuler's analysis is that all negative consequences of Descartes's philosophy arise from the identity between matter and extension. Schuler's dissertation therefore consists mainly of the four topics to which he draws our attention in the *Praefatio*: 1) Descartes's *falsa hypothesis*, namely the identity between body and extension; and its false consequences, that is: 2) the impossibility of the void; 2) the negation of atoms; and 3) the infinity of the world.

The identity between body and extension is discussed in the crucial chapter IV, *De natura, seu essentia corporis in generali*.

Schuler begins by pointing out that the difficulty does not lie in the fact that the human mind cannot know the essence of things. The Cartesian identity between body and extension does not require an adequate knowledge of individual and specific nature, but only a general concept.[61] Now, Descartes correctly teaches that this general concept does not consist in the fact that the body is a hard or a heavy, or a coloured thing. And he also correctly acknowledges that it is impossible for the body to have no qualities: in fact, the body always remains a sensible thing, and the object of our sensations is called quality. But, in spite of his own claim, he wrongly says that nothing which affects the senses belongs to the nature of the body or to the body's essence: indeed, although affecting all senses does not belong to the body's essence, it is nevertheless certain that the senses must be affected by bodies.[62]

However, Descartes is wrong not only in denying that the sensibility belongs to the body's essence, but also in affirming that this essence consists in extension. In fact, if extension belongs to the body's essence, it remains unexplained

---

61   Schuler, *Examen*, ch. 4, 8–9.
62   Schuler, *Examen*, ch. 4, 9.

how what mathematicians claim that a 'line' is a body: as a matter of fact, the line has only one dimension and not three, and nevertheless, it cannot in any way be deemed immaterial. An analogous problem arises if one asks if those three dimensions are divisible or mutually indivisible parts of the body.[63]

Finally, Schuler argues that the body's essence does not consist in extension, because extension can also be attributed to space, place and time, and in a broad sense, even to spirits. Anyone will easily admit, he insists, that spirits, angels and rational souls exist in an indivisible point.[64] One wonders how material or corporeal extension differs from the nature of these immaterial beings. The difference seems to be constituted by impenetrability, that is, the property by which a body prevents the existence of another body in the same space. It is not impossible that two or more spirits exist in the same space; and they can also exist *coextensive* in the same place with the body, also because both reason and faith (Jeremiah 23:24) teach that God is immense, intimately present everywhere, and that he fills the spaces of all bodies. Even the rational soul, according to common opinion, is coextensive with the whole human body. But if, on the one hand, it is not impossible that two or more spiritual substances exist together in the same place, on the other hand it is impossible that two bodies exist together in the same place. In short, it has thus far not been possible to identify any property more essential (*proprietas magis essentialis*) than impenetrability whereby the body or bodily matter differs from other kinds of extension.

It is therefore necessary to replace the Cartesian definition of body as extension with another definition: the body as a material being (*ens materiale*).[65] In this definiton, the genus is being and the specific difference is matter. Now, matter includes sensibility and impenetrability. Sensibility is what makes the body perceptible by some sense; impenetrability removes the coexistence of another body in the same place (*coëxistentiam alterius corporis in eodem loco tollit*).[66]

63   Schuler, *Examen*, ch. 4, 10.

64   Schuler, *Examen*, ch. 4, 10–11: "Denique, in extensione Corporis praecise et generatim sumpti essentiam non consistere ex eo constat, quod etiam spatio, loco et tempori; imo et spiritibus quodammodo, saltem in latiori significatu, attribui possit, (neque enim spiritus, Angelos, et Animas rationales, in puncto indivisibili existere facile quisquam admiserit)."

65   Schuler, *Examen*, ch. 4, 11. Schuler argues that the genus, in this definition, is the Being and not the Substance, for the reasons alleged in *Philosophia*, vol. 1, lib. 1 (*Metaphysica*), ex. 4, 19. Here Schuler contends that Being is twofold, namely, material and immaterial: the one is spirit, the other is body. The difference, rather, between Being and Substance is materiality, and this is what distinguishes a body from all that is non-body and spirit.

66   Schuler, *Examen*, ch. 4, 11.

The criticism advanced by Schuler against the identity between matter and extension does not prevent him from fully accepting Descartes's explanation of condensation and rarefaction, which he considers as conforming to the account of the ancient philosophers against Aristotle.[67] What he cannot accept at all is the idea of the impossibility of the void. This is the first of the erroneous consequences, unacceptable to Schuler, which derives from the identity between matter and extension, and the immensity of God. In fact, as God is infinite and immense, even assuming that there is no corporeal substance in space, there will still at least always be God and, therefore, a being filling space.[68]

This leads to the question of the relationship between the void and material bodies. Schuler claims there is the possibility of a space in which there is no body. Certainly, this is not possible according to the ordinary course of nature. Nevertheless, it has not yet been demonstrated by anyone that the void cannot be produced beyond (or against) the ordinary course of nature, either by a creature or by God. Moreover, as the world has not existed for all eternity, it is necessary that the space in which the world now exists was empty of all bodies before the existence of the world.[69]

Schuler then proceeds to refute Descartes's arguments against the void, which he qualifies as *futilia et inepta*.[70] He dwells at length on Descartes's disputed claim that even God cannot take away every single body contained in the jar without allowing any other body to take their place. According to Descartes, this would be impossibile because, in that case, the sides of the vessel would be in contact; when there is nothing between two bodies, they must touch each other. On these terms, there is an obvious contradiction that there is a distance between the two bodies and, at the same time, this distance is nothing; for every distance is a mode of extension, and therefore it cannot exist without an extended substance. But Descartes is wrong in not distinguishing the different kinds of non-being: in fact, as concerns the negations and the privations, although they are non-beings, they cannot be considered nothing. And this is the case for distance, which is a negation or a privation, insofar as it is a negation of contiguity.[71]

---

67   Schuler only moves away from Descartes in the attribution of the rarefaction to the air, arguing against the Cartesian explanatory model of the sponge; see Schuler, *Examen*, ch. 5, 13.
68   Schuler, *Examen*, ch. 10, 37.
69   Schuler, *Examen*, ch. 10, 37.
70   Schuler, *Examen*, ch. 10,39.
71   Schuler, *Examen*, ch. 10, 42.

The purpose of Schuler's criticisms of Descartes, in fact, is to restore atomism. Chapter 12 marks a crucial point in this task, arguing for the existence of atoms on the basis of the impossibility of an infinite number of particles.[72] According to Schuler, there is nothing truer in philosophy than the doctrine of atoms, not only because of the many errors that depend on their denial, but also because the way to a better knowledge of natural things is precluded by the denial of atoms. In focusing on Descartes's arguments against atoms, one of Schuler's main objections is that atoms, absolutely speaking, are certainly corporeal, but not extended (*nego absolute atomos extensas esse*).[73] The atoms against which Descartes argues are extended, and extension should be infinitely divisible. He appeals to God's infinite power in support of his argument. If it was God's will to create atoms and to further divide the indivisible, by his not having the power to do so he would not be omnipotent.[74] By conceiving of atoms as corporeal but not extended, however, this argument of Descartes that if atoms were a given, the power of God would be finished, does not work.

The main assertion in favour of the existence of atoms, based on the impossibility of a division *ad infinitum*, anticipates the discussion of his chapter 13 in which Schuler proves, against Descartes, that there is no infinite world. Among the arguments that Schuler puts forward is one according to which, if the world were infinite, there would neither be first nor last parts, and there would be no centre of the universe.[75]

According to Schuler, moreover, the Cartesian category of the indefinite is meaningless because between infinite and finite, in fact, *tertium non datur*.[76] I will return to this point soon. In its entirety, Schuler's examination leads to the overturning of the main theses of the first part of the *Principles of Philosophy* and their foundation, that is, the identity between matter and extension. Schuler neither says his analysis is anti-Cartesian, nor does he present it in this way; nevertheless, he undoubtedly observes that he has upset the foundations

---

72   Schuler, *Examen*, ch. 12, 44.
73   Schuler, *Examen*, ch. 12, 46.
74   Schuler, *Examen*, ch. 12, 47.
75   Schuler, *Examen*, ch. 13, 48.
76   Schuler, *Examen*, ch. 13, 49: "Sed videtur nescio quid subindicare Cartesius particula *indefinite*: Neque enim corpus *infinite*, sed indefinite extensum iis contineri spatiis asserit. Quid autem hoc rei est, *indefinite extensum*? Vel enim illud (*indefinite extensum*) infinitum est, vel non infinitum: Si infinitum, corpus ergo revera infinitum erit, quod impossibile, ut jam ostensum est. Si non finitum, ergo finitum. Quorsum ergo nos novo hoc suo vocabulo lusit Cartesius [...] Quod enim non est infinitum et terminorum espers, finitum est [...] Frustra ergo inter *infinitum* et *indefinitum* distinguere velle videtur Cartesius."

of Cartesian philosophy.[77] One year later, the *Exercitationes* will reinforce his criticism of Cartesian philosophy with an analysis of the first part of the *Principles*.

6   The *Exercitationes*

The *Exercitationes* are presented as some 'annotationes' to the first part of the *Principles*. Like the *Examen*, they in fact contain some radical objections to several of Descartes's main theses. In the *Dedicatio*, Schuler explains that, while the *Examen* was still in the process of being published, he had also completed the *Exercitationes*. He decided not to publish the latter at the same time but instead to await the reactions of readers of the *Examen*. In response to any reaction, he could have been able to correct his text. As no answers arrived, the *Exercitationes* could finally be published.

Schuler criticizes, first of all, Cartesian scepticism, by saying that one must not doubt everything,[78] especially not mathematical truths,[79] and that one must not consider false what is doubtful.[80] Then, he observes that the *cogito* is not the first knowledge because it presupposes the knowledge of thought, being and other notions.[81] Moreover, the existence of the *ego* can be demonstrated by any action, not only by thought.[82] Schuler also criticizes the Cartesian negation of final causes, rejecting as false the idea that philosophy must not deal with final causes.[83] The category of the indefinite is not admissible (*tertium non datur* between infinite and finite), and the world is finite. Descartes's suggestion that judgement is the work of the will is also rejected as false, because judgment is the work of the intellect.[84] Descartes does not even give a good explanation of what he proposes, nor establish the clarity of the distinction.[85] Moreover, the Cartesian classification of the *entia* in *Principles* I 48, is defective,[86] where again, the attribute allows for knowledge only of the exist-

---

77   Schuler, *Exercitationes*, *Dedicatio*, n.p.: "Fundamenta Lib. II, Phil. Cartes. posita, subvertisse existimo."
78   Schuler, *Exercitationes*, q. 1, 1–2.
79   Schuler, *Exercitationes*, q. 4, 5.
80   Schuler, *Exercitationes*, q. 2, 2–3.
81   Schuler, *Exercitationes*, q. 6, 7–8.
82   Schuler, *Exercitationes*, q. 8, 9–10.
83   Schuler, *Exercitationes*, q. 26, 34.
84   Schuler, *Exercitationes*, q. 29, 38–39.
85   Schuler, *Exercitationes*, q. 42, 53–55.
86   Schuler, *Exercitationes*, q. 43, 57–68.

ence and not the essence of the substance, and Descartes once more confuses the property with the essence.[87] The essence of the soul does not consist in thought, and the essence of the body does not consist in extension.[88] In addition, within Descartes conception of these principles, the idea of God cannot be clear and distinct.[89]

At the same time, Schuler offers more than criticism alone. As a matter of fact, he openly endorses many of Descartes's theses, some of which he had never accepted before. As regards the mind, he approves of the Cartesian doctrine that the mind can be known more easily than the body.[90] Furthermore, he embraces not only the *a priori* proof[91] and the first *a posteriori* proof of the existence of God, as in the *Philosophia*,[92] but he frankly declares that he accepts also the second *a posteriori* proof.[93] He also affirms the thesis according to which, through the idea of God, it is possible to know not only his existence, but also his nature,[94] and that our knowledge of God is clearer and more distinct than our knowledge of any creature.[95] Finally, he agrees with Descartes on the thesis that everything has been prearranged by God.[96]

Accordingly, not even in the *Exercitationes* is it possible to consider Schuler an anti-Cartesian philosopher. Instead, the *Exercitationes* confirm Schuler as a systematic thinker: the discussion of Cartesian philosophy is indeed a pretext for him to present his own philosophical system and to show its superiority. This is what Schuler does in *quaestio* XXIII, for example. As it is impossible, he claims, to think of anything other than being, metaphysics is the first among all the sciences; then, according to the division of being into material and immaterial, metaphysics can be divided into somatic and pneumatic, and the latter can be divided into theology, angelology and psychology. Psychology is divided into logic and aesthetics, which have as their object two faculties of the human soul, namely, the intellect and the will. By contrast, as the body can be considered in both natural and artificial forms, it is, on the one hand, the object of physics, and on the other, the object of mechanics, mathematics and medicine.[97]

---

87   Schuler, *Exercitationes*, q. 49, 65.
88   Schuler, *Exercitationes*, q. 50, 66.
89   Schuler, *Exercitationes*, q. 51, 67.
90   Schuler, *Exercitationes*, q. 10, 13–14.
91   Schuler, *Exercitationes*, q. 12, 16–17.
92   Schuler, *Exercitationes*, q. 15, 21.
93   Schuler, *Exercitationes*, q. 19. p. 26.
94   Schuler, *Exercitationes*, q. 20, 27.
95   Schuler, *Exercitationes*, q. 16, 22.
96   Schuler, *Exercitationes*, q. 38, 49–50.
97   Schuler, *Exercitationes*, q. 23, 30–31.

7   Conclusion

The evolution of Schuler's reflections on Descartes, from the latter's *Philo-sophia* up to the later Cartesian pamphlets, cannot in any way be characterized in terms of an abandonment of a Cartesianism initially endorsed, because: a) Schuler never was a Cartesian; b) his two works on the *Principles* are not anti-Cartesian since, despite containing strong criticisms of Descartes, they also adopt several Cartesian theses; and c) Schuler's discussion of Descartes is always approached from Schuler's own philosophical perspective. Of course, there is a certain accentuation of an anti-Cartesian tone in Schuler's later writings, but there is no substantial philosophical change in Schuler's reaction to Cartesian philosophy. I think that it would not be inappropriate to liken Schuler's attitude towards Descartes to the concurrent development in the thinking of Henry More in Cambridge, the philosopher who introduced Cartesian philosophy to England.

Even more interesting is the fact that this comparison is not merely extrinsic. A close reading of the *Examen* suggest a strict similarity between Schuler's objections and those that More raised in his correspondence with Descartes in 1648–1649. There are at least three points which prove this. First, consider the object of the objections. Not only are Schuler's objections addressed, exactly like More's, against the second part of the *Principles of Philosophy*, but they focus on the very same points addressed in the four main objections of More's first letter to Descartes: identity of matter and extension; void; atoms; the infin-ity of the world.[98] Secondly, the main claims advanced by Schuler are often the same as those defended by More in his correspondence: the identity between extension and matter is false, since the essence of matter consists in sensibility and impenetrability and extension also belongs to spirits; the Cartesian explan-ation of condensation and rarefaction is true (even if its foundation—identity between matter and extension—is false); the existence of the void and the existence of atoms are not mutually exclusive or contradictory; and, the world is not infinite.[99] Thirdly, there are the arguments. Those used by Schuler to oppose Descartes are often the same as those used by More, as a few examples are sufficient to show: the argument for the identity between matter and sens-ibility (although affecting all senses does not belong to the body's essence, it is nevertheless certain that the senses must be affected by bodies); the arguments against the negation of void (it is possible a space in which there is no body)

---

98   See More to Descartes, 11 December 1648, AT V 236–246, *Lettere* 672, 2592–2603.
99   See, respectively, More to Descartes, 11 December 1648, AT V 238–240, 240–241, 241–242, 242–243, *Lettere* 672, 2596–2598, 2598, 2598, 2598–2560.

and of atoms (if God could not create atoms, he would not be omnipotent); and the argument against the indefinite world (between infinite and finite *tertium non datur*). All clearly echo the arguments used by More in his objections.[100]

In addition to these three particular points, there is a general reason supporting the claim that Schuler's arguments are inspired—directly or indirectly—by those of More. It is well known that all the theses addressed by More against Descartes in his four main objections are actually grounded in the doctrine of spiritual extension, the central issue of More's metaphysics.[101] Now, as we have seen, Schuler strongly endorses this doctrine: extension belongs not only to matter, but also to space, place and time, and in a broader sense, even to spirits. In order to fully appreciate this point, it could be instructive to compare Schuler's attitude towards Descartes with those of other contemporary anti-Cartesian philosophers, like Huet, for example. In his *Censura*, Huet, too, rejects the identity between body and extension, but he never grounds his criticism in the doctrine of spiritual extension.[102] Schuler's strategy must therefore be regarded as distinct from Huet's criticism, a point clearly missed by Petermann in his *Vindicatio*, when, as we have seen, he claims that Huet's arguments were not new, and that they could already be found in Schuler.

In light of this, it is also possibile that More's letters constitute a source for Schuler's objections against Descartes. First, Schuler could have read them in the first volume of Clerselier's *Lettres*, which had been published in 1657. It is well known, moreover, that manuscripts of Descartes' letters circulated, including those belonging to the correspondence with More,[103] so that Schuler may not even have required a reading of Clerselier's edition. Whilst we can cautiously suggest that More's letters constitute a source for Schuler's objections,

---

100   See, respectively, More to Descartes, 11 December 1648, AT V 239, 241, 241–242, 242, *Lettere* 672, 2598.

101   In criticizing Descartes's definition of matter, More claims he is certain that the main theorems of Cartesian physics could stand even if such a definition is rejected, insofar as their truth is independent from the identification of matter with extension; More to Descartes, 11 December 1648, AT V 238, *Lettere* 672, 2594. Descartes will soon express his disagreement on this point; Descartes to More, 5 February 1649, AT V 275, *Lettere* 677, 2622. The development of the correspondence clearly shows that Descartes is right in his claim; see Agostini, "Descartes and More," 5–19.

102   Huet's strategy is grounded on the difference between positive properties (belonging to the body), and negative properties (not belonging to the body); see *Censura*, ch. 4, 192. There is no trace here of the doctrine of the spiritual extension. And yet, not only Huet knew very well More's correspondence with Descartes, but he explicitly follows More in challenging Descartes's claim that the essence of the body does not consist in impenetrability; see Huet, *Censura*, ch. 4, 191–192.

103   See Gabbey, in AT V 669 (Appendix II).

the English edition of Schuler's Cartesian works certainly represented, in fact, a bringing back to England—from where it originated—of a critical reception of Cartesian ideas which had spread to the Continent from More.

Now, if this is true, the question of the reasons for the republication of Schuler's works in England should be reconsidered within a more complex perspective. It seems quite possible to me that the republication was designed to oppose rampant Cartesianism, and in particular, Le Grand's influence. However, also seems reasonable to advance the hypothesis that the English edition of Schuler's Cartesian writings was not merely intended to criticize Descartes's philosophy, but to repropose, and so to support, the criticism advanced against Descartes in More's correspondence, and to show the superiority of the doctrine of spiritual extension in respect of Cartesian principles.

Is it possible to go beyond this hypothesis and suppose a direct role played by More in the Cambridge edition of Schuler's pamphlets? Unfortunately, the available documents do not allow us to confirm this. What is certain, however, is that Schuler's writings were published in Cambridge in the middle of those crucial fifteen years in which More was developing both his doctrine of spiritual extension and his criticism of Descartes—from 1655 onwards, in the *Antidote against Atheism*,[104] through the *Immortality of the Soul* (1659)—until these ideas were fully recapitulated in his *Enchiridium metaphysicum* (1671).

Ackowledgments

I want to thank Erik-Jan Bos, Sarah Hutton, Theo Verbeek and the two anonymous referees for their helpful suggestions.

104   Gabbey, *"Philosophia cartesiana triumphata,"* 202–203.

CHAPTER 12

Descartes by Letter—Teaching Cartesianism in Mid-Seventeenth-Century Cambridge: Henry More, Thomas Clarke and Anne Conway

Sarah Hutton

I think it is the most sober and faithful advice that can be offered to the Christian World, that they would encourage the reading of Des-Cartes in all publick Schools or Universities. That the Students of Philosophy may be thoroughly exercised in the just extent of the Mechanical powers of Matter, how farre they will reach, and where they fall short. Which will be the best assistance to Religion that Reason and the Knowledge of Nature can afford. For by this means such as are intended to serve the Church will be armed betimes with sufficient strength to grapple with their proudest Deriders or Opposers. Whenas for want of this we see how liable they are to be contemned and born down by every bold, though weak, pretender to the Mechanick Philosophy.

> HENRY MORE, *The Immortality of the Soul* (London, 1659), preface, section 15.

∴

1 Introduction

The reception of Descartes at British universities has many similarities to the reception of Cartesian philosophy in Dutch universities and elsewhere in Northern Europe, not least in the challenges for teaching new philosophy in a traditional scholastic academic context. As with Dutch universities, it is possible to track the impact of Cartesian philosophy in English universities through textbooks, reading lists, university disputations, student notebooks, and indeed the records of academic tensions between traditionalists and innovators. However, the insight to be gained from such sources about what actually transpired in tutorials and classes has its limitations. It is easier to track *what* was being taught, than to understand *how* Cartesian philosophy was taught. In this chapter, I focus on another source of evidence for the teaching of Cartesian philosophy—namely correspondence—in order to explore the question of how Cartesian philosophy was actually taught in England. I shall do so by focusing on one of the most important figures in the history of the English reception of Cartesianism, the Cambridge philosopher, Henry More (1614–1687). I shall do so by examining three sources for More's teaching of Descartes's philosophy for what they can reveal about Descartes in the classroom in the mid-seventeenth-century at Cambridge. These are, first, his own writings and, secondly, two manuscript sources, the first of which is a recently-discovered notebook belonging to a Cambridge student by name Thomas Clarke (b. 1633/4).[1] The second manuscript source is More's correspondence with Anne Conway (1631–1679), now published in *Conway Letters*. Some of these early letters are epistolary tutorials recording More's instruction of Conway in philosophy.[2] The letters are important for four reasons. First, they are testimony to the earliest phase of the teaching of Cartesian philosophy in Britain, long before Antoine Le Grand (see Chapter 11, in this volume) or the introduction of Schuler's textbooks (see Chapter 13, in this volume). Secondly, these manuscripts offer insight into how new philosophy could be and was absorbed within an academic set-up where the traditional Aristotelian curriculum was still dominant. Thirdly, and most importantly, these manuscript sources give some insight into the form which the teaching of philosophy took in practice—the interactions between teacher and student. Finally, the letters (especially those of Anne Conway) are important for illustrating the value of Cartesianism for extramural teaching. This is particularly relevant in the case of women since they were denied entry to the university. While I am not directly concerned in this chapter with women and Cartesianism, it suffices to note that there is no little irony in the fact that insight into the university tutorials from which Anne Conway, as a woman, was barred, can be found in the epistolary classroom to which she was confined. I begin with More's claim to being considered a principal actor in the take-up of Descartes at the University of Cambridge, setting this in the wider university context for the reception of Descartes in England.

---

1   See Hutton, "Henry More et Descartes."
2   Nicolson and Hutton, eds., *Conway Letters*, 484–494; Hutton, *Anne Conway*, ch. 2.

2 Descartes at English Universities

Henry More is an important figure in the history of the reception of Cartesianism, especially its English reception.[3] Not only did he have direct contact with Descartes through a brief letter-exchange,[4] but he was one of the first to advocate teaching Cartesian philosophy in universities (see the opening quotation for this chapter) at a time when the university curriculum was still underpinned by Aristotelianism. While not the only Englishman to correspond with Descartes, nor the first to propose including Descartes in the university curriculum, he was the most philosophically important among Descartes's English correspondents.[5] In 1653, in his *Academiarum examen*, John Webster proposed reforming the university curriculum along Cartesian lines. Others at both Oxford and Cambridge studied Descartes in the 1650s and 1660s.

Even though the university statutes stipulated Aristotle as the basis of the teaching curriculum, the collegiate structure at both universities meant that there might be a difference between what was prescribed by statute and what might be taught by an individual college teacher.[6] (Of course, this did not mean there was no opposition or censure of those who promoted new ideas.) As a result of the latitude that this afforded in terms of what could be taught by individual tutors, university regulations are a poor guide to the extent of the uptake of Descartes's philosophy at these universities. The reading list which Richard Holdsworth supplied in his "Directions" written for students at Emmanuel College in the 1640s included works by Descartes.[7] One of the earliest Cambridge Cartesians was Gilbert Clerke (1626–1697), fellow of Sydney Sussex College, who claimed to be the first to have introduced both Descartes's philosophy and mathematics into teaching at Cambridge.[8] Another early student of Descartes was Newton's predecessor as Lucasian Professor of Mathematics, Isaac Barrow (1630–1677). In 1652, Barrow defended a thesis critiquing Descartes's concept of

---

3 See especially Hutton, "Henry More's *Epistola H. Mori ad V.C.*," and Gabbey, "*Philosophia cartesiana triumphata.*"
4 The correspondence consists of four letters by More (11 December 1648, 5 March 1649, 23 July 1649 and 12 November 1649) and two replies by Descartes (5 February 1649 and 15 April 1649). There is also a fragment of a third reply by Descartes, which Clerselier sent to More in 1655, and to which More wrote a response ("Responsio ad fragmentum").
5 Hutton, "Cartesianism in Britain."
6 Hutton, *British Philosophy*, ch. 2.
7 Holdsworth, "Directions."
8 Hutton, "Cartesianism in Britain," 504; see also Gascoigne, "Isaac Barrow's Academic Milieu," 265.

matter (though his references to Descartes were respectful).[9] There were several students at More's own college, Christ's College, who took an early interest in Descartes, among them John Hall (ca. 1627–1656) and Henry Power (1626–1688). Another was John Covel (1638–1722), later master of the College, who delivered an anti-Cartesian MA oration in 1663.[10] Perhaps the most famous student of Descartes was Isaac Newton, who first encountered Descartes as an undergraduate in 1661. Newton's early manuscript notebook ("Questiones quaedam philosophicae") shows that he made a thorough study of Descartes. His study of Cartesianism was extra-curricular, and he appears to have been self-taught.[11]

Other evidence for the study of Descartes at Cambridge comes from Cartesian texts for students that circulated in manuscript copies. Cambridge University Library holds several such student copies of Cartesian texts, though all date from the years after the initial uptake of Cartesian studies: these include a 1668 epitome of Cartesian physics compiled by Jonathan Comer of Christ's College and a much-copied *Physica incipientium, sive cartesiana principia* (1660) by John Felton.[12] But studying Cartesianism at Cambridge was not uncontested. Roger North (1651–1734) reported that when he was a student there in 1667/8, there was, "such a stir about Descartes, some railing at him and forbidding the reading him [...]. And yet there was a general inclination, especially of the brisk part of the University, to use him."[13] Studying Cartesian philosophy became so controversial at this time, that in 1668 the Vice-Chancellor of the university, Edmund Boldero (1609–1679), proscribed debating Descartes's philosophy in university disputations, insisting that Aristotle be used instead.[14] Whether this had any long-lasting effect is hard to determine: when Joshua Barnes (1654–1712), fellow of Emmanuel College, updated Holdsworth's "Directions," sometime after 1678, he added more of Descartes's writings, along with Cartesian

---

9   He would later change his tone. See "Isaac Barrow's Academic Milieu," 264; Gascoigne, *Cambridge*, 54.

10   Gascoigne, "Isaac Barrow's Academic Milieu," 264, 278.

11   Newton, *Certain Philosophical Questions*. On Newton's study of Descartes, see Westfall, *Never at Rest*, 89–101. Newton acquired several of Descartes's works, including Van Schooten's Latin edition of Descartes's *Geometry*, with its copious notes.

12   Luard, *Catalogue*. These are listed and discussed in Kearney, *Scholars and Gentlemen*. As far as I can tell from those which I have examined, it is not possible to reconstruct from these the method of instruction used.

13   North, *Lives*, vol. 3, 15. By the "brisk part of the University," North probably meant students and younger fellows, since the term brisk meant smart or lively, so—by implication—youthful.

14   Gascoigne, *Cambridge*, 34.

texts by the likes of Antoine Le Grand and Jacques Rohault.[15] It may have been in response to the continuing interest in Cartesianism that, in the 1680s, the university press printed Johannes Schuler's anti-Cartesian textbooks *Exercitationes ad* Principorum *Descarti primam partem* (1682), and his *Examinis philosophiae Renati Descartes specimen, sive Brevis & perspicua* Principiorum philosophiae *cartesianae refutatio* (1685).[16]

While there is plenty of evidence that Cartesianism was studied and discussed at Cambridge, the evidence for how it was taught is scant. The indications are that the method of teaching retained the university's scholastic disputation format. Boldero's proscription suggests that Cartesian philosophy was incorporated within this format. Although written in the Netherlands, Schuler's textbooks appear to confirm this, since they tackle their subject in a series of questions which impose a disputatious format. Most of this evidence for Cartesian philosophy being studied and discussed at Cambridge dates from after 1660. For the earliest teaching of Descartes, the main evidence, such as it is, comes from Henry More.

3   Worthington, More and Descartes

The year before Vice-chancellor Boldero proscribed disputations involving Cartesian philosophy, More's friend, John Worthington (ca. 1618–1671), remarked on the enthusiasm for Descartes amongst Cambridge students, suggesting that this could be explained by Henry More's pro-Cartesian stance. In a letter to More on 29 November 1667, he wrote:

> [...] you have as highly recommended Des Cartes, as is possible, and as knowing no better method of Philosophy, you recommend it effectually in some parts of your books whereby you had so fired some to the study of it, that your letter to V.C. (which came long after) could not cool them, nor doth it yet: but they are enravished with it ...[17]

---

15   Holdsworth, "Directions."
16   Although Schuler was critical of Descartes, he did not reject Cartesian philosophy entirely—see Chapter 11, in this volume. Other Cartesian textbooks, such as those by Antoine Le Grand catered for student demand, but most were printed in London, so were not necessarily targeted for the use of students at Cambridge. On Le Grand, see Chapter 13, in this volume.
17   Worthington, *Diary and Correspondence*, vol. 2.1, 254.

The "letter to V.C." mentioned here refers to More's *Epistola H. Mori ad V.C.*, an undated letter to an unidentified recipient, which was first printed in his *A Collection of Several Philosophical Writings* in 1662. More states in "The Preface General" that the original letter was written before the publication of Clerselier's second volume of Descartes's letters.[18] In *Epistola H. Mori ad V.C.*, More presents a measured assessment of Descartes's physics which calls attention to some of the shortcomings in Cartesian philosophy, while at the same time reaffirming his admiration for Descartes and his view that Cartesianism was the best available natural philosophy.[19] If Worthington is correct in stating that More published the *Epistola* with the purpose of reining in student enthusiasm for the study of Cartesianism, this enthusiasm must have been well-established amongst students before 1662, when it was published.[20]

Worthington does not give any details as to which of More's 'books' were the source of the praises which fanned the flames of student enthusiasm for Descartes, apart from the fact that they antedate the "Epistola ad V.C." (which, he says "came long since"). He could be referring to several texts, for example, More's poem *Democritus Platonissans* (1646), or his controversy with Thomas Vaughan published in a series of pseudonymous pamphlets in 1650–1651 (in which he praised and defended Descartes as a "Miracle of Ages" and "incomparable Philosopher"), or perhaps his *Conjectura cabbalistica*, where he praised Descartes "as a man more truly inspired in the knowledge of Nature, then any that have professed themselves so this sixteen hundred years."[21] It is unlikely that the books to which Worthington was referring included More's *Collection of Philosophical Writings* (1662) because it was in this that the *Epistola H, Mori at V.C.* was published. However, these published writings reveal nothing about his pedagogical practices.

Presumably, also, Worthington did not have the Clerselier edition of Descartes's letters in mind as the source of student enthusiasm, since he refers specifically to books (plural) by More. Nevertheless, it was in More's first letter to Descartes, written in 1648, that he expresses his warmest admiration, prais-

---

18   More, *A Collection of Several Philosophical Writings* (London, 1662), "The Preface General," xi. This would date the *Epistola H. Mori ad V.C.* between 1657 and 1659, when the second volume of the Clerselier edition was printed.

19   Hutton, "Henry More's *Epistola H. Mori ad V.C.*"

20   The printing history of *Epistola H. Mori ad V.C.* suggests that it was intended for undergraduate use. After a separate printing in 1664, it was appended to the second edition of More's handbook on ethics, *Enchiridion ethicum*. See Hutton, "Henry More's *Epistola H. Mori ad V.C.*" and Agostini, "Quelques remarques."

21   More, *Observations* (London, 1650), sig A3, 40; More, *Conjectura cabbalistica* (London, 1653), 104. An expanded version of this was published in More, *A Collection*.

ing his correspondent as a giant among philosophers, who are mere "pygmies" compared with Descartes.[22] And he lauds his *Principles of Philosophy, Dioptrics* and *Meteors*, as both beautiful and consonant with nature.[23] So it is likely that More's correspondence with Descartes contributed to student enthusiasm *after* its being made available in print, in Clerselier's edition of Descartes's correspondence in 1657, or in More's own edition of them in his *A Collection of Several Philosophical Writings* published in 1662. But it also appears that the study of Cartesian philosophy was well-established before this—certainly by the time More penned the original *Epistola ... ad V.C.*, which was written earlier, as already noted, before 1659.[24] And when he published it, he linked it clearly with his correspondence with Descartes by placing it immediately after this correspondence in *A Collection*.

There are two slightly puzzling things about Worthington's comments, the first being that More specifically states that the *Epistola* was intended to *defend* Descartes against imputations of atheism, whereas Worthington states that it was intended to dampen student enthusiasm for Descartes. Secondly, More was never an uncritical admirer of Descartes. Even his recommendation that Descartes be studied at universities (quoted above from his *Immortality of the Soul*) was qualified: he recommended studying Cartesian philosophy in order to better understand the *limitations* of his philosophy. More's reservations were based, principally, on his view that "the Mechanical principles" of Descartes could not account for all natural phenomena, and that these could only be explained by positing "some power more than Mechanical" in the operations of nature.[25] By this point in time, More had formulated his own conception of a supra-mechanical power, namely, the Spirit of Nature, whose existence More proposed in order to compensate for the shortcomings of Cartesian "mechanical" natural philosophy, and to reinforce the compatibility of philosophy with religion.[26] The reservations expressed in his *Epistola* are in line with these cautions.

---

22   "Omnes quotquot extiterunt aut etiamnum existunt, Arcanorum Naturae Antistites, si ad Magnificam tuam indolem comparentur, Pumilos planè videri ac Pygmaeos: meque cum vel unicâ vice evoluissem Lucubrationes tuas Philosophicas ..." More, *Epistolae Quatuor ad Renatum Des-Cartes*, 61, in More, *A Collection*.

23   "Omnia profecto tam concinna, in tuis Philosophiae principiis, Dioptrícis & Meteoris, tamque pulchrè sibi ipsis Naturaeque consona sunt, ut mens Ratióque humana jucundius vix optaret laetiusve spectaculum." More, *Epistolae Quatuor ad Renatum Des-Cartes*, 61, in More, *A Collection*.

24   See above, at note 20.

25   More *Immortality of the soul* (London, 1659), preface, section 12.

26   Gabbey, *"Philosophia cartesiana triumphata"*; Reid, *Metaphysics of Henry More*.

It is now well established that More's objections to Cartesianism eventually became more pronounced, leading him ultimately to repudiate Descartes as a "nullibist" or "nowhere-ist" because he acknowledged the existence of the soul, but could not locate it.[27] He was never uncritical of Cartesianism, not even in his correspondence with Descartes, and, as others have observed, many of the points he raises in his letters anticipate basic elements of his later, more-developed critique of Descartes, which ultimately led him to revise his initial assessment of Cartesian philosophy.[28] *Inter alia*, More argued against Descartes that all substance, both corporeal and incorporeal, is extended and, indeed, that God Himself is *res extensa*. For these reasons, the letters are important for the history of More's evolving view of Cartesianism and the development of his own philosophical views. However, in the context of teaching Cartesian philosophy, the letters are also suggestive of a pedagogical scenario, in which the youthful More sought elucidation from the great master.

More's purpose in writing to Descartes was, after all, to ask for clarification about points which More professes to be unable to grasp.[29] And More's third and fourth letters are taken up with detailed queries about different passages in Descartes's *Principles* and *Dioptrics*, such that they serve as commentary and notes on these two texts. Taken together the letters (three from More and two replies from Descartes) can be construed as a set of objections and replies. At the same time, the queries and critical points raised by More are consistent with his reasons for recommending the teaching of Cartesianism in universities, to show the *limitations* of Descartes's philosophy. Arguably, therefore, his correspondence with Descartes has features which make it a useful study aid, especially since he would not have had Cartesian textbooks on which to draw. But, without further corroboration, this is mere speculation.

4   **Thomas Clarke's Notebook**

However, the possibility that the More-Descartes correspondence may have been used in teaching becomes more likely in view of the discovery of a manuscript copy of the More-Descartes letters in a student notebook in the

---

27   Gabbey, "*Philosophia cartesiana triumphata.*"

28   Gabbey, "*Philosophia cartesiana triumphata.*" On More's correspondence with Descartes, see also Anfray, "*Partes extra partes*," and Agostini, "Descartes and More."

29   "quae certe animus meus aut paulò hebetior est quàm ut capiat, aut ut admittat, aversatior." More, *Epistolae quattuor*, 62, in More, *A Collection*.

holdings of Cambridge University Library.[30] The notebook has obvious importance in the manuscript history of the Descartes-More correspondence. But the
fact these letters were circulating in manuscript form among students might
also tell us something about their role in a pedagogical context at Cambridge.
The notebook in question belonged to a Cambridge student, Thomas Clarke,
who inscribed his name clearly on the fly leaf, noting his college (Sydney Sussex), and the date 1654. It contains, in chronological order, copies of the first
three of More's four letters to Descartes and the two replies which he received
from him. At the reverse end of the notebook there are several pages of other
materials, of the kind typically found in student notebooks—pen trials, a few
verses, quotations from the Church fathers.[31] The little that can be discovered
about the owner comes from Venn's *Alumni Cantabrigienses* which mentions
a Thomas Clarke, from Helperby, North Yorkshire, who was admitted as a sizar
to Sidney Sussex College in 1650, matriculating in 1650–1651, and graduating as
a Bachelor of Arts in 1654–1655.[32] But there is no record of who his tutor was,
or what became of him after he graduated. There are many unanswered questions about this manuscript, one being the source which Clarke used to make
his copy, and another how he obtained access to it.[33] The date, 1654, excludes
its having been copied from one of the printed editions of the letters: either
Clerselier's *Lettres de M. Descartes* or More's own in his *A Collection of Several
Philosophical Writings* which were printed later, in 1657 and 1662, respectively.[34]
Preliminary inspection indicates that the letters in the Clarke notebook were
not made from the copies among the Hartlib papers.[35]

---

30   The existence of this notebook was first remarked by Andrew Janiak in a footnote, in
      his *Newton*, 98, without a reference. An error in the Cambridge University Library's
      manuscript catalogue meant it was difficult to trace. See Hutton, "Henry More et Des-
      cartes." Hitherto the only known manuscript copies of More's letters to Descartes are
      contained among the papers of Samuel Hartlib, who acted as the intermediary for both
      correspondents. But the Hartlib holdings are not complete; see Hartlib, *The Hartlib Papers*,
      eds. Greengrass, Leslie and Hannon.
31   For a full description, see Hutton, "Henry More et Descartes."
32   Venn and Venn, *Alumni Cantabrigienses*. A 'sizar' was a poor student at Oxford or Cam-
      bridge who supported himself by acting as a servant to a rich student.
33   Hutton, "Henry More et Descartes."
34   The omission of More's last letter to Descartes may perhaps be explained by the fact that,
      as shown by Alan Gabbey in his seminal article *"Philosophia cartesiana triumphata,"* when
      Clerselier contacted More about the possibility of including his letters in his edition of
      Descartes's correspondence, the autograph copies of More's letters were not in his posses-
      sion at Cambridge. So, they had to be retrieved from Samuel Hartlib, to whom More had
      sent them; see Gabbey, *"Philosophia cartesiana triumphata."*
35   Hutton, "Henry More et Descartes."

One possibility is that, as a student of Cartesianism, Thomas Clarke may have transcribed the More letters from a copy in circulation among undergraduates. Since, as mentioned above, student manuscript recensions of Cartesian philosophy are among the holdings of Cambridge University Library, the Clarke Notebook may have been just such a student copy. That being the case, it must have been made either from More's own copies, or it may have been made from another copy already in circulation among students. If so, it would be the only known example of More's Cartesian correspondence either circulating in manuscript form or being copied among students. There is evidence that More's letters circulated beyond Cambridge.[36] But hitherto, there is no record of them circulating among members of the university, or of More giving anyone at the university access to his manuscripts. The fact that Thomas Clarke was a student at Sydney Sussex College in Cambridge, the same college where the aforementioned Cartesian, Gilbert Clerke, was a fellow, suggests that he very likely encountered them at his own college. Unfortunately, there is no list of Gilbert Clerke's students, so we don't know if Thomas Clarke was one of them. It is very possible that given their shared interest in the new French philosophy, Gilbert Clerke and Henry More were acquainted. If so, it is not beyond the bounds of possibility that More shared his Cartesian correspondence with him, or permitted him to make copies.

As stated previously, however, there is no other evidence that More permitted his correspondence with Descartes to be circulated among students or used in his teaching. Lack of evidence may, of course, be an accident of history. But one thing is certain: in the mid-1650s, a Cambridge student made a copy of a letter of More's which contained his unreserved praise for Descartes in terms which fit Worthington's comment that More had "as highly recommended Des Cartes, as is possible, and as knowing no better method of Philosophy."

5   The More-Conway Correspondence: "Conferences concerning Des Cartes Philosophy"

Leaving open for now the question of the pedagogical status of the Clarke Notebook, I turn to a different source for More's pedagogy, which may enable us

---

36   For example, Clauberg quotes from More's first letter to Descartes in his *Defensio Cartesiana adversus Jacobum Revium* (Amsterdam, 1652). See Hübener, "Descartes-Zitate bei Clauberg." Tobias Andreae, professor of philosophy at Groningen, had copies of the Descartes-More correspondence, which he offered to Clerselier in 1654. My thanks to Erik-Jan Bos for this information.

to ascertain whether there is anything about More's teaching of Descartes's philosophy that suggests he used his correspondence as a teaching aid, or had students copy the letters.

The best evidence for More's teaching practice comes from outside the university, from his correspondence with Anne Conway. As a woman, Anne Conway was not able to enrol as a student at Cambridge, so he taught her philosophy by letter—as it were, by written tutorials. Now published in *The Conway Letters*, the first extant letters from More's side of this correspondence were written shortly after his having undertaken to act as her tutor.[37] Although this tuition was extramural, these letters illustrate the kind of instruction which he might have given her had she been able to study with him in Cambridge. It is reasonable to infer that they reflect the ways he taught his students at Christ's College. This inference is supported by the fact that it was Anne's half-brother, John Finch, who had studied with More in Cambridge, who had persuaded his tutor to give his sister philosophy lessons. And there is at least one example of a Christ's College student who apparently studied Descartes in the way he taught Anne Conway: this was George Elphicke (or Uphicke), whom More recommended for service with the Conways.[38]

Although only a handful of these letters have survived (and only one by Anne Conway herself), there is enough to show that More based his tutorials on the philosophy of Descartes. When considering how Cartesianism was taught, perhaps more significant than their respective arguments is the light which these letters shed on More's teaching practice. The first thing to observe is that it involved reading passages from Descartes's writings. The main text studied was Descartes's *Principles of Philosophy*, in a translation (now lost) which More had evidently made himself for this purpose.[39] She also studied the *Dioptrics* with him. They proceeded through the text, section by section. So, for instance, the earliest letter (dated September 1650) discusses *Principles of Philosophy* I 4; the next extant letter (dated May 1651) discusses book II, sections 18–21. Their discussion is not confined rigidly to the specified sections—for example, More's discussion of *Principles* I 4 refers forward to *Principles* II 30 in order to substantiate his point that Descartes did not "in good earnest affirm that there is no meanes at all to distinguish waking from dreaming."[40] Other topics covered include the argument for the existence of God from the idea of a perfect being

---

37   Nicolson and Hutton, eds., *Conway Letters*, 484–494. See also Hutton, *Anne Conway*, 43–49.

38   Nicolson and Hutton, eds., *Conway Letters*, 145.

39   Nicolson and Hutton, eds., *Conway Letters*, 51, letter 18.

40   Nicolson and Hutton, eds., Conway *Letters*, 484–485.

and the existence of a vacuum. In the third extant letter, More advances arguments to demonstrate the existence of space as non-material extension (as he had done in his first letter to Descartes), and in the one following, he explains the difference between "first and second notions."[41]

The second thing to note is that the approach which More adopted was decidedly non-scholastic, but based on Descartes himself. This is apparent from the letter he wrote in 1650, to accompany his translation of the *Principia*. More gave her his advice on how to approach the work:

> Only let me be bold to commend to you this rule, Though I would have you to habituate yourself, composedly, and steadily to think of anything that you think worth the thinking of, and to drive it on to as clear and distinct approbation as you can, yet do not think of anything anxiously and solicitously, to the vexing or troubling of your spirits at all. What you would force at one time may happily offer itself at another.[42]

This echoes Descartes's own advice to his readers in the preface to the French translation of the *Principles*, where he recommends his reader to read "without straining his attention too much or stopping at difficulties which may be encountered" and then to re-read the text several times, without pausing over difficulties:

> I should like the reader first of all to go quickly through the whole book like a novel, without straining his attention too much or stopping at the difficulties which may be encountered. The aim should be merely to ascertain in a general way which matters I have dealt with. After this, if he finds that these matters deserve to be examined and he has the curiosity to ascertain their causes, he may read the book a second time in order to observe how my arguments follow. But if he is not always able to see this fully, or if he does not understand all the arguments, he should not give up at once. He should merely mark with a pen the places where he finds the difficulties and continue to read on to the end without a break. If he then takes up the book for the third time, I venture to think he will now find the solutions to most of the difficulties he marked before; and if any still remain, he will discover their solution on a final re-reading.[43]

---

41   Nicolson and Hutton, eds., *Conway Letters*, 487–488, 489–490.
42   Nicolson and Hutton, eds., Conway *Letters*, 52.
43   Descartes, *Principes de la philosophie*, *Préface*, AT IX-B 11–12, CSMK I 185.

This was not a special programme for a female student learning at a distance, but directed at all readers, even "those who have never studied."[44] It was evidently also adopted by More with his male students: as already noted, his Christ's College student, George Elphicke, followed this pattern of learning. In 1658, More reported that "he has read Des Cartes Principia over and over, and has a pretty dexterous mechanick wit."[45] Many years later, More gave tutorials to Conway's nephews, Edward and John Rawdon when they were studying at Cambridge. Although More was not their tutor, he agreed to assist in their education. They studied Descartes with him, and More used the same approach of reading through the text which he had used when teaching their aunt. The young Rawdons followed a programme of Cartesian studies, commencing with the *Principles*, then moving on to the *Dioptrics* and *Meteors*. In September 1674, he wrote, "Mr Rawdon [...] comes diligently to his Cartesian lecture, we are just now gott through the 3 first parts of his Principia."[46] On 19 October they began reading "Des Cartes Dioptricks," which they finished by December, as he reported to Lady Conway: "I have now gone quite thorough Des Cartes Dioptricks with Mr Rawdon, and made him understand them from the beginning to the end, the Machine for making glasses not excepted."[47] The subsequent letters (9 and 31 December 1674) report on their progress in study of the *Meteors*.

Another important finding from the "virtual" classroom captured in the More-Conway letters is that More was non-dogmatic and non-prescriptive in his approach to teaching. He proceeded by setting up a series of objections and replies. In so doing, he was in all probability inspired by Descartes's use of "objections and replies" in *Meditations*. Unlike the *quaestiones* and *responsiones* of university disputations, More did not use them in a formulaic or contentious manner. Rather, he expected his pupil to raise queries or objections to the passages prescribed for reading, to which he would then respond. Or he himself raised objections to the passages selected from Descartes and invited Conway to critique his arguments. He then answered her critique and asked her to respond. For example, the first of More's extant letters is a sequel to another (no longer extant) in which he had asked Anne Conway to think of objections

---

44   Descartes, *Principes de la philosophie*, Préface, AT IX-B 8, CSMK I 183. This gender-neutral
     stance would be exploited by Poullain de la Barre to develop his feminist arguments in *De
     l'égalité des deux sexes* (1673) and his *De l'education des femmes* (1674). For Poullain de la
     Barre, see Chapter 18, in this volume.

45   More to Lady Conway, 8 February [1658], Nicolson and Hutton, eds., *Conway Letters*, 145. By
     "mechanick wit," he probably means a geometrical turn of mind, since More exemplifies
     this by the model which he made of his college's New Building.

46   Nicolson and Hutton, eds., *Conway Letters*, 395.

47   Nicolson and Hutton, eds., *Conway Letters*, 397.

to Descartes's argument for the necessary existence of God from the idea of a fully perfect being. We only know anything about her argument from his letter, which reveals that one of the objections she offered was (as quoted by More), "That then the idea of a fully imperfect Being should imply the existence of a Being fully imperfect." More responded to this by pointing out that it amounted to saying,

> the idea [...] of what is fully imperfect, implies a necessary non-existence. And the idea of this fully imperfect, tells us that it is impossible for it to exist or be anything, as the idea of the fully perfect being tells that it does necessarily exist.[48]

Another example of his pedagogical practice comes in his second extant letter, in which the discussion concerns *Principles* II 18, to which Conway "has propounded [...] some objections". The topic at issue was Descartes's denial of the existence of a vacuum, and More's remarks suggest that Anne Conway, apparently, argued in favour Descartes's view against his own:—

> Your Ladyship in courtesy seems to take the strangers part and lean towards his opinion that there cannot be an empty space or any distance but by the interposition of a body or matter.[49]

More follows this with six arguments for the existence of empty space and a vacuum, and asks her to evaluate them: "your Ladyship in your next [letter] shall tell me your judgement which of them is the weakest. For I will not profess them all unconfutable."[50] These six arguments include one based on *consensus gentium* (consent of nations) that "almost all men" hold that the world is finite, and one which turns on a scholastic axiom that there is no motion in an instant. They also include arguments which More had put to Descartes, and arguments he was to repeat in his *An Antidote Against Atheism*, among them that if God annihilated the world and then made a new one, there would be a measurable distance of time between the two events, even though there was nothing in existence. Other points, which echo More's letters to Descartes, include More's rejection of Descartes's claims (*Principles* II 26–27) that the extent of the universe is "indefinite" rather than "infinite."[51]

---

48   Nicolson and Hutton, eds., *Conway Letters*, 484, letter 19a.
49   Nicolson and Hutton, eds., *Conway Letters*, 486.
50   Nicolson and Hutton, eds., *Conway Letters*, 486.
51   *Principia* II 26–27, AT VIII-A 54–55. Nicolson and Hutton, eds., *Conway Letters*, 486–489.

The only extant reply to More by Anne Conway is to the second letter of the group. In this she challenges More's pro-Cartesian position by citing observable evidence which challenges the Cartesian claim (*Principles* I 67–71) that colour is not a property of things but the result of the stimulus of the mind by light,[52] or, as Conway puts it:

> that colours are not re[ally] in the object as they are apprehended by the eye, but that the motion of the optick Nerves stir[s] up a sensation in us w[hi]ch we call colour and make it similitudinary [i.e., similar] so that we think it is without in the object when as nothing but a mere motion was transmitted from the object to our eye.[53]

Against this, she invokes observable phenomena. First, that if colour is not intrinsic to a coloured object, but results from the stimulus of the organ of sense by motion transmitted from an object perceived, how do we explain the appearance of an image of it in a *camera obscura* (or darkened room, as she calls it). She argues that the surface on which the image appears (in this case, a sheet of white paper) is not capable of perceiving motion, or of transmitting motion from the image to the eye. If it were capable of transmitting motion caused by something else, everything which we look at would reflect the image of other things ("yield the eidolum or representation of something else"). This leads her to suspect instead that the object projects an image ("something streames from that object as its Image"). Furthermore, Conway uses the example of a reflection in a window by candlelight, which is not visible in daylight. On Descartes's theory, if images are produced by movement, and if they are not impeded by sunlight, then the reflection should be visible in daylight. Curiously, Conway takes her example of the camera obscura from Kepler (via Sir Henry Wotton[54]), and not from Descartes in *Dioptrics* 5, which suggests that she may not have read it at this point. We do not have More's response. No doubt, he would have drawn her attention to Descartes's arguments in his optics. While Conway's objection to Descartes on the basis of an observable phenomenon may not be "unconfutable" (to repeat More's term), her appeal to observable phenomena is consistent with the fact that much of the attraction of Cartesianism was as an alternative natural philosophy (science). Much of the appeal of Cartesian physics for More and his contemporaries was that it seemed to them to be more consistent with the observable

52   Nicolson and Hutton, eds., *Conway Letters*, 53.
53   Nicolson and Hutton, eds., *Conway Letters*, 494.
54   Her source is Henry Wotton's *Reliquiae wottonianae* (London, 1651).

phenomena of nature. Conversely, it was the growing conviction that Descartes could not account for key phenomena which formed the foundation of More's critique.

It is striking that, in many ways, the master-pupil relationship of his early correspondence with Conway is prefigured in his correspondence with Descartes, especially since More deploys some of the same arguments he had used there. Writing to Descartes, More advanced the kind of critical evaluations of his theories which, later on, he encouraged in Lady Conway. It is difficult to be sure how much he refers to his correspondence with Descartes in his tutorial exchanges with Conway. But on at least one occasion he refers her to his criticism of *Dioptrics*, chapter 2 in his correspondence:

> [...] your Ladiship may see in my last letter to him in the book of letters, page 386. The last lines of the tenth and the first of the 11th paragraph of the 2d chapter are to be understood out of his Principia page 165, 166, where he expands a refraction that is to be made use of pag. 167, n168, of which you may be pleased to read my letter page 393, upon that article ...[55]

The pagination given (p. 386) matches the first volume of the Clerselier edition (1657). Marjorie Nicolson dates the letter and its sequel to 1658. In 1657, More mentions "my designe, concerning Des Cartes Letters"—probably a reference to his intention to send her the newly-published Clerselier edition. A few years later, in 1661, Anne Conway asked him to send her the second volume. There is no indication that More had her make a copy of his correspondence with Descartes as part of the learning process—though it is not beyond the bounds of possibility that he provided her with hand-written copies, as he had done with an English translation of Descartes's *Principles*.

7   Conclusion

Returning to the Clarke Notebook: although there is no evidence to show that More used his correspondence with Descartes as a text for study "in the classroom" at Cambridge, its existence does confirm that Cartesian philosophy was circulating among students as well as professors at the University of Cambridge in the 1650s. Furthermore, the parallels between More's engagement

---

55   Nicolson and Hutton, eds., *Conway Letters*, 145.

with Descartes and his own approach to teaching, as illustrated in *The Conway Letters*, suggest that the correspondence may have served as a model for his pedagogical approach. As his epistolary tutorials with Anne Conway show, this was closer to Descartes than to the disputatious teaching methods of the scholastic tradition. However, while More's correspondence with Descartes may have served More as kind of template for teaching Cartesian philosophy, it seems unlikely that it served as a text for study. But it is not beyond the bounds of possibility that it functioned as ancillary reading, especially after the letters had been printed. Until we know more about the Clarke Notebook, its status in relation to the teaching of Cartesianism text remains speculative.[56] However, although Anne Conway did not have face-to-face tuition with her tutor, her letter exchanges with Henry More gives us important clues as to how Descartes was actually taught in a virtual classroom in the mid-seventeenth-century, from which we can infer how More taught Descartes in a college classroom at Cambridge.

---

56   There is more to the history of the Clarke Notebook which I have not discussed here, since my focus has been on pedagogy. For further information, see Hutton, "Henry More et Descartes."

CHAPTER 13

Teaching Descartes's Ethics in London and Cambridge

Roger Ariew

1 Introduction

Descartes understood that he did not produce a complete curriculum for teaching Cartesian philosophy: he did not construct a complete physics, that is, he did not explain in detail the nature of minerals, plants, animals, and he did not treat humans or deal with medicine and morals. The Cartesians also understood this and some of them tried to fill in the lacunae, constructing complete curricula for teaching Cartesian philosophy based on Descartes's works—whether issued during his lifetime or published posthumously—and on his correspondence (published posthumously in 1657–1667).[1]

For example, Antoine Le Grand (1629–1699) produced several textbooks from which to teach a complete corpus of Cartesian philosophy to his students in London. These include *Philosophia veterum e mente Renati Descartes, more scholastico breviter digesta* (1671), its successor *Institutio philosophia, secundum principia Renati Descartes … ad usum juventutis academicae* (1672 and in numerous editions), and ultimately the English translation and revision of the latter as *Institution of Philosophy*, constituting the first of the two volumes of *An Entire Body of Philosophy according to the Principles of the famous Renate Descartes* (1694)—this last edition, being a "coffee-table" book, was no longer explicitly intended for students. Works such as these allow one to see how Cartesian philosophy was taught in seventeenth-century England, whether one is interested in the teaching of logic, metaphysics, physics, or ethics.

It happens that students in London and Cambridge had another textbook to assist them with their ethics course; a student manual was produced anonymously in London and fashioned out of various statements by Descartes on moral philosophy, taken from his correspondence and from *Passions of the Soul*. This

---

1 The Cartesians also developed Descartes's logic, a standard staple of the curriculum, but a subject that was more likely to be considered by Descartes as an art (*ars cogitandi*), rather than as a science. See Roger Ariew, "The Nature of Cartesian Logic."

textbook, René Descartes, *Ethice: In methodum et compendium, gratiâ studi-osae juventutis concinnata* (1685), became an established reference at Cambridge, being published by the Academic publishers of the University, as part of the compendium called *Ethica* ... (1707). Descartes's *Ethice* makes up the third part of the work, the first being the *Ethica* of the scholastic Eustachius a Sancto Paulo (1573–1640) and the second the *Synopsis Ethices* by the Protestant Etienne de Courcelles (1586–1659), a friend of Descartes and translator into Latin of the *Discourse on Method*.[2]

The tripartite ethics manual gives a fair indication of the structure of an ethics course that could be taught using the compendium, with a clear contrast between the scholastic and Cartesian ethics.[3] Setting aside de Courcelles, who provides a Protestant theological point of view, this chapter discusses the ethics that was taught in seventeenth-century England, that is, the ethics of Descartes and the Cartesians, in contrast with the scholastic ethics of Eustachius a Sancto Paulo and other late scholastics.[4] It begins with a brief account of the ethics found in Eustachius and other late scholastics, focusing

---

2 The texts given in the compendium, other than Eustachius's *Ethica* (part II of his *Summa philosophiae quadripartita*) are de Courcelles's *Synopsis ethices* and Descartes's *Ethice* (both Cambridge, 1702). The first edition of Descartes is the *Ethice* already referred to from 1685. The first edition of de Courcelles is likely the one published in his *Opera theologica* (Amsterdam, 1675), 982–1018, with the second edition being London, 1684—all posthumous publications.

3 There is some evidence that both Descartes's and Eustachius's ethics are discussed in late-seventeenth century England and Scotland; see Heydt, *Moral Philosophy*, 21–28.

4 The *Synopsis ethices* offers a contrast with both Eustachius and Descartes on ethics. In his first chapter (I. *De natura ethices*), de Courcelles argues that ethics is the science of mores, allowing people to strive toward happiness, and differentiating it from wisdom and prudence. The genus of ethics is science (not art) and its difference from other sciences is that its subject is mores and human happiness. The parts of ethics are virtues and happiness. There are two species of ethics, one philosophical, the other theological; the former in conformity with natural light, and about happiness in this life, and the latter dealing with the Christian virtues of faith, hope, and charity, and about our happiness in heaven. But, de Courcelles argues, no one who neglects celestial happiness can be happy in this life. While de Courcelles produces a seemingly standard scholastic account of ethics, the emphasis is clearly on the theological part, that is, on Christian revelation. This is made clear in his final chapter on happiness (xv. *De beatitudine*), where de Courcelles issues the same kinds of provisos as the scholastics about the supreme good and the conditions for it, and argues that it does not consist in the goods of fortune, the body, pleasure, or even knowledge and virtue. He asserts (correctly, see below) that the scholastics consider the supreme good in a dual fashion, that is, both objective, which is God himself, and formal, which is the enjoyment of God. But he states that the distinction is improper: to say that God is our objective supreme good is to make God an object; and if God is the object of the supreme good, he is not himself the supreme good and cannot be enjoyed in himself. For more on de Courcelles's *Synopsis ethices* in the context of protestant theological ethics, see Han Van Ruler, "The *Philosophia Christi*," esp. 253–254.

on their central concept of happiness; it continues with the partial construction of Descartes's ethics from his correspondence, just after the publication of *Lettres de Mr Descartes* (1657–1667) by Claude Clerselier (1614–1684), and ends with the full-blown Cartesian ethics taught in London and Cambridge in the anonymous Ethics student manual and in the textbooks of Le Grand.

2   Ethics in Eustachius and the Late Scholastics

Eustachius's *Ethica* is Part II of his *Summa philosophiae quadripartita* (*secunda pars summae philosophicae quae est ethica, de rebus moralibus; in tres partes divisa*), first published in 1609 with many editions until 1648 (with *Ethica* continuing as a stand-alone work into the 1730s).[5] The immense popularity of the *Summa* can be best explained by its relative conciseness, clarity, and completeness.[6] Whether he is dealing with logic, ethics, physics, or metaphysics, Eustachius rearranges his materials into new conceptual structures and indicates each of these with a table of contents, an index of concepts, and a general analytic table, in which the connection among his topics "can be seen at a single glance."[7] It should also be emphasized that Eustachius is not seeking novelty in his topics, but attempting to reach a consensus with other scholastics about the doctrines he is presenting. These features of Eustachius's *Summa* can be demonstrated for all four parts of his *Summa*; here, it suffices to display them for his ethics and for the portions of ethics dealing with the status of ethics and the account of happiness.

---

5   One might also count the edition by Johannes Adam Schertzer's *Breviarium Eustachianum* (Leipzig, 1663), a 759-page "abbreviation" of Eustachius's *Summa*.

6   Descartes himself praised Eustachius's *Summa* on this account, as "the best book ever made on this subject" (AT III 232). He even decided to write a commentary on it: "My plan is to write, in sequence, a complete course of my philosophy in the form of theses [...]. In the same book, I intend to have printed a textbook course in ordinary philosophy, such as perhaps the one by Brother Eustachius, with my notes at the end of each question, where I will add the various opinions of others, what we should believe concerning all of them, and perhaps at the end I will make a comparison between these two philosophies," (AT II 233). After repeating that he would consider Eustachius "as the best of all the ones who wrote a philosophy" (AT III 234), Descartes indicated that he had begun the project; but the project was soon aborted because of the death of Eustachius: "I am sorry about the death of Father Eustachius; for although it gives me more freedom to write my Notes on his Philosophy, I would nevertheless have preferred to do it with his permission and while he was alive" (AT III 286).

7   For the *Ethics*, this is his "Schema generale, in quo series eorum omnium quae traduntur in hac summa moralis disciplinae ob oculo ponitur." See Ariew, *"Le meilleur livre qui ait jamais été fait en cette matière."*

Eustachius divides his ethics into three parts, generally following the order given to it by the Jesuits of Coimbra in their *Disputations* on Aristotle's *Nicomachean Ethics*, though not entirely agreeing with their conclusions.[8] Eustachius, writing a manual on ethics, deals with all the traditional topics discussed in such texts, not just the good and happiness, but the intellect and will, passions, virtues, and vices as well. He starts with the preliminary questions, that is, the standard questions about the status of the discipline (whatever the discipline); these usually treat the etymology of the term by which the discipline is called, its subject and its end, whether it is a science or art, a theoretical or practical endeavour, often ending with an outline of the divisions and parts of the discipline at stake. He then proceeds with his three main parts, dividing the first, "On Happiness," into treatments of the good, the end, and happiness itself.[9] Part 2, "The Principles of Human Actions," follows with a discussion of the internal principles of human action, such as will and appetite; acquired principles, such as habit; and external principles, including God and Angels.[10] His third part is about "Human Actions themselves, that is, Passions, Virtues, and Vices"; it is further divided into several disputations: concerning the good and evil of human actions; passions, such as love and hate; the virtues in general—prudence, justice, fortitude, and temperance; and ending with a short disputation on vice and sin.[11] All these topics have rich histories and are discussed within the context of a variety of previous and still prevailing ethical views. With respect to happiness, we should, at a minimum, keep in mind: the naturalistic Aristotelian intellectualist view (activity of the soul in conformity with excellence or virtue); the mixed naturalistic and theocentric Thomistic intellectualist view (dividing happiness into two: natural, or Aristotelian and imperfect, and supernatural, or perfect, with supernatural happiness residing in the understanding and consisting in the contemplation of the divine essence); and the theocentric Scotist voluntarist view (consisting in the love of God, which is an act of the will). As we will see, Eustachius's position on happiness, which can be described as the consensus doctrine in seventeenth-century scholasticism, is somewhat different from all of these.[12]

---

8   Conimbricenses, *In libros Ethicorum* (Lisbon, 1593). The contents of the Coimbrans's *disputations* are generally Thomistic.

9   Corresponding to the Conimbricenses, *In libros Ethicorum*, parts 1–3.

10 Corresponding to the Conimbricenses, *In libros Ethicorum*, parts 4–5.

11 Corresponding to the Conimbricenses, *In libros Ethicorum*, parts 6–9.

12 There are, of course, consequences of this view of the good and happiness for the other topics of ethics, such as virtues and vices, but we will not discuss these here.

Eustachius and the late scholastics generally agree that ethics or morals is not an art, but a genuine science. The question about whether ethics is the same as "prudence," that is, advice about how best to behave in particular situations, is usually raised and answered in the negative. The argument is that principles of prudence are particular principles, allowing us to determine what to do in a given circumstance. Principles of morals, in contrast, are universal and certain; because of their universal nature, a person can know the principles of morals and what is good or bad in general, but still choose what is bad in a particular case: we do not always act according to what we know but may be corrupted by our passions or vice.[13] The conclusion is that ethics is a science subordinate to physics, meaning that it bases its conclusions on the—also certain—principles of physics. As with all subalternate sciences, the object of the subalternate science (ethics, in this case) is defined by the addition of some difference to the object of the superior science (physics, in this case), with which it shares a genus. For Eustachius, the object of ethics is human action directed towards a moral good.[14]

There is, in addition, a general consensus among Eustachius and the late scholastics that ethics is a practical science, as opposed to a theoretical science. By this they mean that the aim of ethics is activity, as opposed to contemplation. Eustachius asserts in his Preface that the end of all philosophy is human happiness, but happiness consists partly in the contemplation of the truth and partly in action in accordance with virtue. Hence, in addition to the contemplative sciences there must be a science that provides an account of what is right and honourable and instructs us in virtue and moral probity. Eustachius identifies the latter science as "ethics, that is, moral learning, or the science of morals," which is "traditionally considered to be one of the chief parts of philosophy."[15]

As for happiness, the apparent consensus doctrine, which is shared by Eustachius and other late scholastics—such as the Dominican and Thomist Antoine Goudin (1639–1695), the ex-Jesuit René de Ceriziers (1603–1662) and, in most respects, the Franciscan and Scotist Claude Frassen (1620–1711)—makes

---

13   Eustachius, *Summa* II, *Quaestiones prooemiales, De ipsa morali disciplina*, q. 1: *Quid sit moralis philosophia; et an sit vere scientia*, 2–3; see also Goudin, *Philosophia* (Paris, 1668), III, 4–5.

14   Eustachius, *Summa* II, *Quaestiones prooemiales, De ipsa morali disciplina*, q. II: *Quodnam sit subjectum moralis scientiae*, 4. For Goudin the object of physics is vital motions and human affections insofar as they proceed from a living soul, while ethics studies those same affections insofar as they apply to morality and are considered by reason as good or bad (Goudin, *Philosophia* III, 3).

15   Eustachius, *Summa* II, *Quaestiones prooemiales, De ipsa morali disciplina*, *praefatio*, 1.

particular use of the distinction between objective and formal human blessed-ness. One argues that there is an ultimate end to human life, because there cannot be an infinite chain of final causes without a first final cause that begins to move the will. And if there were not an ultimate final cause for human life, human desires would be in vain. This ultimate end of human life must be sought in and for itself.[16] The good that serves as the object of our happi-ness must be supremely good, absolute, perfect, sufficient, most desirable and most delectable.[17] There are two features in happiness: the object whose pos-session makes us happy and the state that results from possessing this object; thus happiness can be objective or formal, depending upon whether one refers to the object or to the state.[18] Thinking of the object of happiness, we can eas-ily conclude that happiness cannot reside in any external goods or fortune, in the goods of the body, or the goods of the soul—not in riches, honours, glory, power, corporeal pleasures or the operation of the soul.[19] Man's happiness, both natural and supernatural, resides only in God.[20] Referring to the formal happi-ness we can acquire, that perfect happiness cannot be obtained in this life, but humans can obtain an imperfect happiness in this life. Perfect formal happi-ness resides in the intellect, in the vision of the divine essence, and natural formal happiness resides in the activity of the intellect, that is, in the most per-fect contemplation we can have of God in the natural order.[21]

---

16 Goudin, *Philosophia* III, q. 1, art. 1, 8–16; Eustachius, *Summa* II, part 1, *Ethica*, q. IV: *Quotuplex sit bonum*, 14.

17 De Ceriziers, *Le philosophe français* (Paris, 1643), IV, 118; Eustachius, *Summa* II, part 1, *Ethica, quid sit, et qualis esse debeat haec beatitudo*, 22.

18 Eustachius, *Summa* II, part 1, *Ethica, De ipsa felicitate*, 21.

19 Eustachius, *Summa* II, pars 1, *Ethica, De beatitudine hominis objectiva*, q. II: *In quoniam bonorum genere ponenda sit felicitas objectiva*, 23.

20 Goudin, *Philosophia* III, q. 1, art. 2, 16–32; Eustachius, *Summa* II, part 1, *Ethica, De beatitudine hominis objectiva*, q. II: *In quoniam bonorum genere ponenda sit felicitas objectiva*, 14–15.

21 Goudin, *Philosophia* III, q. 1, art. 3, 32–56. Eustachius, *Summa* II, part 1, *Ethica*, q. II. *Quaenam sit supernaturalis felicitas tum hujus alterius vitae*, 28–29. The Protestant, Pierre du Moulin (1568–1658) likewise argues that felicity, or the end of human life, must be praise-worthy and desirable in itself and that the means toward this end must be so as well. Feli-city is the end for man, not qua citizen or king, policeman or student, but qua human—not for a portion of life but for a whole life. Moreover there must be such an end: "God and nature do nothing in vain ... and there is a natural desire in man for felicity, which would be in vain if it were impossible to be satisfied"; du Moulin, *Les elemens de la philosophie morale* (Paris, 1643), 48–49. Similarly, du Moulin discusses the false supreme goods, such as honour and riches, and argues that happiness does not reside in power, pleasure, or habit, but in activity, and this activity must be proper to the noblest of our faculties, meaning the understanding rather than the will; du Moulin, *Les elemens de la philosophie morale*, 51–52.

There are, of course, differences among philosophers regarding these conclusions. De Ceriziers argues that formal human felicity is a most perfect operation of the principal human faculty (allegedly in agreement with Aristotle); the question to be resolved is whether this action belongs to the faculty of understanding, as Aquinas thinks, or that of the will, as Duns Scotus believes. De Ceriziers's answer to this is dual, of course, depending upon whether one is speaking about our future life in heaven or our present life in the here-below. In heaven we cannot perceive God without loving him or love him without perceiving him; nonetheless de Ceriziers argues that the essence of supernatural felicity consists in the action of the understanding, the noblest of our faculties—and in that way he believes that he comes to agree with Aristotle, Plato, and biblical prophecy. He places the felicity for our present life in the love of the supreme being, meaning in our faculty of will, though he admits that something would be missing from our felicity in the present life if we were to love God without tasting the sweetness of the divine object. De Ceriziers summarizes his thought by asserting "eternal beatitude consists in the knowledge of God and temporal beatitude in love of God."[22]

The Scotist theologian Claude Frassen (1620–1711) holds a broadly comparable view. He also divides beatitude into natural and supernatural, perfect and imperfect, and distinguishes between objective and formal beatitude.[23] But he argues against the philosophers, including Aristotle, claiming that objective beatitude does not consist in a good soul or in its habits, nor in the goods of the body, or honour, etc.[24] Objective human beatitude resides in God alone, who is the object of that beatitude.[25] Formal human beatitude consists in the contemplation of God and more principally, in the love of God, the former residing in the understanding and the latter in the will.[26]

Still some, such as the early seventeenth century French textbook authors Théophraste Bouju (fl. 1602–1614) and Scipion Dupleix (1569–1661), maintain a more properly Aristotelian position. Bouju follows the same kind of argumentative path we have seen but comes to radically different conclusions. He argues that if everyone sought for an illusory felicity they could never attain, God and nature would be operating in vain, but that this cannot be the case. Human nature cannot be deceived at all times. A false opinion is only an infirmity of the understanding, and since defects are accidents, they cannot be in us uni-

---

22   De Ceriziers, *Le philosophe français* IV, 124.

23   Frassen, *Philosophia academica* (Paris, 1657), IV, 39–43.

24   Frassen, *Philosophia academica* IV, 43–55.

25   Frassen, *Philosophia academica* IV, 55–57.

26   Frassen, *Philosophia academica* IV, 57–62.

versally and always; thus a judgment held always and by everyone cannot be false.[27] Bouju lists the following conditions for the human happiness we can have in this life (as understood through our "natural light"): it is a good; it is pleasurable and brings the greatest joy; it is something within our power; it can be gotten easily; it is the most desirable of all human goods; it is sufficient, perfect, and desirable in itself, not for something else; it brings tranquillity; and it is the ultimate end of all human actions, though not something fleeting, but for the long run.[28] Given these conditions, it becomes clear that happiness does not consist in external goods, such as riches, power and worldly authority, the favour of eminent people, good fortune, the goods of the body, such as pleasure, health, and beauty, the goods of the mind, such as contentment and pleasure, the affection of the person loved, amusement and diversion, honour, praise and glory, or even the habit of virtue.[29] According to Bouju, human felicity consists in the activity of the soul in accordance with the virtues of perfect wisdom and prudence. This alone, he argues, fits his conditions: wisdom and prudence are goods of the noblest part of our souls; they are accompanied with pleasure and contentment; they are in our power, easy to exercise, the most excellent good for man, the only sufficient, perfect, accomplished goods; they cause in us tranquillity and rest; and they are such that the ultimate end or perfection of man consists in their activity.[30] Like Bouju, Dupleix argues for Aristotle's position that "the supreme good or human felicity is activity of the soul in conformity to virtue in a perfect life," with the addition that this is to be understood for both the active and the contemplative life, and contends that the doctrine is in conformity with, or at least is not repugnant to Christian theology.[31]

To sum up: while some Scholastics argue for a naturalistic, anthropocentric ethics in which happiness resides in the activity of the soul, the general agreement (displayed by Eustachius et al.) is theocentric: happiness is divided into objective happiness, which has God as its object, and formal happiness, whether natural or supernatural, which resides in the intellect and requires both of its faculties: the understanding and the will. There are still some differences as to whether the essence of that formal happiness resides principally in the understanding, that is, in the vision of the divine essence, which entails the love of God, or more principally in the love of God, that is, an act of the will, that requires the contemplation of God as well.

---

27   Bouju, *Corps de toute la philosophie* (Paris, 1614), *Morale*, 6.
28   Bouju, *Corps de toute la philosophie*, *Morale*, 9–12.
29   Bouju, *Corps de toute la philosophie*, *Morale*, 12–19 and 26–34.
30   Bouju, *Philosophie*, *Morale*, 34–39.
31   Dupleix, *L'ethyque* (Paris, 1610), 131 and 135–138.

3   Ethics in Descartes and the Cartesians

Descartes did not say much about ethics or happiness in his published works; there is, of course, his provisional morality in *Discourse* (1637), Part III, and his thoughts in *Passions of the Soul* (1649), Part III, about how best to comport one-self. Descartes did intend ultimately to write a treatise on morals. As he said in the Preface to the 1647 French edition of the *Principles*:

> I believe myself to have begun to explain the whole of philosophy in sequence [...]. To carry this plan to a conclusion, I should afterwards in the same way explain in further detail the nature of each of the other bodies on the earth, that is, minerals, plants, animals, and above all man, then finally treat exactly of medicine, morals, and mechanics. All this I should have to do in order to give to mankind a complete body of philosophy.[32]

Descartes's metaphor of a tree of philosophy, wherein one is depicted as gathering fruits from the tree, requires the establishment of the tree's metaphysical roots, its physical trunk, and its branches constituted by medicine, mechanics, and morals. The problem, however, is that Descartes claims not to have performed sufficient experiments to finish the trunk, so he cannot yet endeavour to venture onto its non-existent branches—that is, he cannot write treatises on medicine and morality—until he has solved a variety of problems concerning animals and humans. Accordingly, he limited himself to some disconnected and unpublished thoughts about morals and happiness in his correspondence. Thus, a more adequate understanding of what might constitute a Cartesian ethics had to wait for the posthumous publication of Descartes's letters (1657–1667) by Claude Clerselier. Further, it is clear that Clerselier's collection of letters was not just a random selection, but the result of Clerselier's wanting to construct new Cartesian texts to fill the gaps in the extant corpus, starting with ethics.

---

32   "Je pense avoir commencé à expliquer toute la Philosophie par ordre [...] Mais, afin de conduire ce dessein jusques à sa fin, je devrois cy-apres expliquer en mesme façon la nature de chacun des autres corps plus particuliers qui sont sur la terre, à sçauoir des minéraux, des plantes, des animaux, & principalement de l'homme; puis, enfin, traitter exactement de la Médecine, de la Morale, & des Mechaniques. C'est ce qu'il faudroit que je fisse pour donner aux hommes un corps de Philosophie tout entier." Descartes, *Principes de la philosophie, Préface*, AT IX-B 16–17, CSMK I 187.

3.1   *Clerselier and the* Lettres de Mr Descartes

Clerselier's first volume of Descartes's correspondence begins with a 1647 letter to Queen Christina on the supreme good, and continues with letters to Princess Elisabeth from 1645 on the happy life. In his preface to the volume Clerselier even argues that his collection of Descartes's letters is equivalent to any other of Descartes's writings, since "one should not fear the public censure of what is written for Princesses and for the most learned people in Europe."[33] What is addressed to such people, who are esteemed for their rank, knowledge, or virtue, will assuredly be well-considered and highly polished. Clerselier then asserts that the highest and most useful subject is the one Descartes examines in his letter to Queen Christina, namely, the topic of the supreme good, which he treated as well in the letters to Princess Elisabeth. He writes:

> Descartes allowed people to see, in these letters, that ethics was one of his most common meditations, and that he was not so powerfully engaged with the consideration of things that happen up in the air, or with the inquiry into the secret paths nature observes in the production of its works here below, such that he failed to reflect frequently on himself, and [...] to regulate the actions of his life, following the true reason. [...] After this, I do not think that any of the people who in their writings accused him of vanity in his studies as being completely engaged with an inquiry into the empty things of which science fills the mind, instead of those that instruct and perfect man, will be able to reproach him in that way.[34]

Following Clerselier's volumes of *Letters* there appeared an anonymous work which was particularly indebted to Clerselier's edition of Descartes's correspondence: *L'art de vivre heureux* (1687) constructs a Cartesian-style ethics from a variety of sources, but especially from Descartes's letters to Christina and to Elizabeth, prominently featured in Clerselier's correspondence.

---

33   Descartes, *Lettres*, ed. Clerselier, vol. 1, preface, n.p..

34   "C'est dans ces lettres ou il a fait voir que la Morale estoit l'une de ses plus ordinaires Meditations, & qu'il n'estoit pas si fort occupé à la consideration des choses qui se passent dans l'air, ny à la recherche des secrettes voyes que la nature observe icy bas dans la production de ses ouvrages, qu'il ne fist souvent reflexion sur luy-mesme, & [...] à regler les actions de sa vie suivant la vraye raison. [...] Aprés quoy, je ne pense pas qu'il y en ait plus aucun, de ceux qui dans leur écrits l'ont accusé de vanité en ses études, comme s'attachant entierement à la recherche des choses vaines, & dont la science enfle l'esprit, au lieu de celles qui instruisent & qui perfectionnent l'homme, qui ose plus luy faire un semblable reproche." Descartes, *Lettres*, ed. Clerselier, vol. 1, preface, n.p..

3.2 *Anonymous and* L'art de vivre heureux

Part I of the treatise discusses human happiness in the here-below. The author sets aside the supernatural happiness of saints in the state of grace, and makes room for a natural and rational kind of happiness that can be attained in this life, in spite of our fallen state. He argues that there are goods to be attained in this life, apart from grace and faith, which, though useless for salvation, permit us to perform morally good acts. These preliminaries allow the author to continue with an extended paraphrase of Descartes's letter to Christina: the only supreme absolute good is God; but there are goods relative to us that depend on us (such as virtue and wisdom) and those independent of us (such as honours, riches, and health), that is, goods of the body and fortune. Happiness or the most solid contentment consists in what is in our power, that is, the goods of our mind: knowing and willing. In Part II of his treatise, the anonymous author continues with a discussion of the nature of the human soul. He calls Aristotle's opinion on the subject "dangerous and obscure,"[35] and adopts the Augustinian-Cartesian view that "the soul is a substance that has only thought as attribute, from which one concludes that it is spiritual and immortal."[36] He follows the discussion of human souls with a few chapters on Cartesian animal-machines, and concludes Part II with chapters on the two faculties of the soul, understanding and will, again in the style of Descartes. Part III of the treatise, on the application and right use of the two powers of our souls, re-joins the discussion of ethics with an extended paraphrase of the letters to Elisabeth, and lists three conditions useful for acquiring felicity: trying always to use our minds as well as possible to discover what we should do in all the circumstances of our lives; having a firm and constant resolution to execute everything advised to us by reason, without allowing our passions or appetites to divert us; and considering that while we are conducting ourselves in this manner, the goods we do not possess are entirely outside our power.[37] In those letters, given our imperfect knowledge, the further truths we need to keep in mind in order to judge well are the existence of God, the nature of our souls, and our distinctness from every part of the universe. Here these are understood as the three principal truths by which to guide our conduct, toward God, the self, and others, namely: there is a God, on which all things depend; know thyself, that is, you should know the nature of your soul; and you should prefer the interests of the whole to your

---

35   Anonymous, *L'art de vivre heureux* (Paris, 1687), ed. Charles, 67; see also 67–73. The work
       is sometimes attributed to the Oratorian Claude Ameline (1635–1706), but it is an unlikely
       attribution; see the introduction by the editor, 7–27.

36   Anonymous, *L'art de vivre heureux*, 73; see also 73–76.

37   Anonymous, *L'art de vivre heureux*, 53–58 and 113–115.

particular interests. For the anonymous author, the passions enter into the discussion only insofar as they can trouble the will, whose constancy constitutes virtue.

Although providing a rather limited perspective, the author of *L'art de vivre heureux* seems to have understood Descartes fairly well (though as an *art* of living well, not as a *science*), delineating an anthropocentric ethics based on the Cartesian view of the soul and its two functions, all in parallel with and apart from a theocentric, supernatural ethics.

3.3   *Descartes's* Ethice

But, arguably, the more interesting offshoot of Clerselier's efforts is the unusual book of Descartes's ethics previously mentioned, the Latin-language student manual printed in London in 1685.[38] Descartes never wrote such a book, but the clever editor was able to put together a three-part treatise out of Descartes's own words from his correspondence with Christina, Elisabeth, Denis Mesland, and Hector-Pierre Chanut, and from the Latin version of *Passions of the Soul*. It may look as though the translator has made a concerted selection from Descartes's letters, but in fact he is just following Clerselier's edition (or, more precisely, the Latin translation of it published in London in 1668). The beginning of *Descartes's Ethics* concerns the supreme good, the happy life, and free will.[39] And the end contains a discourse on intellectual and sensual love.[40] In between these essays constructed from the correspondence is an abbreviated—and at times reordered—version of texts from all three parts of *Passions of the Soul*, with the physiological passages deleted. Completely missing, for example, are the articles on the parts of the body and their functions, the movement of the heart, the animal spirits in the brain, the movements of the muscles, and the sense organs.[41] Also missing are the articles about the

---

38   Descartes, *Ethice*.

39   1. *De summo bono* (from a letter to Christina, 20 November 1647). 2. *De vita beata* (from the following letters to Elisabeth: 4 August 1645, 1 September 1645, 15 September 1645, January 1646). 3. *De libero arbitrio* (from the following letters to Mersenne or Mesland: 27 May 1641, 2 May 1644; and to Elisabeth: January 1646, 3 November 1645).

40   Starting with a fragment of a letter to Chanut (1 February 1647), *Quid sit amor?* and continuing with a discussion of topics such as: *Utrum solo lumine naturali Deum amare doceamur* (To Chanut, 1 February 1647); *Quae sint causae quibus ad hominem unum magis quam alium, etiam incognitis meritis, amandum ferimur* (To Chanut, 6 June 1647); *Uter sit deterior, amoris an odii excessus?* (To Chanut, 1 February 1647); *De laeto animo* (To Elisabeth, October or November 1646); *An satius est laeto esse animo et contento imaginando ea, quae possidemus bona majora et meliora, quam sunt; an vero accuratius pensitare justum utrorumque valorem, atque inde tristitiam contrahere?* (To Elisabeth, 6 October 1645).

41   Also missing are articles 34–38 about the pineal gland.

order and enumeration of the passions and about the physiological effects of the passions on the body.[42] By far the largest section of this middle portion of *Descartes's Ethics* concerns Part III of *Passions of the Soul*, which is the smallest portion—less than a third—of Descartes's treatise, but the one most devoted to moral philosophy. *Descartes's Ethice* provides a broader perspective on Descartes's views of happiness than does *L'art de vivre heureux*, since it adds to Descartes's treatment of happiness (from the letters to Christina and Elisabeth) materials about the passions (from *Passions of the Soul*) and on love (from the Letters to Chanut and Elisabeth).[43]

3.4   *Le Grand's* Ethicks

Descartes's *Ethice* also supplies sufficient texts for Antoine Le Grand to fashion a Cartesian ethics, something he was disseminating at the time as part of his Scholastic textbook intended to teach Cartesian philosophy. In the preface to the last part of his *Institution of Philosophy*, on *Ethicks*, Le Grand states:

> I would also have the Reader take notice, that in this Treatise I follow the Sentiments of DES CARTES: and tho' he hath writ but little concerning *Moral Philosophy*, yet I have a mind to raise this structure upon the Foundation he hath laid, and from what he hath Writ concerning the *Soul of Man*, and the *Passions* to discover his Sense of *Moral Matters*.[44]

As Le Grand intimates, he believes he can represent a complete Cartesian physics, including parts on man, both in respect to his body and in respect to his mind or soul, as a ground for a Cartesian ethics.[45] Thus, after treating man in relation to his body and his soul, and discussing passions of the soul, Le Grand produces a Cartesian ethics, with considerations on such topics as: the greatest

---

42   With the exception of the first sentence from art. 107, and the title and last sentence of art. 112.

43   Part III of the *Passions de l'âme* is just under 27 pp., while parts I and II are 27 and 42 pp., respectively. In the London textbook, ch. I (*De passionibus in genere*) is 19 pp., ch. II (*De passionibus in specie*) 18 pp., and ch. III (*De illis passionibus, quae primitivas sequuntur*), 28 pp.

44   Le Grand, *Entire Body of Philosophy* (London, 1694), vol. 1, 347[b].

45   The *Institution of Philosophy* is divided into ten parts, the first three being Logick, Natural Theology, and Daemonology, and the next six constituting the whole of Physicks, from General Physicks to the World and Heaven, the Four Great Bodies, Earth, Water, Air, and Fire, then Living Things, such as Plants and Animals, and finally Man, both in respect to his Body and in respect to his Soul or Mind. The last part is Ethicks, or Moral Philosophy, treating Man's right Ordering of his Life.

good, the nature of virtue, the usefulness of the passions, their governance, and the more general remedies for them. He begins by arguing that external goods are not the good of man, and then comes to the main question: What is the highest good of the human being in this life, and the ultimate end whereof? He distinguishes between mankind and the private individual, and asserts that the supreme good for mankind is the concurrence of all perfections of which he is capable, the goods of the soul and body and fortune. But for the private individual the supreme good is the right use of reason, which consists in "his having a firm and constant purpose of always doing that, which he judges to be the best."[46] This, of course, is in our power, whereas the goods of body and fortune are not. The proper use of our two main intellectual faculties also produces a satisfaction of mind. The doctrine is encapsulated in the three things we need to observe, which are said to be the foundation of all ethics. The first is that we "strive to attain the *Knowledge* of what we ought to embrace." The second is that

> we stand firm and constant to what we have once resolved upon and purposed; that is, that we retain an immovable *Mind* and *Will*, of doing those things which *Reason commands*, not suffering our *Passions* and corrupt *Inclinations* to lead us aside.[47]

And the third is "that we lay down as unmovable *Ground* and *Principle*, that nothing besides our own *Thoughts* is in our *Power*." Le Grand concludes "that the *Natural Happiness* of *Man* is nothing else but that *Tranquility* or *Joy* of *Mind*, which springs from his Possession or Enjoyment of the *Highest Good*."[48]

Given these ethical foundations, Le Grand examines how to avoid the excesses and ill-use of the passions, but first he argues, against the Stoics and for the Cartesian view, that the passions or affections "are good and contribute to the Perfection of *Human Life*," when the objects of the passions are lawful and the passions proportionate to their objects.[49] According to Le Grand, the passions do not lead humans to vice, but are useful so long as they

> are subject to the command and guidance of *Reason* and proportion'd to their *objects* and *end*; which only takes place when those things are *Loved*

---

46   Le Grand, *Entire Body of Philosophy*, vol. 1, 353[b].
47   Le Grand, *Entire Body of Philosophy*, vol. 1, 353[b].
48   Le Grand, *Entire Body of Philosophy*, vol. 1, 353[b].
49   Le Grand, *Entire Body of Philosophy*, vol. 1, 368[a].

that ought to be *Loved* and when such *Objects* are loved in a higher degree, which because of their greater worth deserve more of our *Love*.[50]

In the chapter on the governance of the passions and the remedies for them, he discusses generosity as another general remedy: it is "the *Key* to all *Vertue*," and "a powerful means to subdue and moderate our *Affections*." Since generosity consists in valuing and esteeming ourselves to the utmost of our worth, we can attain felicity if we find in ourselves a constant resolution to make good use of our will, that is, to undertake what we judge to be best, given that nothing properly belongs to ourselves other than how we dispose of our will and choice.[51] Still, Le Grand does not end his discussion of remedies with generosity. He adds that "the most powerful *Antidote* against our *Affections* is the Love of GOD."[52] While he cannot refer to *Passions of the Soul* for this, he does think it is Descartes's view and concludes by referring his reader to the 35th Epistle of Volume 1 of Descartes's correspondence, which, of course, is the *Dissertation on Love*, that is, the letter to Chanut of 1 February 1647 that made up much of the third part of *Descartes's Ethics*.

In the letter to Chanut, Descartes is answering questions from Queen Christina such as "What is love?" and "Does the natural light by itself teach us to love God?" In the first question, Descartes distinguishes between intellectual or rational love and love as a passion involving the body. According to him, intellectual love "consists simply in the fact that when our soul perceives some present or absent good, which it judges to be fitting for itself, it joins itself to it willingly."[53] But he asserts that in this life, when the soul is joined to the body, rational love is accompanied by sensual or sensitive love:

> These two loves commonly occur together; for the two are so linked that when the soul judges an object to be worthy of it, it immediately makes the heart disposed to the motions that excite the passion of love; and when the heart is similarly disposed by other causes, it makes the soul imagine lovable qualities in objects.[54]

---

50   Le Grand, *Entire Body of Philosophy*, vol. 1, 368[b].

51   Le Grand, *Entire Body of Philosophy*, vol. 1, 376[a]. These, of course, are from the letters to Elizabeth.

52   Le Grand, *Entire Body of Philosophy*, vol. 1, 376[a-b].

53   Descartes to Chanut, 1 February 1647, AT IV 602.

54   Descartes to Chanut, 1 February 1647, AT IV 603.

This intermingling of the two loves makes it difficult for the natural light by itself to be teaching us to love God; in this life the love of God cannot be purely intellectual, but must also have a sensitive aspect: "it must pass through the imagination to come from the understanding to the senses."[55] Only then can knowledge of God through the natural light be said to properly teach us to love God. Descartes produces a number of powerful objections against the possibility of love of God in this life: God's attributes are beyond us and nothing about God can be visualized by the imagination. But he sketches a way for us to attain the love of God by having our mind represent to itself the truths that excite in us the love of God. We should consider that God is a mind, that our soul's nature resembles his; we should take account of the infinity of his power, the extent of his providence, etc. This allows us to communicate this love to the imaginative faculty:

> we can imagine our love itself, which consists in our wanting to unite ourselves to some object. That is, we can consider ourselves in relation to God as a minute part of all the immensity of the created universe. [...] And the idea of such a union by itself is sufficient to produce heat around the heart and cause a violent passion.[56]

Descartes concludes that "our love for God should be, beyond comparison, the greatest and most perfect of all our loves."[57] This final Cartesian doctrine introduces a theocentric element in Cartesian ethics, but it also causes it to be somewhat less intellectualist, since it requires the body and imagination to play a significant role.

The Cartesians developed what they thought were Descartes's views about the ultimate morality, derived from the branches of the tree of philosophy, that is, ethics considered as dependent on the roots and trunk of the tree, or on metaphysics and physics. Scholastic ethics was likewise subordinated to physics, but the similarities between Cartesian and Scholastic ethics seemed few; the break with Scholasticism looked quite definitive. Cartesian ethics was resolutely anthropocentric: good is a perfection belonging to us; the greatest good was not connected with the goods of body and fortune, which do not depend upon us, but rather with the goods of the soul; the supreme good is a "firm and constant resolution to do everything we judge to be best and to use all

55   Descartes to Chanut, 1 February 1647, AT IV 607.
56   Descartes to Chanut, 1 February 1647, AT IV 610.
57   Descartes to Chanut, 1 February 1647, AT IV 613.

our power of mind to know these," and this by itself constitutes all the virtues; happiness and virtue are thus things within our control. While the late Scholastics also held that happiness cannot reside in any created good—not in riches, honours, glory, power, or corporeal pleasures—most of them held that human happiness, both natural and supernatural, has only God as its object. Happiness, whether perfect or imperfect, resides in the intellect, in the vision of the divine essence, or in the will, in the love of God. However, the final twist in Cartesian ethics also involves the love of God; this theocentric element allows Cartesian ethics to re-join traditional accounts and provides a greater convergence between late-scholastic and Cartesian ethics. This is the contrast between the late-scholastic ethics of Eustachius and the fleshed out Cartesian Ethics of Le Grand et al. that might have been taught at Cambridge University at the end of the seventeenth century and beginning of the eighteenth.

CHAPTER 14

Teaching Magnetism in a Cartesian World, 1650–1700

Christoph Sander

1 Introduction

Just before Christmas in 1641, the University of Utrecht held its usual rounds of disputations in philosophy. A candidate and student of Henricus Regius (1598–1679) claimed that every natural phenomenon could be explained by René Descartes's new philosophy without employing concepts such as the 'forms' or 'qualities' of the Aristotelians.[1] His opponent, Lambertus Van den Waterlaet (1619–1678), asked him how he would account for magnetic attraction without referring to 'forms' and 'qualities.' The unknown student was stumped and did not know how to answer. Somewhat feebly, he offered the announcement that Descartes was working on such an explanation. Gisbertus Voetius (1589–1676), who was presiding over the disputation, reacted with sarcasm: feeding the committee with hopes would be convenient for Jewish philosophers, he pronounced, who would give up as soon as they face the smallest difficulties and simply trust on the arrival of Elias, heralding the Messiah.[2]

Three years later, in July 1644, Descartes published his *Principles of Philosophy*. The longest section, namely fifteen per cent of the entire work, is devoted to a single topic: magnetic phenomena.[3] The use of magnetism, e.g., in navig-

---

1 This is reported in Schoock, *Admiranda methodus* (Utrecht, 1643), **** 2ʳ: "opponente [...] inferente, absque formis et qualitatibus praesentis omnia naturalia explicari posse. Cum eius specimen edi peteret Defendens et cum eo praeses vel in uno magnete, excipit arcanum quid inesse, atque ita concederentur qualitates occultae, quas tamen semper negare videntur." Cf. also Verbeek, *La querelle d'Utrecht*, 177; Bos, "Correspondence," 96, n. 9; Duker, *School-gezag en eigen-onderzoek*, 96–97.

2 Schoock, *Admiranda methodus*, **** 2ᵛ: "similes esse tales philosophos Iudaeis seu Rabbinis, qui quotiescumque aqua ipsis haeret aut nodus insolubilis occurrit, dicere solent 'Elias veniet.' Interim spem pretio apud nos non emi." This alludes to Mal. 3, 23–24. On Schoock's and Voetius's anti-Semitic tendencies, cf. Pollmann, "The Bond of Christian Piety," 66.

3 Magnetism comprises 55 paragraphs, cf. Descartes, *Principia philosophiae* IV 133–184, AT VIII-A 275–311, *Principles* 259–293. For a quantitative analysis, cf. especially Meschini, *Indice dei Principia philosophiae*, 407. The lexeme MAGNES counts 198 instances. On Descartes's theory of magnetism, cf. Sander, *Magnes*, 717–743. This theory will not be re-sketched in this

ation with compass needles and the difficulty to explain the alleged magnetic action at a distance in natural philosophy had made magnetism a highly relevant topic by then, both with regard to theoretical and practical matters.[4] Descartes's main goal was the corpuscularian and mechanistic explanation of thirty-four magnetic properties, which Descartes knew mainly through the landmark publication *De magnete* (1600) by William Gilbert (1544–1603), and through his correspondence with the polymath Marin Mersenne (1588–1648).

In his explanation, Descartes first postulates a 'subtle matter' that pervades all seemingly solid bodies.[5] Then, to explain magnetism, he assumes compounds of matter as screw-shaped particles which he calls *particulae striatae*. Any magnetic body, even the earth itself, has a certain sphere of activity limiting its interaction with other magnetic bodies, defined by the radius of the orbital path along which the screw particles travel (see Figure 14.1). These screw particles can enter corresponding threads that run exclusively through magnetic bodies. Alignment of a magnet or a magnetic needle along the north-south axis of the magnetic Earth is caused by the flow of the particles that 'push' the magnetic object in the respective direction. In an equally mechanistic fashion, he explains attraction and repulsion between magnets. The details of this theory do not need to concern us at present, but it is already evident that the Cartesian theory is radically different from the so-called 'occult qualities' of the Scholastic tradition.[6] Descartes's account of magnetism is thus paradigmatic of his more general agenda to account for all physical phenomena by means of nothing but the geometric and kinematic properties of corpuscles— an account that can be labelled as "mechanistic."

Descartes's account of magnetism was highly anticipated; he frequently mentioned it in his correspondence, and it was widely discussed, criticized and celebrated shortly after its publication.[7] With his *Principles of Philosophy*, Descartes's aimed at creating a university textbook following and taking over the role of Scholastic textbooks used in the schools of his day.[8] He did so by implementing original rhetorical and visual strategies and by integrating a lot

---

chapter, as there is various literature dealing with it already. See the references in Sander, *Magnes*, 717–743; Sander, "*Terra AB*."

4   See Sander, *Magnes*.

5   See especially AT XI 24, AT I 176. This subtle matter was deduced from a theory of elements presented earlier in his *Principia*.

6   See note 13.

7   Sander, *Magnes*, 719–728; Strazzoni, "How Did Regius Become Regius?," 374, n. 51; Van Berkel, "Descartes' Debt to Beeckman," 48–59.

8   Cf. as a starting point, Ariew, *Descartes among the Scholastics*. On Descartes's intentions concerning the teaching of his philosophy, see Chapter 1, in this volume.

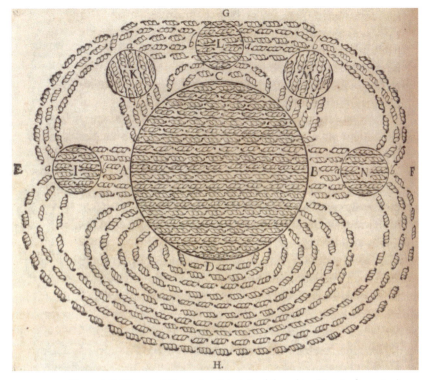

FIGURE 14.1 Descartes's "Terra AB". Descartes, *Principia philosophiae* (Amsterdam, 1644),
          IV 146, 271.
          BERLIN, MAX-PLANCK-INSTITUT FÜR WISSENSCHAFTSGESCHICHTE

of topics—e.g., magnetism—into the canon of textbook natural philosophy.[9]
Moreover, studies have shown how Descartes's role within the Utrecht controversy impacted on his philosophy and shaped some aspects of his *Principia*.[10]
Magnetism was a subject of longstanding controversy in Utrecht university disputations, in which Descartes's magnetism theory was also closely associated
with its advocacy on the part of Regius.

What has not yet been the subject of scholarly work and will thus be the
focus of this chapter is the fate of Descartes's theory of magnetism in the
classroom after the publication of his *Principles of Philosophy*. Many writ-

---

9   His highly persuasive, richly illustrated and up-to-date theory of magnetism is a significant
    example of this pedagogical strategy. Cf. Zittel, *Theatrum philosophicum*, 382–395; Lüthy,
    "Where Logical Necessity," 97–133; Sander, "*Terra AB.*"
10 The role of his magnetism theory has only occasionally been recognized in this regard. Cf.
    as a starting point Verbeek, *La querelle d'Utrecht*.

ten records testify to a wide diffusion of Descartes's thoughts on magnetism and thereby to the reception of his theory of magnetism as an example of the acceptance, transformation and refusal of Cartesian natural philosophy in European institutions of higher education in the 'Scientific Revolution.' It is almost a commonplace in current historiography to point out that this revolution, if it can be so designated, did not happen in an instant and did not simply replace one paradigm, say Aristotelianism, with another, say Cartesianism.[11] Most historians describe the acceptance of Cartesian thought in the seventeenth century as a long and gradual development, as a non-linear, non-monolithic, complicated, multi-layered process, not free of contradictions. As correct as this historiographical picture is, so much too does it need further confirmation based on historical sources. It first, and maybe foremost, needs case studies that provide cross-sectional analyses over longer durations and considering various locations.

This chapter, as with several more in this volume, aims to contribute to this array of case studies on the inclusion and reception of Cartesianism in academic learning in early modern Europe. Its result will not contest the 'big picture' but will add to it and confirm it. Theories of magnetism, as was the case with many other topics, were controversial among different schools of thought. Explaining magnetism touched core elements of natural philosophy and metaphysics, as well as the question of the extent to which natural phenomena could and should be explained at all—a normative question thus. Moreover, by the time Cartesian philosophy emerged, magnetism was often included as a topic in university curricula, being one of the most important 'enabling conditions' for any academic dispute about Cartesianism. Magnetism as a case study is therefore taken to be a reasonable choice.

2   The Institutional Context of Descartes's Theory of Magnetism

The Utrecht controversy occasioned the first criticism of Descartes's physics within an institutional context.[12] While Descartes was not affiliated to this university, his philosophy was eagerly promoted there by Henricus Regius, with Descartes becoming involved in the controversy through letters and pamphlets. Between 1639 and 1644, Descartes and Regius, proponents of a mechanistic and corpuscularian 'new' natural philosophy, were opposed by those

11   Cf., e.g., Garber, "Descartes and the Scientific Revolution."
12   Cf. also Bos, "Correspondence"; Descartes, *The Correspondence, 1643*, 182–192; Van Ruler, *The Crisis of Causality*; Goudriaan, *Reformed Orthodoxy and Philosophy*.

who defended a Calvinist version of Aristotelianism at Utrecht, especially Gisbertus Voetius and Martin Schoock (1614–1669). The physics of the *Principia*
was to a large degree shaped during this controversy. The explanation of magnetism was part of this but, of course, never the central point of the entire
controversy. However, the debate about the cause of magnetism exhibited
the deeper epistemological and metaphysical disagreements between the two
camps: the dispute frequently returned to the question of how Descartes's 'new
type' of natural philosophy was able to cope with phenomena such as magnetic
attraction that were usually accounted for by means of so-called 'occult qualities.'[13]

Sketching the debate on magnetism at Utrecht prior to the publication of
the *Principles of Philosophy* in 1644 shall uncover a general pattern of critique
against Descartes's theory of magnetism which informed also its later reception. The complicated series of events arguably began with Henricus Regius
being appointed as a professor of theoretical medicine at Utrecht in 1638.[14]
One year later, in 1639, the topic of magnetic attraction was brought up in
a university disputation for the first time—planting the seeds of arguments
(*semina contentionum*) yet to come.[15] From a later record of this disputation,
of which the text itself has not come down to us, it can be inferred that Florian
Schuyl (1619–1669) proposed a traditional Aristotelian explanation of magnetic attraction, while his unnamed opponent, presumably a student of Regius,
defended the 'new philosophy' and refused the 'occult qualities' assumed in
magnetic attraction. Regius, personally present at the disputation, claimed that

---

13   Cf., as a starting point, Weill-Parot, "Astrology, Astral Influences, and Occult Properties";
      Hutchison, "What Happened to Occult Qualities?"; Sander, "Tempering Occult Qualities."
14   On Regius's biography, cf. especially Strazzoni, "How Did Regius Become Regius?."
15   See *Testimonium Academiae Ultrajectinae* (Utrecht, 1643), 14: "Quae hactenus semina con-
      tentionum sub glebis delituisse videbantur, primum erumpere coeperunt, occasione dis-
      putationis D. Florentii Schuilii, pro obtinendo Philosophiae magisterio publice institutae
      9 Jul. anno 1639, ubi cum Opponens, secundum sententiam novae Philosophiae, omnes
      qualitates attractrices et qualitatem occultam magnetis oppugnaret, Medicus [sc. Regius]
      stans in subselliis D. Senguerdio, ordinario Philosophiae Professori et Promotori, satis
      indecore insultavit, et contra Doctiss. Candidatum, D. Senguerdii discipulum, triumphum
      ante victoriam cecinit; cum tamen, omnium Professorum judicio, Candidatus perquam
      solide et dextre omnia objecta dilueret, et non inconcinne Opponentem perstringeret,
      atque ad terminos revocaret." Cf. also Verbeek, *La querelle d'Utrecht*, 86; Bos, "Corres-
      pondence," 24; Van Ruler, *The Crisis of Causality*, 33, n. 68. According to Verbeek, *La quer-
      elle d'Utrecht*, 463, this disputation is not to be found. It has been pointed out that the
      record of the event has a bias against Descartes. Senguerdius was appointed ordinary
      professor on 11 March 1639; cf. Voetius and Le Long, *Hondertjaarige Jubelgedachtenisse*,
      77.

the opponent had won the disputation and ended up in a seemingly heated argument with Arnoldus Senguerdius (1610–1667), Aristotelian philosophy professor and Schuyl's promotor.

The report does not tell us about Regius's mechanistic theory of magnetism, nor even explicitly affirm that it was mechanistic, let alone how Descartes would have explained magnetic attraction. It was first in a medical context, namely in a controversy over the physiology of the heart, that Regius in 1640 implicitly referred to his mechanistic account of magnetic attraction.[16] His allusion to magnetic attraction and its alleged mechanistic explanation remained very short and rather elusive.[17] Meanwhile, in December 1641, Regius presided over another medical disputation at Utrecht, which again discussed the powers of the heart. The candidate was Johannes De Raey (1622–1702), and the text for the first time presented a corpuscularian account of Regius's physiology, whilst including also a short passage on the magnet:[18] Regius promises to give an explanation of magnetic attraction by means of certain ties or *vincula* and openly refused 'occult qualities.' He announces that he will show on a future occasion that magnetic action does not work by attraction but by propulsion instead. That his theory still remained so succinct and a mere promise of a full-blown account might have been due to Descartes, who did not approve Regius's account of magnetism.[19]

---

16   See Regius, *Spongia* (Utrecht, 1640). The context is complex: in a disputation on the circulation of blood, presided over by Regius and defended by Johannes Hayman in 1640, an attractive force of the heart is openly denied. Instead, a mechanistic account of the heart and of blood circulation is developed. The magnet is not mentioned, although the 'attractive force' of the heart in Galenic medicine was sometimes compared to the power of the magnet; William Harvey (1578–1657), to whose theory of blood circulation Regius referred, also used this comparison. See the *Disputatio medico-physiologica pro sanguinis circulatione* in AT III 728, 731, 734. Cf. also Schouten and Goltz, "James Primrose," 331–352; Strazzoni, "The Medical Cartesianism"; Fuchs, *The Mechanization of the Heart*, 146–148; Maire, *Recentiorum disceptationes*, 148, 255. In his *Animadversiones*, James Primerose (ca. 1598–1659), an English physician, not only objected to Harvey's theory, but also read the Utrecht disputation and responded very critically to Regius's and Hayman's mechanistic theory in the same year. It is this writing by Primerose against which Regius, in 1640, argued in his *Spongia*.

17   Primerose returned to the question in 1644 and attacked this account of Regius; see *Anditotum adversus Henri Regii* […] *Spongiam* in Maire, *Recentiorum disceptationes* (Leiden, 1647), 9–10, 31.

18   See *De morborum signis* as part of the *Physiologia*, edited in Bos, "Correspondence," 245. See also Regius, *Fundamenta physices*, 186. On the background, cf. also Galen, *On the Natural Faculties*, III, 15, ed. Brock, 325; Aucante, *La philosophie médicale de Descartes*, 312; Grene, "The Heart and Blood," 324–336; McVaugh, "Losing Ground."

19   Regius sent a draft of his *Physiologia* (which was later printed) to Descartes, including the

De Raey's disputation of December 1641 was closely followed by the incident presented in the introduction of this chapter: Lambertus Van den Waterlaet's disputation, in which Regius and Voetius openly clashed on the question of the magnet. Regius informed Descartes about this disputation one month later, in January 1642. Descartes's own account was a subject of some anticipation in Utrecht University, even though he had not yet by this time developed his theory of magnetism. The argument had by then already become a high-stakes—almost ideological—dispute, although Descartes's presence in the debate up to this point was entirely mediated through Regius.

Descartes's role within the Utrecht controversy, however, was not as indirect as it may appear from these findings. Not only was he in close contact with Regius during this time, but he was also directly targeted by the attacks coming from the university establishment at Utrecht. Prompted by Van den Waterlaet's disputation, in December 1641, Voetius publicly defended traditional Aristotelian concepts such as 'occult qualities,' 'substantial forms,' 'final causes,' and the like, against the attacks put forward by Regius.[20] But Voetius was also clearly taking aim at Cartesian philosophy.[21] According to Voetius, magnetism could not be explained by means of geometrical and kinematic concepts (*per motum, quietem, situm, quantitatem, figuram*), and he accused

---

text of the aforementioned disputation by De Raey. Descartes replied to Regius in May 1641, AT III 546, Bos, "Correspondence," 73: "What you have written on the magnet, I would prefer to leave out, because it is not clear" (*quae habes de magnete, mallem omitti; neque enim adhuc plane sunt certa*).

20   Cf. the *Appendix de rerum naturis et formis substantialibus* of December 1641 as printed in Voetius, *Selectarum disputationum theologicarum pars prima* (Utrecht, 1648), vol. 1, 871–881. Cf. esp. Van Ruler, *The Crisis of Causality*, 9–34. For reactions, see Regius, *Responsio, sive notae in appendicem ad Corollaria theologico-philosophica*, and Bos, "Correspondence," 98–118. By 'occult qualities' Voetius meant, following Daniel Sennert, an unobservable quality of a natural body which was nonetheless the cause of its observable effects. For the philosophical background, cf. also Roux, "La philosophie mécanique (1630–1690)," 44–53; Beck, *Gisbertus Voetius*, 67; Goudriaan, *Reformed Orthodoxy and Philosophy*, 113–125. Cf. also the notes on magnetism in Van den Waterlaet, *Prodromus sive examen tutelare* (Leiden, 1642), vol. 1, 50, 88, 108; vol. 2, 9, 31–32. He refers to Voetius's charges and Regius's replies, and puts forward a list of magnetic phenomena that were to be explained by the deniers of 'occult qualities,' publicly or privately (*ad questiones hasce responsum vel publicum vel privatum exspectamus*); vol 2, 31–32. It seems that he also wanted to make good for his silence in the 1641 disputation. We also find Regius's ideas linked to ancient atomism in this chapter.

21   The theologian Voetius claimed already in 1640, in letters to Marin Mersenne, that Descartes's new philosophy showed heterodox tendencies, whilst Aristotelianism was much better harmonized with the word of the Bible; see Beck, *Gisbertus Voetius*, 64. More literature is referenced in Beck, *Gisbertus Voetius*, 60–90.

his adversaries of pansophism and arrogance as they attempted to reduce inexplicable natural phenomena to the mere collision of bodies.[22] Although Voetius's defence of 'occult qualities' is more sophisticated, it will suffice here to emphasize that, while accusing his opponents of arrogance, he invoked magnetism as a typical example of a physical effect which was to be explained by an 'occult quality.' Both Regius and Descartes, to the contrary, considered the notions of 'occult qualities' and 'substantial forms' as pernicious to the study of nature. They argued that these concepts were, by their very definition, inaccessible to human understanding.[23] Descartes even turned the moral accusation of arrogance back on Voetius. As he put it, it would be even more arrogant to define something as 'occult' only because some quality has not yet yielded to human understanding.[24]

Regius's explanatory account of magnetism in the original draft of the *Physiologia*, of which Descartes had disapproved, must have circulated, because in 1643 an extensive argument against it was articulated by Martin Schoock.[25] Schoock summarizes Regius's position as saying that magnetism does not work by attraction but by exhalations passing out at one pole, travelling around the magnetic body by some *circumpulsio* and entering the other pole. Schoock criticizes Regius for virtually eliminating a whole ontological category, namely the power of attraction, and discredits his whole theory as atomistic, invoking all of the heretical implications of the term.[26]

---

22   See especially Verbeek, "From 'Learned Ignorance' to Scepticism," 31–45. Cf. also Van Ruler, *The Crisis of Causality*, 313.

23   For the text of the *Physiologia*, see Bos, "Correspondence," 240. 'Occult qualities' are also refuted in Regius, *Responsio, sive notae in appendicem ad corollaria theologico-philosophica* (Utrecht, 1642), 28–31. Cf. also a February 1642 letter from Descartes to Regius in AT III 505–506, and Bos, "Correspondence," 116. Cf. also Van Ruler, *The Crisis of Causality*, 207, 244.

24   See the letter from Descartes to Regius, [February 1642], AT III 507.

25   See Schoock, *Admiranda methodus*, 228: "Inter lapides opacos admirandus est magnes, cuius operationes non fiunt per attractionem, sed circumpulsione corporum magneticorum vi exhalationis magnetica e tellure versus septentrionem vel austrum exhalantis."

26   Cf. Schoock, *Admiranda methodus*, 229–230: "Sin vero nec de illis studiosus credere debeat agere per attractionem, quod nihil per attractionem agere soleat, liquido ostendatur attractionem entium classe movendam esse." Cf. Schoock, *Admiranda methodus*, 230–231: "Democriti atomi magneticis profluviis seminibusque turgeant [...] Illationis nostrae absurditatem spongiis delebimus simul ac Philosophantibus, definire placuerit quid aut corpus magneticum sit, aut quid magnetici in quoque corpore inveniatur, et ad qualium entium classem hoc referri debeat." Cf. Schoock, *Admiranda methodus*, 231: "Magnetismi malum in novis quibusque opinionibus invenio, sive enim per attractionem sive circumpulsionem agant, multos ad se invitant, quorum ignorantiam scutica expiandam novi commenti palliat. Scilicet qui antiquam philosophiam in pariete nunquam pictam vider-

When Descartes published his theory of magnetism in 1644, it was neither unheralded nor completely new. Not only had he relied on Isaac Beeckman's ideas about the cause of magnetism—treatment of which is beyond the scope of this chapter—but his close contact with Regius certainly informed his theory.[27] More importantly, the Cartesian theory of magnetism was already—*avant la lettre*—employed and invoked in the Utrecht controversy, mostly through Regius's anticipation of it. Descartes's theory of magnetism was drawn into an educational, institutional conflict, and was exploited by both sides, even though it had not yet been published. Many of the assumptions made by both parties, Regius and Voetius, about which direction it would go— a corpuscularian, mechanistic account, rendering 'occult qualities' useless— were more or less correct; Descartes even concluded his account of magnetism with a general refusal of 'occult qualities' whose alleged effects could be perfectly explained by the particles he imagined.[28] However, his theory of magnetism was much more sophisticated and elaborate than either Regius, Schoock, or Voetius probably might have guessed. For the following, it suffices to underline that Descartes's theory was quite extensive, covering thirty-four magnetic effects, and was based on the concept of a certain, screw-shaped type of particle—the *particulae striatae*.

3 The Dutch Follow-Up

With the 1644 publication of the *Principles of Philosophy*, the controversy about the 'correct' explanation of magnetism in Dutch universities was not over.[29] The first to receive and adapt Descartes's theory of magnetism in Utrecht was certainly Regius himself, but his friendly collaboration with Descartes also came to an end at this time. When Regius published his own physics, the *Fundamenta physices*, in 1646, his description of magnetism seemed at first glance to be a copy of Descartes's account.[30] Descartes himself accused Regius of

---

unt, ventis ac sacrae oblivioni eam consecrarunt, quia per occultam circumpulsionem aut circa Fluddi aut Cartesii magnetem haerent, qui duum viri entia eiusdem se cum simplicitas ad se invitando, alarum remigio ad audacis loquentie superbaeque maledicentie sphaeram facile sublevari possunt." Cf. for more details also Sander, *Magnes*, 703, 726–727, 735.

27   On Descartes and Beeckman, cf. Sander, *Magnes*, 721–722.

28   Cf. Descartes, *Principia philosophiae* IV 187, AT VIII-A 314.

29   For the Dutch reception, cf., as a starting point, Verbeek, *Descartes and the Dutch*; Schmaltz, *Early Modern Cartesianisms*; Strazzoni, *Dutch Cartesianism*.

30   As a starting point, cf. Dechange, "Die frühe Naturphilosophie des Henricus Regius"; Ver-

plagiarism, although he did not specifically point to the section on magnetism.[31]

Regius's claimed to have worked out his account years before, and Schoock's
extensive attack against it one year before the *Principia* was printed seems to
prove this.[32] Regius may have developed an elaborate corpuscularian explanation of magnetic phenomena before or independently of Descartes, although
there is very little evidence predating the publication of the *Fundamenta* in
1646 to suggest that this was the case. Featuring many of the same theoretical elements that are crucial also for Descartes's theory, that of Regius was
conceived as an atomistic theory in the manner of Democritus. Having had
access to the *Fundamenta* in draft form since July 1645, Descartes tried to
prevent Regius from publishing his account. While the two men formed an
alliance in the early stages of the Utrecht controversy, this alliance fell apart
shortly thereafter as they came into conflict.[33] Although their quarrels did
not touch upon the issue of magnetism or natural philosophy in particular,
their discord might have influenced Descartes's silence about any influence
from Regius regarding his theory of magnetism. His use of a different vocabulary to that of Regius, may have been a deliberate strategy on the part of
Descartes to disguise this connection.[34] A student of Regius, Petrus Wassenaer (d. 1688), claimed in 1648 that Regius had developed and taught a theory of

beek, "Regius's *Fundamenta physices*"; Strazzoni, "The Medical Cartesianism of Henricus
Regius."

31   This statement is made in the preface of Descartes, *Les principes de la philosophie*.

32   Regius recollected that it was a passage in the commentary by Galen on Plato's *Timaeus*
that had at first given him the occasion, years earlier, to conceive of magnetic attraction as
a *circumpulsio*. See Regius, *Fundamenta physices*, 141–142: "Atque ex his patet, verum esse
illud Platonis, apud Galenum in Timaeo dicentis, magnetem non per attractionem, sed
circumpulsionem agere, quod, ut dicam quod res est, mihi jam ante multos annos occa-
sionem veram magneticarum operationum causam investigandi et proponendi, primum
dedit." Regius refers to Galen, *Aliquot opuscula nunc primum Venetorum opera inventa et
excusa, quorum sequens tibi pagella catalogum indicabit* (Leiden, 1550), 129: "Ipsum per-
fecto dogma per circumpulsionem [...] appellatam, per quam vult respirationem perspir-
ationemque fieri." There is no evidence suggesting that Descartes was inspired by Plato for
his theory of magnetism. In fact, Plato refers to magnetic attraction in a materialistic manner, analogous to the process of breathing. Cf. Plato, *Timaeus*, 80^c. This certainly motivated
Regius's talk of exhalation in the context of magnetic attraction.

33   Bos, "Henricus Regius et les limites."

34   See Verbeek, "Regius's *Fundamenta physices*." Regius's talk of *exhalatio* and *circumpulsio*
was carefully avoided by Descartes, who used these expressions in other places and
replaced them in the context of magnetism by mere synonyms such as *circumfusum*
or *circumiacens*. Cf. Regius, *Fundamenta physices* (Amsterdam, 1646), 130, 133, 140, and
Descartes, *Principia philosophiae* IV 133, 146, 152, 166, 186, 186, AT VIII-A 275–313.

magnetism "years before" Descartes's major publications.[35] This theory, according to Wassenaer's report, was based on "vortex-like exhalations" (*halitus vorticiosi*).

Regius was not the only one impacted by the Cartesian theory of magnetism. His Dutch colleagues engaged assiduously with Descartes's account, without, however, arriving at a clear consensus in terms of approval or criticism. Arnoldus Senguerdius, for example, who presided over the 1639 disputation that ended in a harsh dispute with Regius, in 1643 presided over another disputation in Utrecht and explicitly referred to any corpuscular explanation as *improbabilis*.[36] Although he does not mention Descartes or Regius, this critique was certainly addressed to them amongst others. Those whom Senguerdius named included William Gilbert, a notorious critic of Aristotelian philosophy and major founder of the magnetic philosophy (*De magnete*, 1600), and the Jesuit Niccolò Cabeo (1586–1650), whose *Philosophia magnetica* (1629) could be seen as an updated Aristotelian reaction to Gilbert's magnetic philosophy.[37] In a later disputation (1652) and in the second edition (1652) of his textbook *Introductio ad physicam* (1644), Senguerdius stated that the corpuscular or atomistic account had only recently been refuted (*dudum explosa*), and Descartes is mentioned in a short bibliography amongst those who "ascribe magnetic operations to the movement of corpuscles."[38]

35   See Regius, *Brevis explicatio mentis humanae, sive, animae rationalis* (Utrecht, 1657), 13: "Iam ante multos annos, cum a te nondum quicquam praeter Methodum, Metora, et Dioptricam, in publicam lucem prodisset, docuit, ut plurimis eijus auditoribus constat, [...] magnetis directionem, conjunctionem et excitationem, per geminos et diversos halitus vorticiosos factam." Cf. also Verbeek, "Regius's *Fundamenta physices*," 541. In Wassenaer's *Album amicorum* of 1648, there are benevolent entries not only by Regius, but also by Arnoldus Senguerdius and Gisbertus Voetius, leading to the assumption that the social situation was not as tense as one might expect.

36   Cf. thesis 11 of *De mineralibus ... Abrahamus Roodenburgh, Ultraject. Ad diem 18. Februar ... Anno 1643* as printed in Senguerdius, *Collegium physicum* (Amsterdam, 1652), X1r-Y1r: "Attractio magnetica explicatu difficillima est. Improbabilis est illa opinio, quae illius causam petit ab effluvio minutissimorum corpusculorum; ut et illa quae statuit magnetem ferrum tanquam alimentum suum attrahere, et illa quae dicit illam provenire ab inimicita naturae ferreae et lapideae quae in magnete detur. Probabiliora sunt quae Gulielmo Gilberto et Nicolao Cabeo traduntur; verum nullum illud desclarandi modum hactenus video, in quo animus plane acquiescat."

37   Cf. Cabeo, *Philosophia magnetica* (Ferrara, 1629); Pumfrey, "William Gilbert's Magnetic Philosophy"; Pumfrey, "Neo-Aristotelianism and the Magnetic Philosophy."

38   Cf. *Disputatio XIII. De mineralibus in genere; item de terris, succis et metallis, in specie. Resp. Michaele Eversdijk* as printed in Senguerdius, *Collegium physicum*, 181–206, here 205: "Quantum ad magneticorum attractionem attinet, putarunt veteres ex magnete corpuscula quaedam egredi, quibus agitatis ferrum moveatur ad magnetem. Haec opinio merito

Senguerdius's reference to corpuscular accounts of magnetism in his text-books and disputations testifies to the fact that philosophers increasingly felt the need to take a position in public while Cartesianism gained ground. His disputation of 1652 was held at the Amsterdam Athenaeum, a university-like institution, where he had taken over the chair of philosophy from Caspar Van Baerle (1584–1648).[39] One year earlier, a study on the magnet by Van Baerle was posthumously published in which he proposed a theory of magnetism that clearly employs elements of a corpuscular account, yet without clear allusions to Descartes.[40] Senguerdius's student, Florian Schuyl, who in 1639 was his candidate in the Utrecht disputation arguing against Descartes and in favour of 'occult qualities,' even converted to Cartesianism later in his career. In what has been seen as an example of a "Kuhnian paradigm shift,"[41] in 1667, Schuyl declined to continue calling magnetism 'occult' in a speech given at Leiden University.[42]

---

jam dudum explosa est; consumeretur enim brevi tempore magnes, tot corpusculis ex ipso egredientibus. Praeterquam quod ab illis, non ferrum tantum aut magneticam, sed etam alia corpora eundem subirent motum. Occultum quid hic in rei natura est, in cuius ratione libens fateor, me mihi satisfacere non posse." These disputations do not contain relevant passages: Senguerd (praes.) and Eyndhoven (resp.), *Disputatio physica decima continens quaestiones de mineralibus in specie* (Utrecht, 1645); Senguerd (praes.) and Renesse (resp.), *Disputatio physica nona continens quaestiones de mineralibus in genere* (Utrecht, 1645); Senguerdius, *Physicae exercitationes* (Amsterdam, 1658). The first edition of Senguerdius, *Introductio ad physicam* (Utrecht, 1644), remained silent on mechanistic accounts of magnetic attraction; cf. Senguerdius, *Introductionis ad physicam libri sex*, 2nd ed., 456. He also repeated his remark that this type of explanation had been recently refuted; see Senguerdius, *Introductionis ad physicam libri sex*, 2nd ed. (Amsterdam, 1653), 454–455: "Causam conjunctionis ferri et magnetis, vel raptus unius ad alterum, veteres petebant ab effluviis seu corpusculis, quae ex illis corporibus egrediantur. Prolixe hoc negotium carminibus suis proponit Lucret. lib. 6. Haec sententia dudum explosa est. Quibus tamen recentiorum illa minus displicet, qui etiam a motu exiguorum corpusculorum, hujus et aliorum magneticorum effectuum causam petunt." This part remained identical in the 3rd edition of 1666, 428–430.

39   Cf. Wiesenfeldt, *Leerer Raum in Minervas Haus*, 54.

40   On Van Baerle, cf. Reael and Baerle, *Observatien of ondervindingen aen de magneetsteen* (Amsterdam, 1651); Miert, *Humanism in an Age of Science*, 249; and Sander, "Magnetism in an Aristotelian World."

41   Cf. Van Ruler, "Substituting Aristotle," 160: "It exhibits all the dramatic features of a Kuhnian paradigm shift rather than a simple substitution of one scientific theory for another."

42   See Schuyl, *De veritate scientiarum et artium academicarum* (Leiden, 1672), 22. This is reproduced in Lindeboom, *Florentius Schuyl*, 125–152: "Crediturn fuit antiquitus magnetis facultatem esse qualitatem quandam occultam, quae hominis captum superet, solique magneti propria sit. Jam vero nescio an quicquam magnetis facultatibus sit manifestius,

In various disputations over which he presided before 1660, Martin Schoock clearly did not convert in this way, but kept on defending Aristotelian principles to account for magnetism against Cartesianism in various disputations over which he presided before 1660.[43] Senguerdius's son and student, Wolferdus, who also became a philosophy professor, appears to have been much more sympathetic to Cartesianism than his father.[44] Like Schuyl, Senguerdius *fils* was affiliated to Leiden University, where he taught philosophy. In one disputation (1679), he seems to claim the existence of 'occult qualities,' while in his *Philosophia naturalis* (1681), however, he presents a theory of magnetism that was clearly inspired by Descartes, although Descartes is not mentioned.[45] Senguerdius is still cautious and hesitant to determine (*non ausim determinare*) some of the details in explaining magnetic phenomena—e.g., the concrete shape of the particles—but he explicitly mentions the possibility of particles in the shape of screws, as Descartes had imagined.[46]

A further example of the lack of a clear consensus on the question of the Cartesian theory of magnetism is Jan De Raey (1622–1702), who was a student of Regius and a candidate in the disputation of 1641, in which Regius had for the first time sketched his own theory of magnetism. De Raey later moved to Leiden, where he was appointed philosophy professor.[47] His mission, according to a broad-brush description, was to combine Aristotelianism with Cartesianism. He referred to magnetism as one of the examples that 'subtle matter' would be able to account for and criticized the view of occult qualities as a cause of magnetism.[48] However, he never seems to have explicitly endorsed a mechanistic theory of magnetism à la Descartes or Regius. Later in his career, De Raey

---

  utpote cujus omnia effecta non modo unius aut alterius horae spatio ad oculum demonstrantur."

43 Cf. Schoock (praes.) and various (resp.), *Physica generalis* (Groningen, 1660), 11, 40, 199, 228, 239, 246, 249, 268, 271, 280, 286. Respondents are Johannes Wubbena, Pompejus Venhuysen, Winckt Tonkens, Jacob Duirsma, Marcus De Muinck and Tjapkon Conrad. Descartes is named in some of these, although his theory of magnetism is not discussed.

44 Cf. Wiesenfeldt, *Leerer Raum in Minervas Haus*, 139, 142.

45 See Senguerd (praes.) and Udemans (resp.), *Disputationum physicarum selectarum septima decima; quae est de particulis subtilibus, secunda* (Leiden, 1679), *corollaria*: "Dantur occultae qualitates." See Senguerdius, *Philosophia naturalis*, 1st ed. (Leiden, 1681), 254–259; 2nd ed. (Leiden, 1685), 355–362.

46 See Senguerdius, 1st ed., *Philosophia naturalis* (1681), 259; 2nd ed. (1685), 361.

47 Cf. n. 18. On De Raey, see Chapters 3 and 4, in this volume.

48 See De Raey (praes.) and Crooswyck (resp.), *Disputationum physicarum ad problemata Aristotelis, quintae de materia subtili* (Leiden, 1653), thesis 15; De Raey, *Clavis philosophiae naturalis* (Leiden, 1654), 24–25. De Raey did not develop a clear theory of magnetism himself in either of these works.

became professor in Amsterdam's Athenaeum. At that time, the school was the site of frequent discussions of magnetism, including the aforementioned cases of Van Baerle and Arnoldus Senguerdius. In 1658, Alexander de Bie (ca. 1620–1690) presided over three mathematical disputations on the magnet at the institution (respondents Sibertus Coeman, Johannes Brandlight, Johannes du Pire).[49] As they were primarily mathematical, they dealt with problems of magnetic declination rather than with natural philosophy, and nowhere was there any mention of Descartes. Yet, all three candidates referred approvingly either to the Jesuit Cabeo or to the Jesuit Athanasius Kircher (1602–1680), both of whom were proponents of Aristotelianism and critics of mechanistic or corpuscularian accounts of magnetism.[50]

When Sibertus Coeman (1643–1679), a student of de Bie and Arnoldus Senguerdius, died in Leiden, it fell to another of their students, none other than Burchard de Volder (1643–1709), to write the funeral oration.[51] De Volder described Coeman as trained in the principles of Peripatetic philosophy, which he also claimed for himself and his listeners, but added that "we" in physics also valued Galileo, Gassendi, and especially Descartes, "propagator of mechanical philosophy" (*illustris Mechanicae Philosophiae propagator*) and "light of our era."[52] De Volder is indeed remembered as a progressive scholar and as being sympathetic to Cartesian philosophy.[53] In his inaugural disputation at Utrecht in 1660, formally under the auspices of the rector of the university Gisbertus Voetius, he took an open stand for the Cartesian theory of magnetism in one of the corollaries: "By the *particulae striatae* [i.e., the screw-shaped particles] of Descartes, various effects of the magnet can be demonstrated."[54] He moreover

---

49   De Bie (praes.) and du Pire (resp.), *Disputatio de magnete, quae est de ejus ὀρθοβορεοδείξει* (Amsterdam, 1658); de Bie, (praes.) and Coeman (resp.), *Disputatio mathematica de acus magneticae deviatione* (Amsterdam, 1658); de Bie (praes.) and Brandlight (resp.), *Disputatio mathematica de acus magneticae inconstanti deviatione* (Amsterdam, 1658).

50   On Cabeo, see n. 37. On Kircher, cf. Kircher, *Magnes; sive, De arte magnetica opus tripartitum* (Rome, 1641); Baldwin, "Athanasius Kircher and the Magnetic Philosophy."

51   Cf. Coeman and de Volder, *Orationes duae, quarum altera inauguralis* (Leiden, 1679). See also Chapter 8, in this volume.

52   See Coeman and de Volder, *Orationes duae, quarum altera inauguralis*, E3ᵛ.

53   On de Volder, see Wiesenfeldt, *Leerer Raum in Minervas Haus*; Miert, *Humanism in an Age of Science*; Strazzoni, *Burchard de Volder*.

54   See the *Corollaria physica* in de Bie (praes.) and de Volder (resp.), *Disputatio philosophica inauguralis* (Utrecht, 1660), A4rˑ "9. Vis attractrix, retentrix, expultrix mera ignorantiae asyla sunt. 10. Terra est magnes. 11. Magnetis polus tendit ad polum terrae. 12. Deviationis causa est, quod ab una parte plus terrae, quam ab altera sit. 13. Per particulas striatas Cartesii varia effecta magnetis demonstrari possunt. 14. Cabei qualitas duarum facierum non minus manifesta est quam quatuor primae qualitates."

dismissed the Aristotelian 'attractive force,' but at the same time also affirms that the conception of a 'quality of two faces,' which Cabeo used in order to explain magnetism, was no less manifest than the four primary qualities—in support of an Aristotelian theorem.[55] At the same time in Leiden (1677), de Volder presided over an disputation *De magnete*; while Descartes himself is not mentioned, his concept of *particulae striatae* does feature, this being the clearest shibboleth of the Cartesian theory of magnetism.[56] Yet, as contradictory as it may seem, the Jesuits Cabeo and Kircher are approvingly named within the text as well.

The foregoing analysis of discussions about Cartesian magnetic theory in Dutch institutions in the aftermath of the Utrecht controversy testifies to a tightly woven web of relations between, on the one hand, the scholars involved and, on the other, the different ways of dealing with the Cartesian theory of magnetism. Reactions ranged from outright refusal (e.g., Schoock), through mild critique and neglect (e.g., A. Senguerdius) or syncretistic and diplomatic acceptance (e.g., Van Baerle), to an implicit or even explicit adoption of the Cartesian doctrine (e.g., de Volder and Regius). Many of the scholars were connected to each other through the biographical tracks of their educations and careers, and partly even by family ties. Utrecht, Amsterdam, and Leiden are clearly the most relevant sites of the debate.

But what does this Dutch context prove on a more general level for the of teaching Cartesian philosophy in schools? It seems clear, at least to the extent that the case of magnetism can be taken as a valid indicator, that after the climax of the Utrecht controversy, from 1644 onwards, the situation calmed down within the institutions. Tensions remained and were apparent, but were dealt with much more discreetly. It is striking, for example, that even those who clearly sympathized with the Cartesian theory of magnetism rarely admitted this publicly and often refrained from naming Descartes at all.[57] His name had supposedly acquired a political meaning by that time. In institutions with a traditional curriculum, often still orientated towards the works and doctrines of Aristotle as a point of departure for teaching, openly appearing as a Cartesian might not have always ensured the smoothest or most promising career track. This is not to say that Dutch institutions between 1650 and 1670 did not engage with Cartesian philosophy, but it holds true that they did so in a rather balanced, nuanced and critical way.

---

55   Cf. n. 37.
56   See de Volder (praes.) and Helvetius (resp.), *Disputatio philosophica de magnete* (Leiden, 1677).
57   See note 29.

Focusing on the Cartesian theory of magnetism, it appears significant that empirical and theoretical achievements by two Jesuits, Cabeo and Kircher, were so openly approved within the Dutch institutions—even if the confessional animosities between Calvinists and Jesuits might appear much stronger and more blatant than any suspicion of heresy in Descartes. These Jesuits, too, were operating in an Aristotelian framework, which made them a helpful support in fighting mechanical physics, regardless of any theological disputes. A Jesuit or Aristotelian blend of magnetic theory needed not be taken as contradictory to a corpuscularian account. Van Baerle and de Volder, for example, seemed to be very much in favour of combining elements from both currents of natural philosophy, even if this remained very vague. Yet—and this is an important point—maybe with the exception of Regius's *Fundamenta*, none of the sources analyzed in the Dutch context discussed magnetism with the rigour and sophistication Descartes had exhibited in his *Principles of Philosophy*. Illustrations, for example, which are an important feature in Descartes's account, are lacking in any of the disputations examined. A full-blown support and approval of the Cartesian theory of magnetism as developed in the *Principles of Philosophy* was rare in the Netherlands in the first decades after its publication.

4   The Cartesian Theory of Magnetism Taught in European Schools

Seen from a wider geographical perspective and reaching further into the seventeenth century, the critique and the defence of Cartesian principles in explaining magnetism was livelier, and conducted more openly, in other parts of Europe, as the following spot tests and examples will show.

Although Descartes was not always mentioned, several Dutch scholars included only rather short chapters specifically dedicated to the magnet in their natural philosophical textbooks. Looking at the treatment of magnetism in a selection of seventeen textbooks published (most of them in various editions and even translations) between 1653 and 1727 all over Europe presents strong evidence for the impact of Descartes's theory of magnetism, particularly from the 1670s onwards.[58] This clearly characterizes Cartesianism as

---

58   Cf. especially Mouy, *Le développement de la physique cartésienne*, 82, 160, 225; Clarke, *Occult Powers and Hypotheses*, 136, 146, 160, 168, 216; McClaughlin, "Descartes, Experiments" 333; Roux, "Was There a Cartesian Experimentalism in 1660s France?," 62–63, 74–75, 77 (n. 116), 80, 82; Dobre, "Rohault's Cartesian Physics," 209, 215; Vermeir, "Mechanical Philosophy," 301; Maignan, *Pars secunda philosophiae naturae* (Toulouse, 1653), 1410–1457;

a phenomenon of European scale. Some of these textbooks are clearly and self-professedly Cartesian (e.g., those by Jacques Rohault, Antoine Le Grand, or Pierre-Sylvain Regis), some are clearly anti-Cartesian (e.g., those by Jean-Baptiste de La Grange, or Joannes Vicentius), while others are eclectic and conciliatory (e.g., those by Jean-Baptiste Du Hamel, Francesco Lana de Terzi, or James Dalrymple). All of them deal with the subject in a way whereby there is no question about the influence of Descartes's theory of magnetism, but not all of them name Descartes. Most of them deal with the topic as extensively as it was dealt with in the *Principles of Philosophy*, structuring it in more or less in the same way, namely by discussing various magnetic phenomena one after another. Some of them even add similar illustrations to render the theory more understandable.[59]

4.1   *A Particular Case: Louvain*

Including magnetism in textbooks of natural philosophy made it much more likely that students would come into contact with the Cartesian theory of magnetism, albeit mediated through the views of another author. However, it remains difficult to determine how many of those who refer to Descartes's theory of magnetism in an institutional context had actually read the *Principles of Philosophy*. That his theory of magnetism made it into classrooms, and also into classrooms far beyond the Netherlands, is evidenced by quite a few sources. A particularly well-documented example is the University of Louvain, where Cartesian doctrines were discussed early, found their first supporters around 1650, were condemned in 1662, and served as the basis for an entire curriculum from around 1670 onwards.[60] At the beginning of this process, Louv-

---

Fabri, *Physica* (Leiden, 1671), vol. 4, 1, 16, 23, 129; Rohault, *Traité de physique* (Paris, 1671), 198–236; Rohault, *Tractatus physicus* (Geneva, 1674), 561–603; Rohault, *Tractatus physicus* (Amstedam, 1708), 403–426; La Grange, *Les principes de la philosophie* (Paris, 1675), 243–304; Vincentius, *Discussio peripatetica* (Toulouse, 1677), 356–407; Du Hamel, *Philosophia vetus et nova* (Nuremberg, 1682), vol. 2, 497–507; Le Grand, *Institutio philosophiae*, 2nd ed. (Nuremberg, 1683), 450–458; Du Hamel, *Philosophia vetus et nova* (Paris, 1684), vol. 2, 416–429; Lana de Terzi, *Magisterium naturae et artis* (Brescia-Parma, 1684–1692), vol. 3, 366, 369–375; Dalrymple, *Physiologia Nova Experimentalis* (Leiden, 1686), 486; Bartholin, *Specimen Compendii Physicae* (Copenhagen, 1687), 14–15; Dechales, *Cursus* (Leiden, 1690), vol. 2, 524–527; Regis, *Cours entire de philosophie* (Amsterdam, 1691), vol. 2, 220–257; Le Clerc, *Physica* (London, 1696), 125–133; Hilleprand, *Examen doctrinae Cartesianae* (Vienna, 1707), 98–134; Ode, *Principia philosophiae naturalis* (Utrecht, 1727), 95–109. The majority of these works have not been the subject of study. No examples from the British Islands and the Iberian Peninsula have been included here.

59   See Sander, "*Terra AB.*"
60   For Cartesianism in Louvain, see especially the Chapters 15, 16, and 17, in this volume. See,

ain's professor Libert Froidmont (1587–1653), an early critic of Descartes, had not yet discussed the Cartesian theory of magnetism at the university.[61] His pupil, Vopiscus Fortunatus Plemp (1601–1671), also a professor in Louvain, was critical of many Cartesian doctrines and had discussed magnetic theory in his writings.[62] Plemp's writings on magnetism did not mention Descartes—who in any case had not yet published his *Principia*—but opted for an immaterial cause of the phenomenon, referring to 'sympathy,' and clearly taking for granted 'attractive forces' in nature, as had most physicians operating in a Galenic framework.[63]

Around 1650, there were Louvain professors who championed Cartesian ideas. One such professor was Willem Van Gutschoven (1618–1667). In a disputation presided over by him in 1651, his candidate openly argued along these lines, making magnetism a phenomenon only to be explained by "the laws of statics" and based on the principles of "matter and motion."[64] A 1652 lecture on natural philosophy attributed to Willem Van Gutschoven followed similar

---

as a starting point, in particular Vanpaemel, "Cartesianism in the Southern Netherlands," 221–230; Vanpaemel, *Echo's*; Radelet-de Grave, "Les Jésuites," 72–125. For the student notebooks that are discussed in the following, see especially Vanpaemel, Smeyers and Smets, eds., Ex cathedra; Mirguet and Hiraux, *Collection de cours manuscrits*.

61   Cf. Sander, *Magnes, ad indicem*.

62   See Plemp, *Ophthalmographia*, 1st ed. (Amsterdam, 1632), 78–79, 119, 256, 321; Plemp, *Fundamenta medicinae*, 3rd ed. (Louvain, 1654), 40, 42, 62, 66, 184, 204. Concerning Froidmont, Plemp and the teaching of Cartesianism in seventeenth-century Louvain, see Chapter 15, in this volume.

63   See Plemp, *Ophthalmographia*, 1st ed. (Amsterdam, 1632), 321: "Cuius causa tam mihi manifesta est, quam cur Magnes ferrum trahat. Blandimur vero nobis tantisper, dicendo: ob sympathiam accurrere ad oculos sanguinem." This passage was not included in later editions; see also Plemp, *Fundamenta medicinae*, 184: "magnes attrahit ferrum [...] sine superadita facultate [...] per [...] vim trahentem [...]. Facultas appetens [...] reperitur [...] in magnete [...]." See also Plemp, *Ophthalmographia*, 3rd ed. (Louvain, 1659), 74: "non aliter quam magnes incorpoream quamdam vim ex se fundit, qua ferrum rapit." For background on the theoretical assumptions, see Sander, *Magnes*, 642–743.

64   Van Gutschoven (praes.) and Van Werm (resp.), *Philosophia* (Louvain, 1651), B2ʳ: "II. Praefatis principiis haec verba subsidio sunt: sympathia, antipathia, antiperistasis, virtutes magneticae, influentiae caelestes, qualitates occultae, et tota farrago praepotentum facultatum. Nobis materia et motus sufficiunt, ut ex ipsis tanquam principiis unicuique obviis, effectus naturales secundum leges staticae deducamus. III. Sic nulla potentia superaddita magnes ad polos mundi se dirigit; alterum magnetem et ferrum trahit, et fugat; illico ferro quod attingit, vim suam secundum ferri longitudinem communicat; ferro armatus videsies plus ferri sustinet; et licet magnetis poli contrarie sint virtutis, aeque tamn ad ferrum sustinendum se invicem iuvant, etc." This passage is also repeated in the identical disputation contained in Van Veen and Van Gutschoven, *Philosophia quam praeside*, B2ʳ. Magnetism as *explanandum* for a mechanical account is also briefly referred to

lines, as one of his students' notebooks reports: it argued at much greater length for the Cartesian theory of magnetism, and even mentioned his *particulae striatae*.[65] Descartes's name itself, however, was still not mentioned in this context.

Van Gutschoven's older brother, Gerard (1615–1668), also defended Cartesian natural philosophy. Around 1659, in his *Animadversiones in Ophthalmographiam*, he launched into an attack against Plemp. His critique touched on the magnet only once, and only as an incidental example, but Plemp's reply is more revealing. He ridiculed the mechanistic theory of particles orbiting a magnet as some sort of miraculous dance, and also attacked Regius, a *secator Cartesii*, for his alleged experimental proof of this orbit of particles.[66] In 1666, a Franciscan teacher at Louvain University, Willem Van Sichen (1632–1691), in his *Cursus* also explicitly targeted Descartes and his theory of magnetism.[67] His argument was epistemic: the precise means through which magnetism was caused were uncertain, and Descartes's *particulae striatae* were seemingly contrived without any solid reason.

But times were changing in Louvain, and Cartesian natural philosophy was on the rise. One of his students' notebooks shows that in 1675, Joannes Stevenot (ca. 1640–1718) lectured quite extensively on the Cartesian theory of magnet-

in Van Gutschoven (praes.) and Van Werm (resp.), *Philosophia* (Louvain, 1651), B2[r]. Cf. also Geulincx, *Saturnalia*, 142–143; Vanpaemel, *Echo's*, 82.

65 See MS., Brussels, MS. II 737: anonymous (professor) and Meesters (student), *Physica* (1652), fols. 261[v]–262[v]. The attribution of the lecture to Willem Van Gutschoven is based on circumstantial evidence and argued for in Chapter 15, in this volume.

66 See Plemp, *Ophthalmographia*, 3rd ed. (1659), 249: "Ita est: nam dum omnia naturae opera mechanice, clare et distincte demonstrare conantur, in multos errores labuntur. Vide, quae dicant de motu magnetis ad ferrum, vel ferri ad magnetem; quas comminiscantur particulas ramosas, crassas, striatas; quas exire fingunt, impelli, repelli, redire, et nesco quas choreas agere, nullo praesultore, choraule vel chorago." See also Plemp, *Ophthalmographia*, 3rd ed. (1659), 255: "Illa agentia creata intelligo, quae operantur in distans aliquid a se emittendo; qualia sund calida, frigida, sonora, odora, purgantia, venena. Magneti esse circularem sphaeram activitatis ipse Henricus Regius secator Cartesii scribit in sui Physices fundamentis. Et magnes scobi seu ferragini chalybis immersus eamdem sibi circumquaque seu spherice adfigit, ita ut inde specie erinacei eximatur. Motus absurde huc producitur; quia non est agens, sed vel actio vel passio. Gravitas etiam non est tale agens, quod in distans aliquid a se emittit."

67 See Van Sichen, *Integer cursus philosophicus* (Antwerp, 1666), vol. 2, 58: "Potuisset certe Deus ita naturam instituisse, ut actio illa magnetica fieret per effluvia spirituum magneticorum, vel per particulas striatas, vel per qualitatem toto medio diffusam, aut etiam sic, ut actio inchoetur in distanti, vel denique per alium modum nobis occultum: quis autem de facto a natura adhibeatur, res est plane incerta." On Van Sichen, cf. also Monchamp, *Histoire du cartésianisme*, 430; Coesemans, "Faculties of the Mind," 198. Cf. also Birlens (praes.) and D'Overschie (resp.), *Theses philosophicae* (Louvain, 1660), B2[r]: "Motus corpusculorum Carthesii tam est occultus, quam qualitates occultae."

ism.[68] Around this time, when Louvain had in practice established a Cartesian curriculum, Michael Hayé (d. 1676), a local printer, began producing engravings from some of the woodcuts of Descartes's *Principles of Philosophy*. Among others, he produced an engraving of the main image related to the theory of magnetism.[69] Three students included copies of it in their notebooks on *physica*: Leo Josephus Daco in 1678, Albertus Boone in 1680, and Balthasar Cox in 1687.[70] While Cox only dedicated a few lines to the magnet (fol. 307^r), Boone (fols. 178^r–180^v) and Daco (fols. 324^v–328^v) clearly tried to record a full explanation of the Cartesian theory of magnetism. Daco did not name Descartes here, but referred to his concept of the *particulae striatae*, and also began his section with a critique of 'occult qualities.' Boone mentioned Descartes by name, but more interesting still is the fact that his notes were based on lectures by the Louvain professor Léger Charles De Decker (1645–1723). De Decker is known as the presumed author of an anonymously published anti-Cartesian work, the *Cartesius seipsum destruens* of 1675, which harshly attacked Cartesian philosophy, yet not his theory of magnetism explicitly.[71] This criticism is likewise absent from the student's notes on the Cartesian theory of magnetism, making De Decker's anti-Cartesian polemics appear more as a private battle than the official position taken in the classroom—or, at least this seems to have been the case as far as magnetism was concerned.

One textbook that was especially important in informing the teaching of physics at Louvain was Jacques Rohault's *Traité de physique* (1671, translated into Latin 1674).[72] As has been mentioned, Rohault (1618–1672) followed Descartes in his account of magnetism. In the course of the eighteenth century,

---

68 See MS., Louvain, MS. 261: Stevenot (professor) and Van den Biesche (student), *Physica* (1675–1676), fols. 495^v–501^v. Cf. also Vanpaemel, *Echo's*, 99–102. On Stevenot, see Chapter 16, in this volume.

69 Cf. Vanpaemel, "The Louvain Printers."

70 MS., Brussels, MS. II 106: Wauchier (professor) and Daco (student), *Physica* (1678), fol. 326^r; MS., Louvain-la-Neuve, MS. C165: De Decker and Van Goirle (professors) and Boonen (student), *Physica & Metaphysica* (1680–1681), fol. 179^r; MS., Brussels, MS. 21127: anonymous (professor) and anonymous (student), *Physica* (1757), fol. 306^r. It seems that the page order in Ms. C165 is mixed up. Cf. however Vanpaemel, *Echo's*, 100: "In het laat-cartesiaanse curriculum kreeg het magnetisme weer wat meer ruimte toebedeeld, maar het wist niet uit te stijgen boven het niveau van een curiosum, ergens tussen de behandeling van de fossielen en de mineralen geplaatst. Een afzonderlijk dispuut werd aan het magnetisme niet gewijd."

71 Cf. [De Decker], *Cartesius seipsum destruens* (Louvain, 1675), 10–16; Vanpaemel, "Cartesianism in the Southern Netherlands," 225; Geudens and Papy, "The Teaching of Logic at Leuven University," 376.

72 Cf. Vanpaemel, "Rohault's *Traité de physique*." Cf. also n. 58.

the Cartesian theory of magnetism is dealt with extensively and frequently in Louvain University student notebooks.[73] Also in the second half of the eighteenth century, Hayé's magnetism engraving was reproduced by, amongst many others, the engraver Petrus Augustinus Denique (1683–1746). These prints were included in several notebooks of students at Louvain.[74]

4.2   *Other Parts of Europe*

Student notebooks bear a close witness to what was actually taught in classrooms, and rarely are they preserved as numerously and completely as at Louvain. However, printed university disputations, which survived in much greater numbers, also testify to the wider use of the Cartesian theory of magnetism in teaching. Most of these disputations occurred in German-speaking territories or in Scandinavian or Baltic areas.[75] Disputations *De magnete*, with discussions of the Cartesian theory of magnetism, were held in Weissenfels near Leipzig (1673), Strasbourg (1683), Marburg (1683), Basel (1685, 1686, 1692, 1697), Erfurt (1687), and Zerbst near Magdeburg (1693).[76] The positions taken by the authors

---

73   Referenced manuscripts are held in three libraries: Brussels, Koninklijke Bibliotheek van België / Bibliothèque royale de Belgique (KBB), KU Leuven Libraries (KUL), Louvain-la-Neuve, Archives de l'Université catholique de Louvain (AUL). In brackets, the date of composition is indicated. The folia indicate the section(s) on magnetism. Cf. KBB, MS. II 737 (1652), fols. 261ᵛ–262ʳ; MS. II 106 (1678), fols. 324ᵛ–328ᵛ; MS. II 5444 (1739), fols. 248ᵛ–251ʳ; MS. II 3294 (1763), fols. 172ᵛ–178ʳ; MS. II 3214 (1720–1721), fols. 274ᵛ–279ʳ; MS. II 3703 (1730–1731), fols. 289ᵛ–298ʳ; MS. II 4269 (1754–1755), fols. 343ᵛ–348ʳ; MS. 21127 (1756–1757), fols. 182ᵛ–187ᵛ; MS. II 4523 (1758–1759), fols. 215ᵛ–222ʳ; MS. II 5602 (1760–1761), fols. 371ʳ–378ʳ; KUL, PRECA0021 (1785), fols. 99ʳ–101ᵛ; MS. 247 (1772), fols. 40ʳ–44ʳ; MS. 302 (1774), fols. 178ʳ–187ᵛ, 191ᵛ–192ʳ; MS. 261 (1675–1676), fols. 495ᵛ–501ᵛ; MS. 211 (1686–1687), fol. 307ʳ; MS. 326 (1750–1751), fols. 203ᵛ–212ᵛ; MS. 359 (1754–1755), fols. 158ʳ–164ʳ; MS. 284 (1779–1780?), fols. 122ᵛ–129ʳ; AUL, MS. C210 (1761), fols. 273ʳ–278ᵛ; MS. C202 (1774), fols. 97ʳ–101ᵛ; MS. C163 (1781), fols. 134ʳ–139ᵛ; MS. C59 (1785), fols. 211ʳ–216ʳ; MS. C165 (1680–1681), fols. 178ʳ–180ᵛ; MS. C72 (1714–1715), fols. 257ᵛ–262ʳ; MS. C75 (1738–1739), fols. 286ʳ–291ᵛ; MS. C4 (1755–1756), fols. 566ʳ–574ᵛ; MS. C28 (1746), fols. 287ᵛ–290ᵛ.

74   Cf. Sander, *"Terra AB."*

75   This bibliographic result depends on whether disputations were printed at all; see especially Friedenthal, Marti and Seidel, eds., *Early Modern Disputations*.

76   See Siegfried (praes.) and Pfundt (resp.), *Disputatio physica de magnete* (Weissenfels, 1673); Siegfried, "Curiöse Gedancken vom Magnete" (Dresden, 1704); Scheid (praes.) and Kast (resp.), *Quaestionum decades duae de magnete* (Strasbourg, 1683); Waldschmidt (praes.) and Kursner (resp.), *Disputatio physica, de magnete* (Marburg, 1683); Zwinger (praes.) and Gernler (resp.), *Disquisitionum physicarum de magnete prima* (Basel, 1685); Zwinger (praes.) and Gemuseus (resp.), *Disquisitionum physicarum de magnete secunda* (Basel, 1686); Zwinger (praes.) and Gemuseus (resp.), *Disquisitionum physicarum de magnete quinta* (Basel, 1692); Zwinger (praes.) and Schönauer (resp.), *Disquisitionum physicarum de magnete septima* (Basel, 1697); Zwinger, *Scrutinium magnetis physico-medicum* (Basel,

in these disputations mostly approve the Cartesian theory of magnetism. In Weissenfels, Strasbourg, Marburg, and Zerbst, Descartes's theory of magnetism was welcomed, while the disputation at Erfurt takes a middle position. Theodor Zwinger (1658–1724), in several of the disputations he presided over in Basel, integrated the Cartesian account into his eclecticism, and also referred to disputations on the topic by, amongst others, De Volder, Wolferdius Senguerdius, Du Hamel, and Johann Jakob Waldschmidt (1644–1689).[77]

In a miscellany manuscript of seventeenth-century Gotha, an anonymous German translation can be found of just the section on magnetism from Descartes's *Principles of Philosophy*.[78] The context of the production of this translation, which also reproduces the images, remains unknown, but circumstantial evidence points towards an educational setting, maybe the private tuition of Ernest I, Duke of Saxe-Gotha (1601–1675) and/or his many sons. The Duke was interested in magnetism, and experiments relating to this interest were conducted in Gotha as well.[79] Moreover, Veit Ludwig von Seckendorff (1626–1692), who was educated at Gotha's Ernestine Gymnasium and under the Duke's protection, was acquainted with Cartesian philosophy when Daniel Lipsdorp (1631–1684)—a propagator of Cartesian thought—was the court mathematician in Weimar between 1653 and 1656.[80] As the Gotha manuscript also contains a text by Andreas Reyher (1601–1673), the rector of the Gymnasium at that time, the translation of the part on magnetism has elsewhere, not implausibly, been ascribed to this pedagogue or his circle.[81]

While Cartesianism, and the Cartesian theory of magnetism in particular, seems to have flourished in the German-speaking regions, Descartes did not break through so easily south of the Alps, especially in the Catholic Italian territories. Although the general reception of Cartesian thought in Italy in the

---

1697); Vesti (praes.) and Fischer (resp.), *Disputatio physico-medica de magnetismo* (Erfurt, 1687); Limmer (praes.) and Wolff (resp.), *Dissertatio philosophica, de magnete ejusque effectibus* (Zerbst, 1693). See also Murhard, *Versuch einer historisch-chronologischen Bibliographie des Magnetismus* (Kassel, 1797). Waldschmidt promoted Cartesian ideas in many of his disputations in Marburg; cf. Schlegelmilch, "The Scientific Revolution in Marburg."

77   See especially Zwinger, *Scrutinium magnetis physico-medicum*, 32–37.

78   See MS., Gotha, Chart. A 707: anonymous, *Kürtzliche Erleuterung*, fols. 190ʳ–202ᵛ: "Kürtzliche Erleuterung etlicher Vorgaben von der Würkung des MagnetSteins nach Anleitung und gesetzten Grundstücken von Cartesio." Another part of the manuscript is described in Cooper, "Placing Plants on Paper," 257–277. I thank Jacob Schilling for information on the background of this manuscript and its context. A digital copy of the manuscript is available at https://ch-sander.github.io/raramagnetica (accessed 12 December 2021).

79   Cf. Collet, *Die Welt in der Stube*, 61, n. 114.

80   Cf. Strauch, *Veit Ludwig von Seckendorff*, 150; Lipstorp, *Specimina philosophiae Cartesianae*.

81   See Lotze and Salatowsky, *Himmelsspektakel*, 193.

seventeenth century is not to be investigated here, some examples shall be given of how his theory of magnetism was received in the Roman Jesuit context. In their daily academic life in their many colleges and universities all over the world, Jesuits were, above all, teachers, and the Collegium Romanum was their flagship institute.[82] As is well known, Descartes was himself educated in a Jesuit college.[83] By their statutes and many decrees tied to Aristotelian philosophy, tensions between the Jesuits and Descartes existed already during his lifetime and did not stop after his death. As in the Utrecht controversy, metaphysical issues were at stake, such as the acceptance of 'substantial forms' and 'occult qualities.' Unlike Utrecht's theologians, however, many Jesuits were interested in studying magnetic effects in their own right, not least the aforementioned Cabeo and Kircher.[84] Cabeo's *Philosophia magnetica* preceded Descartes's *Principles of Philosophy* by fifteen years. Kircher's *Magnes*, which was also read by Descartes, also predated the publication of the Frenchman's work.[85] In the third edition of *Magnes* (1653), however, Kircher mentioned Descartes and criticized him as one of Epicurus' followers.[86]

The Jesuit college in Rome also possessed some of Descartes's writings.[87] In the same year of Kircher's reference to Descartes, 1653, Niccolò Zucchi (1586–1670) deals with the Cartesian theory of magnetism in his manuscript study *Philosophia magnetica*.[88] Yet, he does not name Descartes, but only his *particulae striatae* and again considers this theory as atomistic.[89] Zucchi also included a longer digression on the theory of magnetism as put forward by Emmanuel Maignan (1601–1676) in 1653, who himself reworked Descartes's account.[90] In later Jesuit textbooks of natural philosophy, Descartes's theory of

---

82 Cf., as a starting point, Grendler, *The Jesuits and Italian Universities*.

83 Cf., as a starting point, Ariew, *Descartes among the Scholastics*.

84 Cf. Vregille, "Les jesuites et l'etude du magnetisme terrestre"; Sander, *Magnes*, 846–855.

85 See AT XI 635–639; Sander, *Magnes*, 163. Cf. also n. 37 and n. 50.

86 See Kircher, *Magnes*, 3rd ed., 38.

87 Blum, *Studies on Early Modern Aristotelianism*, xii, remarks, "in the handwritten index to the books of the Roman College, Descartes or Cartesius is to be found under letter 'S': de Schartes."

88 Cf. Sander, *Magnes*, 856–857.

89 See MS., Rome, Fondo Gesuitico 1323: Zucchi, *Philosophia magnetica per principia propria proposita et ad prima in suo genere promota* (completed *post* 1653), fol. 70ʳ/66ʳ: "[...] asserentes per diffusas a polis Terrae particulas striatas, quae circa Terram convolvantur, aut per spiritus a centro Terrae provenientes determinari inclinationem magneticorum ad Terram, quod novissime quidam qui cupit novus Author haberi confidentissime protulit, cum tamen asseruisset per meatas magneticorum parallelos exire spiritus determinatinos motuum, qui a magneticis exercentur."

90 See MS., Rome, Fondo Gesuitico 1323: Zucchi, *Philosophia magnetica*, fol. 65ʳ, and n. 58.

magnetism was openly discussed and refuted, mostly on the basis of his having invented particles that have no rational or empirical basis.[91] To which extent this criticism was aired in classrooms is hard to determine, but from other contexts it is known that Cartesian natural philosophy was criticized also in teaching.[92]

Looking to the north of Europe, a different picture emerges.[93] In the Finnish Turku, Cartesianism was discussed in the classroom from 1660, but "did not become *the* new paradigm."[94] Descartes was not mentioned in disputations on magnetism, but, e.g., Daniel Erici Achrelius (ca. 1644–1692) (presiding in 1689) and Petrus Olai Hahn (1651–1718) (presiding in 1689) argued in favour of the emissions of particles or atoms, yet without clear mechanistic commitments.[95] In Sweden too, Cartesianism was discussed from the 1660s.[96] In Uppsala, the first traces of Descartes at the university date to 1663, where Petrus Hoffwenius (1630–1682) was one of Descartes's first promoters.[97] Hoffwenius was educated at Leiden, and the early teaching of Descartes was orientated along the lines set out by Cartesian textbooks of the first generation.[98] In 1678, Hoffwenius

---

91   Fabri, *Physica*, vol. 4, 1, 16, 23, 129; Lana de Terzi, *Magisterium naturae et artis*, vol. 3, 366, 369–375; Dechales, *Cursus seu mundus mathematicus*, vol. 2, 524–527; Hilleprand, *Examen doctrinae Cartesianae*, 98–134.

92   Cf., e.g., Hellyer, *Catholic Physics*.

93   For disputations on magnetism printed in northern Europe, cf. Vallinkoski, *Turun Akatemian väitöskirjat*, vol. 1, 3, 10, 200; Lidén, *Catalogus disputationum in Academiis et Gymnasiis Sveciae*, vol. 1, 68, 71, 248; vol. 2, 132; vol. 3, 10, 12, 76.

94   Quote with italics from Kallinen, *Change and Stability*, 39. Cf. also Kallinen, "Kartesiolaisuus Turun akatemian meteorologisissa väitöskirjoissa 1678–1702," 67–98; Kallinen, "Naturens hemliga krafter," 317–346; Salminen, "Barokin filosofis-teologisen synteesin hajoaminen maassamme," 52–84.

95   Cf. Hahn (praes.) and Procopoeus (resp.), *Disputatio physica amicitiam magnetis cum ferro exhibens* (Abo, 1698), 21: "verum tutissimum videtur statuere cum recensioribus hujus rei investigatoribus, ex Magnete atomos profisci in ferrum, cum ejusdem fere rationis sit utriusque constitutio, ferri nempe ac ipsius Magnetis, adeo ut adhaereant invicem jactis atomis, cum sane nullum corpus detur, quod non halitus et vapores a se emittat." Cf. also Achrelius and Ulnerus, *Contemplationum mundi*, 10. On magnetism in Turku's disputations, cf. Kallinen, "Naturens hemliga krafter," 331–338; Kallinen, *Change and Stability*, 206–209.

96   Cf. Lindborg, *Descartes i Uppsala*; Eriksson, "Framstegstanken"; Dunér, *The Natural Philosophy of Emanuel Swedenborg*.

97   Cf. Lindborg, *Descartes i Uppsala*, 176, 176, 181–183, 251.

98   Among those used were publications by Jan De Raey (1654), Johann Clauberg (1664) and Johann Tatinghoff (1655). Cf. the short comments in Tatinghoff, *Clavis philosophiae naturalis antiquo-novae* (Leiden, 1655), 19–20, 73–74. His endorsement is very implicit, while mentioning "particulae striatae" and refuting 'occult qualities.' Cf. also Lindborg, *Descartes i Uppsala*, 103. Clauberg, *Physica, quibus rerum corporearum*, 229, 279–280, explicitly

published a collection of disputations including one on the magnet, which originated from his activities at the University of Uppsala and which were republished in Pärnu two decades later.[99] Hoffwenius presented Descartes's theory in a succinct form, without polemics against Aristotelians. By 1683, Andreas Drossander (1648–1696) was already using Hoffwenius's disputations as the basis for his own lectures on Cartesian philosophy in Uppsala.[100] His manuscript account on the magnet follows Hoffwenius and is also short, but he explicitly argued against 'occult qualities.'

But the early attacks to which Cartesianism was subjected in Uppsala were not to end soon, even if two further Uppsala disputations and one from Lund make Sweden in the last two decades of the seventeenth century appear a Cartesian stronghold when it came to magnetism.[101] Especially Johan Bilberg (1646–1717), presiding over one of the Uppsala disputations, is a well-known figure in the history of Cartesianism in Sweden.[102] In 1689, the controversy over the acceptance of Cartesian philosophy reached its peak at the university of Uppsala; Bilberg, as professor of philosophy, defended Cartesian philosophy against the theologians of the university. At the time of his *De magnete* disputa-

abstained from a long discussion (p. 221: "quia brevitatis studio doctrinam de Magnete [...] omisi") and referred readers to his *Physica contracta* for the principles, and to Descartes directly, p. 279: "Quomodo magnes ad ferrum moveatur sine ulla attractione, cognitione, appetitu, Physica Cartesiana Princip. parte quarta clarissime docet eum qui non plane est excaecatus." The magnet is only briefly mentioned in Clauberg, *Physica contracta in qua tota rerum universitas*, 196–197. Like De Raey, neither Clauberg nor Tatinghoff discussed magnetism in detail, but they both soon subscribed to Descartes's theory of magnetism in their works. The reception of Descartes's theory of magnetism is not dealt with in Trevisani, *Descartes in Deutschland*. Tatinghoff's textbook is based on disputations held at Wittenberg where the Aristotelian theory of magnetism was criticized early; cf. Sander, "Magnetism in an Aristotelian World (1550–1700)."

99   Hoffwenius, *Synopsis physica, disputationibus aliquot Academicis comprehensa*, 1st ed. (Stockholm, 1678), 112–120; 2nd ed. (Stockholm, 1698), 78–84; 3rd and 4th eds. (Pärnu, 1699 and 1700), 105–112. The second edition also included a copy of a woodcut used by Descartes to depict his theory of magnetism.

100   See MS., Uppsala, A 209: Andreas Drossander, *Prolegomena in physicam Hoffwenii* (1683), fols. 75ʳ–76ʳ.

101   Bilberg (praes.) and Plaan (resp.), *Disputatio physica de magnete* (Uppsala, 1687); Vallerius (praes.) and Linnrot (resp.), *Disputatio physico mathematica de pyxide magnetica* (Stockholm, 1699); Riddermarck (praes.) and Aulaenius (resp.), *Dissertatio philosophica de magnetis ac ferri amoribus & odiis* (Lund, 1692). Cf. also Bilberg (praes.) and Odhelius (resp.), *Specimen cogitationum de magnetismis rerum* (Stockholm, 1683). There, Descartes is mentioned twice but his theory of magnetism is not discussed, nor is any causal theory of magnetism.

102   Cf. especially Nilsson, "Johan Bilberg"; Eriksson, "Framstegstanken"; Lindborg, *Descartes i Uppsala*, 288.

tion (1687), this controversy was already at its height. The text of the disputation shows that Bilberg (and his student Andreas Plaan) explicitly approved the Cartesian theory of magnetism, but also referred, e.g., to the Jesuits Kircher and Cabeo. His dismissal of 'occult qualities' clearly echoes the words of both Descartes and Regius: Only the principles of mechanics safeguard the explanation of natural phenomena and avoid "the figments of 'forms' and 'qualities', concepts more obscure than the phenomena through which they purport to be explained."[103] He, more or less apologetically, concludes his disputation by quoting "some Peripatetic," who makes much the same point.[104] In fact, the quotation is taken from a widely used textbook, the *Philosophia vetus et nova* (1678), which was published anonymously in various editions but is understood to be the work of Jean-Baptiste Du Hamel (1624–1706).[105] Although conciliatory and comparative from the outset, much of this work's natural philosophy, and its theory of magnetism in particular, is more Cartesian than Aristotelian. Bilberg is clearly trying to use this as an example even of a Peripatetic claiming that magnetism is to be explained by mechanical principles, although his chosen source has a clear bias towards Cartesianism.

Another Swedish author of one of the disputations on the magnet, Andreas Riddermarck (1651–1707), had travelled to Dutch cities and later taught both in Lund and Uppsala.[106] His disputation of 1692 is a clear stand for the Cartesian theory of magnetism, although he avoids mentioning Descartes by name, and refrains from attacking Aristotelian concepts.

Harald Vallerius (1646–1716) was another known supporter of Descartes in Sweden, but his 1699 disputation from Uppsala deals more with the magnetic compass and its historical invention than with magnetic theory.[107] Descartes's

---

103   See Bilberg (praes.) and Plaan (resp.), *Disputatio physica de magnete* (Uppsala, 1687), 11–12: "Et cum naturam semper aequaliter operari exinde colligimus, principia Mechanica omnium utilissima cernimus ad opera illa demonstranda, quae si non omnino talia fuerint qualia supponimus, non multum tamen ab illa dispositione abludere possunt. Hac opera scilicet evitare licet fictionem illam formarum et qualitatum, quarum notitia obscurior est rebus ipsius quae per illas solent demonstrari."

104   See Bilberg (praes.) and Plaan (resp.), *Disputatio physica de magnete* (Uppsala, 1687), 18: "Quae omnia et alia huius generis plurima facile ad mechanica principia referuntur. Nec necesse est qualitatem occultam, quae nihil explicat, aut formam substantialem obtenere, non magis quam intelligentiam quandam comminisci. Cum enim effectus naturae explicare volumus, quantum fieri potest, quaerenda est causa, cuius idea clara sit, non obscurior ipso quod querimus."

105   Cf. Du Hamel, *Philosophia vetus et nova* (Paris, 1678), vol. 2, 423–424. Cf. also Du Hamel, *De meteoris et fossilibus libri duo* (Paris, 1660), 202: "Placet mihi Cartesii sententia."

106   Cf. Tegnér and Weibull, *Lunds Universitets historia*, vol. 2, 115–116.

107   This text has to be understood against the background of 'Rudbeckiansim'; cf. especially

theory of magnetism is referred to twice, and it seems that Vallerius is willing to follow Descartes's principles in explaining magnetism.[108] He also refers to propagators of the Cartesian account, like Du Hamel or Rohault, and to critics, like the Jesuits Cabeo, Kircher, Giovanni Battista Riccioli, and Claudius Franciscus Milliet Dechales. Even if these references do not concern magnetic theory, they well illustrate how eclectic these authors were at the end of the century, and that the theoretical disagreements of their referenced sources could easily be smoothed over in the practice of teaching by using them selectively or by focusing on aspects beyond the disputed issues. With Olaus Rudbeck's work, Swedish academia had meanwhile created its own paradigm, whose distinguished antiquarianism could synthesize Cartesianism with Jesuit authors maybe more easily than the natural philosophical curricula in preceding decades and in other countries.

5   Conclusion

What is to be learned from the analysis of the reception of Descartes's theory of magnetism? When the Cartesian theory of magnetism was attacked or defended in controversies of the 1640s and 1650s, the theory of magnetism itself was nothing more than an adjunct to more general issues, such as the acceptance of 'occult qualities' or 'attractive forces' in nature. The dispute was often triggered within the context of medicine, and it was often theological arguments that were invoked to dismiss Cartesian physics and metaphysics at the larger scale opened up by these disputes. The major concepts relevant in magnetism—'qualities' and 'attraction'—were taken to be important features of Galenic medicine and the analogy between magnetism and physiological processes was very common in this tradition.[109] The assumption of 'substantial forms' and 'qualities' was also very much presupposed in rational theological discourses, e.g., in explaining aspects of Christology, creation, and the sacraments.[110] This

---

Roling, *Odins Imperium*, vol. 2, 731–733, 751, 753. He mentions Bilberg, but not his disputation of 1687 or his attitude towards Cartesianism. See Vallerius (praes.) and Linnrot (resp.), *Disputatio physico mathematica de pyxide magnetica* (Stockholm, 1699), 54.

108   Vallerius (praes.) and Linnrot (resp.), *Disputatio physico mathematica de pyxide magnetica* (Stockholm, 1699), 44, 58.

109   Cf. Sander, "Nutrition and Magnetism"; Sander, "Tempering Occult Qualities."

110   Cf., e.g., Hellyer, *Catholic Physics*, 90–114; Beck, *Gisbertus Voetius*, 60–90; Goudriaan, *Reformed Orthodoxy and Philosophy*. See also Goudriaan, ed. and transl., *Jacobus Revius*, 180–183.

held true, more or less, for all Christian confessions at the time: Descartes's theory of magnetism and *a fortiori* his metaphysics were attacked in Lutheran, Calvinist and Catholic institutions. Claiming that magnetic attraction could be explained by nothing else but particles and their motion was a direct attack on Galenic principles, an indirect attack on Aristotelian principles, and potentially a problematic claim in theological contexts, at least in the eyes of few. Such a claim could be said to have undermined or questioned the philosophical underpinning of the major faculties at almost every university at this time. There is little wonder, then, that it met with such resistance, and from there became associated with some subsequent efforts to reconcile Cartesian with Aristotelian thought.

Only in the 1660s did magnetism become a topic in its own right within university learning, with long sections in the curriculum and textbooks discussing causal explanations. Teaching and its literary production were thereby established as the main battleground for these longstanding controversies. Against this background, the medical or broader relevance of concepts such as 'attraction' or 'quality' became less obvious within the discussions on magnetism. These discussions belonged to *physica particularis*, dealing with minerals and the like, and not, as previously, to general natural philosophical discussions about concepts such as 'attraction' or 'quality.' The discussions on magnetism thus became not only more focused, but were also relocated within the system of learning.

Looking more closely at the development of controversies on magnetism theory from a natural philosophical perspective, it seems that their dynamics exhibit more signs of convergence than of radicalization. Authors sympathetic to the Cartesian theory of magnetism did not, for example, always pick up Descartes's exact theory of screw-shaped particles, and regarded the form of the particles as beyond determination by any theory. Aristotelian authors were not keen on promoting 'occult qualities' and often seemed to distance themselves from this concept, which arguably provided the largest attack surface for Descartes and his followers. This is not to downplay the systematic differences between the opposing sides in the debate: elaborate attempts such as that of Van Baerle to combine both theories remained rare. While the adversarial rhetoric remained hostile, camps grew more aware of the weak spots of their own theory as seen from the opponent's perspective. They adjusted accordingly. The Aristotelians, especially, tried to turn the tables by claiming that the invisible particles and their motion were 'occult' at best. More often they claimed that the *particulae striatae* were just dreamt up without any basis in empirical evidence—*fingere, comminiscari,* and *excogitari* are recurrent terms.

It is also noteworthy that sources tended to avoid openly naming Descartes even while unmistakably addressing his theory of magnetism. More often this tendency shows in his followers, especially in phases and contexts of controversy and transition. Critics, on the other hand, tend to frame the Cartesian theory of magnetism as atomistic, alluding to magnetism theories as put forward by Epicurus and Lucretius. These are primarily tactical manoeuvres. Owing to its atheistic connotations, atomism was levelled as a charge more than as the sober description of some philosophical resemblance. Making positive public references to Descartes in the second half of the seventeenth century was also a political statement, given the reputation, or at least the suspicion, of his doctrines as 'heretical', as testified not least by Catholic censorship and condemnation.

On this more political and partly even confessional side of the story, it should also be emphasized that the reception of Descartes's theory of magnetism was not a binary set-up in terms of predictable opponents. Particularly, but not exclusively, Descartes's critics found unlikely allies in the Jesuits Niccolò Cabeo and Athanasius Kircher, and even in William Gilbert. These works were not books typically studied in university classes, and Jesuits were certainly not typical warrantors for Protestant and Calvinist authors, not to speak of Gilbert, who stridently attacked Aristotelianism and traditional university learning.[111] However, they all agreed in their refusal of corpuscularian accounts. On the other hand, Descartes openly acknowledged Gilbert in his *Principles of Philosophy* and his followers also referred to Cabeo and Kircher. While this was mostly for their more empirical findings than for their theories, their theories were not attacked by Descartes's followers in this connection.

On the level of university education, this policy of appealing to named authors across the political or confessional divide arguably led also to a widening of the pool of authors and books that may have been mentioned (and perhaps also discussed) in the classroom. Arguing against Descartes produced new alliances, introducing new authors and thereby new 'knowledge.'

Alongside Descartes's intriguing theory of magnetism adding to the scope and content of a Cartesian curriculum, another effect is probably even more important. The way in which Descartes dealt with magnetism was quite original, both formally and in terms of the length at which he rehearsed his arguments. Both aspects were often adopted along with the theory itself. Many subsequent textbooks describe magnetic effects as extensively, including empirical and experimental findings, and depict both theory and observation visually as

---

111   Cf. n. 37.

Descartes had done.[112] This was equally true for some of his critics. At least structurally, by arguing against Descartes, they also imitated him, e.g., by discussing single magnetic effects one after another; many can also be seen to illustrate their sections on magnetism in ways similar to those of Descartes.[113]

This development holds true for textbooks, for preserved student notebooks, and for printed disputations. The very fact that forty out of forty-nine preserved seventeenth-century printed disputations on the magnet date from after 1644 is not to be attributed exclusively to the influence of Descartes's theory of magnetism, but it testifies to a general development rendering magnetism an increasingly important and established topic for university learning.[114] Descartes had some share in making such a development possible, especially as far as it concerns the way natural philosophy was taught at universities and schools. Even if not all students in the 1680s, and maybe not even the majority of them, were taught the Cartesian theory of magnetism, the fact that they were taught any theory of magnetism, and the manner in which this happened, in one way or another owed much to Descartes and the publication of his *Principles of Philosophy* in 1644.

Acknowledgements

Funding for this research was provided by the Max Planck Research Group "Visualizing Science in Media Revolutions," led by Sietske Fransen (BH-P-19-35).

---

112   Cf. also Lind, *Physik im Lehrbuch*, 87: "Für den Magnetismus wird meist Descartes' Wirbeltheorie angenommen."
113   Cf. Sander, "*Terra AB.*"
114   Cf. Sander, "Magnetism in an Aristotelian World (1550–1700)."

The Anatomy of a Condemnation: Descartes's Theory of Perception and the Louvain Affair, 1637–1671

Mattia Mantovani

"The university of Louvain has long been honoured for not accepting novelties." And so things ought to stay, the papal nuncio of Brussels admonished in a 1662 letter to the faculty of arts.[1] Girolamo De Vecchi's concerns were justified. From the late 1640s onwards, more and more Louvain professors had begun to lecture on the philosophy of René Descartes (1596–1650). Some of their colleagues, especially those of the higher faculties—law, medicine and theology—opposed this new trend, and demanded the intervention of authorities to stop it. Their complaints made it to Brussels, and to Rome. In 1662, in the wake of the nuncio's letter, Descartes's philosophy was condemned by the Louvain faculty of theology. One year later, his works were put on the *Index librorum prohibitorum*. The universities of Utrecht and Leiden had already taken measures against Descartes's philosophy in 1642 and 1648, but the Louvain condemnation of 1662 was the first issued by a Catholic university.[2] It is difficult to determine its exact consequences in countries like Italy and France, but the Louvain condemnation resounded throughout the anti-Cartesian controversies of the years that followed.[3] The 1663 Catholic prohibition on reading Descartes's works *donec corrigantur* slowed down their circulation, and gave a powerful argument to Descartes's opponents.

Léger Charles De Decker reported in his *Cartesius seipsum destruens* that, by 1675, Louvain was still the only Catholic university in which Descartes's philosophy was taught.[4] A proper understanding of how Descartes's philosophy was

---

1 Girolamo De Vecchi to the Faculty of Arts of Louvain, 28 June 1662: "Lovaniensis Academiae antiqua laus est quod novitates non recipiat"; reported in Armogathe and Carraud, "First Condemnation," 97.

2 For Utrecht, see Verbeek, ed., *La querelle d'Utrecht*. For Leiden, see Verbeek, *Descartes and the Dutch*, 46–47.

3 On the reception of the 1662 condemnation in France, see Roux, "Condemnations," especially 757–758.

4 [De Decker], *Cartesius seipsum destruens* (Utrecht, 1675), §19, 137: "nec hatenus ibidem aut

received in Louvain is therefore vital to an overall understanding of the early history of Cartesianism in the academic setting. In this chapter, therefore, I consider the responses to Descartes's philosophy by Louvain professors from 1637 to 1671, that is to say, from the first critique of Descartes's philosophy at Louvain to its establishment at the university faculty of arts.

The main sources at our disposal reveal an intense—almost obsessive—interest in Descartes's theory of perception and, more specifically, of vision, down to the minutest physiological details. In this chapter, I try to account for this apparent oddity, and show what the *fortuna* of Descartes's theory of vision in Louvain reveals about Descartes's own philosophical agenda and the reasons behind its acceptance, or rejection, in the academic milieu. Vision theory is here taken in a broad sense, as to encompass the physics of light, the anatomy of the eye, and the perceptual process proper: *perspectiva* was about all these subjects, and much more.[5]

Contemporaries promptly realized the importance of Descartes's theory of vision: Henry More's (1614–1687) last letter to Descartes—who died, leaving it unanswered—revolved precisely around this subject. By 1650–1651, More had already started an epistolary tutorial given to Anne Conway (1631–1679), based on their shared reading "Des Cartes Dioptricks."[6] Even before his death, Descartes's vision theory had become a topic of controversy also in institutes for higher education, as illustrated by the College of Claremont in Paris; Domenico Collacciani and Sophie Roux have extensively documented the importance granted to the topic by the mathematics professors of this most important French Jesuit college.[7] As it happens, the responses by Paris Jesuits were more flexible and varied than one might expect. For instance, during the 1640s even as strong an opponent of Descartes as Pierre Bourdin (1595–1653)—best known as the "seventh objector" to the *Meditations*—came to adopt theories from the *Dioptrics*, and to have them defended by his students in their disputations, sometimes even presenting Descartes's theory of light and attributing

---

alibi ulla est Schola publica Academiae Catholicae (nostram, de qua statim, non comprehendo) in qua tradatur eiusmodi Philosophia." The work was published anonymously; on its authorship, see Monchamp, *Galilée et la Belgique*, 206. De Decker's claim was almost certainly an overstatement: Louis XIV's condemnation of 1671 strongly suggests that, around those years, this was the case also for Paris. On the diffusion of Descartes's philosophy in French academies and via public lectures already in the 1660s, see Schmaltz, *Constructions*, 94–97 and Chapter 19, in this volume.

5 On the scope and aims of *perspectiva*, see Mantovani, "Roger Bacon on *perspectiva*."
6 On More's teaching—by letter—to Anne Conway, see Chapter 12, in this volume.
7 Collacciani and Roux, "Querelle optique"; Collacciani and Roux, "Times of War."

it to Aristotle. The mathematics professor of the early 1670s—Ignace-Gaston Pardies (1636–1673)—was even accused, not without reason, of being a crypto-Cartesian.

In this chapter, I provide a further example of the diffusion of Descartes's theory of vision in the educational context, and of its various uses around the mid-seventeenth century. Contrary to what happened at the Claremont College, however, Louvain anti-Cartesians proved unyielding in their opposition and, indeed, pushed it to the point of an outright condemnation of Descartes's philosophy.

The chapter identifies five main stages of Descartes's reception at Louvain university, starting from 1637 (section 1), when Descartes dispatched to Louvain copies of his just-published *Discourse on Method*, arguably—as I show—with the hope of having the professors of this renowned university begin lecturing on his philosophy. Descartes's plan backfired: the professor of sacred scripture—Libert Froidmont (1587–1653)—passed a very stern judgement on the *Discourse*, and especially on the *Dioptrics*, in which he was immediately joined by the professor of medicine—Vopiscus Fortunatus Plemp (1601–1671)—an erstwhile friend of Descartes, who, in the years to come, became one of his greatest foes. Up until his death in 1650, Descartes revisited the themes of the *Dioptrics* many times, and expanded upon them in a number of works; in particular, I document Descartes's reactions to Froidmont, and his continual reflections on the larger epistemological and metaphysical implications of his vision theory, which prove to be of special importance for understanding the Louvain affair (section 2). Based on disputations, students' notebooks, archival sources and contemporary correspondence, I correct previous scholarship (section 3) and prove that the teaching of Descartes's philosophy started in Louvain in 1648–1649, by philosophy professors such as Gerard Van Gutschoven (1615–1668), his brother Willem (1618–1667), and Arnold Geulincx (1624–1669). The anti-Cartesians from the higher faculties were not slow to respond, and the controversy somewhat spilled over into something bigger as Plemp and Froidmont managed to convince a number of their colleagues to denounce Descartes in print (section 4). Indeed, the *Iudicia de philosophia cartesiana* of 1652 are an extraordinary—and surprisingly understudied—document of the first reception of Descartes's philosophy. I examine these "assessments" in light of the few surviving *dictata* and *disputationes* of those years, so as to shed new light on the reformation of the philosophy *curriculum* that took place in 1658 and on the process leading to the official condemnations of 1662 and 1663. The chapter concludes (section 5) by considering the aftermath of these condemnations into the 1670s, when the vast majority of philosophy professors came to embrace Descartes's views also on theological matters. At this time—as we shall see—both Cartesianism and anti-Cartesianism underwent significant changes, in Louvain as all over Europe.

1   1637: *Renatus Democritus*

In June 1637 Descartes published his first work, the *Discourse on Method*, followed by three "essays" of this method: the *Dioptrics*, *Meteorology*, and *Geometry*. He circulated the work among fellow intellectuals in the Low Countries and France and eagerly awaited their judgement.[8] By the end of the summer, copies of the *Discourse* was thus sent to "a professor in medicine at Louvain," whom Descartes regarded, first and foremost, as a friend.[9]

Descartes and Plemp had met for the first time soon after October 1629, when Descartes moved to Amsterdam, where he was to live until May 1632.[10] Descartes was the older of the two by over five years, but both were educated in Jesuit colleges and shared a keen interest in medicine and natural philosophy. The two young men soon established a firm friendship, and spent quite some time together discussing *de rebus physicis* and, apparently, performing dissections.[11] Traces of these discussions resurface in the *Ophtalmographia, sive tractatio de oculi fabrica, actione et usu* (Amsterdam, 1632), which earned Plemp an appointment at Louvain one year later. For his part, Descartes too seemed to have Plemp in mind while drafting the *Dioptrics* in the course of the 1630s. Indeed, his famous remark that perception cannot be explained by appealing to some sort of "other eyes within the brain" is easily read as a response to the *Ophtalmographia*.[12] In his tract, Plemp ventured to speculate on why things are not visually perceived upside-down, given the inversion of the retinal image (discovered by Kepler just a few decades earlier). Plemp remarked that the "cognitive faculty" is located in the brain, anterior to the eye, so that it has to apprehend the retinal image from behind, and upside-down—think of

---

8   Descartes to Mersenne, March 1636, AT I 339.

9   Descartes to Vatier, 22 February 1638, AT I 561. Plemp to Descartes, 15 September 1637, AT I 399. All translations, otherwise stated, are mine; as for Descartes, I have taken as my starting point the CSMK translation.

10   Clarke, *Descartes*, 421–423.

11   Plemp, *De fundamentis medicinae*, 1st ed. (Louvain, 1638), II 5, 265: "ipsomet Cartesio, amicissimo viro ac mihi familiarissimo." Plemp, *Fundamenta medicinae*, 3rd ed. (Louvain, 1654), 375: "saepe cum eo de rebus egi physicis."

12   Descartes, *Dioptrique* VI, AT VI 130.

a bat in a cave—thereby compensating for its inversions.[13] The bearded man observing the back of an eye in the well-known engraving of the *Dioptrics* might thus also be regarded as a sort of joking portrait of Descartes's junior comrade (Figure 15.1).

Besides the copy for his long-time friend, another of the copies of the *Discourse* dispatched to Louvain in August 1637 was intended for Plemp's mentor: Libert Froidmont (1587–1653), professor, a few months since, of sacred scripture at the local university.[14] During his previous appointment as a philosophy teacher (1614–1628), Froidmont had written an important treatise on atmospheric phenomena—the *Meteorologicorum libri VI* (Antwerp, 1627)—which explains why Descartes would have solicited his opinion. Descartes appears, furthermore, to have had an even more ambitious goal in mind: in his correspondence with Froidmont, he betrayed the aspiration to have his philosophy taught or, at least, accepted by university professors.[15]

Froidmont immediately undertook the study of the *Discourse*: just three weeks later, he sent back to Descartes a long letter with comments.[16] Whilst praising his ingenuity, Froidmont's criticism of Descartes's natural philosophy overall was quite disparaging, however. He lamented that most of Descartes's explanations were "excessively gross and mechanical," and that they seemed inspired by "the coarse and somewhat bloated [...] physics of Epicurus."[17] Froidmont was appalled by Descartes's presuming to account for all phenomena—other than the operations of the rational soul—by the size, shape and motion of particles of matter. He protested that "so noble operations" such as life and

---

13   Plemp, *Ophtalmographia*, 1st ed. (Amsterdam, 1632), II 19, 135: "Sic igitur existimandum censeo: quamquam objectorum pictura in retiformi tunica repraesentata nobis ex adverso eam videre visis, tota & cernatur & judicetur eversa: facultas tamen cognoscens seu sensus communis non ita apprehendit; ratio est: quia facultas haec versus cerebrum sedem obtinens non habet sibi picturam illam adversam; nam cavum retinae introrsum in caput vergit, & retro cavum illud facultas visa est. Unde eo fere modo concipe facultatem sentientem picturas in oculo dignoscere, quo nos in clausa camera imagines perciperemus retro charta stantes, & desuper in eam despicientes: sic enim omnia ea figura situque, quo foris vere sunt, cognoscerentur."

14   Plemp to Descartes, 15 September 1637, AT I 399. Among Froidmont's other works, the more memorable are the *Ant-Aristarchus* (Antwerp, 1631) and the *Labyrinthus sive de Compositione continui* (Antwerp, 1631). On Froidmont, see Jesseph, "Fromondus."

15   Descartes to Plemp, for Froidmont, 3 October 1637, AT I 411: "sed contra quicquid adeo verum erit et firmum, ut nulla tali demonstration possit everti, non impune, ut spero, *saltem ab iis qui docent*, contemnetur" (my emphasis). AT I 415–416: "aliis materiis fere omnibus potuissem adjungere ad propositiones meas roborandas, quae de industria subticui [...] ne ullis opinionibus in Schola receptis velle viderer insultare."

16   Froidmont to Plemp, for Descartes, 13 September 1637, AT I 402–409.

17   Froidmont to Plemp, for Descartes, 13 September 1637, AT I 406, 402.

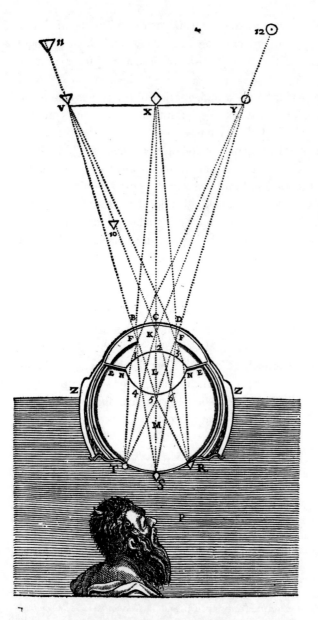

FIGURE 15.1 The eye as a *camera obscura*. Descartes, *Discours de la méthode* (Leiden, 1637), *Dioptrique* V, 37.
WELLCOME COLLECTION, ZC7F33FN

sense-perception "could not" and, indeed, "ought not" to be attributed to so "ignoble," "brute" and "humble causes" as the motions of a few corpuscles, lest one induce atheists to claim that the same held true for the rational soul.[18]

According to Froidmont, as "sublime" an operation as sense-perception demanded causes other than material causes in both inanimate bodies and living beings: namely, and respectively, "real qualities" and "substantial souls." Moreover, in order to mediate between the two and to make these qualities of bodies cognitively present to perceivers, Froidmont objected to Descartes that an additional class of entities was needed: the so-called *species intentionales*.

In general, Froidmont was thoroughly unconvinced by Descartes's account of sense-perception, including his claim that non-rational animals do not sense. No fewer than eight out of the eighteen points raised in his letter to Descartes address precisely this issue, with a remarkable emphasis on neurophysiology. Froidmont was especially puzzled by Descartes's claim that the soul is directly responsive only to the changes occurring in the brain, and reacts to those taking place in the other organs only insofar as they are transmitted to the brain by the nerves. Faithful to Aristotle, Froidmont objected that the soul informs the entire body, thereby giving all limbs sentience.[19]

But when it came to Descartes's theory that the nerves are responsive to nothing but motion, his theory of the visual process in particular was enough for Froidmont to expose the shortcomings of Descartes's philosophy as a whole:

> Who might believe that vision occurs by means of nothing but light's impulse on the nerves of the retina, as he says in the *Dioptrics*? How could the visual power distinguish between, say, purple and yellow, in case it was not determined to see by the quality of an intentional *species*, but by local impulse alone? On account of the fact that the impulse on a certain colour is stronger than another? But also the impulse of one and the very same colour, depending on whether it is more or less intense, would be more or less strong.[20]

---

18   Froidmont to Plemp, for Descartes, 13 September 1637, AT I 403, 404: "Tam nobiles operationes non videntur posse prodire ex tam ignobili et bruta causa [...] non oportet tam excelsas operationes tam humilibus causis tribuere."

19   Froidmont to Plemp, for Descartes, 13 September 1637, AT I 405–406.

20   Froidmont to Plemp, for Descartes, 13 September 1637, AT I 405–406: "Visionem etiam fieri per solum impulsum luminis in fila tunicae vel retinae (ut pag. 50 docet) [*Dioptrique* V, AT VI 128–129] quis credet? Unde enim potentia visura distinguet inter purpureum v. g. et flavum colorem, si non per Speciei intentionalis qualitatem, sed per solum localem

Froidmont's core objection was that mere variations in the degrees of particles' motion could not account for colour-sensations that differ in kind, let alone for opposite colour-sensations, as would be the case for black and white. (The same applied to flavours, as Froidmont argued in relation to the different shapes of the particles in contact with the tongue.) Arguably, Froidmont was not reproaching Descartes as an atomist only in questions of matter theory, as has already been nicely illustrated by Daniel Garber.[21] Indeed, his arguments against Descartes's account of sense-perception seem to rephrase Aristotle's own arguments against Democritus in chapter four of *De sensu*, which—notably—opposed the atomist's account by discussing precisely flavours and colours.[22] Claims about atomism and the composition of matter, and related claims regarding what pertains to sense-perception, had been so fiercely debated in Louvain in the 1630s, that it was almost inevitable that Froidmont and Plemp would approach Descartes's *Discourse* as the work of a "Democritus revived."[23]

According to Froidmont, the shortcomings of Descartes's theory were inevitable consequences of his reducing bodies to chunks of matter alone; once it had done away with the Scholastic apparatus of "intentional species," "real qualities," "real accidents" and "substantial forms," Descartes's philosophy could not but fall short of its task. Without these entities, Froidmont confessed that he was at a loss for any explanation:

---

impulsum, determinetur ad videndum? An quia unius coloris impulsus major est, quam alterius? Sed etiam unius et ejusdem coloris intensioris aut remissioris impulsus major erit aut minor."

21   See Garber, "Revolution." On early-modern readings of Descartes as an atomist, see, more generally, Roux, "Descartes atomiste?"

22   I provide evidence for this claim in Mantovani, *"Renatus Democritus"*. I think it can be proved that Descartes himself developed these claims — already advanced in his *Regulae ad directionem ingenii* XII — by reflecting on this passage from Aristotle's *De sensu*. Froidmont to Plemp, for Descartes, 13 September 1637, AT I 406–407: "Quam etiam paradoxum quod pag. 162 ait [*Météores* I, AT VI 235–236], eadem corpuscula, si languide impellant sensum tactus, gignere frigoris sensationem, et caloris, si fortius impellant! Quasi vero tantum differentiae sit in illo impulsu locali, non in qualitatibus ipsis diversimode afficientibus organum tactus!." AT I, 408: "docet aquam maris apparere salsam, quia partes aquae crassiores punctim potius quam transversim incidunt in poros linguae. Quasi alio sapore tincta appareat, si casu transversim partes illae organo gustus incumbant!." Compare to *De sensu* 4, 442^{b}18–23: "Again, all the sensibles involve contrariety; e.g. in colour white is contrary to black, and in savours bitter is contrary to sweet; but no one figure is reckoned as contrary to any other figure. Else, to which of the possible polygonal figures [to which Democritus reduces bitter] is the spherical figure [to which he reduces sweet] contrary?."

23   On Froidmont's concerns with atomism since the early 1630s, see Palmerino, "Labyrinth of the Continuum." The woodcuts of the *Discourse* seem to have had an important part in this misunderstanding; see Lüthy, "Perplexing Particles."

He [Descartes] hopes to be able to clarify by means of nothing but position, or local motion, quite too many phenomena, which (unless I do not understand anything) cannot be explained without 'real qualities.'[24]

By Froidmont's reckoning, Descartes's philosophy would succeed or fail depending on whether or not it managed to do without any 'real qualities.' And against this measure, it badly failed.

Descartes soon sent a point-by-point response back to Froidmont, but this was not deemed worthy of an answer. As will emerge in what follows, Froidmont remained convinced until the end of his life that Descartes's philosophy was wrong and potentially dangerous; if anything, his assessment of Descartes's philosophy was to become only more negative with the passage of time.

In transmitting Froidmont's 1637 objections to Descartes, Plemp made clear to his friend that, all his affection and esteem notwithstanding, he sided with his erstwhile mentor and current colleague.[25] Slightly embarrassed, Plemp joked that he had been so imbued with Aristotelian philosophy as a schoolboy that he could not change course now.[26] His own objections to Descartes's account of blood circulation and heart motion further testify to this fundamental opposition. Whereas Descartes thought that heat could sufficiently explain blood circulation, Plemp subscribed to the received view, according to which it was a *facultas pulsifica* that explained the beating of the heart.[27]

Undaunted, Descartes gave every indication of being satisfied with his exchange with Froidmont, and went so far as to turn what Froidmont had intended as an aspersion into a title of honour, speaking of *mechanica philosophia mea*.[28] Their punishing critique notwithstanding, Descartes planned to publish these exchanges (together with other) as appendices to a revised, Latin edition of the *Discourse*. The project was superseded by the *Meditations*, but some

---

24   Froidmont to Plemp, for Descartes, 13 September 1637, AT I 408: "Nimis multa sperat se expediturum per solum situm, aut motum localem, quae sine realibus qualitatibus aliis non possunt, aut nihil intelligo."

25   Plemp to Descartes, 15 September 1637, AT I 399: "responsiunculae quaedam sunt, quae *secundum nostra principia* rem aliter explicant" (my emphasis).

26   Plemp to Descartes, 15 September 1637, AT I 400.

27   For an insightful analysis of the topic, see Petrescu, "Descartes on the Heartbeat."

28   Descartes to Plemp, for Froidmont, 3 October 1637, AT I 413–431. Incidentally, this is usually considered to be the first time the term 'mechanical' was used as a qualifier for a philosopher's theory; see, for a recent example, Bertoloni Meli, *Mechanicism*, ix. It can be shown, however, that already five years before Aristotle (!) had been criticized for speaking "mechanically," thereby encouraging "the slaves of the senses" in their materialist convictions. I discuss this evidence and its larger implications in Mantovani, *Mechanical Philosophy: The Origins of a Concept* (in preparation).

unpleasantness occurred in the intervening months. In 1638 Plemp published an abridged version of his exchange with Descartes within his *De fundamentis medicinae*.[29] Almost obsessed as he was with controlling his public image, Descartes felt betrayed when he found out about this publication, believing (not without reason) that his ideas had been misrepresented.[30] His reaction must have been extremely bitter: Plemp was still complaining about Descartes's resentment as late as 1644.[31]

Far from the open door to his philosophy at the university for which Descartes had likely hoped, the copies of the *Discourse* dispatched to Louvain had cost him one of his friendships. On top of their genuine philosophical disagreement and his larger pedagogical and cultural concerns, Plemp's later efforts to have Descartes's philosophy condemned seem also to have been motivated by personal rancour.

2   1637–1648: A Matter of Perspective

But why did Froidmont devote so much attention to the *Dioptrics*? And why, even before that, did Descartes decide to open his *Essays* with a tract on this subject? Froidmont might have been mistaken in reading Descartes as a neo-atomist, but he was certainly right in insisting so much on this subject, and on its importance for Descartes's philosophy as a whole.

Descartes had been engrossed with vision theory since his twenties, and, in this respect, he was not alone: at the time when the *Dioptrics* was published in 1637, optics had already been at the forefront of debate for some decades. In 1604, Kepler's *Ad Vitellionem Paralipomena* had dismantled the century-old account of vision set up in the thirteenth century—the so-called "Baconian" or "Perspectivists' synthesis"—without yet managing to provide an entirely convincing alternative in its stead, especially as regards the perceptual process proper.[32] Plemp's wild speculations about how to account for the inversion of the retinal image discovered by Kepler provide one example among many of a

---

29   Plemp, *De fundamentis medicinae*, 1st ed. (Louvain, 1638), II 5, 265–267.

30   Regius to Descartes, January 1640, AT III 3. On the Descartes-Regius exchange more generally, see Verbeek, Bos and Van de Ven eds., *Correspondence*.

31   *Fundamenta medicinae*, 2nd ed. (Louvain, 1644), II 5, 152. In this second edition, Plemp published the unabridged version of his 1637–1638 exchange with Descartes (152–161). As a reaction to that, Descartes published this exchange himself (remarkably enough, these are the only private letters that Descartes made available for publication in his lifetime); cf. Johan Van Beverwijck, ed., *Epistolicae quaestiones* (Rotterdam, 1644), 125–139.

32   On the "Baconian synthesis" see Tachau, *Vision and Certitude*. On the Perspectivists' the-

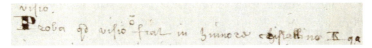

FIGURE 15.2 Louvain student's note on vision. MS., Louvain, MS. 229: Boudart
(professor) and Taijmont (student), *Physica & Metaphsyica* (1665–
1666), fol. 183ʳ: "Proba quod visio *non* fiat in humore crystallino".
MAGISTER DIXIT COLLECTION

puzzlement that lasted for a century and beyond: still in 1709, Berkeley lamented that "the explanation of vision" was bedevilled by a "mighty Difficulty."[33]

University professors were split. Their students found themselves in the
crossfire, as confirmed by their disputations and, graphically (see Figure 15.2),
by a course on natural philosophy delivered at Louvain in 1665–1666. Does vision take place in the crystalline lens, as the Perspectivists had claimed since
the thirteenth century, or does it *not*? The negation added between the lines by
Vincent Taijmont appears to reveal, at a glance, both the difficulty for this new
theory of Kepler (and Descartes) to find acceptance in the curriculum and its
eventual success.[34]

Yet for Descartes, the *Dioptrics* was not just about providing his own solution
to Kepler's conundrum, though this was an issue, and quite a pressing one.[35]
But it is certainly not for his interpretation of the retinal image that Descartes
kept referring his readers back to the *Dioptrics* in all of his subsequent published works.[36] Arguably, Descartes did not devote himself to vision theory only

ory of perception and its aftermath in the early modern age, see Mantovani, "*Visio per
        sillogismum.*"

33   Berkeley, *An Essay Towards a New Theory of Vision* (1709), § 88.

34   MS., Louvain, MS. 229: Boudart (professor) and Taijmont (student), *Physica & Metaphsy-
     ica* (1665–1666), fol. 183ʳ. All student notebooks discussed in this chapter can be consulted
     online thanks to the *Magister Dixit* Project. On Louvain student notebooks, see Mirguet
     and Hiraux, *Collection de cours manuscrits*; Vanpaemel, Smeyers and Van Smets, eds.,
     Ex cathedra. Kepler's theory of the retinal image had also been endorsed by the Jesuit
     Christoph Scheiner (1575–1650), who in his *Oculus hoc est fundamentum opticum* (Oeni-
     ponti, 1619) tried—unlike Kepler and, later on, Descartes—to accommodate the received
     doctrine of *species*; see Gal and Chen-Morris, *Baroque Science*, 203–206; Pantin, "*Simu-
     lachrum,*" 263–267; Dupré, "The Return of the *Species*"; Smith, *From Sight to Light*, 374–375.
     Froidmont and Plemp agreed here with Scheiner; cf. Froidmont to Plemp, for Descartes,
     13 September 1637, AT I 405.

35   The importance of Kepler's theory of vision for Descartes was publicly acknowledge by
     Descartes himself: "Kepler a été mon 1er maître en optique"; Descartes to Mersenne,
     31 March 1638, AT II 86. On the Kepler-Descartes relation, see Simon, *Kepler* and "Théorie";
     Schuster, *Descartes-Agonistes*; and Bellis, "Perception."

36   *Descartes, Responsiones* IV, AT VII 248; *Responsiones* VI, AT VII, 435, 437–438; *Epistola ad*

for its own sake—and, from the 1630s onwards, maybe not even mainly for its own sake—but on account of its epistemological and metaphysical implications.

According to Descartes, the physiology of the visual system had in fact momentous consequences for epistemology, as he made especially clear in arguing against the more materialist-minded account of perception defended by his erstwhile friend Henricus Regius (1598–1676). The mismatch between our sense-perceptions (such as those of colours and light) and their physiological causes (patterns of motions modifying the sense-organs and, hence, the brain) was, for Descartes, a conclusive argument that our "ideas" of sensible qualities have to be innate:

> Nothing reaches our mind from external objects except certain corporeal motions [...] But neither the motions themselves nor the figure arising from them are conceived by us exactly as they occur in the sense organs, as I have explained at length in my *Dioptrics*. Hence it follows that the very ideas of motions and figures are innate in us. *A fortiori* must then be innate the ideas of pain, colours, sounds and the like if, on the occasion of certain corporeal motions, our mind is to be capable of representing them to itself, for there is no similarity between these ideas and the corporeal motions.[37]

Equally significant were the consequences that Descartes intended to draw for metaphysics. He, indeed, believed himself to have shown in the *Dioptrics* and later works that he could account for light and colour perception without having to posit anything in bodies in addition to the geometrical properties

---

P. Dinet, AT VII 582, 602; *Principia* III 130, AT VIII-A 180; III 134, AT VIII-A, 186; IV 189, AT VIII-A, 316; IV 195, AT VIII-A, 319; *Epistola ad Voetium*, AT VIII-B, 162–163; *Notae in programma*, AT VIII-B 359; *Passions de l'âme* I 12–13, AT XI 337–338. See also *Description du corps humain*, AT XI 255. The references in the correspondence are almost innumerable. Indeed, the *Dioptrique*, the first draft of which dates to around 1630, is even referred to in *Le monde* (which Descartes had abandoned in 1633): *Le monde*, AT XI 9, 106; *L'homme*, AT XI 153, 156, 187. There is evidence, however, that Descartes reworked *Le monde*—or, at least, the *Traité de l'homme*—also in the 1640s.

37   *Descartes, Notae in Programma quoddam*, AT VIII-B 358–359. See already Descartes to Mersenne, 22 July 1641, AT III 418: "Et enfin, je tiens que toutes celles qui n'enveloppent aucune affirmation ni négation, nous sont *innatae*; car les organes des sens ne nous rapportent rien qui soit tel que l'idée qui se réveille en nous à leur occasion, et ainsi cette idée a dû être en nous auparavant." On the philosophical implications of Descartes's physiology, see Hatfield, "Descartes' Physiology." On Descartes's changing views on innatism, see n. 88 below.

of extension. "Light and colour," Descartes famously claimed, "are nothing but some arrangements [of particles] consisting in size, shape and motion," and so it was for all sensory qualities.[38] According to Descartes's account, these arrangements then elicit, either by themselves, or by means of other corpuscles and yet other arrangements, some changes in the sense-organs and, thence, in the brain. These brain-states are in turn *institués de la nature* in such a manner as to bring about certain sense-perceptions.[39] The sense-perceptions themselves, however, *need not* be and in fact *are not* similar to the corresponding properties of bodies. Descartes never grew tired of warning that it would be a gross mistake to assume that "in a white or green body is present the self-same whiteness or greenness that I perceive" (as the Aristotelians had been doing for centuries).[40] By so doing, Descartes meant to reject the entire Aristotelian model of cognition *qua* assimilation, and promised his readers to "deliver their minds [...] from all those little images flitting through the air, the so-called intentional *species*"—or *similitudines* (likenesses)—"which fatigue the imagination of Philosophers so much."[41]

Descartes thereby meant his vision theory to prove that he could account for sense-perception without the usual scholastic machinery of "intentional *species*" and "real qualities," which could and should therefore be dismissed as useless. And what had been proved true for the "the noblest and most comprehensive of all senses," the one widest in scope and the most sophisticated of all, was intended to work as a prototype for all the other senses:

> The principal argument that induced philosophers to posit real accidents was that they thought that sense-perceptions could not be explained without them, and this is why I promised that I will give a very detailed account of sense-perception in my *Physics*, taking each sense in turn. Not that I want any of my results to be taken on trust, but I thought that the explanation of vision which I had already given in the *Dioptrics* would

---

38 Descartes, *Principia* IV 199, AT VIII-B 323: "... lumen, color, odor, sapor, sonus, et tactiles qualitates; quae nihil aliud esse, vel saltem a nobis non deprehendi quicquam aliud esse in objectis, quam *dispositiones quasdam in magnitudine, figura et motu consistentes*, hactenus est demonstratum" (my emphasis).

39 See, for example, *Dioptrique* VI, AT VI 130; *Meditationes* VI, AT VII 87; *Passions* I 36, AT XI 357. On this topic, see Mantovani, "The Institution of Nature."

40 Descartes, *Meditationes* VI, AT VII 82: "in albo aut viridi sit eadem albedo aut viriditas quam sentio."

41 Descartes, *Dioptrique* I, AT VI 85: "... Et par ce moyen votre esprit sera délivré de toutes ces petites images voltigeantes par l'air, nommées des espèces intentionnelles, qui travaillent tant l'imagination des Philosophes."

make it easy for the judicious reader to guess what I was capable of accomplishing with regard to the remaining senses.[42]

At stake in Descartes's theory of vision, as Froidmont had perfectly understood already in 1637, were some key notions of scholastic metaphysics, and openly so. Indeed, I think there is a compelling argument to be made that vision theory provided the paradigm argument for Descartes's claim that bodies are nothing but extended substances, with no forms attached.[43] But even without going for so bold an interpretation, the evidence presented here suffices to show that Descartes intended his theory of vision to do more than merely describe the functioning of the eye, or the blending of shadow and light. He meant it to establish a daring new image of the world: a mechanical world, devoid of all *species*, "qualities" and "substantial forms"—except for the thinking mind. Seen through the lens of the *Dioptrics*, the interest in Descartes's vision theory that had so engrossed the Louvain professors thus acquires a completely different meaning and reveals a much wider range of far-reaching implications.

3   1648–1652: Replacing Aristotle

"For almost four years, some people have been struggling to expel Aristotle from our schools."[44] The *incipit* of Plemp's open letter of December 1652 was an open call to arms. His colleagues did not shy away from it, but spoke their minds loud and clear about the impacts wrought by the novel teaching of Descartes's philosophy at the faculty of arts. The six "Assessments of Descartes's philosophy by a few learned men of Louvain Academy," are quite an extraordinary and surprisingly understudied document of the first reception of Descartes's philosophy. I will consider in due course these *Iudicia*, appended by Plemp to the third, 1654 edition of his *Fundamenta medicinae*. For the moment, let us just focus on Plemp's testimony that Descartes's philosophy started being taught in Louvain

---

42   Descartes, *Meditationes, Responsiones* VI, AT VII, 435; *Dioptrique* I, AT VI 81.

43   Descartes believed himself to have established that bodies are in the first place "extended things," based on his "first philosophy" alone. By contrast, I think it can be shown that Descartes's thesis that bodies are *nothing but* extended things—with no forms attached— was not indeed the starting point of Descartes's "natural philosophy," but its crowning achievement. I argue at length for this reading in Mantovani, "The Eye and the Ideas."

44   Plemp, *Doctorum aliquot in Academia Lovaniensi virorum iudicia de philosophia cartesiana*, in Plemp, *Fundamenta medicinae*, 3rd ed. (Louvain, 1654), 375–387, at 375: "Conantur aliqui jam a quadriennio fere pellere e scholis nostris Aristotelem, et nescio quam antiquatam philosophiam inducere."

around 1648–1649, shortly before his death on 11 February 1650. In the late 1640s, Louvain thus became the first—and, for few decades, the only—Catholic university where Descartes's ideas were publicly taught.

Of the extant disputations in philosophy defended in Louvain, the first to display traces of Descartes's philosophy dates to 1651; the first course on 'physics' dates to one year later. There are no reasons, however, to call into question Plemp's testimony: Plemp's appointment as *rector magnificus* in 1647 and 1649 renders him an especially informed and authoritative witness of the academic events of the late 1640s. The very earliest Louvain lectures on Descartes's philosophy, however, seem to have left no traces: the fire of 1914 destroyed the most invaluable sources of Louvain's old university. Of the many courses in natural philosophy taught in those years at Louvain's four colleges (the Castle, the Falcon, the Lily, and the Pig), only two students' notes have survived, none of which seems to display or oppose Cartesian influences. They both come from the college of the Pig, which had the good fortune of being the furthest away from the city centre under siege but was also, it seems, the least open to innovations. The casualties of war seem to have irreparably disfigured our image of the past.

A similar dearth of evidence means that, unfortunately, we do not know who introduced Descartes's philosophy to Louvain University, and why exactly this introduction occurred when it did in the late 1640s. The new appointments in 1646 of Gerard Van Gutschoven (1615–1668) and Arnold Geulincx (1624–1669), as senior professor of mathematics and junior professor of philosophy respectively, seem to signal—or, at least, to have prompted—a novel interest in the "new philosophy." In his monumental and still praiseworthy *Histoire du Cartésianisme en Belgique* of 1886, George Monchamp claimed that Gerard Van Gutschoven had already started to lecture on Descartes's philosophy in 1639, when first appointed as a junior professor of mathematics, but the evidence he provided was quite slim and ran contrary to Plemp's declaration and Monchamp's own indirect testimony concerning the 1648 reaction to Descartes's philosophy by the local Jesuits (see below).[45] Indeed, Gerard Van Gutschoven

---

45   Monchamp, *Cartésianisme*, 143–144. Gerard Van Gutschoven was first appointed professor
      of mathematics in 1639 but, due to frictions with an eighty year old professor of mathemat-
      ics, resigned already in 1641, to come back to Louvain only upon Sturmius's dismissal
      in 1646; Andreas, *Fasti academici studii generalis Lovaniensis* (Louvain, 1650), 249. Baillet
      maintained that, at the time of his first appointed in 1639, Gerard Van Gutschoven had
      already been "domestique de M. Descartes pendant un temps considerable." Monchamp,
      Cartésianisme, 143–144. Baillet, *Vie de Descartes*, vol. 2, 456; see also 399: "il demeuré
      plusieurs années sous luy à copier, et à le servir pour les expériences." Baillet—as well
      known, not always the most reliable source—was relying on a letter by Pierre Bayle to

resigned already from this first tenure at Louvain University in 1641 and immediately matriculated in the study of medicine at the University of Leiden, at a
time when Descartes lived nearby, at castle Endegeest: it was almost certainly
on that occasion that the two men met and—if we are to trust reports—struck
up a close friendship.[46] By the same token, in the absence of independent evidence in its support, I would urge caution in taking on trust Girolamo De Vecchi's
letter to Cardinal Pallavicini, where Descartes's philosophy is said to have "perhaps" been introduced in Louvain by Geulincx: the letter dates to twelve years
after the facts, and is obviously based on second-hand reports.[47]

In addition to Plemp's testimony, moreover, we do possess independent
evidence that Descartes's philosophy was first taught in Louvain in 1648. Monchamp, who had access to materials now lost, provided an unintentional—
and, therefore, even more valuable—piece of evidence in favour of this dating
when he noticed that, in that same year, the theological disputations of the

---

Claude Nicaise (Rotterdam, 28 September 1690), in which Bayle reports a conversation
with a Flemish physician that he had had in the 1680s: a quite late and indirect testimony,
indeed, that all the more makes no mention of the time when Van Gutschoven would
have become so close to Descartes; cf. Cousin, *Fragments philosophiques*, 204–205; Monchamp, *Cartésianisme*, 118–119. A letter by René-François De Sluse to Pascal from 4 October
1659 suggests a more reasonable picture: on De Sluse's account, Van Gutschoven had with
Descartes "des entretiens très particuliers l'espace *de quelques mois en Hollande*"; De Sluse,
Correspondence, XVII, 509 (my emphasis). Moreover, further evidence—see the following footnote—suggests to move Gerard Van Gutschoven's acquaintance with Descartes to
after 1641.

46   *Album studiosorum academiae lugduno batavae*, 325 (the credit for this discovery goes to
Erik-Jan Bos, to whom my heartfelt thanks). Descartes lived at castle Endegeest from May
1641 till May 1642. We already knew that, between 1641 and 1646, Gerard Van Gutschoven
had moved to Antwerp, but later declared himself to have also been practising anatomy
and dissections "in quite some universities," thereby strongly suggesting a stay in the
United Provinces; the information derives from Van Gutschoven's address to the Governor
of the Low Countries to obtain the Chair of Anatomy at Louvain University following the
resignation of Dorlix in 1659; Archives de l'État, Conseil d'État, Faculté des Arts, 1659–1725,
Carton; reported in Monchamp, *Cartésianisme*, 299. At any rate, Gerard Van Gutschoven's
adherence to Cartesianism in the 1650s and 1660s is beyond doubt: as is well known, it
was him, together with and Louis De la Forge (1632–1666), who provided (upon Clerselier's
request) the illustrations for the 1664 posthumous edition of Descartes's *Traité de l'homme*
(I provide evidence of Gerard Van Gutschoven's Cartesianism in the 1650s in what follows).
By contrast, in 1637 Descartes declared himself to be acquainted with only two persons in
Louvain who could understand his geometry, neither of whom was Van Gutschoven; cf.
Descartes to Plemp, 3 October 1637, AT I 411.

47   Girolamo De Vecchi to Cardinal Pallavicini, 29 July 1662: "Il primo che *forse* cominciasse
in Lovanio a fomentare queste novità fu un tale Gulinx [sic] professore ivi assai stimato
in Filosofia" (my emphasis); reported in Armogathe and Carraud, "First Condemnation,"
100.

town's Jesuits devoted special attention to a refutation of *a priori* proofs of God's existence; exactly the sort of proof defended by Descartes.[48] Arguably, the first teaching of Descartes's philosophy at the faculty of arts had already aroused opposition to his views even outside the college walls. (But for marginal remarks of this sort, Monchamp pays little or no attention to courses and disputations: as we are about to see, these sources do in fact permit us to enrich and adjust his readings on a number of points.)

Amidst the destruction and the oblivion wrought by fire and by time, only one 'Cartesian' disputation defended in these crucial years appears to have come down to us intact: a *Philosophia*, defended on 31 July and 1 August 1651 by a student of Falcon College, Hilarius Van Werm, under Willem Van Gutschoven (1618–1667).[49] Little is known of Gerard's younger brother, beyond the fact that he had been appointed professor of philosophy at Falcon College in the 1640s and, in 1650, also of ethics; he seems to have left academia before 1666.

As usual in Louvain, the *Philosophia* disputation of 1651 comprises three main sections—*Logica*, *Physica*, and *Metaphysica*—in keeping with the three-fold structure of the two-year philosophy curriculum. Distinctively Cartesian doctrines are to be found in all three sections and their relative corollaries (*impertinentia*), as in the claim that the prejudices of scholastic philosophy obscure the natural intelligence, or that Tycho's planetary system ends up attributing more motion to the Earth than is found in that of Copernicus; in the purely mechanical explanation of the heartbeat, as well as in the refusal of an appeal to final causes in accounting for the actions of non-rational beings.[50]

It is in the section on natural philosophy, however, that Descartes's influence features most strongly. The tone is set by the second thesis, which lists "sympathy and antipathy [...] magnetic virtues, occult qualities and the entire salmagundi of prepotent faculties"; only to then dismiss them with disdain.

---

48   Monchamp, *Cartésianisme*, 168. On the *a priori* proofs of God's existence in the Scholastics of the time, see Agostini, *Démonstration*; and Mantovani, "A Priori Proofs."

49   The same disputation (corollaries included) was defended, in the very same hours, by another of Willem Van Gutschoven's students, Petrus Van Veen. Compare: Willem Van Gutschoven (praes.) and Van Werm (resp.), *Philosophia* (Louvain, 31 July 1651); Willem Van Gutschoven (praes.) and Van Veen (resp.), *Philosophia* (Louvain, 31 July 1651). Also Willem Van Gutschoven's *Philosophia* disputation of 1655 (see below) was defended by different students.

50   See, successively, Willem Van Gutschoven (praes.) and Van Werm (resp.), *Philosophia* (Louvain, 31 July 1651), *Logica*, *Impertinentia* 1, 2, 4 & *Metaphysica*, *Impertinentia* 2; *Physica* VIII; *Metaphysica* IX. On Copernicanism in seventeenth-century Belgium, see Monchamp, *Galilée et la Belgique*. On *impertinentia* in general, and on Cartesian-minded corollaries in particular, see Chapter 7, in this volume.

"As for us," Willem Van Gutschoven and his student boldly assert, "matter and motion are enough [...] to deduce from them, as from principles self-evident to everyone, [all] natural effects according to the laws of statics."[51] The same is said to hold true for all the operations usually ascribed to the vegetative faculty.[52] Next on the agenda, predictably enough, was the sensitive soul, to which are devoted the longest theses of the whole disputation, occupying a full page of their own.[53]

Significantly, this portion of the disputation opens by endorsing the two points attacked by Froidmont in 1637: Descartes's claim that the mind is ultimately responsive only to the changes taking place in the brain, and his explanation of how the nerves perform their task of delivering to this organ the modifications occurring in the other limbs:

> We only sense in the brain [...]. The function of animal spirits is to stretch out and separate the nerves fibres, so to serve the senses [...]. Hearing is determined by nothing but the vibration of air; smelling by corporeal spirits.[54]

The explanation of hearing and smell sketched in the ninth of the physical "theses" is followed by four long entries entirely devoted to vision. In particular, Willem Van Gutschoven and Hilarius Van Werm defend Descartes's theory (via Kepler) of the retinal image and his understanding of *species* in nonpictorial terms, as well as Descartes's cases against Euclid's principle of magnitude perception and against the universal validity of the cathetus-rule for the location of "images," the exception being hyperbolic mirrors. As it happens,

---

51   Willem Van Gutschoven (praes.) and Van Werm (resp.), *Philosophia* (Louvain, 31 July 1651), *Physica* II: "Praefatis principiis haec verba subsidio sunt: sympathia, antipathia, antipersitasis, virtutes magneticae, influentiae caelestes, qualitates occultae & tota farrago praepotentum facultatum. Nobis materia et motus sufficiunt, ut ex ipsis tanquam a principijs unicuique obviis, effectus naturales secundum leges staticae deducamus." On the Cartesian accounts of magnetism, and their teaching, see Chapter 14, in this volume.

52   Willem Van Gutschoven (praes.) and Van Werm (resp.), *Philosophia* (Louvain, 31 July 1651), *Physica* VIII: "ingeniosa viventium machina [...] non opus habet ulla facultate motrice, concoct. [...] pulsifica, sanguif. ossif. &c." See also *Physica* I: "Determinatae plantarum figurae formas substantiales non adstruunt."

53   Willem Van Gutschoven (praes.) and Van Werm (resp.), *Philosophia* (Louvain, 31 July 1651), *Physica* IX–XIII.

54   Willem Van Gutschoven (praes.) and Van Werm (resp.), *Philosophia* (Louvain, 31 July 1651), *Physica* IX: "In solo cerebro sentimus [...]. Spiritum animalium hic usus est: capillamenta nervorum extendere et separare, ut sensibus superviant [...]. Auditus sola vibratione aëris: odoratus per corporales spiritus determinatur."

Van Gutschoven and Van Werm followed Descartes's *Dioptrics* on virtually all accounts: from the physiology of humours and nerves to the formation of colours; from his general account of the perceptual process to optical illusions. Tellingly, the *Physica* section of the disputation concludes in Descartes's name:

> x. The crystalline humour does not make so that the pyramid [of light rays reflected] from singular points of an object and impinging on the entire pupil is refracted in so many points of the retina, but collects them back into one [point] ...
> xi. Visible *species* are light (*lumen*) which, based on its various reflections and refractions, brings about various colours. *Species* are not the images of objects, and they do not represent their distance, position and size [...] Due to the excessive motion of the nerves [it causes], whiteness damages vision ...
> xii. How many absurdities follow the axioms of some opticians, according to whom the apparent size [of bodies] depends on the proportions of the visual angles [under which they are seen]!
> xiii. Stevin and Descartes are fully right in disapproving of the ancient opticians' statement according to which the locus of the apparent image is to be located, for *all* mirrors, at the intersection of the ray reflected to the eye and the perpendicular drawn from the object to the eye.[55]

The correspondence is so close as to give the impression at times that the *Philosophia* of 1651 is simply translating the *Dioptrics*, or slightly rephrasing its authorized Latin version, the *Specimina philosophiae* of 1644.[56] A disputation is not an obvious kind of writing in which to articulate an argument: this was rather left to its *viva voce* defence. But the order of the propositions provides at

---

55   Willem Van Gutschoven (praes.) and Van Werm (resp.), *Philosophia* (Louvain, 31 July 1651) *Physica*: "x. Humor crystallinus non efficit refracte collectionem pyramidum ex singulis objecti punctis per totam pupillam radiantium in tot puncta tunicae retinae, sed tantum unius in unum [...] xi. Species visibiles sunt lumen, quod pro varia reflexione et refractione varios colores exhibit. Non sunt imagines obiectorum, nec eorum distantiam, situm, magnitudinem aut figuram repraesentant [...] Nimio nervolorum motu albedo visum disgregat [...] xii. Ex hoc quorundam opticorum axiomate, magnitudines apparentes sequuntur proportionem angulorum visualium, quam multa absurda concluduntur! [...] xiii. Cum veteres optici locum imaginis in quocumque speculo statuunt in concursu radij ad oculum reflexi cum perpendiculari ab obiecto ad oculum ducta, summo iure a Stevino et Cartesio erroris arguuntur" (my emphasis). The dissertation is indeed patently attacking the so-called "cathetus rule"; on this rule, see Smith, *From Sight to Light*, 267–268.
56   Compare, for example, Willem Van Gutschoven (praes.) and Van Werm (resp.), *Philosophia*

least a hint as to the line of reasoning at work. It is indeed only at the penultimate proposition, after having laid out Descartes's theory of perception, that Van Gutschoven and his student arrived at the conclusion feared by Froidmont: the rejection of the "real qualities" of the Schools. The proposition is artfully presented as in keeping with Aristotle's most considered statements on the subject, but the manoeuvre could hardly impress Froidmont or Plemp—let alone fool them:

> Aristotle was cautious in defining qualities as "the form based on which something is said to be such and such (*quale*)." Light, colour, odour, savour, sound, heat, cold, humidity, dryness [...] are nothing in bodies but certain dispositions consisting in magnitude, shape and motion.[57]

As was the case for Descartes, the theory of perception appears indeed to have interested Van Gutschoven not only for its own sake, but also—if not mostly—in view of its metaphysical implications for the doctrine of the real qualities. It is no coincidence that Van Gutschoven's list opens with "light and colour," and that he considered them in great detail: in the wake of Descartes, he regarded them as paradigmatic cases, and as one of the main testbeds of the new philosophy.

To be sure, Descartes's *Dioptrics* was about many things, almost too many things: it formulated the law of refraction for the first time, for example, and it concluded with detailed instructions about how to grind hyperbolic lenses that would outperform all known telescopes. Willem Van Gutschoven, for his part, left out these more technical results and focused on the theory of perception proper, also with the intention of exploding the Aristotelians' model of matter and forms.[58] Descartes's *Dioptrics*, an enjoyable piece of writing inten-

---

(Louvain, 31 July 1651), *Physica* XI with *Descartes, Dioptrics* VI, AT VI 145. On the *Specimina philosophiae*, see Vermeulen, "Introduction."

57   Willem Van Gutschoven (praes.) and Van Werm (resp.), *Philosophia* (Louvain, 31 July 1651), *Metaphysica* XIV: "Qualitatem caute definit Aristoteles; est forma a qua aliquid dicitur quale. Lumen, color, odor, sapor, sonus, calor, frigus, humiditas, siccitas [...] nil aliud sunt in rebus quam dispositiones quaedam in magnitudine, figura et motu consistentes." Willem Van Gutschoven pursued this strategy in the *Philosophia* disputation of 1655—with no better success (see below).

58   Willem Van Gutschoven was not insensitive to optics proper, however, as attested by his references to telescopes and the microscope (*Physica* XIII)—apparently, his is the earliest extant reference to the latter instrument in Louvain: see Vanpaemel, "De eerste microscoop." The scant pieces of evidence at our disposal seem to suggest that his brother Gerard was even more engrossed with these more technical aspects, as his publications in the field of practical mathematics would confirm; see Gerard Van Gutschoven, *Regulae muni-*

ded for laypersons as a specimen of a new "method," was thus adapted to the bullet-point form of a disputation in natural philosophy, and the concerns of a professor in the field.[59] What Descartes—out of prudence—had put forward in the *Dioptrics* almost only as a suggestion, was singled out for attention by Van Gutschoven as the punchline of his disputation. His contemporaries complained that no other philosopher had ever relied so heavily on the word "perhaps" or used so many circuitous expressions as Descartes. The criticism was unfair, but captured an important feature of Descartes's argumentative style: his adherence to the *ordre des raisons* sometimes made it difficult to locate his arguments, and to fully appreciate their bearing and implications. Descartes's distaste for disputes and his sly manner did the rest.[60] The stylistic brevity required by the disputation genre forced Van Gutschoven to break from this strategy, and resulted in an apparent radicalization of Descartes's claims or, at least, in making them more patent.[61] *Larvatus prodeo* could work as a motto for Monsieur des Cartes, but "advancing under a mask" would not have gotten Hilarius Van Werm his degree. The disputation adjusted accordingly.

If the theses defended in 1651 under Willem Van Gutschoven are quite extraordinary for their overt endorsement of Descartes's philosophy, Willem Van Gutschoven's commitment to Cartesianism and his keen interest in the theory

---

tionum analogicae, earumque ex methodo Fritagii et Dogenii usus compendiosus (Brussels, 1673) and *Usus quadrantis geometrici* (Brussel, 1674). On the role of optical instruments for the theories of perception of the time, see Hamou, *Mutation du visible*.

59   Descartes to Vatier, 22 February 1638, AT I 560: "un livre, où j'ai voulu que les femmes mêmes pussent entendre quelque chose." On the teaching of Descartes's philosophy to women, see Chapters 12 and 18, in this volume.

60   As an example of the latter aspect of Descartes's works, see the all-important letter to Mersenne, 28 January 1641, AT III 297–298: "I may tell you, between ourselves, that these six Meditations contain all the foundations of my physics. But please do not tell people, for that might make it harder for supporters of Aristotle to approve them. I hope that readers will gradually get used to my principles, and recognize their truth, before they notice that they destroy the principles of Aristotle." The expression *ordre des raisons* comes from Gueroult's *Descartes selon l'ordre des raisons* who, in turn, derived it from *Meditationes, Responsiones* II, AT VII 155–159.

61   On the progressive emancipation of the "contents" of Descartes's philosophy from the "form" in which they had been originally presented (contrary to Descartes's prescriptions to the contrary), see Chapter 1, in this volume. As in some of the cases discussed by Verbeek, teaching practices—or, in the case of the *Philosophia* of 1651, examination requirements—significantly contributed to breaking apart the contents and form. On the adaptation of Descartes's philosophy to the academic context, more generally, see Chapter 2, in this volume, and *passim*.

of perception are further confirmed by a course he gave one year later, which—
to my knowledge—is also the first extant course of the Louvain faculty of art
to discuss Descartes's philosophy.[62]

After providing a primer on Aristotelian physics, with its usual machinery of
prime matter and substantial forms, in 1652 Willem Van Gutschoven presented
his students with a few "more recent" competing theories.[63] The chapter briefly
mentions *Epicurei* and *chimici*, but its greatest share is devoted to Descartes's
philosophy. Even in the absence of his name, the references are unmistakable.
Indeed, the chapter *De corpore naturale secundum recentiores* opens by pointing out that "according to the new philosophers, prime matter is a substance,
or bulk, extended in length, width and depth."[64]

Descartes's name is explicitly mentioned when the course moves to consider
the anti-Aristotelian theories of the elements. In line with the keen interest
in theories of vision amongst the Louvain Cartesians, Descartes's theory of
the elements is presented by Willem Van Gutschoven as primarily intended to
account for how bodies relate to light:

> All bodies can be reduced to three classes: namely, to bodies that shine by
> themselves, those transmitting light, and to opaque bodies. Thus, follow-

---

62   MS., Brussels, MS. II 737: anonymous (professor) and Meesters (student), *Physica* (1652);
      the course was given at the Falcon college. The name of the professor who gave this course
      has not come down to us, but the strong parallels with Willem Van Gutschoven (praes.)
      and Van Werm (resp.), *Philosophia* (Louvain, 31 July 1651), strongly suggest that he is none
      other than Willem Van Gutschoven, who in the 1650s was indeed teaching philosophy
      at the Falcon college. (Were this not the case, the parallelism would prove all the more
      significant, and attest to a yet wider diffusion of Descartes's philosophy among Louvain
      professors). The close parallels with the disputation of 1651 also provide a very compel-
      ling argument for taking on trust the date of 1652 that features at the beginning of the
      manuscript: in fact, it is hard to see what would be the point in, or who could benefit from,
      stating it falsely. My thanks to one of the referees for prompting me to pursue this question.
      The 1652 course is not discussed in Vanpaemel's *Echo's van een wetenschappelijke revolutie*
      that failed also, in my opinion, to appreciate the importance of the *Philosophia* disputa-
      tion of 1651 and of the *Iudicia* of 1652. On the distinction between *professor primarius* and
      secundarius in Louvain, see Chapter 17, in this volume.

63   See, respectively, MS., Brussels, MS. II 737: anonymous (professor) and Meesters (student),
      Physica (1652), fols. 305ᵛ–311ʳ (*Philosophia pars secunda, seu physica*, ch. 1, *De corpore na-
      turali secundum Peripatheticos*); fols. 311ʳ–313ᵛ (*De corpore naturali secundum recentiores*).

64   MS., Brussels, MS. II 737: *Physica* (1652), fol. 311ʳ. Willem Van Gutschoven's disputation of
      1651 already equated matter with quantity; Willem Van Gutschoven (praes.) and Van Werm
      (resp.), *Philosophia* (Louvain, 31 July 1651), *Metaphysica* XII. Distinctly Cartesian theses are
      to be found also in other sections of the course, as in the discussion of radical doubt as
      necessary for grounding philosophy.

ing Descartes, the more recent [philosophers] (*recensiores cum Carthesio*) posit three elements, out of which those bodies emerge which they habitually call the matter of the first, the matter of the second, and the matter of the third element.[65]

The teacher of the course seriously entertained Descartes's explanation, thereby strongly suggesting to students his own stance on the matter; the open support of Descartes's philosophy was apparently found inappropriate for a course, and was thus left to the disputations. The chapter concludes by praising the explanatory power and simplicity of Descartes's account, in this case too leaving little doubt as to Van Gutschoven's personal opinions.[66] Willem Van Gutschoven, moreover, took care that the implications of these theories were not missed by his students. Just as in the dissertation of 1651, he explained Descartes's neurophysiology and its consequences for the theory of the mind.[67] The course format moreover allowed for discussion of Descartes's theory of perception at some greater length, and Van Gutschoven did not squander this chance. In particular, he discussed the famous comparison of light rays to sticks in the hands of a blind man, by which Descartes intended to show that sense-perceptions can be altogether different from their causes, whether remote (external bodies) or proximate (the resulting impressions on the perceiver's sense-organs).[68] Students were also given a taste of Descartes's general account of this mismatch in terms of the "institution of nature" that governs the mind-body union.[69] And, again as in the 1651 disputation, the ultimate conclusion of this complex reasoning was spelled out in the clearest possible terms: "The new philosophers reject those 'real qualities' (1°) as useless, and (2°) as incomprehensible."[70]

---

65   MS., Brussels, MS. II 737: *Physica* (1652), fol. 312ᵛ: "Quia omnia corpora reduci possunt ad tres classes, scilicet ad corpora per se lucentia, lumen transfundentia, et ad opaca corpora, ideo recensiores cum Carthesio tria elementa adstruunt ex quibus illa corpora exsurgunt, et vocari solent materia primi, materia secundi et materia tertii elementi."

66   MS., Brussels, MS. II 737: *Physica* (1652), fol. 313ᵛ: "ex his facile datur ratio cur ignis sit lucidus [...] ex his facile datur ratio cur aqua sit fluida, pellucida."

67   See for example MS., Brussels, MS. II 737: *Physica* (1652), fol. 334ʳ: "recensiores dicunt quod omnis sensatio fiat in cerebro."

68   MS., Brussels, MS. II 737: *Physica* (1652), fol. 333ᵛ: "caecus experitur alias sensationes et affectiones cum baculo quo se dirigit impingit in lapidem, alias dum impingit in aqua [...] ergo idem de anima." Descartes introduced the comparison in *Dioptrique* I, AT VI 83–84.

69   MS., Brussels, MS. II 737: *Physica* (1652), fol. 333ᵛ: "quia deus uniens animam corpori decrevit ut ad similem impulsum oriretur in anima certa affectio." On this topic, see Mantovani, "The Institution of Nature."

70   MS., Brussels, MS. II 737: *Physica* (1652), fol. 333ᵛ: "Recensiores illas qualitates reales rejiciunt 1°: ut inutiles, 2° ut inconcepibiles."

The insistence on this point in both the 1652 course and the disputation of the previous year is especially important, as Willem Van Gutschoven's commitment to Cartesianism is far from unconditional. Indeed, with regard to other topics, he openly opposed Descartes's ideas. For instance, the denial of souls to non-human animals (almost a signature for most of Descartes's followers) was considered in fair detail, but—much to Froidmont's satisfaction—was eventually dismissed as "paradoxical and not without danger to the faith."[71]

It is not important, however, to decide whether Willem Van Gutschoven may fully qualify as a Cartesian. The point is that, in the early 1650s, even professors opposed to key contentions of Descartes's philosophy were open to endorsing his account of the perceptual process, which they saw as Descartes's most valuable contribution to natural philosophy, also in view of the dismissal of *qualitates reales*. This particular use of Descartes's philosophy appears to have been extremely successful in Louvain, to the point of its becoming a sort of accent of the local Cartesian dialect. The sumptuous metaphysics of the Aristotelians was thus to earn them a new title among Louvain Cartesians, who missed no opportunity to tease "the Peripatetics or—to use a fitter name—the Men of Qualities."[72] Louvain *Qualitatisti*, on their part, responded to these jests in a quite different spirit.

4 1652–1663: From the *Iudicia de philosophia cartesiana* to the Condemnations

We do not know how students and the other professors reacted to Van Gutschoven's lectures and disputations of the early 1650s. But we have evidence that, in the same year in which he gave his course on physics, other Louvain teachers who sympathized with the 'new philosophy' felt confident enough to make fun in public of Aristotelian philosophy and the professors who still endorsed it. The occasion arose with the end-of- year *Saturnalia* festivities, when a professor was invited to give a series of speeches, usually on a lighter note, appropriate to a celebratory event. The opening address of 16 December 1652 was delivered by Arnold Geulincx, who had just been promoted senior professor at Lily College.

---

71   MS., Brussels, MS. II 737: *Physica* (1652), fol. 330ᵛ: "paradoxa nec satis tuta in fide."
72   MS., Louvain, MS. 261: Stevenot (professor) and Van den Biesche (student), *Physica* (1675–1676), fol. 1ᵛ: "… a Peripateticis, vel (ut nomen rebus magis adaptetur) a Qualitatistis." On Stevenot, see Chapters 14 and 16, in this volume.

Geulincx can hardly be described as a strict Cartesian: his *Saturnalia* are first and foremost a polemical work, imbued with Francis Bacon's critique of human "idols" in general, and of the Schoolmen's more specifically.[73] Still, in his opening address, it is quite telling that Geulincx took his listeners to be familiar enough with Descartes's philosophy to be able to play along with him.[74] Most of Geulincx' witty speech was intended to ridicule Aristotelians. In order to win over an audience comprised mostly of teenagers, Geulincx (who at the time was not even thirty) tickled their grossest sense of humour, depicting the *materia prima* of the Aristotelians as a happy widow pimping herself to forms, and continued along those lines.[75] Geulincx named no names, but a few passages of his opening address appear to be directed at his Louvain colleagues and their textbooks: the audience was clearly expected to be able to fathom out the targets.[76]

For Plemp this was the last straw. Five days after Geulincx delivered his speech, Plemp wrote the aforementioned letter of 21 December 1652, in which he called for the help of his colleagues to speak out against the new trend at the philosophy faculty. To respond to his call were four professors of theology, plus the university's fiscal lawyer.[77] The *Iudicia de philosophia cartesiana* is an extraordinary document that reveals how Louvain university lecturers reacted to the 'new philosophy': not only to Descartes, but also to Ramus, Campanella, Bacon, Gassendi, and Digby—the last of whom, oddly, is presented here as "the

---

73   On Geulincx's *Saturnalia* and their philosophical significance, see Cassirer, *Erkenntnis-problem* 1; Van Ruler, "Cartesian Disenchantment," 387–388.

74   Geulincx, *Quaestiones quodlibeticae* (Antwerp, 1653); I quote from *Saturnalia*, 2nd ed. (Leiden, 1665), 65–66. Geulincx left Louvain university for Leiden in 1658: his reasons for doing so are still debated, but apparently involved also his intention to marry his cousin, thereby breaching the rule of celibacy; see Van Ruler, "Arnold Geulincx."

75   Geulincx, *Saturnalia*, 101–102: "*Materiam* habere ad instar fœminae; *formam* esse velut marem, eas inter se diligere, concupiscere, num aliud sapiunt? Hinc vero nascuntur longe lateque diffusae quaestiones tribolares: *An Materia Formam appetat, quam amisit?* Dicas eos quaerere, an Materia lugeat maritum vita jam defunctum: *An Forma, qua fruitur Materia, appetitum eius expleat?* ita ut novis nuptiis memoriam exuat veteris connubii: *An plures simul Formas appetat?* et omnibus se pervulget."

76   See, for example, Geulincx, *Saturnalia*, 104–105: "Et vero, si memini recte, id ipsum, quod ridemus, de igne censet Peripateticorum recentium non nemo."

77   Besides Plemp's letter (375–377), the *Iudicia de philosophia cartesiana* comprises feedback by Froidmont (378–382), by three other professors of theology—Petrus Damasus De Coninck (382–383), Christianus Lupus (383–385) and F. Joan Rivius (386)—and by Henricus Vanden Nouwelandt (386–387), fiscal lawyer and attorney-general of Louvain University, ironically, a position previously held by Van Gutschoven's father. Plemp, *Doctorum aliquot in Academia Lovaniensi virorum iudicia de philosophia cartesiana*, in Plemp, *Fundamenta medicinae*, 3rd ed. (Louvain, 1654), 375–387.

leader of the Cartesians."[78] The "Assessments of Descartes's philosophy by a few learned men of Louvain Academy" would certainly deserve a study of their own. For present purposes, however, it suffices to consider Plemp's address and Froidmont's rejoinder.

As was to be expected in light of his profession, Plemp placed great emphasis in his assessment of Descartes's philosophy on its implications for medicine. Descartes's denial of "virtues, or potencies," "substantial forms" and "qualities," as well as of all souls with the exception of the rational soul, threatened to undermine the very principles of received medicine. Plemp contended that phenomena such as the circulation of lymph in plants demanded an "attractive power," and could not be explained by dint of the Cartesians' machinery of "syphons, bellows and pipes."[79] Caustically, Plemp remarked that it was Descartes's blind commitment to his own principles (and his soft spot for Spanish wines) had led him to a premature death. As for his colleagues and their students, by contrast, Plemp wished that they could "live healthier and longer thanks to Aristotelian philosophy."[80] Faithful to his role as professor of sacred scripture, Froidmont doubled down: at stake, he warned, was also the *salus* of their souls.

Plemp's most immediate targets, however, were his colleagues at the arts faculty and, I would suggest, Willem Van Gutschoven specifically. One of the propositions denounced (twice) by Plemp is taken verbatim from the disputation of 1651.[81] In order to discredit Descartes, Plemp also drew an articu-

---

78   Froidmont, *Iudicia de philosophia cartesiana*, 378: "princeps sectatorum ejus." The other
        names are brought up at 377 and 386.

79   Plemp, *Iudicia de philosophia cartesiana*, 375 and 377: "cucurbitulis, siphonibus, antljis"
        (Plemp is apparently quoting here a Louvain Cartesian). On Descartes's understanding
        of living bodies in terms of machines, contrary to the Scholastic theory of his time, see
        Des Chene, *Life's Forms* and *Spirits and Clocks*. Des Chene's insightful analysis is beauti-
        fully confirmed by another passage from Plemp, that seems to have escaped Des Chene's
        attention. In arguing against Descartes's and Gerard Van Gutschoven's account of muscle
        motion, Plemp objected that "Natura non imitatur artem, sed ars naturam: nec possunt
        omnia naturae opera mechanice explicari [...] Non adducunt musculi partes, quemad-
        modum nos funiculo pondus aliquod attrahimus: *nam musculi ipsi agunt vitaliter, non
        horologialiter*"; *Animadversiones*, in Plemp, *Ophthalmographia*, 3rd ed. (Louvain, 1659),
        247 (my emphasis). I analyse the passage, and its relation to Gerard Van Gutschoven's
        illustrations to the 1664 edition of Descartes's *Traité de l'homme* in Mantovani, "Tale of
        Two Images."

80   Plemp, *Iudicia de philosophia cartesiana*, 377: "ego auguror vos per Philosophiam Aristotel-
        icam saniores et diutius victuros."

81   Compare "Nobis materia et motus sufficiunt, ut ex ipsis tanquam a principijs unicuique
        obviis, effectus naturales secundum leges staticae deducamus" to "Materiam & motum
        sufficere, ut ex ipsis tanquam principijs unicuique obvijs, effectus naturales secundum

lated comparison between Descartes's and Democritus' lives and doctrines. The affinities between the two, Plemp claimed, had always been so striking that, upon their first meeting, he had renamed his erstwhile friend *Renatus Democritus*.[82]

In his response, Froidmont wrote that he especially enjoyed this comparison, which he reworked so as to compare Descartes with Epicurus, in this case again to warn against the possible implications for theology. Froidmont is quite explicit that he was not so much troubled by Descartes's own statements—however problematic he found them—as by the consequences that someone "with a spirit less honest and Catholic than his" could draw.[83] A good share of Froidmont's letter is thus devoted to denouncing Averroes, Wyclif and Pomponazzi on the basis of the Bible, Augustine and the Church Councils, and thus reflects Froidmont's general concern with maintaining orthodoxy rather than with Descartes's thought specifically.[84]

In his tirade, Froidmont also found a chance, however, to speak of Descartes's own ideas and their reception at Louvain. Froidmont's attack against some unnamed persons who had publicly made fun of Aristotle's philosophy confirms that the *Iudicia de philosophia cartesiana* were prompted by Geulincx's opening address.[85] But even more significant is Froidmont's presentation of Descartes's philosophy. Most of his condemnation was again reserved for Descartes's account of the perceptual process; further evidence of the importance of this topic for Louvain professors, whether they were for or against it:

---

leges staticae deducantur"; respectively Willem Van Gutschoven (praes.) and Van Werm (resp.), *Philosophia* (Louvain, 31 July 1651), *Physica* II; Plemp, *Iudicia de philsophia cartesiana*, 375 (see also 376: "materia et motus sufficunt."). By contrast, the expression "static" and derivatives are never attested in Descartes, but for the references to Jean De Beaugran as the "Géostaticien"; cf. Descartes to Mersenne, 31 March 1638, AT II 82, 84, 98; cf. AT I 479; AT II 146, 190, 269, 270, 331, 435, 502; AT III 81, 83, 131, 207.

82   Plemp, *Iudicia de philosophia cartesiana*, 376.

83   Froidmont, *Iudicia de philosophia cartesiana*, 379.

84   See, in particular, Froidmont, *Iudicia de philosophia cartesiana*, 380–381. Just three years earlier, Froidmont had published a *Philosophia christiana de anima* (Louvain, 1649) where, in conformity with a rigorously Thomist model, he attacked the understanding of the soul-body relation advanced by the physicians—the "Bodinistae" and "Campanellistae" of his time—whom he accused (them too) of following Democritus and Epicurus; see Cellamare and Bakker, "Christian Psychology." Froidmont's criticism of Descartes in the *Iudicia* is clearly to be interpreted in light of a much ampler agenda: it suffices to notice that Descartes is never mentioned in the work of 1649.

85   Froidmont, *Iudicia de philosophia cartesiana*, 380: "Eliminant enim facultates et potentias omnes animae, et universam illam inutilium (ut ipsi putant) instrumentorum supel-

And here comes Mr. Descartes, who establishes that the threads of the optic nerves are pushed and moved by the *species*, and that this tickling and transmission of motion warns the soul, which hides deep in the brain, so as for it to know about the colours or light that are outside [...] It seems that Mr. Descartes arrived at this theory after looking once at a spider, seated in the middle of its web and waiting for a thread to tremble to pursue its prey.[86]

Froidmont's account of how Descartes came to develop this view is, of course, purely conjectural, and almost certainly derives from a passage of the *Dioptrics* where Descartes claimed that the nervous filaments need not be thicker than a spider's web (but the metaphor of the soul as a spider had a much longer history behind it).[87] This piece of criticism was a novel from Froidmont, and there are other indications that he was not just parroting his letter of 1637. Indeed, in his "assessment" of 1652, Froidmont also commented on Descartes's claim that colours and all sensory ideas are innate: a theory that Descartes only articulated in the 1640s.[88] It is clear, therefore, that Froidmont had kept abreast of Descartes's

---

lectilem diffringunt, et fragmina eorum novis & ridiculis insignita nominibus inter fabulas et risus platearum spargunt."

86   Froidmont, *Iudicia de philosophia cartesiana*, 378: "Nam ecce D. Cartesius fila nervi optici impelli et moveri a speciebus sciscit; ista vellicatione & continuatione motus, animam, quae in recessu cerebri delitescit, velut admoneri, ut colores vel lumen, quod foris est, cognoscat .... D. Cartesius autem hac cogitatione videtur in eam sententiam adductus: quod aliquando vidisset araneum in centro retis sui sedentem & expectantem, fili vellicatione admoneri ut ad praedam exurgeret."

87   *Descartes, Dioptrique* IV, AT VI 111–112: "Car, sachant que ces filets sont ainsi enfermés en des tuyaux [...] il est aisé à entendre qu'encore qu'ils fussent beaucoup plus déliés que ceux que filent les vers à soie, *et plus faibles que ceux des araignées*, ils ne laisseraient pas de se pouvoir étendre depuis la tête jusqu'aux membres les plus éloignés, sans être en aucun hasard de se rompre, ni que les diverses situations de ces membres empêchassent leurs mouvements" (my emphasis). On the comparison of the soul to a spider holed up in the brain, see Summers, *Judgment of the Senses*, 92. Schuyl will employ it in the address to his 1662 *De homine* edition, possibly in response to Froidmont; *Ad lectorem*, n.p. [xxxii].

88   Faithful to his theological background, Froidmont understood and presented this thesis though the lenses of Aquinas's theory of angelic cognition: "Quod facilius intelligi posset, si anima hominis et bruti, species phantasticas ab ortu suo, ut S. Thomas de speciebus Angelorum existimat, infusas haberent, quibus impulsu quodam & vellicatione nervorum uti admoneretur"; Froidmont, *Iudicia de philosophia cartesiana*, 378. On angelic cognition as a model for Descartes's theory of knowledge, see Scribano, *Angeli e beati*. On Descartes's evolution on innatism, see Machmer and McGuire, *Descartes's Changing Mind*, which has however (in my opinion) pushed the issue too far. For a more balanced assessment of the matter, see Schmaltz, "Review Essay." I think it can be proved that Descartes modified his views on innateness around 1638–1639, after reading Herbert of Cherbury's *De*

works or, at least, someone had kept him informed to this effect. Until the end of his life—he died the following year, in 1653—Froidmont remained convinced, and with very sound reasons, of the importance of perception theory for Descartes.

The forcefulness of the reaction contained in the *Iudicia* against Descartes led Louvain Cartesians—or, at least, Willem Van Gutschoven—to somewhat tone down their Cartesian pronouncements, arguably in the hope of not exacerbating the clash with the colleagues of the higher faculties. The evidence for this comes from yet another disputation: one more *Philosophia*, presided by Willem Van Gutschoven and defended on 6 August 1653 by the Falcon student Egbertus Van der Gheest. From having featured in the main text of the 1651 *Philosophia*, Descartes's name was moved to the corollaries in 1653 which, for their part, no longer criticized the Ptolemaic system and final causes. Yet, the Cartesian aspects remained prominent insofar as a number of the theses defended in 1653 can be traced back to the *Philosophia* of two years earlier. Significantly, this was especially true for the theses on vision, which Willem Van Gutschoven either repeated verbatim—e.g., "species visibiles sunt lumen ..."—or rephrased, if anything, so as to stress their Cartesian bearings, as in the refusal of the assimilation model of sense-perception: "the image painted on the retina is usually dissimilar to the object, and it is not by looking at this image that we are made to see."[89] The 'homunculus-model' criticized here was exactly the one endorsed by Plemp and ridiculed in the *Dioptrics*: Willem van Gutschoven could hardly miss the target of Descartes' allusion.

Willem Van Gutschoven was even ready to turn Froidmont's criticism on its head, and to fully endorse the analogy Froidmont had used in his repudiation:

---

veritate—both the second Latin edition of 1633 and Mersenne's French translation of 1639; see Mantovani, "Herbert of Cherbury, Descartes and Locke."

89   Willem Van Gutschoven (praes.) and Van der Gheest (resp.), *Philosophia* (Louvain, 6 August 1653), *Physica* XVII. For the passage quoted above on the *species visibiles*, see *Philosophia* (1653), *Physica* XVIV, which corresponds to Willem Van Gutschoven (praes.) and Van Werm (resp.), *Philosophia* (Louvain, 31 July 1651), *Physica* XI. This holds true not only for the section on *Physica*: compare for example *Philosophia* (1651), *Metaphysica* XIV with *Philosophia* (1653) *Logica* I, on Aristotle's "cautious" definition of "quality." It is also worth noticing that passage on the "leges staticae" of the 1651 disputation (*Physica* II) was rephrased, two years later, in terms of "mechanicae ... principia"; *Philosophia* (1653), *Physica* III. Willem Van Gutschoven had come closer to Descartes's original terminology. For the Cartesian account of tides presented in this disputation, see Chapter 16, in this volume.

The threads of the nerves, spread from the brain throughout the body [...] determine the soul to sense-perceptions by virtue of their various trembling, just as the motions of the threads make a spider flee, or are alluring to it.[90]

Willem Van Gutschoven's students continued to defend theses in favour of Descartes in the years to come. In 1655, in fact, he attempted to present Descartes's natural philosophy as not inconsistent with that of Aristotle, although he had little success with this endeavour: ten years later, Van Sichen was still calling his bluff.[91] In other respects, Van Gutschoven's disputations seem to have enjoyed some measure of success: in 1657, the Utrecht Cartesian physician and philosopher Lambertus Van Velthuysen (1622–1685) mentioned having at hand four of these disputations, as proof that "de natuerlijcke Philosophie van des Cartes" was publicly taught at Louvain.[92]

The fight between supporters and opponents of Descartes's philosophy in general, and his theory of vision in particular, endured throughout the 1650s and beyond. A single episode may suffice to give a sense of the acrimony involved in these disputes. Gerard Van Gutschoven's first act upon being promoted professor of anatomy, surgery and botany in 1659 was to attack Plemp's treatise on the eye. Plemp immediately hit back with a point-by-point rejoinder. Plemp's response to Gerard Van Gutschoven's *Animadversiones in Ophthalmographiam* occupies more than fifty pages: its sheer size attests to the importance assigned to the topic by both parties.[93] This is further confirmed by its tone. Plemp referred to Descartes's philosophy as a "pestiferous doctrine,"

---

90   Willem Van Gutschoven (praes.) and Van der Gheest (resp.), *Philosophia* (Louvain, 6 August 1653), *Physica* xv: "A cerebro per corpus dispersa nervorum fila, vario tremore [...] animam ad sensationes determinant: veluti araneam filorum motus fugant invitantve."

91   Cf. Willem Van Gutschoven (praes.) and Buntinx (resp.), *Philosophia* (Louvain, 18 August 1655); Willem Van Gutschoven (praes.) and Lamberti (resp.), *Philosophia* (Louvain, 11–12 August 1655). But for the dedication, the two disputations are identical, down to the corollaries. To get a sense of the disputation strategy, it suffices to quote *Physica* VIII: "Recte cum Arist. formam naturalem explicamus per figuram artificialem." The disputation is especially interesting, also because it expands on Descartes's "laws of nature" in relation to Aristotle's physics; I hope to study this disputation further in the future. Van Gutschoven's disputation was expressly criticized by Van Sichen, *Integer cursus philosophicus* (Antwerp, 1666), II 20.

92   Van Velthuysen, *Nader bewys* (Utrecht, 1657), *Voor-reden* n.p. [xii]. I am grateful to Erik-Jan Bos for this information, as well as for informing me on Van Werm's acting as rector in 1664 and 1669.

93   The full title, as given by Plemp, reads *Doctissimi viri D. Gerardi Gutischovii Doctissimi viri M.L. Mathematices, Anatomices, Chirurgices et botanices in Academia Lovaniensi pro-*

cursing the day it had entered Louvain University.[94] For his part, Gerard Van Gutschoven was so keen on the *Dioptrics* that he undertook to compose a commentary on this work. This was incomplete at the time of his death in 1668, and the drafts were soon lost.[95]

As predicted by Froidmont in 1637, vision theory had moved to the vanguard in the battleground of the new philosophy. Accordingly, classes on Aristotle's *De anima* took on greater importance throughout the 1650s, as they presented professors with a natural occasion to address Descartes's innovations in the field—whether to endorse or to oppose them. This quite peculiar feature of Louvain Cartesianism might contribute to explaining the important modifications of the philosophy *curriculum* that took place in 1658, when the arts faculty decided to significantly increase the teaching hours devoted to Aristotle's psychology (as also attested by the extant courses).[96] The motivations behind this decision were likely to have been as mixed as they were manifold: arguably, anti-Cartesians thought that prolonged training in Aristotle could aid their cause, whereas Cartesians took advantage of this change to discuss in even greater detail Descartes's own views on the matter.[97] Certainly, many

---

fessoris Animadversiones in Ophthalmographiam: Ad easque Plempii responsio. They were published in Plemp, *Ophthalmographia*, 3rd ed. (Louvain, 1659), 247–299.

94   Plemp, *Animadversiones*, in Plemp, *Ophthalmographia*, 3rd ed. (Louvain, 1659), 252.

95   According to Lipstorp's testimony, Van Gutschoven was working on this commentary already in the early 1650s; cf. Lipstorp, *Specimina philosophiae cartesianae* (Leiden, 1653), 16: "Primum in genere inventionis est dioptrica immane quantum! ab aliis, quibus antehac usi sumus, [Descartes] discedens, praesertim quod Naturam Luminis, Visionis, & Refractionis concernit, quam mira brevitate et suubtilitate ita exornavit, ut quaecunque de reflexionibus, refractionibus, caeterisque rebus ad perfectionem visus pertinentibus intelligi possunt, persecutus fit [...] Qua in re, si lucem visurae sint istae eruditae commentationes, quas Subtilissimus Dn. Gudscovius, Lovaniensis Mathematum Professor in eam adornat, haud scio, num quid amplius in eo genere desiderari possit." But Van Gutschoven had already written a paper on the matter around 1650—of which yet nothing seems to have survived: in 1659 Plemp wrote that "about nine years previously" Van Gutschoven had presented him with a piece of writing on optical matters; cf. Plemp, *Ophthalmographia*, 3rd ed. (1659), *Animadversiones*, 259: "Ab annis plus minus novem obtulisti, si memineris, mihi scriptum, quo acriter defendis specierum illam in pariete et charta repraesentationem fieri per reflexionem velut a speculo." In a letter to Oldenburg (quoted in Monchamp, *Cartésianisme*, 473), Sluse wrote that Van Gutschoven was working on a treatise on catoptrics when he died in 1668, whose drafts were yet to be found.

96   Louvain Decan and Art Faculty (Ulrich Randaxhe) to Girolamo De Vecchi, Apostolic nuncio in Brussels, 5 July 1662: "Jam per quadriennium integrum [...] iisque (quod multorumet primariorum pridem votum fuit) substieremus exactam librorum *De Anima* pertractationem." Reported in Monchamp, *Cartésianisme*, 611–612. See also Armogathe and Carraud, "First Condemnation," 98.

97   On the *curriculum* reform of 1658, see Vanpaemel, *Echo's van een wetenschappelijke revolutie*, 43–53.

other reasons converged to yield the decision, but once again, unfortunately, the loss of most of the documentation seems to forbid us from drawing any definitive conclusions. At any rate, the teaching of Descartes's philosophy was not confined to the courses on psychology, nor to the philosophy faculty alone. Gerard Van Gutschoven was not the only professor of medicine with Cartesian leanings: by the 1660s, it was instead Plemp who represented the minority view. Cartesian students, moreover, started to make a name for themselves also at the other higher faculties: Hilarius Van Werm—the student who had defended the *Philosophia* disputation of 1651—was soon to become professor of law and, by 1664, rector. The university of Louvain, which had "long been honoured for not accepting novelties," was become a harbour for the 'new philosophy.' It was about time for its opponents to take countermeasures.

Monchamp has carefully reconstructed the different stages that led to the condemnation of September 1662.[98] The documents later brought to light by Armogathe, Carraud and Collacciani have permitted us to add further detail to this picture. In particular, the documentation from the Archives of the Congregation of the Doctrine in Rome enables us to follow, almost week by week, the exchanges between Louvain, Brussels and Rome between May and September 1662. Plemp and others must have been plotting to this end for years. Already in 1652, Plemp had urged his Louvain colleagues to follow the examples of Utrecht and Leiden, and he had announced as early as in 1659 that the theologians "were about to thrust [Descartes's philosophy] back into hell."[99] Arguably, during his five years in Rome, the Louvain professor Christianus Lupus (Christiaan De Wulf)—who had already railed against Descartes in the *Iudicia* of 1652—became a figurehead for Louvain anti-Cartesians, and managed to relay their complaints to important members of the Curia, among whom was the cardinal Francesco Albizzi (1593–1684). Albizzi must have shared De Wulf's concerns and must have urged him to take action. Shortly after De Wulf's return to Louvain, Descartes's philosophy was condemned. The wheels had already been set into motion when, under the supervision of Pierre Dorlix on 29 August, Theodore Aerts defended a *Repetitio thesium omnium medicarum* favourable to Descartes.[100] The nuncio did not manage to forbid the defence

---

98   Monchamp, *Cartésianisme*, 337–370.

99   Plemp, *Iudicia de philosophia cartesiana*, 377; see also 387. See Plemp, *Ophthalmographia*, 3rd ed. (1659), 252: "brevi audiemus censuram Theologorum primatis, qui illam et S. Scripturae repugnantem, ss. Patribus dissentientem, et ab Ecclesiae sensu aberrantem, ad inferos usque dejiciet."

100   The disputation has been recently identified by Collacciani, "Censure." In 1662 the four professors of medicine in Louvain were: Plemp (theoretical medicine), Pierre Dorlix

but, a few days later, on 7 September 1662, the faculty of theology condemned these disputations and, taking advantage of the occasion, also censured five statements taken from Descartes's *Meditations* and *Principles of Philosophy*, as notable "specimen[s] of his errors" in metaphysics. In particular, they condemned (i) Descartes's definition of substance and (ii) his denial of real accidents, as well as Descartes's claims that (iii) extension is the essence of bodies, that (iv) the world has no boundaries and, finally, that (v) there cannot exist but one world.[101]

The theology professors expounded especially on the second point: Descartes's argument that 'real accidents' are a contradiction in terms. And this time, at last—as their associates in Rome were likely expecting them to do for quite a few years—they officially condemned Descartes on account of transubstantiation. Were he right, they argued, it would "follow that, in the Eucharist, the accidents of wine and bread do not remain without the substance," and this ran contrary to Catholic dogma.[102] Already in 1652, Plemp had left it to his colleagues of the theology faculty to determine whether Descartes's account ran against orthodoxy on this point.[103] Their few protestations, lost in the appendix of a medical treatise, proved insufficient, however: Descartes's philosophy kept on steadily spreading throughout the 1650s. In those years, the anti-Cartesians themselves were instead objecting to Descartes's account of the structure of the nerves and to his theory of *bêtes machines*, and Plemp is no exception. In the mid-1650s his main writings revolved around Descartes's explanation of the heartbeat.[104] All in all, transubstantiation does not appear to have been an issue at the forefront of the Louvain polemics of the late 1640s and of the 1650s,

---

(practical medicine), Guillame Philippi (institutions), and Gerard Van Gutschoven (anatomy).

101　On this condemnation and on the seventeenth-century condemnations of Descartes's philosophy more generally, see Ariew, *Descartes among the Scholastics*, 241–265; Ariew, "Censorship." By contrast, I have reservations about Collacciani's claim ("Reception") that the 1662 condemnation was motivated by Philippi's alleged commitment to occasionalism. The issue is never raised in the condemnation dossier, and the episodic use of the expression *causa occasionalis* is not by itself decisive, or especially controversial: the concept was well-established in Counter-Reformation metaphysics, suffice it to mention Suárez; cf. Perler, "Occasional Causation."

102　*Deliberation of the faculty of theology of Louvain*, "Ex quo consequens est non remanere accidentia panis et vinis sine subjecto in eucharistia"; reported in Armogathe and Carraud, "First Condemnation," 105. Descartes had already objected to this conclusion, and so did quite a few of his followers; see Armogathe, *Theologia*; Ariew, *Descartes among the Scholastics*, 217–240. For Louvain, see Willem Van Gutschoven (praes.) and Van der Gheest (resp.), *Philosophia* (Louvain, 6 August 1653), *Metaphysica* x.

103　Plemp, *Iudicia de philosophia cartesiana*, 376.

104　See Petrescu, "Heartbeat on the Heartbeat." On concurrent debates about Descartes's

especially as compared to the two decades that followed.[105] For the Roman Curia, by contrast, transubstantiation was hugely important, and they appear to have expressly required the theology professors to take firmer measures to uphold Catholic orthodoxy. This is indirectly confirmed by the 1663 *Censura* of Descartes's *Meditations* commissioned by the Congregation of the Holy Office to Giovanni Agostino Della Natività, in which the problem of the transubstantiation features very prominently. (It was this censure, together with a consonant report by Stefano Spinula, that led to the placement of Descartes's works on the *Index librorum prohibitorum* on 20 November 1663.)[106] The dispute over Descartes's philosophy was no longer solely the business of the professors of Pig college and their colleagues downtown: it had become a Roman affair, and Rome demanded official censures and countermeasures on a much larger scale. And so, in 1662, Louvain became the first Roman Catholic university to officially condemn Descartes's philosophy.

Plemp gloried in this outcome. In 1664, he went so far as to re-issue his *Fundamenta medicinae* for the third time, also with the intention—I would suggest—of offering in print both censures and a detailed report of the condemnation procedure, as some sort of grim preface to his own work.[107] He also had other, more local concerns, however. Little or no attention has been paid to the way in which Plemp presented these condemnations—their 'paratext,' as it were. As it turned out, what the nuncio in Brussels and the cardinals in Rome had found most troubling in Descartes's philosophy was not quite what Plemp and his colleagues had been debating since the 1630s. Plemp had no intention of downplaying their accusations; nor is this my intention here. Yet, it is important to realize that, for over two decades, Louvain anti-Cartesians had been fighting quite a different war to the one in which Rome perceived them to be engaged. To the best of my knowledge, for example, the indefinite extension of the world, or its unicity (the fourth and fifth propositions censured in 1662)

---

account of blood circulation and the heartbeat more generally, see Mantovani, "Descartes's Man under Construction."

105   It is crucial to remark, moreover, that three of the theses condemned in 1662—(i), (iv) and (v)—have no bearing on transubstantiation, not even tangentially.

106   *Censura*; reported in Armogathe and Carraud, "First Condemnation," 80–81. These two "censures," however, address many other topics, some of which have no direct bearing on theological matters. As this chapter intends to prove for Louvain, the early modern condemnations of Descartes's philosophy are extremely complex affairs, which resulted from a cluster of philosophical, political, sociological and educational concerns and from a number of actors, each with their own agendas. The Roman condemnation of 1663 is no exception.

107   Plemp, *Fundamenta medicinae*, 4th ed. (Louvain, 1664), vi–xx.

had never been brought up by Plemp or Froidmont, nor by the Van Gutschoven brothers. The Louvain-Rome condemnations of 1662 and 1663 thus provide only a partial—if not plainly misleading—insight into what had been going on in Louvain's four colleges since the late 1640s.

Accordingly, in presenting the condemnation of 1662, Plemp made sure to include and focus on the topics dearest to him. For instance, Descartes's claim that non-human animals lack sentience—already condemned by Froidmont in 1637, but barely mentioned by the official censure—was carefully analysed and fulsomely rejected by Plemp.[108] *A fortiori*, he could not pass over in silence the disputes that had bothered so many members of his university for so many years. Taking his cue from Schuyl's posthumous publication of Descartes's *De homine* in 1662, Plemp reiterated his objections to Descartes's physiology, added a few new ones, and drew from each their culminative import. The list of imputations he produced, almost worthy of Borges, was Plemp's last public statement in a war of nerves he had been waging for almost thirty years:

> Descartes's philosophy cannot be held by a Catholic.
> Many reasons and pieces of evidence prove that beasts are endowed with cognition (the counterarguments to this effect end up in nothing).
> Descartes ignored the composition of saliva.
> He falsely introduced valves within the nerves.
> The pineal gland is not the seat of the soul, nor the organon of the common sense: the rectum is nobler than this gland.
> By standing firm in his principles, Descartes killed himself in Stockholm. His arrogance was appalling.[109]

108　Plemp, *Fundamenta medicinae*, 4th ed. (Louvain, 1664), x–xv.

109　Plemp, *Fundamenta medicinae*, 4th ed. (Louvain, 1664), iii–iv: "Philosophia Cartesii non potest teneri a Viro Catholico. Bestias praeditas esse cognitione multis rationibus & experimentis probatur, argumentaque in contrarium allata solvuntur. Salivae materiam ignoravit Cartesius. Falso valvulas in nervis comminiscitur. Glandula pinealis non est animae domicilium, nec sensorium commune: intestinum rectum est ipsa nobilius. Principijs suis insistens Cartesius se ipse interemit Holmiae. Miranda ejus arrogantia." Schuyl's edition is expressly quoted at *Fundamenta* x, xv–xvii. Almost thirty years later, Plemp still had Froidmont's 1637 letter in mind, and may indeed still have had it on his desk: compare, for example, "tot & tam nobiles in corpore nostro actiones non posse calori elementari aut spiritibus attribui, nisi ab anima directi" and "videtur dicere quod calor, qualis in fœno calefacto, possit exercere omnes operationes animalis in corpore humano [...] Ergo calor fœni, sine alia anima sensitiva, potest videre, audire, etc. Tam nobiles operationes non videntur posse prodire ex tam ignobili et bruta causa"; respectively, Plemp, *Fundamenta medicinae*, 4th ed. (Louvain, 1664), xv and Froidmont to Plemp, for Descartes, 13 September 1637, AT I 403–404.

5   1662–1671: The Aftermath

The impact on Louvain Cartesians of the 1662 condemnation should not be
overstated. Much to his disappointment, Girolamo De Vecchi's injunctions
were substantially ignored by the faculty of arts. Whilst officially, on 28 June,
the professors of philosophy denounced the public, widespread diffusion of
works contrary to Catholic dogmata, for their faculty, they limited themselves
to admitting that Descartes's natural philosophy was widely received therein,
and not always in line with the doctrines previously taught.[110] Monchamp had
a point when he described this decree as a "quasi-consécration de la réforme
cartésienne."[111]

The case of Guillaume Philippi (1600–1665), the erstwhile teacher and promoter of Geulincx, is instructive.[112] While one would have expected that his
summa philosophiae, published between 1661 and 1664, should have been ma-
jorly affected by the condemnation of 1662, as it happens, Philippi sided with
Descartes in the third and last volume—the *Medulla physicae* (1664)—almost
as clearly as he had in the first, the *Medulla logicae*, published three years
earlier. To be sure, in writing two years after the condemnation, Philippi took
care to temper his tone. But, for all the lip service paid to the censors' remarks,
even the most casual reader could not fail to notice that he had not quite
abandoned his Cartesian convictions, especially as regards the theory of perception.[113]

And yet, from the 1660s onwards, the issues of orthodoxy in general, and
of transubstantiation in particular, seem to have taken on greater importance
among Louvain professors—all the more so when the *Journal des Sçavans* of

---

110   *Decree of the Faculty of Arts*, 28 August 1662: "Porro cum Renati Carthesii scripta multorum
         nunc manibus terantur; noverit quoque praeedicta juventus, quod etsi author ille in multa
         quae naturae experimenta concernuntur non infœliciter videatur incidisse, nihilominus
         quaedam in illis reperiantur sanae et avitae dictae facultatis Artium doctrinae non satis
         consona"; reported in Armogathe and Carraud, "First Condemnation," 103.
111   Monchamp, *Cartésianisme*, 348–349. Also, some of the "condemnations" of Descartes's
         philosophy in the United Provinces prove in fact to be legitimations in disguise; cf. Ver-
         beek, *Descartes and the Dutch*.
112   Monchamp, *Cartésianisme*, 368–369. Unfortunately, Philippi's only extant courses date to
         before the introduction of Descartes's philosophy in Louvain; MS., Louvain, MS. 202: Phil-
         ippi (professor) and Van der Goos (student), *Metaphysica* (1639); MS., Louvain-la-Neuve,
         MS. C97: Philippi and Stockmans (professors) and De Wauldret (student), *Logica* (1640);
         MS., Louvain-la-Neuve, MS. C96: Philippi and Stockmans (professors) and De Wauldret
         (student), *Physica & Metaphysica* (1640–1641). On the distinction between *professor pri-
         marius* and *secundarius* in Louvain, see Chapter 17, in this volume.
113   Monchamp, *Cartésianisme*, 375–378. But cf. 418–425.

January 1666 published an article which discussed the condemnations of 1662 and 1663, thereby making them known to a much larger readership.[114]

It is not a matter of all or nothing: already in the 1650s there had been voices opposing Descartes's vision theory on account of its incompatibility with the mysteries of Catholic religion. And, by contrast, the *Dioptrics* continued to be discussed for its own sake beyond the 1670s, with some Louvain professors actually expounding on its more technical aspects, with a special attention paid to its treatment on telescopes.[115] But the balance between these two strands seems to shift around the mid-1660s: arguably, as a result of the events set into motion by the condemnation of 1662.[116] A Louvain course on physics of the mid-1680s provides a very visible example of these mounting preoccupations. Descartes famously compared the eye to a *camera obscura*, wherein external objects project inverted images. In the *Dioptrics*, as we have seen, he illustrated the matter by means of little rhombi and triangles (Fig. 15.1). In his illustrations for the *Traité de l'homme*, Gerard Van Gutschoven was to use apples and arrows.[117] They were light, playful images: the woodcut of the *Dioptrics* even contained a friendly joke. More often than not, post-condemnation Louvain had graver concerns. Under the appearance of an innocent game, the putto of the engraving *On visible species* was intended to display the optics at play in the Eucharist and, more subtly, the possibility for the new vision theory to accommodate this miracle: the grapes represented in so great detail, the table with an elegant tablecloth that gives it the appearance of an altar, and the solemn staging of the scene are all, I think, unmistakable allusions (Figure 15.3).[118]

---

114   *Journal des sçavans* (4 January 1666), 61. Truth be told, the *Journal* complained of Plemp's unfair treatment of Descartes: the piece started out as a review of Plemp's 1664 *Fundamenta*.

115   This progressive emancipation—especially progressive in the seventeenth-century—of the study of light from the study of sight is at the centre of Smith, *From Sight to Light*. On the teaching of Descartes's theory of perception into the 1680s in the United Provinces, see Chapter 9, in this volume.

116   The debate over Cartesian philosophy became more heated also in France at that time: because of the Louvain and Rome condemnations, to be sure, but also due to so unexcepted an event as the passage of two comets in 1664 and 1665, and the discussions they elicited; see Roux, "Two Comets"; see also Collacciani and Roux, "Times of War." This episode should dispel any doubts as to the complexity of seventeenth-century (anti)Cartesianism, and the impossibility of reducing it to any single set of problems.

117   On Gerard Van Gutschoven's images on visual perception for Clerselier's edition of the *Traité de l'homme*, see Hatfield, "Images as Interpretations"; Mantovani, "Tale of Two Images."

118   On the iconography of Louvain disputations, see De Mûelenaere, *Early Modern Thesis*. On the iconography of Louvain courses, see Vanpaemel, Smeyers and Smets, eds., *Ex cathedra*;

FIGURE 15.3 The optics of the Eucharist. Lambert Blendeff, *De speciebus visibilibus*, in
         MS., Louvain, MS 211: Muel and Werici (professors) and Cox (student), *Meta-
         physica & Physica* (1686–1687), fol. 85ʳ.
         MAGISTER DIXIT COLLECTION

The debate over the Eucharist had already been raging in Paris for a few decades, as attested also by a battery of letters on the subject by Clerselier, who
had tried to convince as many people as possible that Descartes's theory was
in keeping with orthodoxy.[119] He failed: in 1671 Louis XIV banned the pub-

---

       Berger, *Art of Philosophy*. On Louvain "Cartesian" engravings, see Vanpaemel, "Louvain
       Printers." In Chapter 16, in this volume, Palmerino analyses some of these "Cartesian"
       engravings as for cosmology.
119   On Clerselier's correspondence, with special focus on the Eucharist question, see Agostini,
       Lettres de Clerselier. On the different positions of French Cartesians on the subject of the
       Eucharist, see Schmaltz, "Cartesianism in Crisis."

lic teaching of Descartes's philosophy, on the grounds of religious scruples.[120] Meanwhile, most Louvain professors of philosophy were trying to convince their students of the tenability of Descartes's theory; and with better success, it seems. In his *Entretiens sur la philosophie*, published that very same year, Jacques Rohault (1618–1672) could object to French anti-Cartesians that

> Descartes's doctrine [of the transubstantiation], which had formerly been rejected by the School of Louvain, is now so well received, that of sixteen philosophy professors, there are fourteen who teach it.[121]

This was almost certainly an overstatement, and it was promptly contested by Louvain anti-Cartesians such as Léger Charles De Decker (1645–1723).[122] In fact, Christiaan De Wulf was still warning students against Descartes's monstrous doctrines in 1675.[123] Yet, the documents at our disposal confirm that Cartesian ideas continued to hold sway in the colleges nevertheless. An evaluation committee visited the faculty in 1673 and reported that "nobody would admit of teaching Descartes [...] but several professors do indeed teach his doctrine": "nowhere in the Southern Netherlands," they complained, "is philosophy less well-founded than in Louvain."[124] Little by little, the Cartesians' debates on the Eucharist spilled over even into the classes of logic, especially as it came to defining the category of "quantity" and its relation to substance.[125] (Arguably, Arnauld's long stay in nearby Brussels from 1679 to 1694 might have played a role in this: early-eighteenth-century Louvain courses of logic display traces of *The Port-Royal Logic* on other subjects.)[126]

---

120   As it turns out, the reasons for the 1671 condemnation cannot solely be reduced to concerns for orthodoxy either; see Roux, "Condemnations."

121   Rohault, *Entretiens sur la philosophie* (Paris, 1671), 112: "Ces avantages, Monsieur, vous paroissent-ils peu considerables, & vous étonnerez-vous aprés cela, de ce qu'un de mes amis me made que cette Doctrine qui avoit esté autrefois comme rejettée par l'Ecole de Louvain, y est maintenant si bien receüe, que de seize Professeurs en philosophie, il y en a quatorze qui l'enseignent." On Rohault's Cartesianism, see Chapter 19, in this volume.

122   [De Decker], *Cartesius seipsum destruens* (Utrecht, 1675), §19, 137–138. On De Decker's endorsement of other aspect of Descartes's philosophy—e.g., magnetism—see Chapter 14, in this volume.

123   De Wulf, *Censura*, in [De Decker], *Cartesius seipsum destruens* (Utrecht, 1675), n.p. [xi].

124   Vanpaemel, "The Louvain Printers," 244, quoting from Reusens, *Documents*, 672–673.

125   See Coesemans, "Facultative Logic." On the teaching of logic in Louvain, see also Papy, "Logicacurussen"; Geudens and Papy, "Teaching of Logic."

126   Arnauld had indeed been the first to ask Descartes in print about the Eucharist; see *Meditationes, Objectiones* IV, AT VII 217–218. For Descartes's reply, see *Responsiones* IV, AT VII 374–375; see also Descartes to Mesland, 9 February 1645, AT IV 165–170; Descartes

In the aftermath of the condemnation, the faces of Louvain Cartesianism and anti-Cartesianism were changed, and not just metaphorically. The generation of Louvain professors who had personally known Descartes died off: Gerard Van Gutschoven in 1668, one year after his brother Wilhem; Plemp in 1671. Meanwhile, in the United Provinces and in France, a new generation of thinkers—taking their lead from Descartes's works—were about to change, once and for all, the story of Cartesianism and the entire philosophical landscape: in 1674, there appeared the first volume of Malebranche's *Search after Truth*; three years later, Spinoza's *Ethics* made it into print. Defending and opposing Descartes became quite another business, both inside and outside the classroom.[127] The *Kartesiomania* of the earlier decades turned into something more daring, and the opponents of the 'new philosophy' had now to face yet mightier foes.

Descartes and Plemp had started as good friends in 1630s Amsterdam, discussing together, referring to each other in their works, performing dissections shoulder to shoulder. In the rift that followed between the two, Louvain classrooms became their battleground. Plemp eventually triumphed, but his victory was short-lived: the Louvain University faculty of arts basically ignored the condemnation of 1662, only to become Cartesian through and through. By the end of his life, Plemp was surrounded by Cartesians including amongst his colleagues in medicine. With the wisdom of hindsight, Voltaire was perhaps right in remarking that these condemnations helped Descartes's cause, spicing up novelty with the taste of the forbidden. Seventeenth-century Cartesians were in no position to appreciate this irony. On the whole, however, they ought to have been content: although quite later than he had wished, Descartes's philosophy had finally made it on to the Louvain *curricula*.

Acknowledgments

The research for this chapter has been financed by the Fonds voor Wetenschappelijk Onderzoek—Vlaanderen as part of the FWO-NWO Lead Agency Project *The Secretive Diffusion of the New Philosophy in the Low Countries: Evidence on the Teaching of Cartesian Philosophy from Student Notebooks, 1650–1750*

---

to Arnauld, 4 June 1648, AT V 194. On this issue, see also Menn, "Stumbling Block." On the influence of *The Port-Royal Logic* on Louvain teaching, see Chapter 17, in this volume.

127   On the promotion of Descartes's philosophy by Spinoza's circle, and on the role of Spinoza's philosophy in the anti-Cartesian polemics of those years, see, respectively Chapter 5 and Chapter 10, in this volume.

(3H160697). The final stages of this research have been financed within my FWO Project *From Truth to Ecology: Reassessing the History of the Senses, 1250–1750* (12ZY922N). My thanks to this institution, and to KU Leuven, for their generous support. I would also like to thank here Erik-Jan Bos, Davide Cellamare, Steven Coesemans, Vincenzo De Risi, Christoph Lüthy, Carla Rita Palmerino, Jan Papy, Christoph Sander, all participants of the Nijmegen conference and the two referees of this volume for their perceptive comments to various drafts of this chapter. I dedicate this chapter to Theo Verbeek, with my heartfelt thanks for his advice and support since the beginning of my time at the Descartes Centre in Utrecht, and for paving the way for this entire volume.

CHAPTER 16

Descartes's Theory of Tides in the Louvain Classroom, 1674–1776

Carla Rita Palmerino

1   Introduction

In 2011, Geert Vanpaemel drew attention to a set of engravings that were produced around 1670 by the Louvain printers Michael Hayé and Lambert Blendeff, and that were sold to the students of the Faculty of Arts, who inserted them in their notebooks. The engravings, most of which were derived or drew inspiration from Descartes's works, entered the Louvain classroom "at a time when the Faculty was experiencing a major reform of its curriculum," as a result of which professors gained more freedom to teach what they saw fit.[1] While in 1662 the Louvain faculty of theology condemned Descartes's philosophy following an admonition of the papal nuncio of Brussels, as we learn from Mattia Mantovani's contribution to this volume (Chapter 15), the professors of the Faculty of Arts substantially ignored the decree. In 1673, an evaluation committee visited the faculty and concluded, in an official report, that "nobody would admit of teaching Descartes [...], but several professors indeed do teach his doctrine."[2]

The Cartesian engravings that the two publishers Michael Hayé and Lambert Blendeff put on the market are a tangible proof of the Cartesian orientation of Louvain teaching at the time. But what role exactly did these illustrations play in the classroom? On the basis of a preliminary analysis of the notebooks, Vanpaemel came to the conclusion that the Louvain pictures were not necessarily meant as illustrations of points addressed in class, but served rather as "embellishments or memorizing aids."[3] Students bought the engravings and inserted them in their notebooks, but the course texts generally did not refer to them, "and certainly the didactic aspects of the engravings (e.g., the use of letters to

---

1   Vanpaemel, "The Louvain Printers," 244.
2   Vanpaemel, "The Louvain Printers," 244. Vanpaemel, in turn, borrows his quote from Reusens, *Documents*, 672–673.
3   Vanpaemel, "The Louvain Printers," 252.

indicate certain parts of the objects shown) are not embedded in the lectures of the professors."[4]

More recently, Susanna Berger has devoted some attention to visual representations in the notebooks of Paris and Louvain philosophy students, stressing their function "as critical tools in the organization and the exploration of difficult questions."[5] Berger also remarked that while the texts of philosophy notebooks were *verbatim* transcriptions of the lessons, students enjoyed some freedom in the selection and production of images. In the Parisian and Louvain notebooks, one finds in fact not only engravings and etchings, but also drawn copies of these prints as well as original sketches, with which students often emphasized their endorsement of new philosophies.[6] As a nice example of the latter practice, Berger reproduces the title page of a notebook (Figure 16.1) compiled by Georgius Jodoigne, a student of the Falcon (one of the four pedagogies of the Louvain Faculty of Arts, together with the Castle, the Lily, and the Pig), who paid homage to Descartes by drafting in the upper left medallion the inscription "CartesIVs MagnVs phILosophIae DVX" ("Descartes the great leader of philosophy") and an original chronogram of MDCLXXVIII (1678), the year in which the course was taught.[7]

Vanpaemel and Berger shed an interesting light on the production and function of visual elements in the Louvain notebooks, but do not dwell on the relation between the content of the lessons and the accompanying images. In the following pages, I will try to say more on this by focusing particularly on one engraving, the *Explicatio Epicycli Lunaris et Aestus marini* (Explanation of the lunar epicycle and of the tides), which was produced by both Hayé and Blendeff, and later reprinted by Giles-Peter Denique. The *Explicatio* is interesting for a number of reasons. First, in contrast to other Louvain engravings, it is not copied from Cartesian works, but is rather an original image, albeit inspired by the visual representation of ebb and flow in Descartes's *Principles of Philosophy* (1644). Moreover, the *Explicatio* seems to contradict Vanpaemel's hypothesis about the marginal didactic importance of the Louvain engravings. As I will try to document in this chapter, professors devoted considerable attention in their physics courses to the Cartesian theory of the tides, which they explained by making explicit reference to the *Explicatio*. The fact that some

---

4   Vanpaemel, "The Louvain Printers," 248.
5   Berger, *The Art of Philosophy*, 116.
6   Berger, *The Art of Philosophy*, 127, 134.
7   MS., Louvain, MS. 250: De Novilia and Snellaerts (professors) and Jodoigne (student), *Logica* (1678), fol. 2ʳ.

FIGURE 16.1   Homage to Descartes in the factotum frontispiece of a student's notebook. MS., Louv-
               ain, MS. 250: De Novilia and Snellaerts (professors) and Jodoigne (student), *Logica*
               (1678), fol. 2ʳ.
               MAGISTER DIXIT COLLECTION

notebooks do not contain the original engraving, but rather pen drawings, more or less faithful to the original, is an additional proof of the fact that students did not regard the illustration as a mere "embellishment" or memorizing aid, but rather as a valuable didactic tool that could guide them in the understanding of Descartes's account.

The *Explicatio* is reproduced in 25 of the 561 handwritten lecture notes digitized in the "Magister Dixit" project of Louvain University.[8] The first occurrence dates from 1674 and seems to coincide with the first treatment of the tides in the Louvain classroom. Before that date, no mention of the phenomenon is found in the notebooks of a physics course, at least not in those included in the "Magister Dixit" database. By 1720, the engraving must have been sold out, as the printer Giles-Peter Denique produced a new copy that we encounter in several notebooks drafted between 1720 and 1773. Interestingly, the Burndy Library of the Dibner Institute hosts a manuscript entitled *Tractatus in quo de vero partiu[m] universi situ et motu, seu de vero mundi systemate inquiritur*, produced around 1770 at Louvain University, which also includes Denique's engraving.[9]

The Louvain notebooks, in which the *Explicatio* repeatedly reappears between 1674 and 1773, seem to confirm Aiton's claim that Descartes's explanation of the tides "remained virtually unchanged for a century" and that most Cartesians were blind to the obvious flaws of the theory.[10] It is interesting, in this respect, to see that Louvain professors endorsed and taught Descartes's account of the tides phenomenon, which was based on the vortex theory of planetary motion, without even mentioning alternative explanations, such as those of Galileo, who in the *Dialogue Concerning the Two Chief World Systems* (1632) identified the cause of ebb and flow in the combination of the diurnal and annual motion of the Earth, or Newton, who in the *Principles* (1687) ascribed the phenomenon to the gravitational attraction exerted on the ocean waters by the Moon and the Sun.

Yet, although Louvain professors remained faithful to the Cartesian theory for a century, their approach to the explanations of the tides did change over

---

8   *Magister Dixit Collection.* For a description and an analysis of the Louvain student notebooks, see Mirguet and Hiraux, *Collection.* A very interesting analysis of the Louvain notebooks is provided in the collective volume Vanpaemel, Smeyers, Smets and Van der Meijden, eds., *Ex cathedra.*

9   MS., Washington, MSS. 001296 B: Verheyden (professor) and anonymous (student), *Tractatus in quo de vero partiu[m] universi situ et motu, seu de vero mundi systemate inquiritur* (1770 ca.).

10   Aiton, "Descartes's Theory of the Tides," 337.

the course of time. As we shall see, in the notebooks produced between 1674 and ca. 1710, the ebb and flow of the sea is usually addressed in the treatise *De elementis*, as a part of a discussion of the properties of water.[11] Around 1720, professors started discussing the tides in a cosmological context, more precisely in the treatises *De sphaera* or *De cosmographia*. Although Descartes, contrary to Galileo, did not consider ebb and flow to offer a direct proof of the Earth's motion, he accounted for this phenomenon in terms of his vortex theory of planetary motion.[12] It is therefore no surprise that the tides became a more prominent topic in the Louvain classroom around 1720, when the professors of the Faculty of Arts started to endorse heliocentrism more openly and to teach Descartes's cosmological system. As we shall see, the later notebooks contain not only a summary of Descartes's theory of the tides, but also a discussion of possible objections, with detailed answers.

In the following sections, I will first analyze the *Explicatio* and its sources (section 2) and then discuss some examples of its use in the Louvain classroom, for the period ca. 1674–1710 (section 3), and 1720–1776 (section 4).

2   The *Explicatio Epicycli Lunaris et Aestus marini* and Its Cartesian Sources

The *Explicatio Epicycli Lunaris et Aestus marini*, contains a representation, accompanied by a long caption, of the Cartesian theory of the tides (Figures 16.2a & 16.2b).

As one reads in the first column of the caption, ABCD represents an oval vortex with the Earth at the centre and the Moon moving along the circumference from A (full Moon) to D (half Moon) to C (new Moon) to B (half Moon). When the Moon is located on either extremity of the shorter axis AC (which the caption calls "diameter," following Descartes), i.e., at the perigee, it moves more

---

11   As Vanpaemel pointed out, "the content of the courses was determined by the statutes [of] 1567–1568, stipulating strict adherence to the traditional Aristotelian corpus of treatises. The statutes were renewed in 1639 without major changes." Vanpaemel, "The Louvain Printers," 243.

12   According to Descartes's vortex theory of planetary motion, the Earth is at rest with respect to the surrounding subtle matter and hence, strictly speaking, does not move. However, as Tad Schmaltz noticed, this "is in some tension with the explanation of the tides that Descartes offers in the Fourth Part of the *Principia*," according to which "the motion of the vortex responsible for the tides is more rapid than that of the earth and the moon." Schmaltz "Galileo and Descartes on Copernicanism," 79.

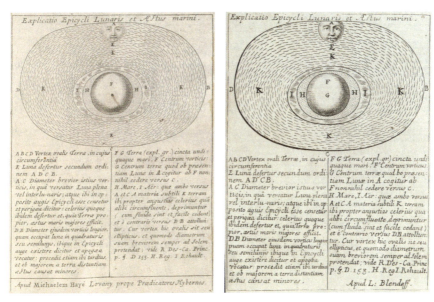

FIGURE 16.2A–B
Michael Hayé's and Lambert Blendeff's engraving *Explicatio Epicycli Lunaris et Aestus marini*.
MAGISTER DIXIT COLLECTION

quickly, causing the strongest tides. The weakest tides occur instead when the Moon is further removed from the Earth, i.e., at the apogee (points B and D), where its motion is slower. The second column of the caption explains that the Moon's presence in A causes a slight displacement of the centre of the Earth from F towards G. As for water (H) and the air (I) surrounding the globe, the figure shows that they take on an oval shape as a consequence of the pressure exerted by the motion of the subtle matter, which accelerates when passing through the narrow channel EK. Finally, the reader is referred to Descartes's *Principles of Philosophy* as well as to Henricus Regius (1598–1679) and Jacques Rohault (1618–1672) for an explanation of the oval or elliptical shape of the Earth vortex and of the fact that the Sun is always located on the prolongation of the shorter axis of the vortex.[13]

Vanpaemel drew attention to the original character of the *Explicatio* and stressed that the illustration "has no apparent model in the Cartesian textbooks, although there is some resemblance to the picture in Descartes's posthumous *Le monde*."[14] If one looks at the three sources cited in the caption, one

---

13   On Rohault's cosmological theories, see Chapter 19, in this volume.
14   Vanpaemel, "The Louvain Printers," 251, 249.

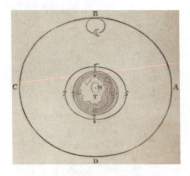

FIGURE 16.3
The representation of the tides in the *Principia*.
Descartes, *Principia philosophiae* (Amsterdam, 1644),
IV 49, 219.
LIBRARY OF CONGRESS

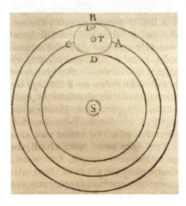

FIGURE 16.4
The representation of the tides in Regius's *Fundamenta physices*. Regius, *Fundamenta physices*
(Amsterdam, 1646), IV, 91.
BOOKSGOOGLE.COM, PUBLIC DOMAIN

sees, however, that the *Explicatio* shows a stronger resemblance to the illustration contained in Descartes's *Principia philosophiae* (1644) (Figure 16.3) than to the more schematic diagrams found in Regius's *Fundamenta physices* (1646) (Figure 16.4) and in Rohault's *Traité de physique* (1671) (Figure 16.5). The illustration contained in Chapter XI of Descartes's *Le monde* (Figure 16.6), which was posthumously published in 1664 and then republished in 1677, bears a clear resemblance to that of the *Principia*.[15]

In *Principia* IV 49 (*Concerning the Ebb and Flow of the Ocean*), Descartes uses the representation shown in Figure 16.3 so as "to visualize that small vortex of the heaven which has the Earth as its center."[16] He points out, however, that the centre of the Earth (T) does not always coincide with the centre of the vortex (M). Given that the Earth is kept in its place by the pressure of the circling

---

15   On the cosmological illustrations of *Le monde* and *Principia philosophiae*, see Dobre, "Depicting Cartesian Cosmology."

16   *Principia* IV 49, AT VIII-A 233, *Principles* 206. The translations from the *Principia* are all borrowed from Descartes, *Principles*, transl. Miller and Miller.

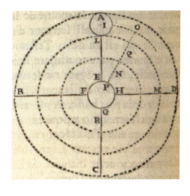

FIGURE 16.5
The representation of the tides in Rohault's *Traité de physique*. Jacques Rohault, *Traité de physique* (Paris, 1671), II 16, 134.
BOOKSGOOGLE.COM, PUBLIC DOMAIN

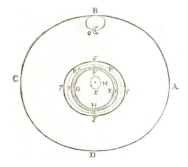

FIGURE 16.6
The representation of the tides in *Le monde*.
Descartes, *Le monde* (Paris, 1677), 477.
BOOKSGOOGLE.COM, PUBLIC DOMAIN

subtle matter, when the Moon is at B, the centre of the Earth descends slightly towards D, so that the pressure remains equal on all sides.

> In this not only is the space through which the heavenly matter flows between B and T made narrower by the Moon at B, but so is the space through which the heavenly matter flows between T and D. It follows that this heavenly matter flows more rapidly in those spaces and therefore presses more upon the surface of the air at 6 and 8, and upon the surface of the water at 2 and 4, than it would if the Moon were not on diameter BD of the vortex. And since the bodies of air and water are fluid and easily obey this pressure, these bodies must be less deep above parts F and H of the Earth than if the Moon were not on diameter BD; and, on the contrary, [these bodies] must be deeper at G and E, so that the surfaces of the water 1, 3, and of the air 5, 7, swell there.[17]

In Descartes's view, then, the pressure exerted by the circling subtle matter causes a displacement of the solid Earth globe and a change of shape of the

---

17   *Principia* IV 49, AT VIII-A 234, *Principles* 206.

fluid bodies of air and water, thereby causing the tides.[18] Having accounted for
the main cause of ebb and flow, Descartes devotes three separate articles to
the daily, monthly, and yearly tidal cycle. In article 50 he explains why the daily
rotation of the Earth around its axis results in the occurrence of two high tides
and two low tides a day, and how the rotation of the Moon around the Earth
causes a daily delay of around fifty minutes in the beginning of the tides. In
article 51 he tries to account for the fact that the tides are stronger around new
Moon and full Moon, by noticing that "the diameter on which the full or new
Moon is situated is shorter than the diameter which intersects it at right angles."
This explains why "the ebb and flow of the ocean must be greater when the
Moon is full or new than during the intervening times."[19] In article 52, finally,
Descartes gives the reason "why they are greatest at the equinoxes."[20] This has
to do with the fact that the plane of the ecliptic at the equinoxes intersects the
plane of the equator and hence the Moon, which is always close to the plane of
the ecliptic, has a more direct influence on the Earth.

If one compares the account of tides contained in the *Principles of Philosophy* with the caption of the Louvain engraving, one sees that in the latter
the explanation of the monthly tidal cycle precedes that of the main cause of
the tides and that no mention is made of the daily tidal cycle (i.e., the occurrence of two high tides and two low tides a day), and of the yearly variations
in the tidal ranges. As far as the figures are concerned, the *Explicatio* adds a
few elements to the woodcut of the *Principia*, namely: i) a representation of
the circling subtle matter; ii) an indication that the Moon appears full when
in A; and iii) a clearer distinction between the sea (H) and the air (I). What is
missing in both the Cartesian and the Louvain figures is a visual explanation
of the last point mentioned in the caption, namely the fact that the shorter
axis of the vortex always stretches out towards the Sun. This element is instead
visible in the *Fundamenta physices* (Figure 16.4), where Regius represents the
revolution of the Earth-Moon vortex around the Sun, thereby making clear
that the latter is aligned with the axis BD. It is also worth mentioning that the
Louvain caption contains two elements that may look Keplerian at first sight,
namely the designation of the shape of the vortex as 'elliptical' and the reference to the variations in the orbital speed of the Moon. As a matter of fact,
both claims are borrowed from the *Fundamenta physices* of Regius, wherein the
shape of the Earth's vortex is also referred to as 'elliptical,' and where Regius

---

18   For Descartes's tidal theory, see among others, Aiton, "Descartes's Theory of the Tides";
      Gaukroger, *Descartes' System*, 18, 169–171; Schuster, *Descartes-Agonistes*, 484–487.

19   *Principia* IV 51, AT VIII-A 235–236, *Principles* 207.

20   *Principia* IV 52, AT VIII-A 236, *Principles* 207.

explains that the Moon moves faster when it is closer to the Earth (perigee) than when it is further away (apogee), due to the varying speed of the subtle matter.[21]

We have just seen that the author of the *Explicatio* modified the original figure of the *Principles of Philosophy*, incorporating elements borrowed from other sources. As Vanpaemel rightly pointed out, Blendeff and Hayé must have received input from one or more members of the Faculty of Arts or of Medicine, who helped them select and design the images for their engravings. The most likely source of this kind of input is Gerard Van Gutschoven (1615–1668), first professor of mathematics in the Faculty of Arts and then of anatomy, surgery and botany in the Faculty of Medicine, who produced the illustrations for Descartes's *Treatise on Man* upon Clerselier's invitation. Gerard Van Gutschoven is likely to have encouraged the use of images in the Louvain classroom, but as Vanpaemel observes, he died in 1668, when Hayé and Blendeff had not yet started to produce the engravings.[22]

Gerard Van Gutschoven had a younger brother, Willem (1618–1667), professor of philosophy and of morals at Louvain University, who together with Arnold Geulincx (1624–1669) was responsible for the reform of the curriculum of the Faculty of Arts.[23] On 6 August 1653, Willem Van Gutschoven's student Egbertus Van der Gheest defended a philosophical *disputatio* which briefly addresses the phenomenon of the tides. In the *Physica* section, Van Gutschoven and his student criticize those who identify the "fictitious attraction of the Moon" (*fictitium Lunae tractum*) on the Earth as the cause of ebb and flow, and state that the tides are due to the pression exerted by the Moon on the circling aether, which in turn presses the underlying seawater causing ebb and flow.[24] As we shall see in the following section, this is exactly the explanation of the tides phenomenon that Louvain professors taught to their students with

21   Regius, *Fundamenta physices* (Amsterdam, 1646), 74–75.

22   Vanpaemel, "The Louvain Printers," 253. On Gerard Van Gutschoven, see Chapter 15, in this volume and the sources cited there.

23   Vanpaemel, "De verspreiding van het cartesianisme," 261. Not much is known about Willem Van Gutschoven. See Monchamp, *Histoire du Cartésianisme*, 295–297.

24   Willem Van Gutschoven (praes.) and Van der Gheest (resp), *Philosophia* (Louvain 1653). I thank Mattia Mantovani for bringing this *disputatio* to my attention. The criticism of the hypothesis of lunar attraction is formulated, in this *disputatio*, in *Physica* II: "Marini aestus causam quaeris? aqua vento impulsa, fluit, refluitque: intra hoc genus consiste, potius quam ad fictitium Lunae tractum confugias." The alternative Cartesian explanation is briefly presented in *Physica* VII: "Aestum maris Luna pressione causat: nam deorsum protuberans aetherem praeterfluentem in mare subjectum premit, quod pressioni cedens diffluit, Lunaque recedente refluit."

the help of the *Explicatio*. Willem died one year before his elder brother and he too, therefore, cannot have supervised the production of the engravings. The 1653 *disputatio* indicates, however, that the *Explicatio*, like other images printed by Hayé and Blendeff, was a vehicle through which Cartesian theories, already popular among Louvain professors, reached and were disseminated in the classroom.

3   The Ebb and Flow of the Sea in the Treatise *De elementis*, 1674–1710 ca.

As mentioned in the introduction, the *Explicatio Epicycli Lunaris et Aestus marini* is included in 25 notebooks digitized in the *Magister Dixit* project. If one looks at the content and structure of these notebooks, one sees that from 1674 until around 1710 the phenomenon of ebb and flow was usually addressed using the treatise *De elementis*, which treated also of the saltiness of the sea (*De salsedine maris*) and the origin of springs of water (*De fontibus et eorum origine*).

Unsurprisingly, one of the first professors to discuss the *Explicatio* in his courses was Joannes Stevenot (ca. 1640–1718), who was among the most outspoken supporters at Louvain University of Descartes's philosophy. In the courses taught in the Faculty of Arts, Stevenot compared the Ptolemaic, Tychonian and Copernican systems, siding explicitly with the last one of these. One of his central arguments in favour of heliocentrism boiled down to the claim that the rival worldviews could not provide a convincing explanation of important physical phenomena, notably gravity and the tides.[25]

Given this premise, it comes as no surprise to encounter in the notes of a physics course taught by Stevenot in 1675–1676 at the pedagogy of the Pig, a section *De coelis*, in which the professor dealt at length with the arrangement of the celestial vortices, and a section *De elementis*, in which both gravity and the tides are accounted for in terms of vortex theory.[26] Particularly interesting

---

25   Vanpaemel, "*Terra autem in aeternum stat*," 111.

26   MS., Louvain, MS. 261: Stevenot (professor) and Van den Biesche (student), *Physica & Meta-physica* (1675–1676), fols. 715^v–718^r (gravity), fols. 723^r–726^r (tides). Usually the courses at the Faculty of Arts were co-taught by a *professor primarius* and a *professor secundarius*, but I this case, Stevenot seems to have been the only teacher of the course. In her book Susanna Berger does not mention Stevenot's course, but she does reproduce two drawings from Van den Biesche's notebook and observed that, although they are unsigned, the drawings are likely to have been made by the student himself "for his own personal edification." Berger, *The Art of Philosophy*, 136.

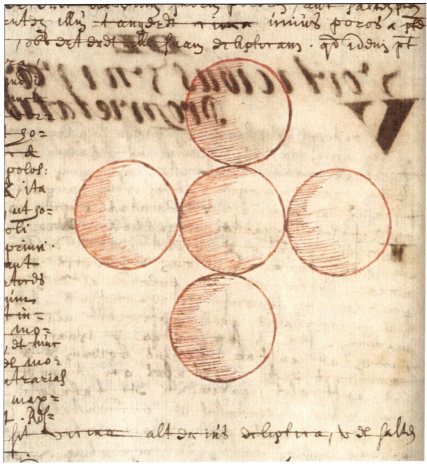

FIGURE 16.7   The arrangement of the Cartesian vortices. MS., Louvain, MS. 261: Stevenot
(professor) and Van den Biesche (student), *Physica & Metaphysica* (1675–
1676), fol. 512ᵛ.
MAGISTER DIXIT COLLECTION

from our point of view are the pen drawings with which one of Stevenot's students, Michael Van den Biesche, illustrated Descartes's theories.[27] The sketch
in Figure 16.7 represents the mutual arrangement of the celestial vortices and
also serves to explain how material particles can flow from vortex to vortex;
Figure 16.8 depicts the emission of light particles from the Sun; Figure 16.9 is

---

27   The manuscript contains notes of two sets of Physics lessons, almost identical in structure
and content, followed by one set of lessons *De coelis*.

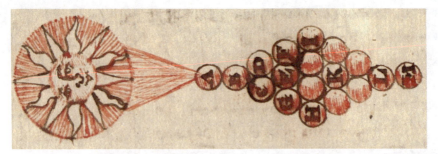

FIGURE 16.8   The emission of light particles by the Sun. MS., Louvain, MS. 261: Stevenot
                     (professor) and Van den Biesche (student), *Physica & Metaphysica* (1675–
                     1676), fol. 357ᵛ.
                     MAGISTER DIXIT COLLECTION

a faithful copy of the famous illustration of the formation of a rainbow from
Descartes's *Météores*; and Figure 16.10 reproduces the *Explicatio Epicycli Lunaris et Aestus marini*.

    If one looks at the text accompanying the drawing shown in Figure 16.10, one
sees that in his lessons Stevenot discussed an element of the Cartesian theory
of the tides which is not mentioned in the caption of the *Explicatio*, namely
the daily delay in the timing of ebb and flow. The explanation he gave is based
on the *Principia* IV 50, where Descartes argues that while the Earth makes a full
rotation around its axis, the Moon also advances slightly, so that it takes twenty-
four hours and twelve minutes before a given point of the Earth surface (F) is
placed under the Moon.[28]

    While the presence of Cartesian illustrations and theories in Van den Bies-
che's notebook confirms what we already knew about the orientation of Stevenot's teaching, it is much more surprising to encounter a pen drawing illustrating the phenomenon of the tides in the notes of a course taught in 1680
at the pedagogy of the Castle by Léger Charles De Decker (1645–1723), as *professor primarius*, and Gaspar Van Goirle as *professor secundarius*.[29] As we learn
from Vanpaemel, the division of tasks between the two professors entailed
that the *primarius* would teach the treatises *De causis, De corpore naturali* and
De anima, whereas the *secundarius* would be responsible for the *De motu, De
sphaera, De elementis* and *De meteoris*.[30]

28   MS., Louvain, MS. 261: Stevenot (professor) and Van den Biesche (student), *Physica & Meta-
      physica* (1675–1676), fol. 724ᵛ.

29   MS., Louvain-la-Neuve, MS. C165: De Decker and Van Goirle (professors) and Boonen (stu-
      dent), *Physica & Metaphysica* (1680–1681), fol. 148ᵛ. On the distinction between *professor
      primarius* and *secundarius* in Louvain, see Chapter 17, in this volume.

30   Vanpaemel, "The Louvain Printers," 250.

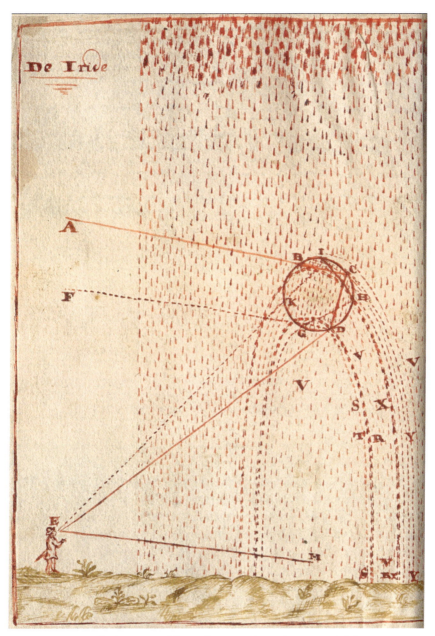

FIGURE 16.9   The formation of the rainbow according to Descartes's *Météores*. MS., Louvain, MS. 261: Stevenot (professor) and Van den Biesche (student), *Physica & Metaphysica* (1675–1676), fol. 550[av].
MAGISTER DIXIT COLLECTION

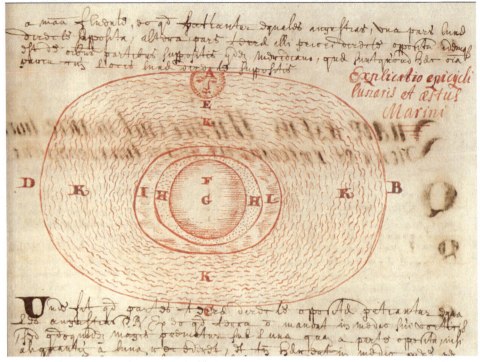

FIGURE 16.10   A pen drawing of the *Explicatio*. MS., Louvain, MS. 261: Stevenot (professor) and Van den
Biesche (student), *Physica & Metaphysica* (1675–1676), fol. 724ʳ.
MAGISTER DIXIT COLLECTION

Léger Charles De Decker is known as a staunch anti-Cartesian and as the
probable author of a polemical work significantly entitled *Cartesius seipsum
destruens sive Dissertatio brevis in qua Cartesianae contradictiones, & hallucinationes*, published in 1675 under the initials I.T. Philosopho Lovaniensi.[31] Yet,
the notebook redacted in 1680 by the student Albertus Boone contains many
Cartesian elements, such as the *Explicatio iridis iuxta Carthesium* (fol. 173ʳ), a
reproduction of Michael Hayé's illustration *Modus quo particulae striatae per
terram et magnetem fluunt* (fol. 179ʳ), and a detailed summary of Descartes's
explanation of tides, accompanied by a pen drawing (Figures 16.11).

---

31   Paquot. *Mémoires*, vol. 13, 157; Vanpaemel, "Kerk en wetenschap," 184. In a footnote of his
Galilée et la Belgique, Monchamp quotes an unpublished manuscript by Jean-Noël Paquot,
where the authorship of the *Cartesius seipsum destruens* is instead attributed to Jacques
Tacquenier, professor of philosophy at Louvain from 1644 until 1652. However, Monchamp
himself was inclined to believe that the author of the book was De Decker; Monchamp,
Galilée et la Belgique, 206, n. 1.

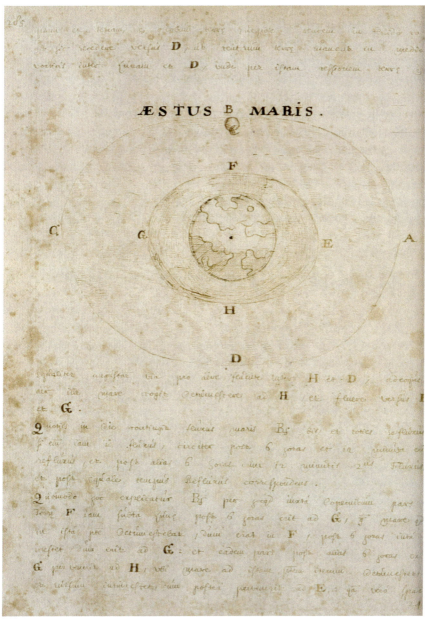

FIGURE 16.11   A pen drawing illustrating Descartes's tidal theory. MS., Louvain-la-Neuve, MS. C165: De Decker and Van Goirle (professors) and Boonen (student), *Physica & Metaphysica* (1680–1681), fol. 148ᵛ.

MAGISTER DIXIT COLLECTION

In introducing the explanation of the tidal phenomenon, the professor made explicit mention of Descartes: "Quomodo Cartesius explicat" (fol. 148^r). Significantly, the direct source of the drawing is not the Louvain *Explicatio epicycli lunaris et aestus marini* (Figure 16.2), but rather the illustration of Descartes's *Principia* (Figure 16.3), from which the notebook borrows not only pictorial elements, such as the representation of the Moon and of the Earth's continents, but also the position of the letters. As one can see in Figure 16.11, the notebook's detailed account of the tidal phenomenon contains references to specific elements of the pen drawing, which seem to indicate that the professor lectured with Descartes's *Principia* at hand. The example contradicts Vanpaemel's claim that the course texts did not make mention of the engravings and that Louvain professors did not actually refer to the Cartesian illustrations, nor discuss "the didactic aspects of the engravings (e.g., the use of letters to indicate certain parts of the objects shown)."[32]

Should we interpret Boone's lecture notes as a proof that by 1680 De Decker had converted to Cartesianism? This would probably be too hasty a conclusion, as the chapter *De aestu maris* in the notebook is part of the treatise *De elementis* that was usually taught by the *professor secundarius*, in this case Gaspar Van Goirle.[33] It is natural to assume, however, that De Decker knew and approved of the content of the lessons given by the *secundarius*. Moreover, as we learn from Christoph Sanders's contribution to this volume (Chapter 14), Boone's notebook bears witness to De Decker's endorsement of Descartes's account of magnetism, in the light of which a parallel endorsement of the Cartesian theory of tides does not appear unlikely.

The notebook under scrutiny nicely illustrates how, in the Louvain classroom, "Aristotelian and Cartesian doctrines were taught next to each other, and [...] the professor himself was free to comment on which doctrine seemed to him the best suited."[34] In introducing the Cartesian theory of the tides to his students, Van Goirle draws attention to the fact that it presupposes the validity of the Copernican system ("iuxta Copernicum") which is nevertheless criticized in another section of the course.[35] The chapter *De systemate mundi*,

---

32   Vanpaemel, "The Louvain Printers," 248.

33   MS., Louvain-la-Neuve, MS. C165: De Decker and Van Goirle (professors) and Boonen (student), *Physica & Metaphysica* (1680–1681). The notebook clearly indicates that the *Metaphysica* was taught by De Decker (fol. 1^v) and that Van Goirle taught the *Physica pars secunda* (fol. 214^r), which includes the *Tractatus in sphaeram*. It is not clear, however, who taught *De elementis*, but this was usually one of the tasks of the *secundarius*.

34   Vanpaemel, "The Louvain Printers," 248.

35   MS., Louvain-la-Neuve, MS. C165: De Decker and Van Goirle (professors) and Boonen (student), *Physica & Metaphysica* (1680–1681), fol. 148^v.

which is part of the treatise *De sphaera*, also taught by the *secundarius*, first discusses the Ptolemaic system, then the Copernican system, including a number of objections, and ends with a presentation of the Tychonic system.[36] We thus see a significant difference with the above-mentioned course by Stevenot, who in the *Tractatus de coelis* dwelled at length on Descartes's vortex theory of planetary motion and explicitly endorsed the Copernican system, which he discussed after the Ptolemaic and Tychonic systems.

Later notebooks attest that, in eighteenth-century Louvain, professors continued to teach Descartes's theory of the tides, ignoring Newton's alternative explanation. In 1708 Ursmarus Narez (1678–1744), who was to become professor of botany in 1710, of institutions of medicine in 1717, and finally of anatomy and surgery in 1719, taught an afternoon course of physics as *professor secundarius* at the pedagogy of the Pig.[37] An extant notebook, compiled by the student Jacob Steijart, confirms the testimony of the eighteenth-century historian and bibliographer Jean-Noël Paquot, according to whom Narez followed Descartes's mechanical philosophy and rejected the theories of the "ancient school."[38] In the treatise *De elementis secundum mentem antiquorum*, which opens with an explanation of the difference between the elements as conceived by ancient philosophers and by the chemists (*de elementis chÿmichorum*), one encounters, under the element water, the usual discussion of Descartes's account of the flux and reflux of the sea. What is interesting about Steijart's notebook, is that it contains both Hayé's engraving of the *Explicatio* (fol. 254^r) and an original pen drawing (see Figure 16.12), obviously derived from the representation of the terrestrial vortex found in Rohault's *Traité de Physique* (see Figure 16.5), which seems to indicate that the sources quoted in the caption of the *Explicatio* were actually discussed in the classroom.

Steijart's notebook abounds with original—often playful—drawings of mechanical, pneumatical, and astronomical objects, which provide evidence of Narez's adherence to the theories of the *recentiores*. One recognizes, among others, inclined planes, barometers, air pumps, the Moon's spots seen through

---

36   MS., Louvain-la-Neuve, MS. C165: De Decker and Van Goirle (professors) and Boonen (student), *Physica & Metaphysica* (1680–1681), fols. 241^r–241^v.

37   MS., Louvain-la-Neuve, MS. C90: Narez (professor) and Steijart (student), *Physica* (1708–1709). The name of the *professor primarius* in this case is not known.

38   Paquot, *Mémoires*, vol 1, 638: "Il suivoit la Physique de Descartes, il regardoit le Méchanisme & les principes de l'Hydraulique, comme la base de l'Economie animale, & ne vouloit ni des *fermens*, ni des *orgasmes*, encore moins des *facultés* & des *puissances* de l'ancienne Ecole, don on n'étoit pas encore bien révenu à Louvain, lorsqu'il commença d'y enseigner."

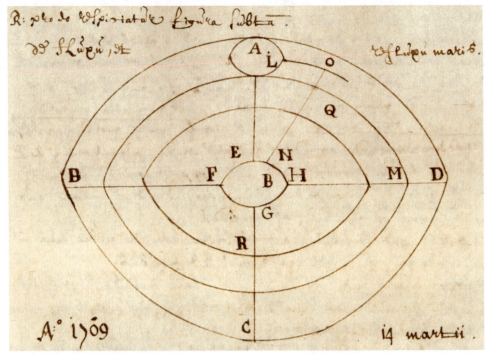

FIGURE 16.12   Steijart's pen drawing based on Rohault's representation of the tides. MS., Louvain-la-
                 Neuve, MS. C90: Narez (professor) and Steijart (student), *Physica* (1708–1709), fol. 255ᵛ.
                 MAGISTER DIXIT COLLECTION

the telescope, and Saturn's rings. Given this abundance of imagery, there is a
striking absence of visual representations of Descartes's vortex theory, which
as we have seen were very prominent in Van den Biesche's notes of Stevenot's
course.[39] As we shall see in the following section, it was around 1720 that Louvain professors started to establish a clear link between the Cartesian account
of the tides and the vortex theory of planetary motion and to emphasize, in
their lessons, Descartes's contribution to the improvement of Copernicus's theory.

---

39   In the treatise *De coelis*, the three world systems are presented according to the standard
      chronological order. First, that of Ptolemy, then Copernicus's system (with objections),
      and finally Tycho's system.

4   Descartes's Theory of the Tides in its Cosmological Context, 1720–1773

In January 1691, the Louvain Faculty of Arts took disciplinary measures against Martin Van Velden (1664–1724), a professor *primarius* in philosophy at the Falcon, who was responsible for having defended the validity of the Copernican system in the classroom. This decision marked the beginning of a complicated legal controversy between Van Velden and the faculty board, which has been well documented before.[40] However, the "trial" did not have long-term negative consequences for Van Velden, who in July 1695, having become royal professor of mathematics in the meantime, published a list of philosophical theses publicly defended by his student Carolus Allegambe.[41] One of the theses concerned the heliocentric system or, to be more precise, the Cartesian version if it. Van Velden praised Descartes's ingenuity in conceiving vortices and in describing the origin and formation of the universe, and then mentioned the essential ingredients of the heliocentric worldview, namely the double motion of the planets around the Sun and around their own axes, the satellites of Saturn and Jupiter, and, finally, the Moon, "companion and follower of the Earth," which causes the tides.[42] On the basis of this thesis, Monchamp concluded that Van Velden was the first in Louvain to have introduced the Copernico-Cartesian system, which was later endorsed by other professors, notably Jean-François Grosse, Franciscus Josephus Engelbert (1696–1764), and Gulielmus Walricus Leempoel (1750–1815).[43]

The Louvain notebooks reveal that the supporters of the Copernico-Cartesian system were more numerous than Monchamp assumed. Particularly instructive, in this respect, are the notes that Petrus Josephus Brunin, a student of the Falcon, took of a physics course taught in 1720 by the *professor primarius*

---

40   Stevart, *Procès de Martin-Etienne Van Velden*; Monchamp, *Galilée et la Belgique*.

41   Monchamp, *Galilée et la Belgique*, 319.

42   Van Velden (praes.) and Allegambe (resp.), *Theses philosophicae*, (Louvain, 1695); quoted in Monchamp, *Galilée et la Belgique*, 320: "Ingeniose suos vortices Des Cartes effinxit, et sane quam eleganter hujus universi corporei fabricam quasi ab ovo orsus est. Dignum certe suo ingenio systema dedit. Planetae duplicem potissimum prae se ferunt motum: annuum et diurnum. Annuus est quo circa solem per orbes volvuntur ellipticos; diurnus, quo circum axem aguntur proprium. Quem motum diurnum in Saturno, Jove, Marte, Venere, telescopium palam facit; nec causam videmus cur idem non obtineat in Mercurio. Saturnum quinque, Jovem quattuor esse stipatum satellibus, notius est quam ut adnotari debeat. Luna, quid est aliud quam Terrae comes aut assecla? Aestum ciet illa marinum, nautis optatissimum."

43   Monchamp, *Galilée et la Belgique*, 319–328.

Peter Joseph Van Tieghem and the *professor secondarius* Michael Bessemers (1690–1751). In the treatise *De sphaera*, the professor, presumably Bessemers, presented and refuted Ptolemy's system, then expatiated upon Tycho's system together with a number of objections, and finally turned to the heliocentric system, discussing first Copernicus's original version, and then Descartes's supposed improvement on it. Finally, he reviewed the most common theological, astronomical and physical objections against heliocentrism, proposing a solution for each of them.[44]

Interestingly, in his treatment of Descartes's improvement of the Copernican system ("Prosecutio systematis Copernicani prout a Carthesio est perfectum"), Bessemers put the accent on the same two elements that Van Velden had emphasized in his 1695 thesis, namely Descartes's vortex theory of planetary motion and his account of the genesis of the universe. In the notebook, one reads that while Copernicus had supposed the Sun to be located at the centre of a finite cosmos, Descartes assumed that the universe was composed of an "indefinite" number of vortices ("infinite" being an attribute reserved only for God), each of which had a sun at its centre. Following the genealogical account of the formation of the universe in the *Principles of Philosophy*, Bessemers went on to explain how God, at the beginning of times, had created a big mass of matter which he then divided into equal particles and set in motion. God also decreed that the total quantity of motion should remain constant, so that where matter came into collision, the quantity of motion lost by one particle would exactly correspond to that acquired by another. On the basis of this principle, Descartes could account for the differentiation of the original particles of matter into three elements, that is to say the very subtle particles of the first element, of which the Sun and the stars are composed, the round particles of the second element, which make up the vortices, and the bigger parts of the third element, which constitute the Earth, the other planets, and the comets.[45]

---

44   MS., Brussels, MS. II 3214: Van Tieghem and Bessemers (professors) and Brunin (student), *Physica* (1720–1721). See fols. 311ᵛ–315ᵛ for the discussion of the Ptolomaic system; fols. 315ᵛ–317ʳ for the Tychonic system; fols. 317ᵛ–322ʳ for Copernicus's heliocentric theory; fols. 323ʳ–324ᵛ for Descartes's improvement on the Copernican system; and fols. 325ʳ–327ᵛ for the discussion of the objections to the Copernican system.

45   MS., Brussels, MS. II 3214: Van Tieghem and Bessemers (professors) and Brunin (student), *Physica* (1720–1721), fols. 323ʳ–323ᵛ. For Descartes's distinction between infinite and indefinite, see Descartes, *Principia* I 27, AT VIII-A 15; for the account of the origin of the cosmos, see *Le monde* VI, AT X 34, and *Principia* III 46, AT VIII-A 46; for the conservation of the quantity of motion after a collision, see Descartes's third law in *Principia* II 40, AT VIII-A 65; *Principia* III 90–92, AT VIII-A 144–145; *Principia* IV 146, AT VIII-A 287.

Bruning's notebook also attests to the fact that the professor dwelt at length on the genesis of the three types of matter, insisting in particular on the formation of the oblong and screw-shaped particles (*particulae striatae*) of the first element, which Descartes considered responsible for terrestrial and magnetic attraction.[46]

Equally detailed was Bessemers's summary of Descartes's account of the structure of the cosmic vortices, in which the smallest and fastest particles of the second element occupy the lower regions, whereas the biggest and slowest particles fill the upper regions, where they are freer to realize their centrifugal tendency. Focusing in particular on the solar vortex, the professor explained how the planets, which are chiefly composed of particles of the third element, are kept in stable orbits thanks to the fact that their centrifugal tendency is equal to that of the surrounding particles of the second element.[47] Interestingly, in this context, Bessemers also anticipated a central point of the Cartesian account of the tides, namely the fact that

> the space between the Earth and the Moon in the terrestrial vortex becomes narrower due to the presence of the body of the Moon and hence the globules of the aether [= second element] in passing through a narrower space cause, with their force, a slight displacement of the Earth towards the Sun [...] and some of them also move more quickly.[48]

We see here, for the first time, a Louvain professor embed the Cartesian explanation of the main cause of the tides in its original cosmological context.

A few pages later, in a section of the treatise *De sphaera* devoted to the influence of the stars on sublunary phenomena, one encounters a chapter significantly entitled "the tides caused by the pression of the Moon" (*De aestu*

---

46   MS., Brussels, MS. 11 3214: Van Tieghem and Bessemers (professors) and Brunin (student), *Physica* (1720–1721), fols. 323^r–323^v. For Descartes's account of the formation of grooved particles, see *Principia* III 90–92, AT VIII-A 144–145; *Principia* IV 146, AT VIII-A 287. On Descartes's screw-shaped particles, see among others, Gaukroger, *Descartes' System*, 175 ff.; for the proliferation of corpuscular shapes in Descartes's *Principia*, see Lüthy, "Where Logical Necessity."

47   MS., Brussels, MS. 11 3214: Van Tieghem and Bessemers (professors) and Brunin (student), *Physica* (1720–1721), fols. 324^r–324^v. Cf. *Principia* III 82–86, AT VIII-A 137–142.

48   MS., Brussels, MS. 11 3214: Van Tieghem and Bessemers (professors) and Brunin (student), *Physica* (1720–1721), fol. 324^v: "Spatium inter terram et lunam in vortice terrestri propter praesentiam corporis lunaris fiat angustius adeoque globuli aetherei per angustius illud spatium transeuntes vi sua terram paululo a suo loco dimovent versus Solem (...) unde oportet aliquos globulos celerius moveri."

maris causato a pressione lunae). The professor here gave a detailed summary of the Cartesian explanation of the daily, monthly, and yearly tidal cycle by making reference to the figure of the *Explicatio*, which appears in a new version, identical to those by Hayé and Blendeff, and signed by the Louvain printer Petrus Denique. In conformity with what he had done in the case of the heliocentric system, the professor listed a number of objections to the Cartesian account of ebb and flow, which he then refuted by denying either the premise or the consequence.[49]

A survey of later notebooks reveals that, around 1730, it became standard practice, in the Louvain classroom, to present Descartes's theory of the cause of the tides as the most probable explanation, and to try and respond to some common objections. This is what we see, for example, in a 1730 notebook of a physics course taught by Philip Van Billoen (1704–1775), *professor primarius* at the Falcon, who devoted the very last of these classes to a thorough discussion of the tides. With the help of Denique's engraving of the *Explicatio*, Billoen tried to make clear to his students that Descartes had provided the most probable explanation of this puzzling phenomenon ("Causa probabilissima videtur illa Cartesii") and that the objections raised against his theory were in fact unfounded.[50]

We encounter an even more extensive treatment of the tides in a philosophy course that the *professor primarius* Hubertus Cornet (d. 1763) and the *professor secundarius* Franciscus Graven taught in 1746 at the Lily. Interestingly, the notebook compiled by Judocus Henricus Grondel contains no single engraving or etching, but a great number of pen drawings with which the student marked the most important topics discussed in the classroom. Particularly impressive is Grondel's drawing of the Copernican system (Figure 16.13), based on an engraving by Denique which, in turn, was a mirrored copy of a famous image designed by Sébastien Leclerc and engraved by Juan Olivar for Fontenelle's *Entretiens sur la pluralité des mondes* (Figure 16.14).[51]

In Grondel's notebook we find an interesting tension between the professor's cautious claim that "the Copernican system could be advocated as a hypo-

---

49   MS., Brussels, MS. II 3214: Van Tieghem and Bessemers (professors) and Brunin (student), *Physica* (1720–1721), fol. 338^r for Denique's version of the *Explicatio*; fols. 340^v–341^v, for the discussion of the objections to the Cartesian theory of the tides.

50   MS., Brussels, MS. II 3703: Van Billoen (professor) and Marij (student), *Physica* (1730–1731). The presentation of Descartes's theory of the tides is at fols. 476^r–479^r. For the objections and replies, see fols. 479^r–480^r.

51   For Denique's engraving, see Vanpaemel, Smeyers, Smets and Van der Meijden, eds., Ex Cathedra, 284.

FIGURE 16.13   Grondel's representation of the Copernican system. MS., Louvain-la-Neuve, MS. C28:
Cornet and Graven (professors) and Grondel (student), *Philosophia & Metaphysica*
(1746), fol. 18ʳ.
MAGISTER DIXIT COLLECTION

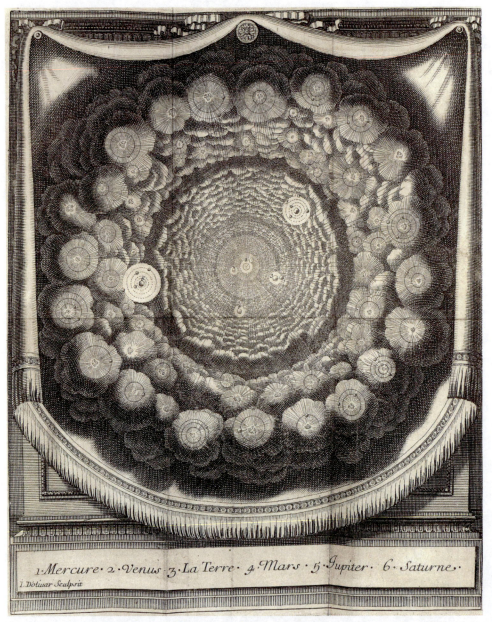

1. Mercure · 2. Venus · 3. La Terre · 4. Mars · 5. Jupiter · 6. Saturne·

I. Dôliuar Sculpsit

FIGURE 16.14   Juan Olivar's engraving of the Copernican system. Based on a design by Sébastien
Leclerc for Fontenelle's *Entretiens sur la pluralité des mondes* (s.l., 1686).
THE SKY ATLAS, FACEBOOK.COM, PUBLIC DOMAIN

thesis," and the great pains he took in answering the theological, physical, and astronomical objections against it.[52]

A pen drawing (Figure 16.15a), which only reproduces the pictorial elements of the *Explicatio*, marks the section devoted to the flux and reflux of the sea, which is embedded in the treatise *De cosmographia* alongside the treatment of astronomical phenomena such as the stellar parallax and the lunar and solar eclipses (Figure 16.15b). The fact that the lecture notes contain references to the letters of the *Explicatio* (see Figure 16.2), which are not included in the student's drawing, is an obvious sign that by the middle of the eighteenth century, Louvain professors still made use of the *Explicatio* in the classroom. In Grondel's notebook, the short summary of Descartes's theory is followed by a long list of objections and replies, the most important of which concern the general explanation of the cause of the tides, and more specifically the question of whether the mass of the Moon, being so small in comparison to the area of the vortex, can exert a pressure on the subtle matter sufficient to generate ebb and flow. The professor answered to this latter objection by pointing out that the tides were not so much the effect of the displacement of the vortex's particles produced by the Moon, as of their acceleration.[53] The notebook also addresses and tries to respond to some more specific objections that signal a contradiction between Descartes's theory and the occurrence of ebb and flow in specific geographic areas.[54]

As a final example of the central role assigned to the Cartesian theory of tides in the Louvain curriculum, I want to refer to the richly illustrated notebook by Dominicus Josephus Lodewijckx, a student of the Pig, who in 1760 attended the *Physics* lessons taught by *primarius* Joannes Franciscus Verberght and *secundarius* Gerardus Deckers. Thanks to Lodewijckx's clear handwriting and diligent notes, some of which carry a date, we know that Louvain professors

52   MS., Louvain-la-Neuve, MS. C28: Cornet and Graven (professors) and Grondel (student), *Philosophia & Metaphysica* (1746), fol. 25ʳ (*Conclusio: Systema copernicanum propugnari potest ut hypothesis*). For the discussion of the objections to the Copernican system, see fols. 25ᵛ–36ʳ.

53   MS., Louvain-la-Neuve, MS. C28: Cornet and Graven (professors) and Grondel (student), *Philosophia & Metaphysica* (1746), fol. 197ʳ: "Licet Luna in quodam vorticis parte versatur, non tamen idcirco a loco illo expellitur tanta materiae aethereae quantitas quantum est corpus lunare prout supponit argumentum, sed quia materia aetherea celerius fluit quam luna, hinc per idem spatium in eadem quantitate transit, ac sic luna in alia vorticis regione versaretur cum autem spatium illud ob lunae presentiam sit coarctatum evidens est quod materia illa angustias istas pertransiens multo celerius fluere debeat ab ante et consequenter fortius premere aquas sibi subjectas."

54   MS., Louvain-la-Neuve, MS. C28: Cornet and Graven (professors) and Grondel (student), *Philosophia & Metaphysica* (1746), fols. 197ᵛ–200ᵛ.

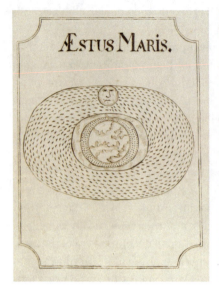

FIGURE 16.15A–B
Grondel's representation of the Copernican system. MS., Louvain-la-Neuve, MS. C28: Cornet and Graven (professors) and Grondel (student), *Philosophia & Metaphysica* (1746), fols. 195ʳ–196ᵛ.
MAGISTER DIXIT COLLECTION

devoted several lessons to discussing the cause of ebb and flow. Lodewijckx's notebook contains the usual summary of the Cartesian explanation of ebb and flow, accompanied by Denique's engraving of the *Explicatio*, and then, on a page dated 20 August 1760, one reads the following thesis: "It is probable that the pressure of the Moon is the cause of the tides" (*Thesis. Probabile est pressionem Lunae esse causam aestus maris*).[55] The discussion of the thesis, which was based on the presentation and refutation of a long list of objections to the Cartesian theory, similar to those we have encountered in Grondel's notebook, lasted several lessons and was terminated on 6 September 1760.

As far as I was able to ascertain, the Louvain professors, whose intention was to convince the students of the validity of the Cartesian explanation of the tides, only addressed those objections for which they had an answer and failed to tackle a crucial problem that could easily be spotted with the help of the *Explicatio*. As the engraving clearly shows, if the tides were due to the pressure exerted by the subtle matter on the ocean waters located below the

---

55   MS., Brussels, MS. II 5602: Verberght and Deckers (professors) and Lodewijckx (student), *Physica* (1760–1761); see fol. 279ʳ for the *Explicatio*, fol. 281ʳ for the thesis.

Moon, then the low tide should occur when the Moon is at the Meridian and the high tides (I) when it is at the quadrature. However, as the German geographer Bernard Varenius observed in his very influential *Geographia generalis* (1650), the opposite is the case, as the sea bulge occurs under the Moon's vertical.[56] Interestingly, Varenius proposed a way to amend the Cartesian theory, which however continued to be endorsed by Descartes's followers in its original form.[57]

Varenius's *Geographia* went through several reprints and appeared in 1733 in an English translation "improved and illustrated" by Isaac Newton and by his pupil Jacob Jurin. In two long footnotes appended to the chapters on ebb and flow, Jurin stressed the inadequacy of Descartes's explanation of the tides, which was based on the fictitious theory of vortices, a fiction "in no way agreeable to nature and motion," and reminded the readers that "Sir Isaac Newton most successfully explains as well the *Flux* and *Reflux* of the Sea, as most other appearances of nature, from his universal Principle of Gravity or attraction."[58]

Newton's explanation of the tides, based on the principle of universal gravitation, did not reach the Louvain classroom, where professors continued to teach the "fictitious" Cartesian theory of vortices. So far as I could ascertain, the last occurrence of the *Explicatio Epicycli Lunaris et Aestus marini* in the notebook of a course taught at the Faculty of Arts dates from 1773, that is to say, exactly one century after its first appearance in the Louvain classroom.[59]

5   Conclusion

My analysis in this chapter of the uses that Louvain professors made of the engraving *Explicatio Epicycli Lunaris et Aestus marini* had a dual purpose. First, my aim was to shed light on the relation between text and images in the students' notebooks. Vanpaemel's hypothesis that the engravings were not mentioned in the course texts and that in some cases they even "went against the

---

56   Varenius, *Geographia generalis* (Amsterdam, 1650), 182.

57   For Varenius's criticism of Descartes's theory of the tides, see Palmerino, "'Ils se trompaient d'ailleurs tous deux'."

58   Varenius, *Complete System of General Geography* (London, 1733), 237 n. *b*, 243 n. *f*.

59   MS., Brussels, MS. II 5342: Vrammout and Forgeur (professors) and Jonckers (student), *Physica* (1773), fol. 39^r. Descartes's theory is discussed at length also in two later notebooks, in which the *Explicatio* is not reproduced, however. MS., Louvain-la-Neuve, MS. C158: Verheyden (professor) and Staumont (student), *Physica* (1776), fols. 110^r–129^r. MS., Louvain-la-Neuve, MS. C148: anonymous (professor) and Lahaye (student), *Physica* (1777–1778), fols. 60^v–71^r.

explanations" provided in the classroom, does not hold true for the *Explicatio*.[60] As we have seen, professors relied on the image and on the accompanying caption for their treatment of Descartes's theory of the tides, and those students who had not bought the engraving made pen drawings that functioned as an essential complement to the lesson's text. It is interesting to see, moreover, that while most of these drawings are faithful copies of the *Explicatio*, others are based on the representations of ebb and flow found in Descartes's *Principia Philosophiae* and in Rohault's *Traité de Physique*, two sources quoted in the caption of the engraving, which were obviously also discussed in the classroom.

The other goal of this article was to shed light on the way in which Descartes's theory of the tides was integrated into the teaching of natural philosophy at the Louvain Faculty of Arts. We have seen that prior to 1670, the year in which Blendeff and Hayé put their engravings on the market, Louvain professors did not address the phenomenon of ebb and flow in their courses. This topic entered the Louvain classroom together with the *Explicatio*, which figures prominently in the students' notebooks composed between ca. 1670 and 1770. Significantly, professors of the Faculty of Arts taught their students only the Cartesian theory of the tides, the notebooks being totally devoid of any evidence of discussions around Galileo's and Newton's alternative explanations.

A survey of the notebooks, however, reveals an important shift in the way in which Descartes's account of ebb and flow was integrated into the teaching of natural philosophy. In the first fifty years (1670–1720), Louvain professors presented ebb and flow as a phenomenon useful to understanding the properties of water, which were discussed in the treatise *De elementis*. The vortex theory of planetary motion, in which Descartes's account was originally embedded and which was essential to understanding his explanation of the main cause of the ebb and flow, was usually not discussed in the classroom and no explicit link was established between Descartes's theory of the tides and his cosmological views. A notable exception here is Joannes Stevenot, one of the earliest supporters of the Cartesian philosophy in Louvain, who, as early as 1675, lectured on Descartes's cosmology and matter theory. As we have seen, the notebook drafted by Stevenot's student, Michiel Van den Biesche, abounds with visual representations of Cartesian corpuscles.

It was only around 1720 that most Louvain professors started to address the Cartesian theory of the tides in its original cosmological context. This went hand in hand with the introduction of the Copernico-Cartesian system in the teaching of natural philosophy. Most notebooks redacted between 1720 and

---

60   Vanpaemel, "The Louvain Printers," 252.

1770 present Descartes's vortex theory of planetary motion as an improvement over the Copernican system, and treat the flux and reflux of the sea as a phenomenon that provides tangible evidence of the existence and causal efficacy of rotating subtle matter. Like other Cartesians, Louvain professors seemed blind to the obvious shortcomings of Descartes's theory of ebb and flow and continued to ignore Newton's explanation of the phenomenon in terms of lunar attraction. The discussion and refutation encountered in the notebooks of possible objections against the Cartesian account of the tides do not represent a real exercise in critical thinking, but rather a dialectical tool used by professors to persuade their students that it was Descartes who had provided the most plausible explanation of the puzzling phenomenon of the tides.

Acknowledgments

I wish to thank Davide Cellamare and Mattia Mantovani, as well as the two anonymous referees, for their insightful comments and suggestions.

CHAPTER 17

Traces of the *Port-Royal Logic* in the Louvain Logic Curricula

Steven Coesemans

1 Introduction

In René Descartes's scientific project, logic occupied a tenuous position. Descartes was very dismissive of scholastic logic as a scientific discipline. Scientific disciplines should broaden our knowledge and serve to solve intellectual problems, and logic—on Descartes's account—failed to hold up to this criterion. The syllogisms of scholastic logic, both demonstrative and dialectical, served mainly as a didactical tool; they were well-suited to transferring knowledge, but rather less apt for acquiring new knowledge.[1]

When we wish to trace the ramifications of this conception of logic for the curriculum of early modern universities, the conflict is immediately apparent. Logic—scholastic, syllogistic logic—made up a significant portion of what was taught in the philosophy curriculum.[2] When Cartesian philosophy started to gain traction at early modern universities, this conflict needed to be resolved in a satisfactory manner. Education is traditionally rather resilient towards sudden structural changes, and this was no different for the case of logic in Cartesian philosophy.

Even so, the well-established scholastic logic evidently needed to be adapted in order to continue to function in a curriculum that would be Cartesian in its inspiration. Descartes considered his own logic to be briefly articulated in his *Discourse on Method*, according to the preface of the French translation of his *Principles of Philosophy*.[3] In the *Discourse*, he first criticizes traditional logic for not being any use in "learning what one does not know". Syl-

---

1 Descartes, *Discourse on Method*, AT VI 18–19. See, e.g., Ariew, *Descartes and the First Cartesians*, 107–112.
2 For an example of a fifteenth- and sixteenth-century curriculum in Louvain, see Roegiers, "Het Leuvense artesonderwijs," 26–29. For seventeenth- and eighteenth-century Paris, see Brockliss, "Natural Philosophy at the University of Paris," 36–37. An in-depth treatment of the seventeenth- and eighteenth-century logic curriculum in Louvain can be found in Coesemans, "Faculties of the Mind," 41–192.
3 Descartes, *Principles of Philosophy*, *Préface*, AT IX-B 15. This passage is discussed in Ariew, "La logique de Port-Royal," 29–30.

logisms can be useful in explaining to others what one has already learned previously through other methods; as such, it is a valuable didactic tool. Since Descartes is, however, not concerned with writing a treatise on pedagogy, traditional logic falls outside of his intellectual project. He expresses the same sentiment in his *Rules for the Direction of the Mind*: a scholastic logician (*dialecticus*) can only devise a syllogism if they know the truth of it before actually deducing it in the syllogism.[4] Similar objections to traditional syllogistic logic were raised by Descartes's contemporaries, such as Pierre Gassendi (1592–1655).[5]

The project of Descartes would require a logic which *did* lead to original knowledge. His logic is a heuristic project as much as an analytic one, and is chiefly concerned with method.[6] The task of logic is, in this project, to identify the correct order of problems and to provide the methodological tools for actually solving these problems.[7] Descartes's method consists of analysis and synthesis. In his view, the scholastic philosophers used syllogisms mostly in a synthetic way, neglecting the role of analysis in problem-finding and problem-solving.[8] Synthesis moves from general notions to specific notions, whereas analysis does the reverse.[9] In contrast to the scholastic logic, a scientifically responsible logic must provide a set of rules which serve as methodological guidelines in the development of science.[10]

Descartes summarizes his method in his four (in)famous rules:[11]

---

4   *Dialecticus* has two distinct meanings in this timeframe: it is either synonymous with *logicus*, or deals specifically with someone versed in topical logic. In the passage from the *Regulae*, it is likely that Descartes means the former, but the point we wish to make stands in both cases. See also Ariew, "La *logique de Port-Royal*," 30. For the passage in Descartes, see *Regulae ad directionem ingenii* X, AT X 406.

5   Capozzi and Roncaglia, "Logic and Philosophy of Logic," 96.

6   Beck, *The Method of Descartes*, 272–273; Lolordo, "Descartes's One Rule of Logic," 52, Roux, "Logique et méthode."

7   Gaukroger, *Cartesian Logic*, 7.

8   Newman, "Descartes on the Method of Analysis," 76. Descartes's exact position on the use of analysis and synthesis is not always straightforward, as Newman explores in the rest of his article. The generalized account I have given will suffice for this chapter, however.

9   Coesemans, "Faculties of the Mind," 189–190. There is also a complex relation with *a priori* and *a posteriori* argumentation, for which I refer to Larmore, "Descartes's Psychologistic Theory of Assent" and Garber and Cohen, "A Point of Order."

10   Descartes's view on traditional syllogisms, topical logic and his own logic as method is not always consistent in its details. I refer again to Ariew, *Descartes and the First Cartesians*, 107–112 for a more nuanced discussion.

11   Descartes, *Discourse on Method*, II, AT VI 18–19: "Le premier estoit de ne receuoir iamais aucune chose pour vraye, que ie ne la connusse euidemment estre telle: c'est a dire, d'euiter soigneusement la Precipitation, & la Preuention; & de ne comprendre rien de plus en mes

- To not consider anything to be true, which has not been evidently recognized as such;
- To divide all difficulties into as many small divisions as possible, to ease their solving;
- To always proceed from the simplest difficulties to the more complex;
- To always make complete overviews of problems, so that we would not omit anything by mistake.

While these rules provide a useful framework for Descartes's own intellectual project, they are rather limited to serve as the basis for a university logic curriculum. In the second half of the seventeenth century and the first half of the eighteenth century, other Cartesians remedied this deficit and produced handbooks of logic which they considered to be in line with the Cartesian project. Not all Cartesians supported Descartes's criticism of scholastic logic to the same extent, but all of them put a strong emphasis on method as an important part of the logic curriculum.[12]

In this chapter, we focus on what is undoubtedly the most influential of these handbooks, i.e., the *Port-Royal Logic*, published as *L'art de penser* (1662) by Antoine Arnauld (1612–1694) and Pierre Nicole (1625–1695). Studies on the *Logique* are in no short supply, but little has been made of its integration in actual university curricula.[13] This is all the more surprising for a book that was conceived as a student textbook: undoubtedly its significance, from the perspective of intellectual history, lies in its impact on students of philosophy in university classrooms. For generations of students, their first foray into philosophy were classes based on some version of the *Logique*. It falls to us, then, to consider and

---

iugemens, que ce qui se presenteroit si clairement & si distinctement a mon esprit, que ie n'eusse aucune occasion de le mettre en doute. Le second, de diuiser chascune des difficultez que i'examinerois, en autant de parcelles qu'il se pourroit, & qu'il feroit requis pour les mieux resoudre.

Le troisiesme, de conduire par ordre mes pensées, en commençant par les obiets les plus simples & les plus aysez a connoistre, pour monter peu a peu, comme par degrez, iusques a la connoissance des plus composez; et supposant mesme de l'ordre entre ceux qui ne se precedent point naturellement les vns les autres.

Et le dernier, de faire partout des denombremens si entiers, & de reueuës si generales, que ie fusse assuré de ne rien omettre."

12   Ariew, *Descartes and the First Cartesians*, 159; Coesemans, "Faculties of the Mind," 12.

13   See, e.g., Ariew, "La *logique de Port-Royal*"; Finocchiaro, "Theory of Argument"; Gardies, "La *logique de Port-Royal*"; Marin, *La critique du discours*; Martin, "Distributive Terms"; Pariente, *L'analyse du langage*; Pousseur, "Bacon et La *logique de Port-Royal*"; Robinet, "Leibniz et La *logique de Port-Royal*." None of these studies devote any significant attention to the teaching practice in which the textbook was actualized: in this regard, this sample of the existing literature is indicative of the whole of *Port-Royal Logic* scholarship.

examine in which ways the professors of these classes adopted and adapted this textbook and presented it to their students, if we are to gain any insight at all into the foundation of the further intellectual development of these young minds.

Ideally, this sort of research would proceed in the following manner. We would make a comparative study of two curricula. One the one hand, there is the logic curriculum as it is laid out in the *Port-Royal Logic*. We have numerous excellent editions of the text, as well as a wealth of secondary literature available.[14] As such, we are unlikely to encounter large methodological problems on this side of the comparison. On the other hand, however, we have the logic curriculum of early modern universities as it was actually taught in classrooms. The overview of this curriculum would need to be sufficiently detailed for us to compare it to the *Port-Royal Logic* not only at the structural level, but also on the treatment of specific details. For example, it is not enough to know that both spend some time treating both the analytic method and the synthetic method: we need to know *precisely* what is taught about analysis and synthesis in order to make a meaningful comparison to the treatment in the *Logique*. Any less detailed presentation of the university curriculum would be insufficient to accurately gauge the real influence of the handbook. At this point, we can scarcely produce such a detailed overview for any one university, let alone a more general overview on a regional or national level. Such general overviews would have to be constructed by induction from specific curricula, none of which we can accurately assess at this moment.[15]

The reconstruction of any specific curriculum requires careful study of primary source material *other than printed textbooks*. The reason for leaving textbooks out of our reconstruction is self-evident: if we are to use the reconstruction afterwards for gauging the influence of a specific textbook, we cannot presuppose the textbook's influence on specific doctrines beforehand. Official documents and university statutes tend to discuss the curriculum in none but the most general terms, which is not nearly sufficient for the project I propose to undertake. What is required, then, is the study of texts produced by teachers and students themselves, often in the forms of manuscript lecture notes.

---

14   Arnauld and Nicole, *La logique ou l'art de penser*, eds. Clair and Girbail (Paris, 1981) is an excellent recent edition.

15   Even then, the question remains whether the regional or national level is the best way to induce general overviews from specific curricula. One might argue that, e.g., a generalized curriculum of Jesuit schools in one given period might be more meaningful than a generalized curriculum of French schools.

For the University of Louvain in the Low Countries, I have mapped out the curriculum of logic for the relevant period.[16] A digital database for notebooks originating here is readily available online.[17] If we wish to gain some insight into what was actually going on in early modern classrooms, similar projects need to be started for other universities. The vast majority of philosophical, and indeed intellectual activity, at least quantitatively, took place inside the classroom; in order to acquire a meaningful understanding of the early modern intellectual climate, it is absolutely crucial that this gap in the scholarship be filled. A tremendous number of relevant manuscripts is extant around the globe; all that is required is to actually start reading them. The case of Louvain is aided here by the digital database called *Magister Dixit*. When the material for other universities is digitized, this type of curriculum reconstruction can take place for other early modern institutions as well.

The case of the University of Louvain can then serve as a first foray into the study of the impact of the *Port-Royal Logic*. It is my hope that, as studies advance and we learn more about curricula at other universities, we might expand our viewpoint and achieve a more global understanding of the impact of this and other logic textbooks.[18]

2   The Louvain Logic Curriculum

The course of philosophy at the Faculty of Arts in Louvain comprised two years, the first nine months of which were devoted to logic.[19] As was often the case in early modern curricula, the curriculum consisted of logic, physics, metaphysics and ethics.[20] Courses were taught by two professors per student group. The senior professor, or *primarius*, taught the morning courses, which contained the *magis principalia*, i.e., the most important parts of the curriculum. The junior professor, or *secundarius*, taught the afternoon courses, which contained the *minus principalia*, i.e., the less important parts of the curriculum.[21] Generally speaking, the *magis principalia* consisted of *logica realis*: logic insofar as it considers its object in itself (e.g., the discussion of categories, syllogisms,

---

16   Coesemans, "Faculties of the Mind."
17   *Magister Dixit* Collection.
18   See also Geudens and Coesemans, "Introduction," 2 for more notes on the necessity of combined research of both textbooks and manuscripts.
19   Papy, "Logicacursussen," 107.
20   Ariew, *Descartes and the First Cartesians*, 11.
21   Coesemans, "Faculties of the Mind," 38–40.

method, etc.). The *minus principalia* consisted of *logica sermocinalis*: logic insofar as it is represented by language (e.g., signs, words, sentences). By the second
half of the seventeenth century especially, *logica sermocinalis* was considered
to be a non-essential part of logic, as noted in a 1677 student notebook:

> Words and things distinct from the operations of the mind are not the
> object per se of logic, but only its accidental object. Logic does not treat
> of words for their own sake, but only for the sake of the operations of the
> mind.[22]

The *logica sermocinalis* curriculum as found in the manuscripts, is based for the
most part on two printed textbooks by local Louvain authors, i.e., Augustinus
Hunnaeus's *Prodidagmata* (1552) and Laurentius Ghiffene's *Prodidagmata*
(1627), especially for the treatment of signs and words. The treatment of sentences which follows appears to be mostly based on a local tradition of commentaries on Aristotle's *De interpretatione*. Finally, the *minus principalia* contain the treatment of topical argumentation and fallacies, even though it is far
from obvious that these subjects would fall under *logica sermocinalis*. They had,
however, been a traditional part of the *minus principalia* since at least the sixteenth century, when the divide between *logica realis* and *sermocinalis* was not
yet articulated in the courses.[23]

From the early seventeenth century up until at least 1671, the *logica realis*
curriculum was structured according to the books of Aristotle's *Organon*, supplemented by local (often idiosyncratic) treatises. The 1660s are characterized
by curricular change: about a dozen Louvain logic manuscripts are extant
for this decade, some of which leave out certain treatises or expand upon
others.[24] The curriculum stabilizes into a more definitive form from 1674 at
the latest.[25] It is now structured according to the three 'operations of the

---

22   MS., Louvain, MS. 250: De Novilia and Snellaerts (professors) and Jodoigne (student),
     Logica (1678), fol. 10ᵛ: "[V]oces et res distinctae ab operationibus mentis non sunt obiec-
     tum per se logicae sed tantum obiectum per accidens, nam logica non tractat de vocibus
     gratia ipsarummet sed tantum tractat de illis gratia operationum mentis."

23   Coesemans, "Faculties of the Mind," 99–117 and 192.

24   Depending on the precise method of counting. Some courses are split up into two or more
     manuscripts, and in some cases, we only have some very short exercises which provide
     little information about the structure of the curriculum. See Coesemans, "Faculties of the
     Mind," 227–229.

25   No manuscripts have been identified between 1671 and 1674, save for MS., Liège, MS. 2538:
     Stam (professor) and anonymous (student), *Logica* (1673), which has not yet been digit-
     ized or studied.

mind', with one book being devoted to each operation. Drawing on James Buickerood's characterization of the logic of this period, we will call this curriculum 'facultative logic'—a term which refers to the important place of the various faculties of the mind (reason, the will, memory) in the logic of this period.[26]

The three operations of the mind, according to which this facultative curriculum is structured, are known from medieval Thomism.[27] In the last quarter of the seventeenth century, the curriculum counts three such operations, in line with earlier seventeenth-century scholastic thought, such as Eustachius a Sancto Paulo's *Summa philosophiae quadripartita* (1609).[28]

The first operation of the mind is *simple apprehension*, which we might also call a *concept* or an *idea*.[29] Here, as with the other operations, there is a clear distinction between the formal idea on the one hand (i.e., the mental act of conceiving) and the objective idea on the other hand (i.e., the concept insofar as it is conceived).[30] The focus throughout the course lies on the formal idea, the mental act. For example, when I conceive a tree, this book of logic principally deals with the act of conceiving the tree, rather than with the tree insofar as it is conceived by me. The latter, the objective idea, is the subject of logic only indirectly, according to the Louvain professors. In focusing on the idea as an act of representation, the theory of ideas they subscribe to is closely related to Arnauld's theory of ideas as mental acts.[31]

---

26   Buickerood, "The Natural History of the Understanding."

27   Schmidt, *The Domain of Logic*, 175–301; Aalderink, *Philosophy*, 299–302.

28   Eustachius a Sancto Paulo, *Summa philosophiae quadripartita* (Paris, 1609). On this type of late scholastic handbook, see Ariew, *Descartes and the First Cartesians*, 43–57.

29   MS., Louvain, MS. 250: De Novilia and Snellaerts (professors) and Jodoigne (student), *Logica* (1678), fol. 29ᵛ.

30   This distinction stems from medieval Thomism, and was probably introduced in the Louvain courses through Suárez. See Andersen, *Metaphysik im Barockscotismus*, 270–271; Ashworth, "Petrus Fonseca"; Ayers, "Ideas and Objective Being," 1064–1065; Coesemans, "Faculties of the Mind," 107–108; Forlivesi, "La distinction," 10–15; Martin, "The Structure of Ideas," 4–5. It should be noted that the manuscripts use *idea* and *conceptus* synonymously.

31   Coesemans, "Faculties of the Mind," 143; Kremer, "Arnauld on the Nature of Ideas"; Nadler, *Arnauld*, 4–5. For more general notes on 'ideas' in seventeenth-century philosophy, see Ariew and Grene, "Ideas." Note that in recent scholarship, Pearce has attempted to dislodge Nadler's interpretation of Arnauld's ideas as mental acts; see Pearce, "Arnauld's Verbal Distinction." Pearce's argumentation, however, partly relies on the fact that he sees the object of an idea insofar as it is conceived as the modification of an external object, rather than of the mind conceiving it; this interpretation leads to various difficulties, and remains unconvincing.

The second operation is *judgment*. A judgment combines two concepts through the act of assent or dissent.[32] For example, I can join the concepts of *leaf* and *green*; when I give my assent to this combination, I now have the judgment that *the leaf is green*. The manuscripts eschew the term *proposition* here, and tend to use it for a judgment insofar as it is represented in a sentence. This affirmation or assent is guided by reason, but ultimately effected by the will. In this, the manuscripts are similar to the account of judgment and the will Descartes provides in the *Meditations*.[33] While some seventeenth-century manuscripts still raise the question whether the will is ultimately responsible for judgement before deciding that it is, almost all eighteenth-century manuscripts consider it obvious that it is.[34]

The third operation of the mind is *discourse* or *reasoning*. This operation deduces one judgment from multiple others. Typically, this occurs through syllogisms. When I have previously judged that Socrates is a man, and that all men are mortal, the third operation of the mind allows me to conclude that Socrates is mortal. The treatment of the third operation of the mind in the Louvain curriculum is almost entirely based on a tradition of discussions of the *Prior Analytics*.[35]

In the first quarter of the eighteenth century, a fourth operation of the mind was added to the Louvain curriculum: method.[36] Method had been a part of logic textbooks since the late sixteenth century, with Zabarella's *De methodis*

---

32 This is not a given in seventeenth-century philosophy. Gassendi viewed judgment as an affirmation or denial of an idea rather than of the connection between ideas; see Nuchelmans, "Proposition and Judgement," 120–121; Lolordo, "'Descartes's One Rule of Logic," 61, n. 20.

33 *Meditationes*, AT VII 56–57. See also Stevenson, "Freedom of Judgement," 225–229; Aalderink, *Philosophy*, 65–91.

34 See in the seventeenth century, e.g., MS., Louvain, MS. 250: De Novilia and Snellaerts (professors) and Jodoigne (student), *Logica* (1678), fols. 117ᵛ–118ʳ. MS., Louvain-la-Neuve, MS. C174: Vandenhove and De Quareux (professors) and Wagemaeckers (student), *Logica* (1698), fols. 18ʳ–19ᵛ. In the eighteenth century, see, e.g., MS., Louvain, MS. 241: Van Billoen and Lilivelt (professors) and Seiger (student), *Logica* (1737), fols. 53–55. MS., Louvain, Ms. 236: Van Horick and Thijsbaert (professors) and Du Mont (student), *Logica* (1761), fols. 74–81.

35 Coesemans, "Faculties of the Mind," 180–184.

36 At this stage, it is difficult to say precisely when method was added to the curriculum. No manuscripts have been dated between 1699 and 1714. Cinck' and Du Bois's *Logica*, was dictated in 1714 and contains a treatise on method, Goffoul' and Audenaert's was dictated in 1720–1723 and does not contain one; MS. II 4272: Cinck and Du Bois (professors) and Dues (student), *Logica* (1714–1715); MS., Louvain, MS. PM0326: Goffoul and Audenaert (professors) and Devisscher (student), *Logica* (1723). All later manuscripts, starting in 1734, have a treatise on method. It is certainly possible that in the period 1700–1734 some professors

(1578).[37] In the late scholastic textbooks, the notes on method were usually included in the treatment of either judgment or discourse.[38] Starting with the *Port-Royal Logic*, however, logic was increasingly seen as a distinct operation of the mind.[39] Indeed, the *Logique* was the first textbook to explicitly treat method as a fourth operation. As such, its structural impact, directly or indirectly, on the Louvain curriculum of the eighteenth century is significant, if only through its separate treatment of method. In what follows, I will gauge its influence in a more direct manner, through a comparison of the treatment of specific subjects in the *Logique* and the Louvain logic curriculum.

3   The Influence of the *Port-Royal Logic* on the Louvain Logic Curriculum

3.1   *The First Operation of the Mind*

For the discussion of the first operation of the mind, we will look at two cases where the curriculum can be connected to the *Logique*: the question of substance and accident, and the Aristotelian categories. Other parts of the curriculum show little sign of being directly indebted to the *Logique*. Another study might be made on the influence of another of Arnauld's major works: his *Des vraies et fausses idées*, which was present in at least some libraries of the Louvain Arts Faculty.[40] Many eighteenth-century notebooks contain a small treatise *De veritate et falsitate idearum* which is at least partly indebted to Arnauld.[41] In this chapter however, I will limit myself to the discussion of the *Logique*.

3.1.1   Substance and Accident

One of the first treatises in the book on simple apprehension is a treatise *De substantia et accidente*. Note that the titles of these treatises vary between manuscripts, and that some manuscripts might split some treatises up—even

---

were quicker in adopting the new treatise than others. A new study of three undigitized manuscripts from 1724–1725 might reveal more in this regard: see Coesemans, "Faculties of the Mind," 234.

37   Dear, "Method," 147–150; Ong, *Ramus*, 16; Schuurman, *Ideas*, 13–14 and 40–44; Vasoli, *La dialettica*, 604–606.

38   Ariew, *Descartes and the First Cartesians*, 53–55.

39   Ariew, "Descartes, les premiers cartésiens et la logique," 68–71.

40   Coesemans, "Faculties of the Mind," 256.

41   Coesemans, "Faculties of the Mind," 160–163.

in those cases, the core text generally remains very similar.[42] Seventeenth-century notebooks tend to give an account of substance and accident that is mostly in agreement with late scholastic notebooks, such as Eustachius a Sancto Paulo's *Summa philosophiae quadripartita*. From the early eighteenth century onwards, the manuscripts include both an Aristotelian and a Cartesian account of substance and accident, explicitly in contrast to one another.

The Aristotelian account defines substance as an *ens per se subsistens*, and an accident as an entity which requires a subject in which to exist. These accidents are, generally speaking, real beings.[43] The Cartesian account, on the other hand, is a substance-mode metaphysics, which is a lot sparser on an ontological level. A manuscript from 1735 describes it as follows:

> The Cartesians and most of the more recent philosophers, however, say that substance is an entity in itself, i.e., a being which subsists according to itself, in such a way that it does not have to be in anything in order to be an entity or to be conceived as an entity. [...] According to the Cartesians, however, and most of the more recent philosophers, an accident is nothing else than a mode of a being in such a way, that accident considered in itself is not some entity, but only some affection of a being or a determined mode according to which substance exists.[44]

---

42   Compare MS., Louvain, MS. 250: De Novilia and Snellaerts (professors) and Jodoigne (student), *Logica* (1678), fols. 38ᵛ–48ᵛ, which contains a treatise *De substantia* and a separate treatise *De accidente*, which discusses more or less the same subject matter with the same conclusions as MS., Brussels, MS. 19376: Lamine and Van den Hove (professors) and Van Riethoven (student), *Logica* (1689), fols. 7ʳ–7ᵛ, which contains one treatise *De substantia et accidente* which is significantly shorter, mostly due to small handwriting and significantly less discussion of examples and exceptions. On the other hand, MS., Louvain, MS. PM0326: Goffoul and Audenaert (professors) and Devisscher (student), *Logica* (1723), fols. 126–140, titles its treatise *An accidentia realiter distinguantur a substantia*.

43   There is a curious and highly problematic idiosyncratic doctrine in Louvain manuscripts about so-called *pure modes*, which are considered to be ontologically inferior to other accidents (*accidentia strictae entitatis*). The criterion is that pure modes cannot be supernaturally separated from their subject by God. See Coesemans, "Faculties of the Mind," 117–130.

44   MS. Louvain, MS. 241: Van Billoen and Lilivelt (professors) and Seiger (student), *Logica* (1737), fol. 39: "Cartesiani vero et recentiores plerique dicunt substantiam esse entitatem in se, hoc est ens secundum se subsistens, ita ut nulli ali [*sic*] inesse debeat ut sit entitas vel tanquam entitas concipiatur. [...] Iuxta Cartesianos autem et plerosque recentiores accidens quodlibet nil est nisi modus rei ita ut accidens secundum se spectatum non sit entitas aliqua sed tantum quaedam entis affectio seu determinatus modus quo existit substantia."

Substance is still a being unto itself. A rock, a chair, a tree: these things still have no need of anything else, ontologically speaking, to exist. The greenness of the tree, however, is no longer a real quality which *inheres* into the tree. It is rather a *way of existing* of the tree, a modification or simply a *mode*.[45] In this account, the mode of a substance is no longer ontologically independent. It is no longer a separate entity, as real accidents were in the Aristotelian account; it is meaningless to speak of greenness without a substance, e.g., a tree, that is actually green.

This type of substance-mode metaphysics is precisely what we find in the *Port-Royal Logic*:

> I call the following manner of being, or mode, or attribute, or quality: that which, being conceived in the thing, and being conceived as not being able to exist without it, determines it to be in a certain way, and causes it to be named as such.[46]

The fact that accidents are relegated to being *ways of being* does not necessarily mean that they become simply beings of reason, but they at least lose the claim to being a real entity.[47] This shift was relatively common in seventeenth-century metaphysics and is certainly not original to the *Logique*.[48] Still, the approach of the *Logique* is similar enough to the manuscripts to make it a plausible source. Yet, it would be unwarranted to conclude, based on this evidence alone, that Louvain professors would have taken their notes directly from the *Logique*. In any case, they are in agreement with each other, since all manuscripts reject the Aristotelian account and subscribe to the Cartesian view.[49]

---

45   Pasnau, *Metaphysical Themes*, 180–187 calls this a *deflationary* account of accidents. On the vanishing distinction between *substantia*, *res* and *ens* in seventeenth-century thought, see Blum, *Studies*, 307.

46   Arnauld and Nicole, *La logique*, 47: "J'appelle maniere de chose, ou mode, ou attribut, ou qualité, ce qui étant conçû dans la chose, & comme ne pouvant subsister sans elle, la détermine à être d'une certaine façon, & la fait nommer telle."

47   Pasnau, *Metaphysical Themes*, 274 speaks of reducing one's ontological commitment.

48   Pasnau, *Metaphysical Themes*, 258–262.

49   It remains to be studied to what degree the manuscripts' metaphysics of modes is actually Cartesian in the sense that it is consistent with Descartes's theory of modes; cf. Pasnau, *Metaphysical Themes*, 262–266.

3.1.2   The Aristotelian Categories

The Louvain curriculum of facultative logic (cf. *supra*) discussed the Aristotelian categories in the book on the first operation of the mind.[50] In seventeenth-century manuscripts, the treatment of the categories is clearly a reworked and abbreviated version of the commentaries on Aristotle's *Categories* in the early seventeenth-century curriculum. They are considered to belong to the first operation of the mind because they are the (traditionally ten) forms a concept or idea could take. Most manuscripts limit themselves to discussing the three or four most important categories: substance, quality, quantity and sometimes relation. This was part of an ongoing trend of diminishing focus on relation as an Aristotelian category. In 1615, each of the six minor categories still received a handwritten folio of treatment; by the 1660s, they were only mentioned to exist; and from 1674 onwards, they were omitted altogether.[51]

In the last quarter of the seventeenth century, the major categories were still briefly treated. From the first quarter of the eighteenth century onwards, Louvain professors adopted an increasingly sceptical attitude towards them. Around 1720, it was still common to explain the system of Aristotelian categories only to refute it; from 1735 onwards, Aristotle's categories were only quoted as being a useless relic of the old philosophy. The following rejection from 1741 is quite telling:

> Here, then, the ten categories of Aristotle, which to some are so full of mysteries! Even though they are of little use; not only are they useless for the correct forming of judgment; they are very often even harmful to it. This is due to two causes, which it will be worthwhile here to note.[52]

Let us now compare this rejection of Aristotle's categories with the same rejection in the *Logique*. The resemblance is so strong, that the notebook here clearly draws directly on the *Logique*:

> Here, then, are the ten categories of Aristotle, of which one makes so much mystery, even though there is very little use in them; not only are

---

50   Coesemans, "Faculties of the Mind," 152–155.

51   Coesemans, "Faculties of the Mind," 95.

52   MS., Brussels, MS. II 4352: Verlaine and Page (professors) and De Libotton (student), *Logica* (1743), fol. 57: "En decem Aristotelis categorias, tot mysteriis ut quibusdam placet plenas! Tametsi illa parum utiles sint; tantumque abest ut ad rite formandum iudicium prosint, ut etiam saepissime noceant; idque duas ob causas, quas hic notare opere praetium erit." Note that this is one of the very few places in these manuscripts where we actually find an exclamation mark.

they useless for the forming of judgment, which is the goal of the true Logic, but they are often harmful to it, for two reasons, which it is important to note.[53]

For the above-mentioned rejection, both the manuscript and the *Port-Royal Logic* give the following reasons. One considers these categories as being actually existing, with a solid reasonable foundation; whereas in reality they are completely arbitrary. A person could come up with any number of categories to explain reality. They are the product of one person's imagination, who decided to arrange reality just so, without any special authority—anyone else could devise a more or less equally valid system of categories. In fact, recent philosophers have constructed a set of seven categories, listed in the following verses, originally composed by none other than Henricus Regius:[54]

> *Mind, measure, rest, motion, configuration and figure,*
> *Are, with matter, the beginnings of all things.*[55]

The second reason is that this type of philosophy will make students (and people in general) think they know more than they really do: this not only makes them feel self-important, but is actively harmful for their studies.

In the above passage, the notebooks read almost like a literal Latin translation of the *Logique*. It is very clear that professors were actively reading and using the *Logique* at least in the mid-1730s. Some thirty years later, the treatment of the subject matter was largely the same, although it is paraphrased to a large degree.[56] Some elements remained the same, such as the distich with the

---

53   Arnauld and Nicole, *La logique*, 51: "Voilà les x. Categories d'Aristote dont on fait tant de mysteres, quoiqu'à dire le vrai ce soit une chose de soi très peu utile, & qui non seulement ne sert gueres à former le jugement, ce qui est le but de la vraie Logique, mais qui souvent y nuit beaucoup pour deux raisons qu'il est important de remarquer."

54   Verbeek, *Descartes and the Dutch*, 13; Lüthy, *David Gorlaeus*, 146; Strazzoni, *Dutch Cartesianism*, 23–38. The professors also include a distich for the classic Aristotelian categories, which appeared in some late seventeenth-century textbooks: "Arbor sex servos ardore refrigerat ustos / rure cras stabo, sed tunicatus ero." Cf. Coesemans, "Faculties of the Mind," 153–154. For an overview of other Cartesian adaptations of the categories, cf. Blum, *Studies*, 308–310.

55   MS., Brussels, MS. II 4352: Verlaine and Page (professors) and De Libotton (student), *Logica* (1743), fol. 57: "Mens, mensura, quies, motus, positura, figura / Sunt cum materia cunctarum exordia rerum." See Bos, "Een kleine geschiedenis."

56   See, e.g., MS., Louvain, MS. 244: Petit and Ghenne (professors) and Van Aerschot, *Logica* (1761–1763), fols., 78–91.

seven 'modern' categories quoted above. In 1764, one professor notes that modern philosophers took these seven categories from Plato.[57] It is unclear whether this professor and his contemporaries still made direct use of the *Logique*, but given the high degree of paraphrasing in these manuscripts, it seems relatively unlikely.

Even the professors who directly used the *Logique* were not wholly uncritical of it. While Arnauld and Nicole concluded that the study of categories was useless as a whole, the Louvain professors still maintained that there was a correct and useful division of reality into (depending on the professor) three or four categories. These four categories are God, matter, created mind and the animal soul.[58] The criterion the Louvain professors observe is the following: any set of categories should be large enough to represent all the objects of our ideas, and to discern them from one another, but no larger than is needed to do so.

From that perspective, the divide between matter and (created) mind seems relatively straightforward, at least from a Cartesian dualist perspective. God must also have his own category, due to his ontological distance from creation. Everything depends on him, and he depends on nothing: He is too ontologically different to belong to either created mind or matter. Angels, for that matter, belong to the category of created mind, just as the rational soul.[59] I will not elaborate here on the debate over the animal soul.[60]

We can surmise that the Louvain professors of at least the late 1730s directly read and actively used the *Port-Royal Logic* for this part of the curriculum. Even so, they still reached different conclusions on the study of categories. Later generations of professors were also heavily impacted by the *Logique*, either directly through their own reading or indirectly through the courses of their predecessors.

---

57   MS., Louvain, Ms. 236: Van Horick and Thijsbaert (professors) and Du Mont (student), *Logica* (1761), fol. 48.

58   An example of a notebook accepting the animal soul as a category is MS., Louvain, Ms. 236: Van Horick and Thijsbaert (professors) and Du Mont (student), *Logica* (1761), fol. 48. An example of a notebook rejecting it is MS., Louvain-la-Neuve, MS. C145: Page and Bultot (professors) and Lambillion (student), *Logica* (1757), fol. 44.

59   MS., Louvain, Ms. 236: Van Horick and Thijsbaert (professors) and Du Mont (student), *Logica* (1761), fol. 53.

60   Coesemans, "Faculties of the Mind," 155; for notes on this debate in early modern philosophy in general, see Barth, "Leibnizian *Conscientia*," 227–228; Clarke, *Descartes's Theory of Mind*, 32–37; Cottingham, "A Brute to the Brutes"; Des Chene, *Spirits and Clocks*; Harrison, "Animal Souls"; Osler, *Divine Will*, 71; Park, "On Animal Cognition," 60–71; Wilson, *Epicureanism*, 156.

3.2   *The Second Operation of the Mind*

The treatment of the second operation of the mind shows much less direct influence from the *Logique*. The second operation of the mind is judgement, which is represented by a proposition. According to the Louvain professors, judgment is not effected by the intellect alone. The intellect can provide motives for judgment, but the ultimate decision to affirm or deny the connection between two ideas lies with the will rather than the intellect. This account differs from the treatment in the *Port-Royal Logic*, where the will does not ultimately determine the operation of judgment.[61] On top of this, the Louvain curriculum is also structurally very dissimilar to the *Logique* in the book on the second operation of the mind.

Most of the discussion in the manuscripts deals with the notion of assent, i.e., the act of the will to make a certain judgment based on the motives provided by the intellect. There are many kinds of assent: firm and weak, certain or uncertain, prudent or imprudent, evident or not evident, etc. This discussion is completely absent from the *Logique*, and we will therefore not elaborate on it. Only two elements from the book on judgment in the Louvain curriculum are also discussed in the *Logique*: the foundation of our assent, and the causes of error in judgment. The *Port-Royal Logic* discusses both of these in other places than the book on the second operation.

3.2.1   The Foundation of Assent

In the fourth book of the *Port-Royal Logic*, which discusses method, we find a short treatment of two possible foundations for assenting to a judgment. We can judge something based on our own rational capabilities, as we generally do in science or philosophy. Alternatively, we can judge something based on our trust in a worthy authority, as we do in matters of faith. The *Logique* quotes St. Augustine here: "*Quod scimus, debemus rationi; quod credimus, autoritati.*"[62]

The Louvain manuscripts make a more complex, threefold subdivision: assent can be given not only in science and faith, but also in opinion. Assent in science is here considered to be only applicable to necessary propositions, including propositions which result from demonstrative syllogisms with necessary premises, such as mathematical propositions. Assent in opinion is applicable to uncertain but probable propositions: the strength of the assent should

---

61   Cassan, "La logique." The position that the will is not ultimately responsible for judgment was held by various other seventeenth-century philosophers, such as Gassendi; cf. Lolordo, *Pierre Gassendi*, 58–59.

62   Arnauld and Nicole, *La logique*, 335–336; Augustine, *De utilitate credendi*, 11.

be proportionate to the probability of the proposition here.[63] An example of such a proposition would be, in seventeenth-century natural philosophy, that the tides are caused by the moon—a much debated topic in seventeenth-century natural philosophy.[64] Assent in faith coincides with the *Logique*'s notion of authority, but the Louvain curriculum makes a further subdivision between divine (i.e., revelation), angelic (the testimony of an angel), or human authority. Both divine and angelic authority are considered to be more certain than science.[65]

All of this discussion contains too many differences from the *Port-Royal Logic* to prove any influence. The discussion concerning assent found in the Leuven manuscripts is presented in a completely different moment when compared to the *Port-Royal Logic*; the similarities between the *Logique* and the manuscripts are too few, in these cases, to draw any meaningful link between them.

3.2.2   The Cause of Our Errors

The Louvain curriculum deals at some length with the causes of our errors. The cause of our errors always lies with the will; it makes us prone to judge about things which the intellect has not yet perceived well enough.[66] The principal question is here precisely what makes the will judge erroneously. The manuscripts identify four main factors: precipitousness in judging, ambiguity of terms, strong emotion, and imagination.[67]

---

63   Coesemans, "Faculties of the Mind," 175.

64   MS, Brussels, MS. II 5412: Engelbert and Cornet (professors) and Guilielmus Borghs (student), *Logica* (1735), fol. 65^r.

65   MS, Brussels, MS. II 5412: Engelbert and Cornet (professors) and Guilielmus Borghs (student), *Logica* (1735), fols. 70^r–70^v. The professor notes here that there is an exception to this rule: science as it is practised by the blessed in heaven is not necessarily less certain than divine or angelic revelation. This should be judged on a case-by-case basis, apparently. This type of argument, which seems far-fetched to the modern reader, is relatively common in these manuscripts. See, e.g., on the subject of innate ideas, the question whether infants sometimes contemplate God; MS., Louvain, Ms. 236: Van Horick and Thijsbaert (professors) and Du Mont (student), *Logica* (1761), fols. 31–32. (They do.)

66   MS., Louvain, MS. 250: De Novilia and Snellaerts (professors) and Jodoigne (student), *Logica* (1678), fol. 127^v.

67   The account differs between manuscripts. Some, especially the seventeenth-century manuscripts, restrict themselves to precipitousness in judgment, while others, such as De Neve, list up to ten causes; MS., Liège, MS. 1264: De Neve (professor) and anonymous (student), *Logica* (1674), fols. 58^r–59^r. Most of them are instances of one general cause in judgment, which is the fact that we judge about things we have not perceived clearly and distinctly enough. See Coesemans, "Faculties of the Mind," 176–177.

The discussion of precipitousness in judging hearkens back to Descartes's discussion of erroneous judgment in the *Third Meditation*.[68] Precipitousness, which amounts to a predisposition of the will to judge without having perceived the object of the judgment well enough, may arise from different factors. Examples include bad epistemological habits which we have retained from our childhood, powerful emotion, or the reverence we have for untrustworthy human authorities.

Ambiguity of terms causes us to erroneously attribute the properties we discover in one thing to another thing with the same or a similar name. For example, I might say that an animal that I will eat later is healthy, meaning that the animal is not sick, and thus suitable for human consumption. Another person might think that I mean that the consumption of the meat will be beneficial to my health, based on my ambiguous use of 'healthy'.

Passion or strong emotion might cloud our judgment, increasing the likelihood that we judge a person without perceiving her clearly. We might conclude that all of her qualities are good based on our love, or bad based on our hate, without allowing our intellect to objectively evaluate these qualities. We might, for example, judge that a politician with whose ideology we disagree issues almost exclusively ill-conceived policies, which, intellectually speaking, is highly unlikely.

Our judgment may also be clouded by our imagination, causing us to attribute certain qualities to subjects that are incapable of these qualities. A classic example is the scholastic *horror vacui*: our imagination erroneously combines the idea of fear with the idea of matter, creating the false judgment that matter fears the void.[69]

Of all these causes, only precipitousness is mentioned as such in the *Port-Royal Logic*. I refer to the following passage from the first of two *Discourses* which serve as an introduction to the whole work:

> Everywhere, one encounters only false minds, who barely have any judgment about truth, who understand everything with false prejudice, who convince themselves with the worst sort of reasonings, and would convince others with them; who let themselves be taken away by the slightest of appearances; who are always in excesses and extremes; who have no firm hold on the truths they know, because it is rather coincidence that

---

68   *Meditationes* III, AT VII 35–36. Cf. also Aalderink, *Philosophy*, 101–113.
69   MS, Brussels, MS. II 5412: Engelbert and Cornet (professors) and Guilielmus Borghs (student), *Logica* (1735), fols. 83ᵛ–84ʳ. For the example of the vacuum specifically, see Blum, "*In fugam vacui*," 431–438.

binds them, than firm light [...] who only judge about the truth of things based on the tone of voice used: he who speaks with ease and severity is right, he who has some trouble explaining himself, or shows some emotion, is wrong. They know no better.[70]

It is relatively unlikely that the treatment in the curriculum was directly influenced by this passage from the *Logique*, for two reasons. First, this section in the *Logique* does treat precipitousness of judgment, but hardly in the systematic way the student notebooks do, namely as characterized by enumeration of causes, subtle distinction between those causes, etc. The *Logique* delivers more of a general dismissal of the thoughtless way most people lead their lives, rather than a detailed account of the cause of our errors in judgment. Secondly, the treatment of precipitousness in judgment as a major cause of our errors appears in the Louvain curriculum from 1677 onwards, which is about forty years before any of the other major influences we can find in the *Logique*. It is more likely that, in this case, the professor had a passage from Descartes himself in mind, such as the Third Meditation.

3.3   *The Third Operation of the Mind*
The book on discourse in the Louvain curriculum deals entirely with the various kinds of syllogisms and the rules for constructing them. There is very little change or evolution here throughout the seventeenth and eighteenth centuries.[71] The eighteenth-century curriculum still treats this as a book on syllogistic logic in the same vein as the early seventeenth-century curriculum. This is also what the *Port-Royal Logic* does—but as there is so little internal evolution in the curriculum, this provides no evidence of an influence of the *Logique*.

3.4   *The Fourth Operation of the Mind*
As I have noted before, the *Port-Royal Logic* had promoted method to the status of fourth operation of the mind in logic curricula. Other handbooks would

---

70   Arnauld and Nicole, *La logique*, 16: "On ne rencontre par-tout que des esprits faux, qui n'ont presqu'aucun discernement de la verité, qui prennent toutes choses d'un mauvais biais, qui se payent des plus mauvaises raisons, & qui veulent en payer les autres; qui se laissent emporter par les moindres apparences; qui sont toûjours dans l'excès & dans les extremités; qui n'ont point de serres pour se tenir fermes dans les verités qu'ils savent, parceque c'est plutôt le hazard qui les y attache, qu'une solide lumiere [...] qui ne jugent de la verité des choses que par le ton de la voix: celui qui parle facilement et gravement a raison: celui qui a quelque peine à s'expliquer, ou qui fait paroître quelque chaleur, a tort. Ils n'en savent pas davantage."
71   For some minor evolutions, see Coesemans, "Faculties of the Mind," 180–184.

follow, such as the popular and unequivocally Cartesian *Cours entier de philosophie* (1690) by Pierre-Sylvain Régis (1632–1707).[72] The Louvain curriculum followed suit in 1714 at the latest. The notebooks divide the questions method should deal with into questions about words (i.e., hermeneutical questions) and questions about things (i.e., about their existence or their nature).[73] This question can also be found in the *Port-Royal Logic*.[74]

I will limit myself here to the discussion of the analytic method in the notebooks. The synthetic method is treated only very summarily in the Louvain curriculum, and mostly consists of about one folio with some didactic notions.[75] The whole treatment of synthesis is too short to pinpoint any specific influences from the *Logique*, and thus is of little use to this chapter.

The analytic method can be summarized in four rules, according the Louvain notebooks:[76]

1. We must establish the *status quaestionis*. This involves careful consideration of what the question is about, establishing what the unknown part is we are meant to find out, and what the terms of the question mean.

2. The difficulties of a question can be subdivided into smaller difficulties, each of which will be easier to solve than the question itself.[77]

3. Within this subdivision, we must order the parts, beginning with the parts that are the easiest to comprehend and to explain. Only after solving those, we can move on to the more difficult parts.

4. When we have found a certain principle, on which the solution of the problem depends, the solution lies in the comparison of this principle with the question at hand.[78]

---

72 A good treatment of Pierre-Sylvain Régis's philosophy in general can be found in Schmaltz, *Early Modern Cartesianisms*, 300–314. A discussion of the Cartesian logic handbooks immediately following the *Port-Royal Logic* can be found in Ariew, *Descartes and the First Cartesians*, 157–165.

73 MS., Louvain-la-Neuve, MS. C145: Page and Bultot (professors) and Lambillion (student), *Logica* (1757), fols. 178–181. Discussion in Coesemans, "Faculties of the Mind," 187–189.

74 Arnauld and Nicole, *La logique*, 300.

75 For MS., Louvain-la-Neuve, MS. C145: Page and Bultot (professors) and Lambillion (student), *Logica* (1757), fol. 182. MS. Louvain, MS. 241: Van Billoen and Lilivelt (professors) and Seiger (student), *Logica* (1737). Discussion in Coesemans, "Faculties of the Mind," 189–190.

76 Coesemans, "Faculties of the Mind," 185–187.

77 This, of course, is the hard part. The professors are cognizant of this, and make an appeal to the student's own ingenuity in this step. See MS., Louvain, MS. 241: Van Billoen and Lilivelt (professors) and Seiger (student), *Logica* (1737), fol. 124.

78 The manuscripts use the phrase *principium aliquod certum*, though it is not entirely clear what they mean precisely.

The first of these rules is more or less a summary of the chapter on analysis in the *Logique*.[79] The second and third rule are very similar to the second and third rule of Descartes's *Discourse* which we mentioned at the beginning of this chapter. Since the *Logique* also enumerates Descartes's four rules, it is plausible that professors would have copied these from the *Logique* rather than the *Discourse on Method*. There is however another candidate which also contains the first three of these rules, and where the wording is also strikingly similar: Malebranche's *De la recherche de la vérité* (1675).[80] All of these works were extant in libraries at the Louvain Arts Faculty, and it remains difficult to determine which of these works professors had to hand when composing their course.[81] Moreover, it remains possible, and even likely, that professors would consult more than one of these books.

4   Conclusion

The Louvain curriculum shows definite signs of influence from the *Port-Royal Logic*, especially from the first quarter of the eighteenth century. However, it is clear that, while the *Port-Royal Logic* was an important source after this period, it is far from being the only determining influence in the formation of the curriculum. Scholastic texts left their legacy, of course, as did the handbooks of other Cartesians and the works of Descartes himself, which were abundantly present in the Louvain college libraries. Where the *Port-Royal Logic* was clearly consulted and even copied, professors did not always follow its conclusions, as we have seen in the case of the *Categories*.

Directly or indirectly, the most significant influence of the *Port-Royal Logic* on the Louvain curriculum was the introduction of method as a fourth book of logic. This is a structural overhaul of the curriculum, which carries a significant interest for the history of ideas in this period.

---

79   Arnauld and Nicole, *La logique*, 299–306.
80   Malebranche, *Recherche*, 380–381. See also Schuurman, *Ideas*, 48–50.
81   Cf. Coesemans, "Faculties of the Mind," 187; 256–268.

CHAPTER 18

Cartesianism and the Education of Women

Marie-Frédérique Pellegrin

A book of which I wanted that the women themselves would be able
to understand something.

DESCARTES to Vatier, 22 February 1638.[1]

∴

1 Introduction

The education of women in the early modern age is a fundamental issue that
can be associated with, among other things,[2] the great movement of self-
control and the control of others described by Michel Foucault in his essay on
criticism. During the early modern age, the art of governing that stemmed from
Christian pastoral care gradually extended itself to all areas, including secular
ones. Foucault drew attention to the governing of children, indigents, beggars,
families and homes as well as of armies, cities and states, extending through
to the control of bodies and minds.[3] Textbooks and treatises about the education of women multiplied during the sixteenth and seventeenth centuries
and almost all of them were aimed at a strict control of women's will to know
and to act. These writings may be positioned at the intersection of the different types of the government of human beings discussed by Foucault; what is at
stake is the control of women's bodies and minds for purposes which are eminently social. These writings deal with the education of a specific category of
the population, and merge with other prescriptions towards good government:

---

1   Cottingham's translation of the opening quotation is "in a book which I wished to be intel-
     ligible in part even to women," which is less close to the French text: "Un livre où j'ai voulu
     que les femmes mêmes pussent entendre quelque chose"; AT I 560, CSMK I 86. All translations
     of Descartes are from the Cottingham edition unless otherwise indicated. Unless otherwise
     stated, all translations from other authors are mine.
2   See Timmermans, *L'accès des femmes à la culture.*
3   Foucault, "Qu'est-ce que la critique?," 37. He speaks of a "veritable explosion in the art of gov-
     erning men" from the fifteenth century onwards.

from the direction of the soul, to economy (in the original sense of the term, namely, as the orderly management of the household), through to instructions for a harmonious marriage or for the education of children by the mother. Even if Foucault does not mention the example of treatises and manuals for the education of women, these writings corroborate his thesis of an increasing emphasis on self-government and the government of others in the early modern age. When reading most of these manuals, one may see that girls and women were generally subject to injunctions that confined them to a fairly limited learning.[4] The government of females perfectly exemplifies Foucault's thesis of population control with regard to knowledge and power. Educational textbooks were never intended for both sexes: their moral and social aims were different for girls and for boys.[5] There is a need, therefore, of a separate chapter, or book, to deal with the education and instruction of women for any full study of Descartes in the classroom.

A portion of the literature on women's education is about women's instruction proper. The stakes are extremely high: what can women know? What culture, what knowledge is necessary and beneficial to them? Instruction—the part of education that more specifically concerns intellectual and cultural learning and, therefore, acquired knowledge—poses a general problem: the education of girls is not only different, but quite restricted in comparison to the education of boys.

Everywhere in the context of education, a choice needed to be made in terms of what subjects should be taught and what readings recommended. In the textbooks that were addressed to women, philosophy was generally ignored, or explicitly discouraged.[6] Yet, the philosophical education of girls was far from a non-existent issue in modern times. It might even be argued that the education of girls became a *topos* in the early modern age, and that Cartesianism

---

4   For the moment, no distinction is made between women and girls, even though from a Cartesian point of view, true education (that of reason) can only begin in adulthood.

5   A late seventeenth-century illustration of this statement can be found, for example, in Locke's *Some Thoughts Concerning Education* (London, 1693): the manual is explicitly about educating a young man and sometimes, very rarely, the author notes this or that prescription which might be suitable for girls, with the clear implication that the main content of the manual does not concern them.

6   See, for example, Fénelon, *Traité de l'éducation des filles* (Paris, 1687), ch. 7: "Will it be by throwing a young girl into the subtleties of philosophy? Nothing could be worse"; "Sera-ce en jetant une jeune fille dans des subtilités de philosophie? Rien n'est si mauvais." Or again, "For children in whom one sees a spirit capable of going further, one can [allow this], without throwing them into a study that smells too much of philosophy"; "Pour les enfants en qui on apercevra un esprit capable d'aller plus loin, on peut, sans les jeter dans une étude qui sente trop la philosophie."

is one of the main reasons for its emergence as such. With Cartesian philosophy, the education of women changed its means and objectives. With this type of education, Cartesian philosophy changed its aims in turn. The idea of a female appropriation of Cartesianism may seem to run up against two glaring difficulties: the apparent absence of any real teaching of Cartesianism in the early modern age, at least in most schools in France, and the impossibility for women to access a philosophical culture.[7] These difficulties, which may be seen as threatening to forestall any investigation before it even begins, should, to the contrary, be seen as added incentives to pursue it. It is precisely the conjunction of the two difficulties that makes it possible to understand the peculiar link between Cartesian philosophy and women's education in the seventeenth century.

2   The Function of Philosophy: Inclusion or Exclusion of Women?

The idea that there existed a female audience for philosophy may seem dubious when one considers the state of the education of women in the early modern age. In fact, education for women and girls was provided at home, by a tutor or in confessional schools run by nuns. In both cases, one was highly unlikely to run into a philosophy teacher.

Yet, women's interest in philosophy was a social phenomenon widespread enough for Molière to write two plays on the subject.[8] He would not have done so if this motive had not been likely to produce in his audience an effect of recognition and familiarity: whether defined as a "precious ridiculous woman" or a "learned woman," the woman in love with wisdom constituted a *topos* enabling her to become a comic figure. Even if they may have been greeted with laughter and ridicule, there were women with an avid interest in philosophy at the time.

The question then is what kind of science and philosophy early modern women wanted to, or could learn about. The university character of scholastic education is not, in itself, an argument for making Cartesianism more attractive to the female public. The fact that the learning of philosophy for women, as for girls, essentially took place in the private sphere does not mean that scholasticism was disdained. A woman could therefore be initiated into Cartesianism as much as to Aristotle, according to the books and tutors at her disposal.

---

7   See Schmaltz, ed., *Receptions of Descartes* for a detailed presentation of this contrasting reception of Cartesianism in France.

8   Molière's *Les précieuses ridicules* dates from 1659, *Les femmes savantes* from 1672.

One may thus find textbooks for educating girls in philosophy that defended the learning of Aristotelianism. In *Les avantages que les femmes peuvent recevoir de la philosophie et principalement de la morale, ou l'Abrégé de cette science* of 1667, Louis de Lesclache (ca. 1620–1671), for example, a rhetorician and popularizer of Aristotle, offers two arguments for teaching Aristotelianism to women. First, he tried to convince men in general, and husbands in particular, that philosophy is a discipline that should be taught to women. More specifically—and this is Lesclache's second point—the study of Aristotelian philosophy should be promoted amongst women as a corrective to Cartesian philosophy, of certain moral and intellectual theses of which he was highly critical. Lesclache's first point clearly shows that the question of teaching philosophy to women was thought to be of importance. The teaching of philosophy was contested in the first place because of its intrinsic emancipatory force. Indeed, Lesclache first addressed husbands: they are the ones who must be convinced that philosophy will not make their wives too "intellectual and talkative" (*raisonneuses et bavardes*), and that they themselves will retain their marital authority. Lesclache does this with humour, noting that if many husbands do not want their wives to learn philosophy, it is because they are afraid that the wives will realize how mediocre their husbands are. Philosophy challenges authority and prejudices about what is appropriate for women and what is appropriate for men. The one pitfall to be avoided, according to Lesclache, is introducing women to Cartesian philosophy:

> Those who accustom their listeners to talk about many things in a problematic way, think that they have found the art of well exercising their minds; but they have found the art of making them despise the sciences, for he who has acquired a habit of doubting believes that there is nothing certain in the sciences, hence he despises them and is not attached to them. This method is very pernicious, because it risks to push those who follow it to fight against the most important truths of Religion and to fall into impiety.[9]

---

9   Lesclache, *Les avantages que les femmes peuvent recevoir de la philosophie* (Paris, 1667), 18: "Ceux qui accoutument leurs auditeurs à discourir problématiquement de plusieurs choses, pensent qu'ils ont trouvé l'art de bien exercer leurs esprits; mais ils ont trouvé celui de leur faire mépriser les sciences, car celui qui a contracté une habitude de douter, croit qu'il n'y a rien de certain dans les Sciences, d'où vient qu'il les méprise et qu'il ne s'y attache pas. Cette méthode est très pernicieuse, car elle met en péril ceux qui la suivent de combattre les vérités les plus importantes de la Religion et de tomber dans l'impiété."

Taken back to its most spectacular operation—the methodical doubt that extends to theological questions such as divine goodness and truthfulness—Cartesianism is totally unsuitable to be recommended for women. In effect, according to Lesclache, Cartesianism tends towards a free examination which extends to the truths of faith and Christian morality. All these different elements are important to a proper understanding of the disqualification of Cartesianism in women's education.

First of all, let us start by addressing the identification of Cartesianism with a philosophy that calls everything into doubt. In order to understand how Descartes was taught at that time, one would first have to decide what makes Descartes's a specific doctrine. This, of course, is not self-evident, but depends on whether one is favourably or unfavourably disposed towards his philosophy.[10] In any case, the answer of an Aristotelian like Lesclache is clear: Cartesianism is a philosophy that calls into question all received truths, both scientific and theological; it is the calling into question of this second category of truths in particular that poses a problem.

Secondly, this incomplete and prejudiced understanding of Cartesianism allows the author to judge that Descartes's philosophy has disastrous intellectual and moral consequences. Intellectual consequences, in the first place, since the habit of doubt and problematic discourse makes Cartesianism tend towards scepticism. Science, having become doubtful and undermined in its entirety, is thereby despised. Morals, in the second place, are undermined also, because uncertainty about what is true can be exported from science to religion. A Cartesian—Lesclache continues—is not that far from a Pyrrhonian or an atheist. It is obvious that, within an educational project that primarily trains girls to become Christian wives, it is necessary to keep them away from this type of philosophy. Moreover, the very title of Lesclache's work indicates the essentially *moral* aim of women's instruction. Philosophy for women is above all moral philosophy: philosophy is "very useful to women since they, just like men, must avoid vice and practise virtue."[11] Once the spectre of Cartesianism has been rejected in the first pages of his work, Lesclache is thus able to lay out the main points of a broadly Christian-Aristotelian philosophy which forms the core of his educational programme.

Lesclache's case shows that, in the early modern age, it was possible to speak in favour of women's access to philosophy, while the philosophy deemed suitable for this purpose could vary. If girls could benefit from the teaching of

---

10   On this question, see the indispensable book edited by Kolesnik-Antoine, ed., *Qu'est-ce qu'être cartésien?*.

11   Lesclache, *Les avantages que les femmes peuvent recevoir de la philosophie*, 22.

Aristotelian philosophy, this would be by means of textbooks or tutors precisely because they could not access it through public schools. It is worth noting that some attribute Lesclache's book to his wife, who might have published it under her husband's name, as did many women at the time in order to contribute to a discipline where they were generally denied legitimacy.[12]

In principle, there is no reason to grant to Cartesianism any advantage over the so-called vulgar philosophy when it comes to addressing women. However, Lesclache's work indicates a reaction to a certain feminine fondness for Descartes's philosophy. Lesclache intended to bring women back to moral philosophy, by turning them away from questions belonging to metaphysics or physics, which (he thinks) are domains not within their competence. Depending on one's point of view, philosophy for women as proposed by Lesclache may thus appear as an opening or as a delimitation. He wanted to include women, to be sure, but he did so by restricting the fields of philosophy to which he believes them to be suited.

Yet, Descartes's philosophy has certain traits that were likely to be attractive to a female public. The first of these traits is sizeable and immediately striking: its language. It is well known that all of Descartes's published works were either originally issued in French, or were translated into French during Descartes's lifetime.[13] Cartesianism was therefore more easily accessible to readers who did not study at school and who did not know Latin. As Descartes himself pointed out:

> If I am writing in French, my native language, rather than Latin, the language of my teachers, it is because I expect that those who use only their natural reason in all its purity will be better judges of my opinions than those who give credence only to the writings of the ancients.[14]

---

12   See Briquet, *Dictionnaire historique des Françaises*, 177. She writes that Madame de Les-clache "lived in the seventeenth century. She composed works of philosophy, which she published under the name of her husband"; "[elle] vécut au XVIIe siècle. Elle composa des ouvrages de philosophie, qu'elle fit paraitre sous le nom de son mari."

13   Thus, if the *Meditationes* and the *Principia philosophiae* were first published in Latin, in 1641 and 1644, respectively, a French translation followed in 1647. Both translations were validated by the author. Descartes writes, for instance, *Principes de la philosophie, Lettre de l'Autheur a celuy qui a traduit le livre*, AT IX-B 1, CSMK I 179: "The version of my Principles which you have taken the trouble to make is so polished and so thorough as to make me hope that the work will be more widely read in French than in Latin, and better understood."

14   Descartes, *Discours de la méthode* I, AT VI 77.

Descartes is here explicitly opposing the traditional type of education he himself had received. The educational stakes are high, for in making a contrast between his own language and that of his preceptors, Descartes also places himself firmly within a new manner of teaching. Moreover, the vernacular language largely renounces the often jargon-like conceptualizations of scholastic philosophies. The clarity and distinction of the concepts offered by Descartes were defended by all Cartesians, and became almost a signature of their movement. When it comes to pedagogy, a philosophy like Descartes's thus lightens the cultural prerequisites for its comprehension. Entering Cartesianism is— according to Descartes—an autonomous decision which requires the exercise of the mind rather than a baggage of knowledge previously acquired.[15] One might recall here Hegel's judgement, according to which the success of Descartes's philosophy ultimately resulted from its simplicity (whether apparent or real):

> The influence that Descartes exerted on his time and on the formation of philosophy in general is mainly due to the free and simple and at the same time popular way in which, leaving aside all presuppositions, he started from popular thought itself and from quite simple propositions.[16]

Of course, it would be naïve to take such assertions at face value. Cartesianism, like any philosophy, requires a difficult and laborious work of the mind. It involves a certain technicality, and it demands a specific understanding of concepts. Still, as a show of great clarity, the intended and stated aim of which is to broaden the audience for philosophy, this claim is quite remarkable in itself.

Finally, the opening quotation of this chapter provides a final piece of evidence for conceiving a strong connection between women and Cartesian philosophy: Descartes himself indicates in effect that he addresses himself equally

---

15   Descartes to Regius, January 1642, AT III 492, CSMK III 206: "I readily admit that since I employ only arguments which are very evident and intelligible to those who merely have common sense, I do not need many foreign terms to make them understood. It thus takes very little time to learn the truth I teach and find that one's mind is satisfied on all the principal difficulties of philosophy." Cf. Ibid., AT III 499: "Je veux bien confesser que, d'autant que je ne me sers que de raisons qui sont très évidentes pour les faire entendre; et ainsi, qu'on peut bien plutôt avoir appris les vérités que j'enseigne, et trouver son esprit satisfait touchant toutes les principales difficultés de la philosophie."

16   Hegel, *Vorlesungen über die Geschichte der Philosophie*, 123: "Die große Wirkung, die Cartesius auf sein Zeitalter und die Bildung der Philosophie überhaupt gehabt hat, liegt vornehmlich darin, auf eine freie und einfache, zugleich populäre Weise mit Hintansetzung aller Voraussetzung von dem populären Gedanken selbst."

to women as to men.[17] The inaugural text of exoteric and mature Cartesianism, properly speaking, the *Discourse on the Method*, aimed to reach a wide audience. Descartes intention that women would be included in his readership was unusual at the time, and especially remarkable, since the *Discourse* is not specifically intended for an exclusively female readership—by contrast with Lesclache's work, for instance. Descartes's *Discourse* was a general work, presenting a method for reforming the mind—potentially, *all* minds—so as to achieve truth. The methodical formation of the mind is thereby coeducational in a novel way, since women are now included among the readers of a philosophical work accompanied by scientific essays, yet without being singled out for special condescension or treatment allegedly better adapted to their gender. This is undoubtedly a key moment for women in terms of the inclusiveness of philosophy. Descartes wrote the *Discourse* with this readership in mind.

3   The Women Promoters of Cartesianism

As is well known, the teaching of Cartesianism was an extremely delicate issue throughout the seventeenth century. By publishing the *Principles of Philosophy* in 1644, Descartes had provided a manual explaining the full features of his thought. This work was intended to become the pedagogical support for the teaching of his doctrine for any teacher or tutor who wished to do so. The very structure of the book, unusual for Descartes, indicates this: its short articles preceded by explicit titles for the reader; the division by philosophical field; the clear attempt at exhaustiveness—all indicate a project of which one of the dimensions is eminently pedagogical.[18] Descartes's textbook was not adopted by French universities, however. When some of Descartes's disciples were appointed in colleges or universities in France and abroad, their teaching

---

17   See the introductory quotation to this chapter from a letter written by Descartes to Vatier, 22 February 1638, AT I 560.

18   The preface contains an appeal to "people of moderate intelligence" who "neglect to study". Descartes insists, *Principes de la philosophie, Préface*, AT IX-B 12–13, CSMK I 185: "I should like to assure those who are over-diffident about their powers that there is nothing in my writings which they are not capable of completely understanding." The pedagogical challenge is explicit: "I would wish to explain here the order which I think we should follow when we aim to instruct ourselves." On Descartes's own intentions concerning the teaching of philosophy in the universities, see Chapter 1, in this volume. On Henry More's teaching—by letter—to Anne Conway, see Chapter 12, in this volume.

was usually contested or even censored.[19] Traditional educational institutions were reluctant to disseminate Cartesian theses during Descartes's lifetime and throughout the latter half of the seventeenth century. The listing of his works on the *Index* in 1663 did little to dispel this reluctance. Cartesianism would thus need to find different channels through which to spread in France.

Even if the present volume shows that Cartesianism found a foothold in the classroom, women had to find other ways to access Descartes's ideas. First and foremost, these were accessed by reading, of course, but also by means of public conferences and private meetings. Unlike colleges and universities, the venues for these meetings and conferences were usually mixed-gender. By attending public meetings in their town, or in the living room of some local personality, members of the bourgeoisie from a variety of professions could learn the rudiments of Cartesianism. The peculiar material conditions of its dissemination rendered the audience of Descartes's philosophy more diverse. And, among these new audiences, there were also women from well-to-do backgrounds. The explicit address to women as potential readers by Descartes was not without considerable effects, and resulted in an especially feminine interest in this 'new philosophy,' as it was often called. Harth could speak, for example, of Descartes's philosophy as a "university without walls" for upper-class seventeenth-century women.[20] Some of these women professed themselves "Cartesians"—like Madame de Grignan, daughter of Madame de Sévigné, for instance[21]—and Molière's *femmes savantes* were said to fall into a trance whenever they heard about *ses tourbillons*, the famous vortices of Descartes's cosmology.[22] But where did they hear about these vortices? In their readings, but also during demonstrations and experiments organized in what would later be called salons. Several of them are reputed to have been run by women defending Cartesianism, such as the salon of Madame de Gueudreville, documented notably by Christiaan Huygens.[23]

---

19   In addition to the well-known Cartesian quarrels such as that in Utrecht, one can give the example of the setbacks encountered by someone like Ambrosius Victor throughout his teaching career. See Pellegrin, "Martin, André."

20   Harth, *Cartesian Women*, 3.

21   See Pellegrin, "Cartesianism and Feminism," 570–571. See also the analyses of Sophie Roux on "the actors" of Cartesianism, showing the difficulty involved in defining them; Roux, "Pour une conception polémique du cartésianisme," 316.

22   Molière, *Les femmes savantes*, III, 2, 882–883: "Descartes pour l'aimant donne fort dans mon sens. / J'aime ses tourbillons.—Moi ses mondes tombants" ("Descartes for the magnet gives strong in my sense / I like its whirlpools.—I like its falling worlds."). The Cartesian physical theory of vortices is set out in Descartes, *Principia* III 65 ff.

23   See Clair, "De Gueudreville interlocuteur de Rohault," 49.

Women are therefore both recipients and promoters of Cartesianism, insofar as they actively contributed to its diffusion. Descartes and women were, in a
certain sense, natural allies: Cartesianism was forbidden in many universities,
just as women were not allowed to attend university. Women's role in spreading
Cartesianism finds its best exemplar in the efforts made throughout her entire
life by Elisabeth of Bohemia, Princess Palatine, to promote Descartes's thought.

The correspondence between the two from 1643 to 1649 is well known as an
invaluable correspondence in the field of moral philosophy, and in the underlying psychophysiology. It is also well known that Descartes dedicated the *Principles of Philosophy* to the princess. A dedication to a woman of high rank was
not unprecedented at the time, nor were the praises with which he celebrated
the dedicatee. His homage is striking, however, because it singles out for praise
the intellectual qualities with which Descartes judges Elisabeth of Bohemia to
be endowed. These were what made her unique. Descartes claimed that Elisabeth of Bohemia was special insofar as she had an excellent mind for both
metaphysical and mathematical subjects, and because she possessed both an
extraordinary understanding and an extraordinary will.[24] Owing to these qualities, Descartes concluded that Elisabeth was the only one who had perfectly
understood his entire system.[25]

With this description, Elizabeth of Bohemia is raised to the rank of a perfect
and accomplished disciple of Descartes. The historiography of Descartes has
long tended to adopt this perspective, making of Elisabeth the ideal pupil for
the emergence of a pedagogically excellent master, as evidenced in this assertion by Ernst Cassirer, for example:

> It was only through his friendship with Princess Elisabeth Palatine that
> Descartes became a master not only in the art of thinking, but also in the
> art of teaching. It was then that he discovered not only the profession, but
> also the passion for teaching.[26]

The essays collected in the book *Elisabeth de Bohême face à Descartes: deux
philosophes?* show a different reality.[27] Elisabeth was not a disciple of Descartes.
Moreover, she cannot be said strictly speaking to be a Cartesian, even if this

---

24   Descartes, *Principes de la philosophie, Epistre*, AT IX-B 23, CSMK I 192.
25   Descartes, *Principes de la philosophie, Epistre*, AT IX-B 22, CSMK I 192: "You are the only
      person I have so far found who has completely understood all my previously published
      works."
26   Cassirer, *Descartes*, 85.
27   Kolesnik-Antoine and Pellegrin, *Elisabeth de Bohême face à Descartes*.

does not prevent her from playing an essential role in the diffusion of Descartes's philosophy.[28] To begin with, during the peregrinations that marked her life as a princess in exile, Elisabeth noticed the resistance to Cartesian philosophy in the philosophical milieus in which she moved. This was undoubtedly the main stumbling block to any teaching of Descartes's philosophy: masters of philosophy usually having neither the desire nor the openness of mind to consider turning to a new doctrine. Addressing herself to Descartes, Elisabeth explicitly contrasted her will to learn with the intellectual laziness of the scholars she had met:

> It is to this will that I owe the intelligence of your works, which are obscure only to those who examine them by the principles of Aristotle, or with very little care, as the most reasonable of our doctors in this country have confessed to me that they do not study them, because they are too old to begin a new method, having used the strength of body and mind in the old one.[29]

She lamented the lack of knowledge of Cartesianism in Germany, where all philosophers seemed to be in thrall to scholasticism and, therefore, to pedantry.[30] She noticed with dismay how very few school philosophers were willing to restart their education from scratch in adulthood, as recommended by Descartes in both the *Discourse on Method* and in the *Meditations on First Philosophy*. The situation was quite different for women, most of whom had not benefitted from this type of education during adolescence.[31]

Elisabeth of Bohemia did her best to promote the teaching of Descartes's philosophy. There is substantial evidence of her desire to spread Cartesian philosophy wherever she resided, both in Holland and in Germany. For instance, Giulia Belgioioso has reconstructed the details of a Cartesian circle that became established around Elisabeth in The Hague, a circle which spread

---

28   Not to mention her role in the formation of the moral side of Cartesian philosophy, which is not elaborated upon here because it is beyond the scope of the current argument.

29   Elisabeth to Descartes, 1 August 1644, AT IV 132.

30   Elisabeth to Descartes, December 1646, AT IV 579–580, Shapiro, *Correspondence*, 152: "There is no one else here who is reasonable enough to understand them [Descartes's works], even though I have promised this old duke of Brunswick, who is at Wolfenbuttel, to give them to him in order to adorn his library. I do not believe that he will use them to adorn his feeble brain, as it is already throughout occupied by pedantry." On Descartes's own consideration of 'pedantry,' see Chapter 1, in this volume.

31   Though, of course, this was not at all the case for Elisabeth of Bohemia, who was educated in many languages and sciences from an early age.

within the Illustrious School of Breda, founded by the Prince of Orange.[32] But one can also recall the significant information collected by Sabrina Ebbersmeyer concerning the private lessons on Cartesianism that Elisabeth very likely gave in Heidelberg.[33] We know from a letter from the mathematician Joachim Jungius (1587–1657) that a group of students who were reading the *Principles of Philosophy* in the city had benefitted from Elisabeth's instruction on the subject. This diffusion is also attested to by Elisabeth herself allowing the circulation in various learned circles, notably the English Royal Society, of two handwritten letters addressed to her by Descartes on mathematics (before these were made available in the Clerselier edition of 1667). Ebbersmeyer adds that Elisabeth seems to have done the same for certain letters on the passions. When her brother restored the University of Heidelberg in 1652, she advised him on the choice of teachers.[34] Finally, when Elisabeth was appointed abbess of a Lutheran community in Herford in 1667, she made the place a refuge for heterodox thought, and a haven for intellectual exchanges of all kinds. According to Descartes's biographer, Adrien Baillet:

> She made the Abbey a philosophical Academy for all kinds of people of spirit and literature, without distinction of sex or even religion [...]. It was enough to be admitted that one was a philosopher, and above all a lover of the Philosophy of M. Descartes [...]. This Abbey was considered to be one of the first Cartesian schools.[35]

One can undoubtedly wonder about this Cartesian academy in Hervord. At that time, Elisabeth's intellectual preoccupations took a more spiritual and theological turn.[36] But the importance of Cartesian thought remains. In his *Vie du*

---

32   Belgioioso, "Descartes, Elisabeth et le cercle cartésien de La Hague," 13–44. On the University of Breda and Cartesianism, see Chapter 11, in this volume.

33   Ebbersmeyer, "Elisabeth of Bohemia."

34   Ebbersmeyer, for example, names the German philologist and historian Johannes Freinsheim (1608–1660), who was also acquainted with Descartes during his stay in Sweden. He dedicated his *Orationes* (Frankfurt, 1655) to Elisabeth. And one year later, he became councillor of the electorate and honorary professor at the University.

35   Baillet, *Vie de Descartes*, vol. 2, 235: "Elle fit de l'Abbaye une Académie philosophique pour toutes sortes de personnes d'esprit et de Lettres, sans distinction de sexe ni même de Religion [...]. C'était assez pour y être admis que l'on fut philosophe, et surtout amateur de la philosophie de M. Descartes. [...]. Cette abbaye fut considérée comme une des premières écoles Cartésiennes."

36   This description of Baillet, behind its somewhat surprising and excessive aspect, is revealing: Elisabeth has always been an essential element of the European republic of letters. She has always actively contributed to the intellectual exchanges between all participants

père Malebranche, father André specifies that at the time at the end of her life, when Elisabeth begins to exchange letters with Malebranche, she testified that "since M. Descartes, she had not read anything so beautiful."[37]

Elisabeth's constant commitment to the spreading of Cartesianism is now well documented. This dissemination took various forms, all of which have to do with pedagogical and educational activities: for instance, research networks based on Descartes's mathematical or moral works, private courses in Cartesianism, the appointment of professors with knowledge of this philosophy, and the transformation of an abbey into an intellectual centre favourable to Cartesians in particular. This case—which is certainly exemplary, because Elisabeth is not just any woman initiated in Descartes's philosophy—shows that Descartes's teaching took many different paths at that time, to which women notably contributed.

4   Stasimaque and Eulalie: Common Sense as a Philosophical Guide

The educational project of the Cartesian feminist philosopher François Poulain de la Barre (1648–1723) could be seen as another part of the tradition already represented by Lesclache. The latter wanted to persuade fiancés and husbands that a woman who practises philosophy is not a bad wife. In *De l'éducation des dames*, his second feminist work dating from 1674, Poulain de la Barre brings four characters into dialogue: Stasimaque, a convinced Cartesian; Timandre, an "honest man" often steeped in prejudice; Sophie, an accomplished scholar; and Eulalie, a young woman anxious to learn.[38] In these dialogues, Timandre starts from premises quite similar to those of Lesclache, recalling that "learned ladies are in such bad odour that it would be a severe punishment to an eligible man to give him one as a wife."[39] The Cartesian perspective adopted in this classical setting, however, totally changes the discourse on teaching philosophy to women. In his scathing judgement of women scholars, Timandre is merely mouthing a cliché of which Poulain de la Barre wanted to get rid.

---

in this arena. See Parageau, *Les ruses de l'ignorance*, 150, where the author discusses an "eclectic approach," with reference to Elisabeth of Bohemia.

37   Malebranche, *Oeuvres*, vol. 18, 131.

38   De la Barre, *De l'éducation des dames* (Paris, 1674), ed. Pellegrin, 159. Even if the term is, of course, anachronistic, we use the term "feminist" in a broad sense, to define any argument in favour of the intellectual, social and political emancipation of women.

39   De la Barre, *De l'éducation des dames*, ed. Pellegrin, 163, transl. Welch and Bosley, 144.

In this book on women's education, Sophie and Stasimaque guide the debates, which advocate the Cartesian method and principles for education. When she is asked to suggest some books to Eulalie, we find Sophie recommending the reading of Descartes, as follows:

> If you wish to read Descartes' *Principles*, and the first volume of his letters written to the Queen of Sweden and the Princess of Bohemia, that is even better. You will realize from these letters that he did not consider women incapable of the highest sciences.[40]

Descartes's textbook is thus to be recommended, alongside his correspondence with these two women. There was indeed a Cartesian pedagogy intended for women: his teaching in general, as presented in the *Principles of Philosophy*. This book was intended for schools, but from these schools women were, of course, excluded. Nevertheless, for a 'second-generation' Cartesian like Poulain de la Barre, this manual, translated into French, promised to become the official manual of a pedagogical Cartesianism for women.

But no less important were Descartes's personal exchanges with women. Indeed, these offer concrete evidence to suggest that, for Descartes, women's minds could be as excellent as those of men. As examples, they are encouraging, but Poulain de la Barre insists on the fact that there have always been women scholars in all areas. Indeed, knowledge is a natural and spontaneous impulse in every human being, whether or not they are allowed to study.

Recognizing the value of the native capacity of the mind is the touchstone of reasoning: it is a sufficient argument to explain the right and the need to educate women. Human nature is made in such a way as to favour the development of all the faculties that nature gives it; this starts with the development of the mind. The famous opening sentence of the *Discourse on the Method* already contains an educational programme that seems to hold out the promise for the emancipation of women:

> Good sense is the best distributed thing in the world [...] the power of judging well and of distinguishing the true from the false—which is what we properly call 'good sense' or 'reason' is naturally equal in all men.[41]

This equality in intellectual capacity extends the possibilities of philosophy to everyone. According to Poulain de la Barre, the "jurisdiction of common sense"

---

40   De la Barre, *De l'éducation des dames*, ed. Pellegrin, 272, transl. Welch and Bosley, 237–238.
41   Descartes, *Discours de la méthode* I, AT VI 1–2, CSMK I 111.

must concern all fields.[42] This possibility is more likely to become an effective reality for girls, the author explains, because boys have in fact the immense disadvantage of being educated in prejudices and vulgar philosophy from an early age.[43] Their spirit is therefore corrupted. Girls, by contrast, having been neglected by the traditional educational programmes, keep this original common sense in its authentic, pristine state.

Girls' education must therefore not be modelled on that of boys. If girls were put in boys' schools, they would lose their native intellectual qualities.[44] Moreover, when comparing girls and boys, there are remarkable intellectual differences, which Poulain de la Barre is reluctant to take for "natural". Quite to the contrary, he insisted that these differences are linked to the different ways in which they are treated and are therefore learned differences according sex. With little or no education, girls are better at understanding what is hidden from them. They are forced into a form of self-education which keeps their common sense alive and reveals the natural character of wisdom.[45]

It should be noted that, in the same way that Poulain de la Barre pits girls against boys, he pits scholars against women in general.[46] The criterion remains the same—common sense:

Not only do we find that a lot of women judge things as well as if they had had a better education, without either the prejudices or the confusion of ideas endemic among learned men, but there are many who have

42   De la Barre, *De l'éducation des dames*, ed. Pellegrin, 273, transl. Welch and Bosley, 238.

43   In the *Préface* to the *Principes de la philosophie* (AT IX-B 12–13, CSMK I 185) Descartes refers to "those who have studied bad sciences the most [...] are the greatest victims" of "preconceived opinions."

44   De la Barre, *De l'éducation des dames*, ed. Pellegrin, 165, 166, transl. Welch and Bosley, 146–147: Timandre says: "It would not do women much good to become as clever as men. On the contrary [...], it would be disastrous for them to be given the opportunity [...]. You know better than I what changes meditation and study can bring about in the student"; Stasimaque adds: "Your fear would be justified [...] if ladies were taught science in the way it's taught in our Schools, but if they follow the method laid out in the second part of the book you mentioned, they will get along fine".

45   In Descartes' *La recherche de la verité*, AT X 506, Eudoxe evokes "a good mind, even one brought up in a desert and never illuminated by any light but the light of nature could not have opinions different from ours if he carefully weighed all the same reasons."

46   We can draw a parallel between the opposition Poulain de la Barre sets up between women and scholars and that set up by Descartes (and by Elisabeth herself) between Elisabeth of Bohemia and the other scholars: see Pellegrin, "Feminine Body."

such wise good sense that they can speak on scientific topics as if they had always studied them.[47]

Common sense alone can therefore compensate for all want of learning, even specialized and technical learning. The operation of judgement rests entirely on common sense and, as such, it is best preserved from school prejudices. Two groups are opposed to each other here: uneducated women and scholars; and it is clear that the science of the latter has resulted in their loss of the ability to use common sense. Poulain de la Barre here takes up the Cartesian motif of minds vitiated by bad education as opposed to minds full of common sense.[48] The novelty of Poulain de la Barre's account is in his imposition of gendered characteristics on this opposition between these two types of mind. Some of Descartes's remarks on Elizabeth of Bohemia or Christine of Sweden already went in this direction: highlighting a few women who served as a counter-example to the mass of ordinary scholars. In the case of Poulain de la Barre, this is an opposition between two clearly discernible groups: those whose "good sense" is still intact, and those whose common sense has been corrupted by inappropriate or formulaic studies. The boys brought up with scholastic principles show signs that received education not only fails to form them but rather, in fact, deforms them. Poulain de la Barre explains that the exercise of common sense, by contrast, makes it possible to participate in real science, even without prior instruction. Importantly, there is a double parallel at work here (between boys and girls and between women and scholars): once again, Cartesian education privileges adulthood over childhood, and Poulain de la Barre will draw all of the attendant conclusions from this. His education manual is aimed at women rather than girls.

5   Women, Girls, Mistresses and Disciples

An education that is based on the development of common sense is an education that can only take place at a certain age, when the mind is free from traditional outside influences: nannies, parents and other badly educated tutors. Descartes argued for the need of a significant degree of intellectual autonomy

---

47   De la Barre, *De l'égalité des deux sexes* (Paris, 1673), ed. Pellegrin, 77, transl. Welch and Bosley, 65.

48   "Among those who have studied whatever has been called philosophy up till now, those who have learnt the least are the most capable of learning true philosophy." *Principes de la philosophie, Préface*, AT IX-B 8–9, CSMK I 183.

in order to begin to reason properly on one's own. Childhood, according to Descartes, is a time of training that one does not master and to which one must therefore passively submit. This is a remarkable feature of the Cartesian educational project, to the extent that it is not primarily pedagogical in the etymological sense of the term, i.e., it is not primarily intended to guide children. This brings us to the distinction between girls and women. This Cartesian theme undoubtedly explains Poulain de la Barre's choices in this field. In the first place, it is necessary to educate adults. Moreover, the title of his work *De l'éducation des dames* (unlike, for example, François Fénelon's *Treatise on the Education of Girls* in 1680) refers to women and not to girls.

Poulain de la Barre's educational project thus encompasses two stages: a first stage, during which some adults become aware of the need to free themselves from prejudices and received learning; and a second stage, during which they turn to teach to others what they have learnt. There are therefore two distinct educational plans, which correspond to two different chronological moments. Ideally, they should be linked together:

> Lest the reader be surprised that we are speaking here only of women who have reached the age of discretion—although by the word "education" we normally mean the art of raising and instructing children—we should note that we can attribute to the word a broader meaning than we ordinarily do, and by showing what we should learn at a more advanced age, we show at the same time what can be taught to us in childhood. And since education depends largely upon those who are responsible for it, and there are few who know what they should know in order to be able to teach well, we thought it best to begin with what is useful for forming mistresses, before saying what we think of the method of forming the minds of the disciples according to the maxims we are proposing.[49]

---

49   De la Barre, *De l'éducation des dames*, ed. Pellegrin, 157, transl. Welch and Bosley, 141: "Pour n'être pas surpris de voir que l'on ne parle ici que des femmes qui sont en état de discernement, quoique par le mot d'Éducation l'on entende communément l'Art d'élever et d'instruire les enfants, il faut considérer que l'on peut donner à ce mot une signification un peu plus étendue que l'ordinaire, et qu'en marquant ce que nous devons apprendre dans un âge avancé, on marque en même temps ce que l'on peut nous enseigner dès l'enfance. Et comme l'Éducation dépend principalement de ceux qui en ont le soin, et qu'il y en a peu qui sachent ce qu'il faut savoir pour s'en bien acquitter, ou qui le sachent de la manière qu'il le faut, l'on a jugé à propos de commencer par ce qui peut servir à former les Maîtresses, avant que de dire ce que l'on pense sur la méthode de former l'esprit des disciples suivant les maximes que nous proposons."

In the education and family system, from a Cartesian perspective, it is easier to reform the mind in adulthood than it is when the student is still a child. Reason may well be native, but it is only at the end of childhood that one finds oneself "in a state of discernment," all the more so when the first teachers are almost always bad masters.

The first task of Cartesian education is to train adults. Only at a later stage will the educators take on responsibility for teaching children. They are first and foremost those whom the child trusts and emulates on the point of being converted to Descartes's science and method. The principles of education—as Poulain de la Barre makes clear in the excerpt quoted above—are not fundamentally different whether one is an adult or a child. If educational reform has to take place, however, it necessarily begins with teachers. Without this there will be no new kinds of pupils.

But who will train these teachers? Are masters needed to train the masters? The risk of falling into an infinite regression here is avoided by Poulain de la Barre's work itself, which is intended to provide the training that educators need. Its dialogical character has a pedagogical significance in its own right, like that which one can also see at work in Descartes's own dialogue *The Search for Truth*. As mentioned, Poulain de la Barre's two characters, Sophie and Stasimaque, are perfectly wise. They are responsible for initiating the other two, Eulalie and Timandre, into Cartesianism, which is the simplest and most natural way to wisdom. The book itself contains the model of an educational setting that can be adopted and followed by its readers. The regression stops there, or more precisely at the previous generation. Stasimaque learned Cartesianism at the source, that is to say, during conferences by convinced Cartesians. Having taken place in venues of knowledge alternative to the universities, he claims that these lectures have made him despise the university and scholasticism. In all rigour, therefore, we should say the following things: that Descartes had many disciples (we can think here, in particular, of all those who are classified under the category of "petits Cartésiens"); that these disciples taught Cartesianism as best they could;[50] and that among their audience was Stasimaque—the character in the dialogue representing Poulain de la Barre himself—who had assigned to himself the mission of writing a book for initiating a female following into a true understanding of Cartesianism. Once initiated, these women would train students in the new philosophy. It would take three generations of this kind of knowledge transfer before an effective education of the women could be realized, and four before an effective education of the girls was possible.

---

50   As is shown in many of the other contributions in this book.

In addition to the Cartesian impetus, there is another reason for giving priority to adult education over that of children; a very concrete one, in fact, which shows that we are confronted here with wide-ranging pedagogical considerations in the early modern age. The project of educating women was taking place in a society that was reluctant to admit—or was even openly hostile to—their intellectual emancipation. If this project was to be effective, it had to propose a realistic plan of action. Poulain de la Barre, therefore, favours a tactic of progressive diffusion of light and knowledge starting from only a few *enlightened* persons: it will be enough that "two or three highly placed people should give their daughters an education, and for that a well-known convent and an individual teacher would have set an example."[51] The new educators will therefore spread the principles of a new teaching in the dual setting of some family homes and convents (the two places where girls are educated).[52]

These two venues underline the fact that these educators of young women are always necessarily in fact themselves women. This aspect of Poulain de la Barre's educational project is important: the education of girls will not take place in a context of mixed education. Not, at least, in its early stages. The teachers will be women; all classmates will be girls.[53] The reasons given for this choice are largely pragmatic. Poulain de la Barre considers that men will resist coeducational learning as much as possible. They will not be able to bear shar-

---

51   De la Barre, *De l'éducation des dames*, ed. Pellegrin, 178, transl. Welch and Bosley, 156: "[que] deux ou trois personnes de qualité et de considération fissent bien instruire leurs filles, et ce serait assez qu'une maison notable de Religieuses et quelque maitresse particulière commençât."

52   De la Barre, *De l'éducation des dames*, ed. Pellegrin, 177–178, transl. Welch and Bosley, 156: "How can we get an education without going to college," persisted Eulalia. "The same way as most men, who never go there either," replied Stasimachus. "Whatever you want to learn can be taught by tutors, the way writing and dancing are". "There something else that could be done that would be much more convenient," interjected Timander. "Given that there are already mistresses perfectly versed in various subjects, they could teach girls, who in turn would be trained as teachers, just as our male teachers are trained in universities and other places." One can think here of the educational work carried out by Jacqueline Pascal in Port Royal, or that of Madame de Maintenon for the girls' school of Saint Cyr.

53   Poulain de la Barre insists in his first treatise that women are fully capable of teaching: "the easiest and most natural public use of what one has mastered thoroughly is to teach it to others, and if women had studied at a university with men, or at other universities set up exclusively for them, they could receive degrees and become Doctors or Masters of Theology or Medicine or both kinds of Law. Moreover, the natural gifts which make them so good at learning would also make them excellent teachers." De la Barre, *De l'éducation des dames*, ed. Pellegrin 178, transl. Welch and Bosley, 156.

ing knowledge with girls. It is necessary therefore to educate girls separately and protect them from their male classmates. Moreover, the desire to disseminate or not to disseminate one's knowledge is a mark, respectively of its veracity or falsity. Knowledge that one does not want to share is not real knowledge, Poulain de la Barre argues, insofar as it must contain something obscure and even aberrant that we do not wish to reveal. What is at stake is not a question of emancipation and harmony: it is a question of power. Knowledge, especially abstruse and reserved for the initiated, is an exclusive property, separating its holders from the uninitiated from whom the holders intend to keep it jealously guarded. Authentic wisdom, by contrast, contains an educational dimension in itself: the strongest proof that a person is wise is when that person knows how to make known to others what she or he knows.[54] The ability to spread this *enlightenment* is the very criterion of true science. Now this scientist carries out his or her pedagogical work in a perfectly Cartesian way, since she or he "brings clear and distinct ideas into our minds."[55]

As for the content of this teaching, this too is essentially Cartesian because "from excellent knowledge, Descartes undertook the discovery, and succeeded in it with such happiness, that it seems that he left to those who came after him only the task of studying it." Notwithstanding the nuances he adds to it immediately thereafter,[56] this judgement of Stasimaque shows the substance of the teaching advocated by Poulain de la Barre: "few principles" and to "know oneself perfectly"—these are what Cartesianism allows. Doing philosophy requires, first and foremost, the study of Descartes. But by studying Descartes, the student acquires a science of himself—and of herself. Descartes's philosophy can be seen as encompassing all the more specific sciences and, even more importantly, as teaching who we are and what we must do. Elsewhere, I have already described the detail of the order of sciences advocated by Poulain de la Barre, and noted the specificities of his Cartesianism, with its minimal focus on metaphysics but directed instead towards action, morals and

---

54　De la Barre, *De l'éducation des dames*, ed. Pellegrin 168, transl. Welch and Bosley 148: "'My greater satisfaction comes from learning about things and communicating them to other people', said Sophia".

55　De la Barre, *De l'éducation des dames*, ed. Pellegrin, 191, transl. Welch and Bosley, 167: "qui porte dans l'Esprit des idées claires et distinctes."

56　De la Barre, *De l'éducation des dames*, ed. Pellegrin, 278, transl. Welch and Bosley, 243: "But please note that I am not claiming Descartes is infallible or that everything he proposed is true and unproblematic, or that one has to follow him blindly, or that others couldn't find something as good or even better than he has left us. All I am saying is that I believe him to be one of the most reasonable philosophers we have, whose method is the most universal and the most natural."

politics.[57] While Poulain de la Barre's philosophy for women might thus seem to concern morality above all—just as was the case for Lesclache—in reality, their two philosophies are quite different. For Lesclache, moral philosophy is directly linked to the future role of daughters as wife and mother. According to Poulain de la Barre, on the contrary, moral philosophy aims at self-determination of judgement and, as such, it is vital for men and women alike. His argument is that philosophy—by which he meant Descartes's philosophy—emancipates women, insofar as it gives them the power to claim a place in the professional sphere, in the same way as men. Philosophy therefore does not aim to improve domestic life, but to extricate oneself from it.[58] Philosophy, for Poulain de la Barre, is thus both an intellectual and a professional training, and is the only means for revealing to women that they are suitable for all jobs.[59] The science of oneself, acquired by philosophy, proves that there is no basis for inequality of any kind between women and men.

6   Conclusion

In early modern education textbooks, girls were always treated separately from boys, and in contrast to boys. There do not appear to be treatises on mixed forms of education at the time. At first glance, Poulain de la Barre's work is not exceptional in this regard, since it seems to be exclusively devoted to girls and always to contrast their experience with that of boys. However, this impression is misleading. First of all, if there is any difference, it is in highlighting the short-comings of boys by comparison with girls. The latter are more generally better endowed with the Cartesian qualities of the mind in its natural state, that is to

---

57   See Pellegrin, "La science parfaite," 377–392. The programme of instruction of women as
       a science of oneself is detailed by showing the gaps in relation to Cartesian science.

58   De la Barre, *De l'égalité des deux sexes*, ed. Pellegrin, 121, transl. Welch and Bosley, 100.
       Again, it should be noted that Poulain de la Barre does not underestimate the difficulties
       of thinking about educating women to participate in public life. He knows that they are
       currently denied access to most jobs: "They would go into business, husbands could hardly
       get out of leaving to them the running of the household and of following their advice."

59   In *De l'égalité des deux sexes*, ed. Pellegrin, 116 ff., transl. Welch and Bosley, 95 ff., Poulain de
       la Barre lists all the professions to which women can have access. He demonstrates why
       they are able to do so, profession by profession, in a specific part of the second part of the
       book entitled "That women are no less capable than men of state positions." They can thus
       be army generals, ministers in the Church, or queens (a function which he distinguishes
       from that of regent). It should be noted that the women with whom Descartes philosoph-
       izes are not unemployed but hold eminent political posts: one, Elizabeth of Bohemia, is a
       princess (then abbess); the other, Christine of Sweden, is a queen.

say, gifted with common sense and capable of good judgement. Secondly, boys are not in fact excluded from this education programme. They must follow the same principles as those presented by Poulain de la Barre in his manual for women: but while for girls it is a question of education, for boys, it is a question of re-education. Starting with women and extending to men, for the first time we are presented with an education manual that addresses both equally in an explicit and philosophically-founded way:

> This is why they [these writings] have been given the title, *On the Education of Women*, although they are no less useful for men, just as books intended for men are also used by women, there being but one way of teaching both, since they are of the same species.[60]

Acknowledgments

I warmly thank Mattia Mantovani for the translation of this paper.

60   De la Barre, *De l'éducation des dames*, ed. Pellegrin, 157, transl. Welch and Bosley, 238.

CHAPTER 19

Rohault's Private Lessons on Cosmology

Mihnea Dobre

1 Introduction

The cosmology of the French Cartesian Jacques Rohault (1618–1672) is peculiar. It is built on core Cartesian elements (i.e., vortices), yet the name René Descartes (1596–1650) is nowhere mentioned. It includes a comparison between the traditional Ptolemaic system and the Copernican one, with the explicit endorsement of the latter. However, manuscript material derived from Rohault's lectures in the 1660s reveals a greater emphasis on the traditional account, which might be linked to his activity as a private teacher. This chapter examines the development of Rohault's thought and aims to contextualize the cosmological explanation presented in his *Traité de physique* (1671).

To get a more accurate picture of such development, I compare the manuscript sources with Rohault's published work, but I also take into account indirect evidence related to his pedagogy. For example, several entries in the Accounts of the House of Conti for the years 1668 and 1669 refer to Rohault.[1] In 1668 and 1669, Rohault received three payments for his services to the famous Conti family.[2] Two of these were made for the mathematical lessons offered in four-month batches each year. The other payment was for two globes, and the book records date it to 1668. These must have been the famous—and widespread—armillary spheres depicting both celestial and terrestrial globes. The inventory list made after Rohault's death includes a reference to "Item deux Globes L'un Terrestre et L'autre Celeste," so it would seem that Rohault was devoting time and resources to delivering lessons in cosmology.[3] Rohault's Conti connection and the existence of the two globes raise more problems than they solve, however, as it is unclear what he was doing with these items and how he may have used them in his private lessons. On the one hand, we lack information about Rohault's private lessons (we only know that such lessons

---

1 See McClaughlin, "Rohault and the Natural Sciences," 34.
2 The episode is discussed in McClaughlin, "Rohault and the Natural Sciences," 34. For an overview of Rohault's pedagogy, see Clair, *Bio-Bibliographie*, especially ch. XI ("Préceptorats"), 40–42.
3 For the inventory made after Rohault's death, see McClaughlin and Picolet, "Un exemple," 17.

took place); on the other, the episode predates the publication of his philosophy. Therefore, we need to look more closely at his wider engagement with cosmological issues in both his published work and the manuscript sources.

Rohault was a successful Cartesian experimentalist who offered public conferences on a variety of natural philosophical topics.[4] In private, he had been teaching mathematics since the mid-1640s. He published little and only later in life, but this did not impede an unofficial dissemination of his public lectures through manuscripts circulating in Paris in the 1660s. For modern scholars, there is something curious about the generally good reception of Rohault's writings, the timing of their publication, his Cartesian commitments, and his intellectual development during the 1660s.[5] By working through the questions about the content and development of his cosmology, we can shed some light on these bigger conundrums, and ultimately return to our starting point concerning the Conti connection: what was Rohault likely teaching about cosmology in private and why?

Only in 1671 did Rohault publish the *Traité de physique*, a comprehensive volume on general natural philosophy (Part I), cosmology (Part II), the terrestrial and meteorological phenomena (Part III), and living bodies (Part IV).[6] The general outlook of the book is Cartesian, and most of the scholarly attention has been devoted to the study of the first part of the treatise.[7] The current chapter aims to examine the second part of the book, but instead of revisiting some of the traditional Cartesian themes—e.g., the vortex theory—I will focus on the mix between traditional cosmology and Rohault's own explanation of celestial phenomena, including, most notably, the comet of 1664–1665. I compare the structure and content of the *Traité* with two manuscript versions of Rohault's public conferences: the Ms. 2225 of the Bibliothèque Sainte-Geneviève in Paris and the *Physique nouvelle (1667)* recently published

---

4 See Clair, *Bio-Bibliographie*; McClaughlin, "Descartes, Experiments"; Roux, "Was There a Cartesian Experimentalism?"; Dobre, "Rohault's Cartesian Physics"; Dobre, "Jacques Rohault."

5 See, for example, Ariew, *Descartes and the First Cartesians*; Des Chene, "Cartesian Science"; Dobre, "Rohault and Cartesian Experimentalism"; Dobre, "Rohault's Cartesian Physics"; Dobre, "Rohault's Mathematical Physics"; Hoskin, "Mining All Within"; McClaughlin, "Was There an Empirical Movement?"; Roux, "Was There a Cartesian Experimentalism?"; Schmaltz, *Early Modern Cartesianisms*; Schüller, "Clarke's Annotations"; Spink, "The Experimental Physics."

6 See Rohault, *Traité de physique* (Paris, 1671). The treatise was reprinted numerous times in the early modern period, in French, but also in Latin and English translations. For modern editions of the book, see Rohault, *A System of Natural Philosophy*; Rohault, *System of Natural Philosophy*; Rohault, *Traité de physique*.

7 For the discussion of Rohault's treatise in the secondary literature, see the list of studies provided in the bibliography.

by Sylvain Matton.[8] A word of warning is perhaps in order here: each of these two sources is difficult to date, such that an accurate chronology might be difficult to achieve. However, in the current study, what I want to suggest is that even an attempt to establish a more precise record for the development of Rohault's cosmological views can enrich our understanding of his natural philosophy. This approach allows for an open-ended answer to our particular question about Rohault's private lessons and broadens the perspective on his Cartesianism.

2   A Biographical Introduction

Rohault is rightly considered to be one of the most significant Cartesians of the early modern period.[9] He is presented as a promoter of Cartesianism, both through his public conferences in Paris and through the social network coordinated by his father-in-law, Claude Clerselier (1614–1684).[10] This is not the place in which to enter into extensive details about his life, but in order to develop an argument about his cosmology, I need to highlight several biographical elements, as well as to outline his philosophical views.

Jacques Rohault was born in 1618 into a family of merchants in Amiens. He began his studies in Amiens, but completed them in Paris. As of 1644, he was known as a "professeur de mathématiques" in Paris.[11] Unfortunately, details about his early life are difficult to reconstruct due to the lack of sources and only scattered information is available from administrative documents. One can only speculate about the possible connections between Rohault and the libertine thinkers of the period, as well as about his private teaching. For example, he was a close friend of Cyrano de Bergerac (1619–1655) and Molière (1622–

---

8   The manuscripts referred to here are Rohault, MS., Paris (Bibliothèque Sainte-Geneviève), MS. 2225: Rohault, *Conférences sur la physique, faites en 1660–1661, par Jacques Rohault, et recueillies par un de ses auditeurs qui, dit-il, y a ajouté du sien* (1660); Rohault, *Physique nouvelle* (1667).

9   There seems to be a general agreement between scholars on this point; see, for example Balz, "Clerselier and Rohault"; Mouy, *Le développement*; Clair, *Bio-Bibliographie*; Mazauric, "Préface."

10   See McClaughlin, "Rohault and the Natural Sciences"; Clair, *Bio-Bibliographie*. Most of the details presented below about Rohault's life follow Clair.

11   See the documents in MS., Paris (Archives Nationales, Châtelet de Paris), MS. Y//184-Y//187: *Insinuations* (15 October 1644–2 March 1651): "Jacques Rohault, professeur de mathématiques" (9 October 1644), "professeur dès sciences de mathématiques demeurant à Paris" (10 February 1646).

1673).[12] Notarial documents present him as a "professor of mathematics" in Paris, but without further qualifications. This means there is no evidence about the status (and the number) of his students, but a safe conjecture about his early practice can be derived from reports about the audiences of his public lectures. These audiences were formed of people of all sorts—social and intellectual status, gender, and nationality—willing to pay a fee, which was collected by Rohault's first wife, Nicole Fillassier.[13] By the mid-1650s, Rohault's name starts to appear as a speaker in other Parisian gatherings, such as the famous conferences at the residence of Habert de Montmor (1600–1679). Around 1660, he offered lectures at his own house—the "Wednesday" conferences (Les Mercredies)—but he continued to entertain the visitors of other salons, as at those hosted by Montmor and Melchisédech Thévenot (ca. 1620–1692).[14] The early 1660s seem to have been his most productive years, at least with respect to his public conferences.[15] The correspondence of Christiaan Huygens (1629–1695) from this period reveals the interest of the Dutchman in committing to the construction of a pneumatic device for Rohault and in discussing experimental results with him. When a new Academy was established by the French King in 1663, Henry Oldenburg (ca. 1619–1677) listed Rohault among the figures who were not invited to join the institution.[16]

The following year (1664) marks the transformation of Rohault, the public lecturer and professor of mathematics, into one of the chief defenders of Cartesianism. He prepared a brief lecture that was printed (anonymously) together with Descartes's *The World, or Treatise on the Light*—Rohault lectured on that topic in one of the meetings held by Montmor—and he married

---

12 Such friendship is indicated, for example, in the preface by Henry Le Bret (1618–1710) to Cyrano de Bergerac, *Histoire comique, contenant les états et empires de la Lune* (Paris, 1657). For the connection between Rohault and Cyrano, see especially Alcover, *Cyrano*. For Rohault's relation to literary and libertine figures in early modern Paris, see Milhaud, "Rohault et Molière"; McClaughlin, "Quelques mots sur Rohault et Molière"; McClaughlin, "Rohault and the Natural Sciences"; Clair, *Bio-Bibliographie*.

13 See especially chapter XII, "Les 'conférences' Rohault," in Clair, *Bio-Bibliographie*. For a comparison between Rohault's conferences and those hosted by other Parisian academies of the time, see Roux, "Was There a Cartesian Experimentalism?"

14 For the spread of private and public conferences in the second half of the seventeenth-century in Paris, see Brown, *Scientific Organizations*; McClaughlin, "Les rapports"; Brockliss, *French Higher Education*; Roux, "Was There a Cartesian Experimentalism?"

15 This has been detailed in Dobre, "Rohault's Cartesian Physics"; Roux, "Was There a Cartesian Experimentalism?."

16 See Oldenburg, *The Correspondence*, vol. 2, 73. This would suggest that Rohault was expected to be part of the savant society, just like other figures on Oldenburg's list: Roberval, Fermat, Bernard Frénicle de Bessy, Auzout.

Geneviève Clerselier (1640–1681), one of Claude Clerselier's daughters.[17] The new family connection opened fresh possibilities for the mathematics professor, who most probably became more absorbed in his private teaching than in public conferences. In a letter written in 1670, Rohault explains how busy he is with his regular teaching duties, and judging from the decreased number of reports about his public lectures, one can infer that Rohault reduced their frequency.[18] It was only in 1671 that he published his *Traité de physique*, where he explains in the preface that he was forced to do so due to the spread of unofficial versions of his conferences, one example being the *Physique nouvelle* (1667), published under Rohault's name.[19] Neither the date included in the title nor the author's name are to be trusted in this case.[20] Even if the notes were taken during Rohault's Wednesday lectures, he did not authorize their publication, and the reader is left wondering about the accuracy of the text.[21] The book is nonetheless a useful source for tracing the evolution of Rohault's thinking, especially if studied in connection with the Ms. 2225 Paris manuscript. I will return to these sources in the next section of the chapter, but before doing so, I shall add some details about the final years of Rohault's life.

Rohault died at the end of 1672. All of the books that had been printed under his name during his lifetime appeared in 1671. The first was the famous *Traité de physique*, followed at the end of the year by the *Entretiens sur la philosophie*.[22] The *Traité* seems to have been the only text that was long in the making, since the *Entretiens* were written in the second half of 1671.[23] This latter book resulted from a conversation between Rohault and an unnamed theologian from the Sorbonne—with the treatise structured as a dialogue between de two—and initially circulated in the form of a letter. More metaphysical in nature, and dealing with some of the controversial topics of the time (e.g., the problem of transubstantiation and the Cartesian elimination of substantial forms), the

---

17   See Rohault, "Discours de la fiévre," in Descartes, *Le monde* (Paris, 1664). For Rohault's family, see chapter VI, "L'homme," in Clair, *Bio-Bibliographie*, 26–31.

18   The letter is from 6 August 1671 and it was published in Clair, *Bio-Bibliographie*, 169. However, for a couple of years already, the number of reports about Rohault's conferences decreased significantly.

19   See the unpaginated preface in Rohault, *Traité de physique*.

20   Sylvain Matton tackles these problems In a preliminary study; Matton, "Remarques sur le manuscrit."

21   An example is in Dobre, "Rohault and Cartesian Experimentalism," 391, n. 1.

22   Rohault, *Traité de physique*; Rohault, "Entretiens."

23   Rohault's correspondence with Nicolas-Joseph Poisson (1637–1710), reproduced at the end of Clair's monograph, explains some of the context; see Clair, *Bio-Bibliographie*, 167–169. The letter to Monsr. Guyard, which forms the basis for the *Entretiens*, is dated 10 June 1671; see Clair, *Bio-Bibliographie*, 170–178.

Entretiens raised immediate concerns about the orthodoxy of Rohault's religious position, which some theologians questioned. The episode reflects the constant trouble of Cartesians facing censure and explains Clerselier's apologetic tone in the preface to Rohault's *Oeuvres posthumes* of 1682.[24]

The collection prepared by Clerselier consists in a series of mathematical treatises, which were offered by Rohault as lessons in his private teaching.[25] They were meant for publication around the same time with the *Traité*, as stated in the publication rights of the treatise:

> Our dear and well-beloved Jacques Rohault, who devoted his entire life to the study of Philosophy and Mathematics, [has] showed us with great modesty that he had composed several Treatises, which he wants to print & offer to the public, if he is granted with our Letter; among these, *a Treatise on physics, or of the Science of nature & that of Cosmography, seen by Mr. de Mezeray, our Counselor & Historiographer; the Fifteen Books of the Elements of Geometry by Euclid, the Practical Arithmetic; the Resolution of Plane & Spherical Triangles; the Practical Geometry; the Fortifications, the Mechanics & the Perspective.*[26]

---

24   See Rohault, *Oeuvres posthumes*, unpaginated; McClaughlin, "Clerselier's Attestation of Descartes's Religious Orthodoxy"; Dobre, "Rohault and Cartesian Experimentalism." In the 1660s, Cartesian philosophy was subject to condemnations and censure; see, for example, Ariew, "Censorship, Condemnations, and the Spread of Cartesianism"; Armogathe and Carraud, "Première condamnation"; McClaughlin, "Censorship and Defenders"; Roux, "Condemnations of Cartesian Natural Philosophy Under Louis XIV"; Schmaltz, *Radical Cartesianism*. About the role played by frictions between Cartesian physics and the Catholic doctrine of transubstantiation in the teaching of Cartesianism, in Louvain, see Chapter 15, in this volume.

25   The context of this publication and some of the consequences derived from the strategy outlined by Clerselier in the preface to Rohault's posthumous works are discussed in Dobre, "Rohault"; Dobre, "Rohault's Mathematical Physics." The former study explores the reception of Rohault's treatise on mechanics, which was part of the mathematical works printed in the collection of Rohault's posthumous works. The latter details Clerselier's role in attracting Rohault into the dispute with Fermat.

26   "Nostre cher & bien amé Jacques Rohault, s'estant toute sa vie appliqué à l'étude de la Philosophie & des Mathematiques, Nous a tres-humblement remontré qu'il auroit composé plusieurs Traitez qu'il desireroit faire imprimer & donner au public s'il en avoit nos Lettres sur ce necessaires; entr'autres *un Traité de Physique, ou de la Science naturelle, & celuy de Cosmographie, veus par le sieur de Meseray nostre Conseiller & Historiographe; Les quinze Livres des Elemens de Geometrie d'Euclide, L'Arithmetique Pratique; la Resolution des Triangles Rectilignes, & Spheriques; la Geometrie Pratique; les Fortifications, les Méchaniques, & la Perspective.*" The "Privilege du Roy" is printed at the end of the book; see Rohault, *Traité*. For the edition of the posthumous works, Clerselier uses the same public-

In the preface, Clerselier downplays the significance of these mathematical writings, mostly due to their technical content, which did not appeal to a large audience. However, he adds a brief note that is of interest to our present concerns: these courses were usually offered by Rohault to his students among the French elites.[27] One is left wondering what the content may have been of Rohault's private lessons in natural philosophy. The case of the two celestial globes offered to his employer, the Conti family, will help us uncover some nuances in Rohault's cosmology.

3   Cosmology in Rohault's *Traité de physique*

Rohault's cosmology is featured in the second part of his sizable *Traité de physique*, a book divided in four parts. In the early modern period, the first part of the book was usually printed as a separate volume. It deals with general natural philosophy, covering topics such as the notion and principles of physics, the method of studying particulars, a new explanation of the so-called sensible qualities, pneumatic experiments, the theory of motion, optics and the theory of vision. All explanations are based on matter and local motion and follow contemporary developments, including some famous experiments and the use of instruments in natural philosophy, as attested most notably in the chapter on the fear of a vacuum (ch. 12). The section on general natural philosophy would remain the most popular part of the book, as is shown both by the early modern annotations to the treatise made by Antoine Le Grand (1629–1699) and Samuel Clarke (1675–1729), and by the more recent scholarly debates around Rohault's work.[28]

---

ation right; Rohault, *Oeuvres posthumes*, unpaginated. For a discussion of the double use of the "Privilege," see Dobre, "Rohault and Cartesian Experimentalism," 394.

27   The relation with Conti's family is well established, but in the unpaginated preface, Clerselier claims that Rohault was about to become tutor of the Dauphin; see Rohault, *Oeuvres posthumes*.

28   The Latin edition of Rohault's treatise published in 1682 includes comments made by Antoine Le Grand. The format was very popular and, by the end of the century, several reprints of this edition were published. In 1697, Samuel Clarke—who will later become one of the main expounders of the natural philosophy of Isaac Newton (1642–1727)—made a new Latin translation. It was annotated, but what is peculiar to Clarke's editorial work on Rohault is that in subsequent editions, he changed the annotations from those in the initial edition. For Clarke's annotations, see Hoskin, "Mining All Within"; Schüller, "Clarke's Annotations"; Dobre, "Rohault's *Traité de Physique*"; Dobre, *Descartes and Early French Cartesianism*.

The second part of the *Traité* is about cosmology or "cosmography." Rohault provides an account of the various hypotheses available to explain the system of the world, and indicates his support for the Copernican system. The third part of the book deals with earthly phenomena, from the three elements of Cartesian natural philosophy to minerals and meteorological phenomena. Finally, the least studied section of Rohault's book is the fourth part, which is devoted to medicine. Traditionally, these three final sections—from book two to book four—were printed as the second volume of Rohault's *Traité de physique*.

The section on cosmology opens with a general presentation of the discipline, which is rendered under the traditional name of "cosmography."[29] In subsequent chapters, Rohault presents the general aspects related to this topic, including a brief account of the technical apparatus for the study of the celestial globe.[30] All these are rather introductory and resemble traditional astronomical discussions but without the technical details (e.g., astronomical tables, mathematical explanations, etc.). After announcing the availability of several hypotheses to explain celestial phenomena, Rohault proceeds with the examination of the most common account: the Ptolemaic hypothesis.[31] He devotes many pages to the traditional astronomical system, even if he will not support it in the end. The reader is carefully introduced to the circles of the world (as explained in the title of ch. 4). In succeeding chapters—usually organised as a set of observations and conjectures—he discusses the Sun, the fixed stars, the

29   For example, the title of the first chapter reads "Of the meaning of the Word Cosmography, and the Usefulness of the Science." The definition is standard, "What I now propose, is to give a general Idea of the World, that is a Description of the Nature, Situation, Magnitude, Figure, and some other Properties of the principal Parts of which the visible World is composed: And that Science which treats of these several Particulars, is called *Cosmography*"; Rohault, *System of Natural Philosophy*, vol. 2, 3. All quotations from Rohault's *Traité* are given from the early modern English translation made by John Clarke (1682–1757) and printed in 1723–1724 (2nd edition in 1728–1729; 3rd edition in 1735); the modern edition of 1987 reproduces the text of the 2nd edition (1728–1729).

30   Chapter 4 concerns "Of the Figure of the World. Of the principal Points, Lines and Circles, which are imagined to be upon the Superficies of it." Rohault, *System of Natural Philosophy*, vol. 2, 8. However, unlike technical astronomical treatises of the period, Rohault's chapter is only preparatory (i.e., defining the key terms), and does not include any mathematical calculations. For an example of the astronomical treatise of the period, see the two astronomy books identified in the inventory of Rohault's library: Durret, *Nouvelle theorie des planets* (Paris, 1635); Riccioli, *Almagestum novum* (Bologna, 1651). For the inventory of Rohault's library, see McClaughlin and Picolet, "Un exemple."

31   The Ptolemaic system is discussed up to chapter 16, with the "Conjectures how to explain the apparent Motions of the Stars" introduced in chapter 3; see Rohault, *System of Natural Philosophy*, vol. 2, 7.

Moon, eclipses, the planets, starting with Mercury and Venus, and then Mars, Jupiter and Saturn.[32] The first part of Rohault's cosmography ends with an illustration of the Ptolemaic system.[33]

The second part of his cosmography examines the hypothesis that the Earth turns on its own axis.[34] The order is similar: first, some technical details about the poles and the circles of Longitude and Latitude, then the "explanation" of particular celestial phenomena. There is a subtle shift of language, from the "conjectures" of the Ptolemaic system presented in the first part in conjunction with "observations," to "explanations" in the discussion of the Copernican theory. However, the table of contents reveals a similar order: the Sun, the fixed stars, planets (in the same order: Mercury and Venus, then Mars, Jupiter, and Saturn). All these are preparatory for a comparison among the three astronomical systems (Ptolemaic, Copernican, and Tychonian). The system of Tycho Brahe (1546–1601) is presented only briefly, and allows Rohault to conclude that Copernicus's system exclusively explains all of the observed phenomena, or at least offers better explanations.[35] The final five chapters of his "cosmography" are meant to illustrate the advantage of the modern cosmological explanations. They cover the nature of the stars, comets, the rejection of judicial astrology, a chapter on gravity, and an explanation of the tides. This, in a nutshell, is the outline of Rohault's cosmology. Let us now turn to an analysis of its content, which will be done through a comparison of the text of the *Traité* with the two manuscripts mentioned above.

As previously indicated, details about Rohault's early philosophical commitments are hard to establish, not least because, before the publication of the *Traité*, he authored only two short texts. One was a lecture on fevers, delivered at the Montmor salon but printed (anonymously) together with Descartes's

---

32 Most of the time, a chapter dealing with "conjectures" follows immediately after a chapter about "observations" about a given celestial object or phenomenon. In the other cases, both observations and conjectures are treated in the same chapter.

33 In the original French edition, images are reproduced in the text. In later editions, images are added in a separate table, at the end of the book. For the purpose of this chapter, I disregard these differences, although the availability of the images with the text must have facilitated the understanding of the argument. For a brief comparison between Rohault's cosmological illustrations and those of Descartes, see below.

34 "The Second Part of Cosmography or An Explication of the Phaenomena, upon Supposition that the Earth turns about its own Center in twenty-four Hours," Rohault, *System of Natural Philosophy*, vol. 2, 46. This section starts from chapter 17.

35 Chapter 23 gives an account "Of the System of Tycho-Brahe"; Rohault, *System of Natural Philosophy*, vol. 2, 58–59. The comparison is made in chapter 24—"Reflections upon the Hypotheses of Ptolemy, Copernicus, and Tycho"—but the discussion continues in the subsequent chapters; see Rohault, *System of Natural Philosophy*, vol. 2, 59–63.

The World, or Treatise on the Light in 1664.[36] The other was a letter of 1658 to Pierre de Fermat (1601–1665), in which Rohault defended Descartes's method in the study of the refraction of light, a letter printed only in 1667, in the third volume of Descartes's correspondence.[37] Except for these two documents, Rohault's activities and his natural philosophical views can only be inferred from other sources. Travellers attending Parisian conferences and fellow natural philosophers have noted in their correspondence and private journals some considerations with respect to Rohault's lectures. Such scattered sources cover a timespan from the end of the 1650s up to the mid-1660s. What is more important for our current topic is that whenever they refer to particular issues addressed by Rohault, they mention themes included in the first part of the *Traité*. The notable exception is the explanation of the magnet, which is provided in the third part of the book.[38]

It is quite curious that no reference to cosmology appears in any of these sources. There are three possible explanations for this. First, Rohault was not discussing this topic in his conferences. Second, his views were rather standard—with respect both to the traditional world system and the modern explanations—and hence were disregarded by those recollecting Rohault's lectures. Third, unlike the cases of other phenomena in which he had an interest (e.g., magnetic phenomena, pneumatic experiments, optics), Rohault's cosmology was still being developed at that time, and he therefore lectured on this subject in private lessons only, not in his public conferences.

In order to decide between these alternatives, closer attention to the unofficial manuscripts is in order: the Ms. 2225 of the Bibliothèque Sainte-Geneviève in Paris and the *Physique nouvelle* (*1667*). As mentioned above, details included in these two sources can be misleading if too much interpretative weight is placed upon them—the differences between terms used in the two manuscripts and those used in the *Traité*, could, for example, be due to copyists' errors, without denoting a change of view—yet the general image derived from them should support a stronger conclusion.

The *Physique nouvelle* (*1667*) was published in a modern edition in 2009. It includes most of the content of the *Traité* and—due to the year included in the title of the manuscript (1667)—is considered an early version of the treatise

---

36   See Rohault, "Discours de la fiévre"; Descartes, *Traité de la Lumiere*. Rohault's lecture on fevers was printed in a slightly modified form as the last chapter of part IV of the *Traité*. For Rohault's medicine, see Dobre, "Rohault on Medicine."

37   See Descartes, *Lettres* III. For Rohault's role in the dispute with Fermat, see Dobre, "Rohault's Mathematical Physics."

38   On Rohault and magnetism, see Chapter 14, in this volume.

published by Rohault just one year before his death.[39] The fourth part, on living beings, is missing, but the other three sections are presented in quite a similar manner to the way in which they appear in the *Traité*. There are, however, some significant differences, including in the "Second partie de physique ou le système du monde."[40] A quick inspection of the chapter titles might lead to the conclusion that most of the material is present; after all, the examination of the "fabrique du monde" or "cosmographie" is structured in a similar way. Rohault makes some general comments about the observed phenomena, and then examines the two competing hypotheses, the Ptolemaic and the Copernican systems of the world.[41] Technical concepts are briefly discussed in a separate chapter.[42] The parallelism between the *Physique nouvelle* and the *Traité* extends further still, with the traditional explanation (i.e., the Ptolemaic system) presented in the same manner: chapters concerning "observations" or *apparances* of a celestial body or phenomenon are immediately followed by a "conjecture" to describe *la cause*. Only when the Copernican system is discussed do chapter titles indicate an *explication* of such observed celestial phenomena.[43] The general landscape of moving celestial bodies—the Sun, the so-called fixed stars, and all the known planets, including the Moon—is completed by a final chapter on the tides.

Compared with the *Traité*, the final part of the corresponding section on cosmography in the *Physique nouvelle* is shorter. Moreover, it skips some other topics, which conclude the second part of the *Traité*: "De la Nature des Astres" (ch. 25), "Des Cometes" (ch. 26), "Des Influences des Astres, & de l'Astrologie Judiciaire" (ch. 27), and "De la Pesanteur, & de la Legereté" (ch. 28). The final chapter on the tides is included in both, which seems to suggest its availability from an earlier date. This conjecture is further supported by the other manuscript—Ms. 2225 of the Bibliothèque Sainte-Geneviève in Paris—where

39   See Blay, "Introduction" and Matton, "Remarques sur le manuscript," in Rohault, *Physique nouvelle*.

40   Rohault, *Physique nouvelle*, 241. See pp. 241–331 for the second part of the manuscript.

41   Neither in the *Traité* nor in the *Physique nouvelle* does this preliminary discussion of the available hypotheses feature the name of Tycho Brahe. However, Tycho's system is discussed in Rohault's cosmology, where it is considered as similar to the Copernican system.

42   The two quotations come, respectively, from *Physique nouvelle*, 241, and *Traité de physique*, vol. 2, 3. Just as in the published *Traité*, the technical chapter is more of a glossary of terms for the Ptolemaic system. They are explained in the fourth chapter, "Des principaux points et cercles qu'on imagine dans le monde"; Rohault, *Physique nouvelle*, 249.

43   See, for example, the chapters on "Explication du mouvement apparent des étoiles fixes," "Explication des mouvements apparents de Mercure et de Vénus," "Explication des mouvements apparents de Mars, Jupiter et Saturne," and "Explication des apparences de la Lune," in Rohault, *Physique nouvelle*, 312, 313, 315, 319.

the topic is listed as the subject of a "Wednesday" conference from November 1669. But a more difficult question about these two sources is suggested by the latter manuscript, which includes, at the very end and without any date, a discussion about comets.[44] If the question of the tides appears in all three sources—the *Physique nouvelle (1667)*, the conference from November 1669 (Ms. 2225), and in chapter 29 of the second part of the *Traité*—the comets are not at all featured in the *Physique nouvelle*. This contradicts the standard reading which places Ms. 2225 as an early version (1660–1661, with the curious addition from November 1669), followed by the *Physique nouvelle (1667)* as an intermediary stage in the writing of the *Traité*, and finally the official publication of Rohault's treatise on natural philosophy in 1671.[45] Would it be possible to reverse the order of the first two sources? In the argument developed further in this chapter, I situate the *Physique nouvelle* before some parts of the Ms. 2225, with the caveat that neither manuscript was authored by Rohault.[46] The uncertainty surrounding the authorship, and the problems in ascribing precise dates to the conference notes and to the manuscript of the *Physique nouvelle*, force us to take other evidence into account.

The most obvious place to begin our investigation is the chapter on comets in the *Traité* and its corresponding section in the manuscript from the Bibliothèque Sainte-Geneviève in Paris. The topic of comets gained increased visibility in the public debates of the period, mostly due to the famous astronomical observations of 1664 and 1665. These triggered a huge number of disputes and they encouraged even more astronomical observations at a time when the astronomical observatory was established in Paris. Curiously, Rohault seems to have been out of step with all of these developments. He acknowledges the new apparition of a comet in December 1664, but he is not interested in astronomical observations, and seeks instead to explore a plausible explanation for the phenomenon. His account is offered as an alternative to the scholastic explanation. In fact, the names that appear most often in his writings are those of Aristotle and Ptolemy. Both in the manuscript and in the *Traité*, Rohault argues that comets are not fiery exhalations, as Aristotle had mistakenly believed. The basis of this claim is in the works of the modern astronomers and mathematicians: it is in view of the recent advances in

---

44 See MS., Paris (Bibliothèque Sainte-Geneviève), MS. 2225: Rohault, *Conférences sur la physique*, fols. 82ᵛ–91ʳ.

45 See, for example, Blay, "Introduction."

46 A similar, more nuanced approach is suggested by Sophie Roux. The *Fragment de physique* (1662), attributed to Cyrano de Bergerac, but linked to Rohault's *Traité* is also included in the discussion; see Roux, "Was There a Cartesian Experimentalism?," 73–74.

mixed-mathematics that Aristotle's view about comets should be changed.[47] However, as Roger Ariew convincingly argued, the late-Scholastics were already quite flexible in adapting the Aristotelian explanation to the new astronomical phenomena.[48]

The best example in the context of Rohault's discussion of comets is that of the Jesuit Jacques Grandamy (1588–1672), professor at the Collège de Clermont in Paris, who had already published a book in 1665 on *Le cours de la comete qui a paru sur la fin de l'année 1664, et au commencement de l'année 1665. Avec un traité de sa nature, de son movement, et ses effets.* For Grandamy, comets move above the Moon, in the cosmic fluid of Tycho's system, and they are produced when celestial matter condenses in a manner that allows the reflection of Sun's rays. It is an accidental alteration explained through optics, but the general Aristotelian framework is still present (e.g., the search for causes and the discussion about the substantial form), despite the adherence to Tycho's cosmology.[49] Grandamy's selective approach to Aristotle—rejecting the cosmological separation between the sublunar and supralunar realms, but preserving the general explanatory principles—is indicative of the adaptability of seventeenth-century scholastic viewpoints in the face of new observations and experiments. Grandamy is keen to collect observations about the comet and to discuss them.[50] In chapter 6 of his book, he refers to a famous assembly gathered on 10 January 1665 at the Collège de Clermont, a meeting which was also reported upon in the *Journal des sçavans* of 28 February 1665. According to the French journal, notable figures of the aristocracy attended the event: "Monsieur le Prince [Prince de Condé], M. le Duc [the Duke d'Enghien, the eldest son of the Prince de Condé], & M. le Prince de Conty [the son of the Prince

---

47   The reference to modern astronomers is in *Traité*, II, ch. 26, art. 13, p. 108; for the mathematicians, see MS., Paris (Bibliothèque Sainte-Geneviève), MS. 2225: Rohault, *Conférences sur la physique*, fol. 85. This includes a discussion of the stellar parallax; see Rohault, *Traité*, II, ch. 26, arts. 13–14, 107–108; see also Rohault, *System of Natural Philosophy*, vol. 2, 83.

48   See Ariew, "Theory of Comets at Paris."

49   See Grandamy, *Le cours de la comète* (Paris, 1665). See also Ariew, "Theory of Comets at Paris," 366–369; Roux, "The Two Comets of 1664–1665," 110–111.

50   Most of the observations included in Grandamy's analysis are from other Jesuit astronomers; Grandamy, *Le cours de la comète*, 10: "Les observation que j'ay faites avec beaucoup de soing s'accordent parfaitement avec celles de ces trois sçavants hommes [Ignace Gaston Pardies (1637–1673), Adrien Auzout, and Gilles François de Gottignies (1630–1689)], aussi bien que celles qui m'ont este envoyées d'Aix en Provence par le Pere Regis de Lyon, par le Pere Bertet, de Langres par le Pere de Billy, de Poitiers par le Pere Verdier, de Doüay par le Pere Seneschal, & de quelques autres Provinces par d'autres Peres de nostre Compagnie curieux de ces connoissances." Some of the figures mentioned in this list were authors of similar treatises on comets, e.g., Jacques de Billy (1602–1679).

de Condé], suivis d'un grand nombre de Prelats & de Seigneurs de la Cour."[51] The speakers were the Jesuit Nicolas d'Harouys, Gilles-Personne de Roberval (1602–1675, mathematics professor at the Collège Royal), Vincent Phelippeaux (a physician from Louvain), and Jacques Grandamy. The extensive note in the *Journal des sçavans* complements the general report of what was discussed in the conference at the Collège de Clermont with new observations made by Adrien Auzout (1622–1691) and Jacques Buot (d. 1678).[52] A far-reaching debate emerged, and the report of the journal seems to have been the catalyst. Against the scholastic account, Jean-Baptiste Denis (1640–1704) published (anonymously) a Cartesian explanation entitled *Discours des comètes selon les principes de Mr. Descartes*, a book in which the January conference hosted by the Collège de Clermont is discussed in the light of the report in the *Journal des sçavans*.[53] Sophie Roux labels the opposition between Denis and Grandamy as a "'confrontation' rather than as a 'controversy'," and shows very convincingly that—at least in the case of the Jesuits of the Collège de Clermont—the scholastic arguments marked a more general attack against Descartes's philosophy.[54] The battle, she writes, was soon considered as having been brought to a close:

> [...] the confrontation between Denis and d'Harouys did not develop [further] after the June theses. Denis left the ground and there was no Cartesian to take over the fight against the Jesuits. Indeed, the notes taken by a lawyer on the lecture on comets that Jacques Rohault gave in his *Mercredis* that very year, contain neither allusion to the January conference nor to any kind of controversy about the nature of comets between Cartesians and Jesuits or, more generally, "Ancient philosophers," as they were called. Rohault was apparently happy to establish that comets are neither stars nor planets, to put forward overused arguments against Aristotle, and, finally, to briefly introduce the theory of comets that Descartes presented in *Principia philosophiae*.[55]

---

51 See *Journal des sçavans* (28 February 1665), 41.
52 For the four points presented in the aforementioned conference, see *Journal des sçavans* (28 February 1665), 41–44. From p. 44 to p. 48, the editor, Denis De Sallo ["Sieur de Hedouville"] (1626–1669), refers to the observations made by Adrien Auzout and Jacques Buot. The consequences of this interest in the nature of the comets are discussed in Roux, "The Two Comets of 1664–1665."
53 See especially Denis, *Discours sur les comètes* (Paris, 1665), 37–38.
54 See Roux, "The Two Comets of 1664–1665," 111–114.
55 Roux, "The Two Comets of 1664–1665," 116–117.

The reference to Rohault is rather surprising in this context and needs some further clarification. As a leading Cartesian, Rohault would have been expected to provide a thorough defence of Descartes's views about the comets, and yet he is described as not having involved himself in the disputes. The problem is further increased by the missing date in the conference notes of the lawyer "F.," which fails to offer any indication about his involvement in the debate.

The manuscript version of Rohault's views on this topic opens with a definition of the comets as luminous celestial bodies of a temporary nature, with a particular shape (the tail) and motion with respect to the Sun.[56] According to Rohault, the comet is an object of "l'admiration aux philosophes," raising "leur curiosité." Moreover, since Aristotle's explanation in terms of a fiery exhalation produced in the sublunar realm has several shortcomings, one should rely on the discoveries made by "les Mathematiciens du dernier siècle."[57] It is remarkable that Rohault makes an appeal to a general knowledge—less specialized and rather common, as one can find in the pages of the *Journal des sçavans*—in the sense that his discourse lacks any technicalities of the astronomical treatises of his time. However, he preserves the general traits of such treatises: the comet is presented as an extraordinary astronomical object; he rejects the Aristotelian explanation; he discusses the problems related to the shape, the temporary character, the place, the motion, and the effects of the comet. The received knowledge is simply rejected as failing to provide a good explanation for comets.[58] Grounding his explanation on the theory of the vortex ("tourbillon"), Rohault places comets in the fluid matter of the world and sees them as objects capable of traveling through the vortices, adding into the mix Descartes's three elements, subtle matter, and the explanation of the motion of light. However, the lawyer "F." notes at the very end of the manuscript: "But it is impossible to understand all these well, if one had not read the principles of Mr. Descartes, and seen the images of his book."[59] The intricacy of Rohault's explanation is in stark contrast to the clarity of Descartes's text in the third

---

56   See MS., Paris (Bibliothèque Sainte-Geneviève), MS. 2225: Rohault, *Conférences sur la physique*, fols. 82ᵛ–83ʳ.

57   Respectively, See MS., Paris (Bibliothèque Sainte-Geneviève), MS. 2225: Rohault, *Conférences sur la physique*, fol. 83ᵛ, 84ʳ, 85ʳ.

58   See especially the conclusion reached on MS., Paris (Bibliothèque Sainte-Geneviève), MS. 2225: Rohault, *Conférences sur la physique*, fol. 87ʳ.

59   See MS., Paris (Bibliothèque Sainte-Geneviève), MS. 2225: Rohault, *Conférences sur la physique*, fol. 91ʳ: "Mais il est impossible de bien entendre tout cecy, si on n'a lû les principes de Monsr. Des Cartes, et veu les figures de son livre." For the "tourbillon," see fol. 87ᵛ.

part of the *Principles of Philosophy*, but especially to the use of illustrations.[60] There is no reference to figures accompanying the discourse on comets, and unfortunately, the visual representations of the *Traité* are no improvement in this respect. The very visual character of Descartes's cosmology is drastically reduced in Rohault. Instead, the reader of the *Traité* stumbles upon the two classic representations of the world-systems: the Ptolemaic (PII.16.18) and the Copernican (PII.22.7).[61] Rohault adds the famous illustration of the ring around Saturn, which was published by Huygens at the end of the 1650s (PII.16.11).[62] Four other diagrams complement the set of cosmological images, all dealing with slightly more technical details, but not implying very complex mathematical calculations: the calculation of the distance between the Earth and the Moon (PII.12.5); the explanation of the movement of the planets and why they were traditionally ascribed a retrograde motion (PII.21.6); the heaviness of bodies and why bodies tend to descend towards the middle of the Earth (PII.28.12); and the system between the Earth and the Moon, which is supposed to explain the tides (PII.29.13 and PII.29.16).[63]

In order to complete the analysis of his cosmology, I will now turn to Rohault's chapter on comets in the *Traité de physique*. The first article ("Why we treat of Comets in this place") offers a justification for discussing this topic:

> When I gave an Account of the Observations of the several caelestial Bodies; I should have mentioned those made from Time to Time about Comets; but I purposely forbore this, because I know that they are not, in the common Opinion of Philosophers, reckoned amongst the heavenly Bodies; and because I was unwilling to increase the Difficulty of the Subject I was handling, by adding a Thing which requires much Attention, and which is but little understood hitherto. But now, seeing Men have always had a great Curiosity to understand the Nature of Comets, I think

---

60   For Descartes's use of images in his natural philosophy, see Lüthy, "Where Logical Necessity Becomes Visual Persuasion"; Zittel, "Conflicting Pictures"; Mantovani and Cellamare, eds., *Cartesian Images*. For a comparison between Descartes's illustrations in the third part of the *Principia* and in *Le monde*, see Dobre, "Depicting Cartesian Cosmology." A digital exhibition of Descartes's cosmological images is available online at Dobre, Babeş, and Bujor, eds., "Cartesian Cosmological Illustrations."

61   In the original French edition—see Rohault, *Traité*, vol. 2, 64 and 81,—PII.16.18 refers to the second part of the *Traité* (PII), followed by the chapter number (16), and the article (18). This notation will be used further in the chapter.

62   See Rohault, *Traité*, vol. 2, 61. See Huygens, *Systema saturnium* (The Hague, 1659).

63   The illustration in the chapter on the flux and reflux is repeated twice; see Rohault, *Traité*, vol. 2, 50, 77, 124, 132 and 134.

I ought not so far to lay aside this Matter, as not to say at least what is most certainly known about it; leaving it to them who shall come after, to philosophize in a different Manner, if any new Observations that shall at any time be made, oblige them to alter our Hypothesis, or to mend our Opinion.[64]

Rohault seems willing to preserve an ambiguity about his own Cartesian explanation. Just like in other parts of his natural philosophy, he introduces the explanation by way of a hypothesis, which he is ready to alter, if new observations will make it untenable. However, he is not performing astronomical observations himself, and those he includes in other parts of his cosmology are either very famous (e.g., the telescopic observations of Galileo Galilei (1564–1642) and Huygens's observations of Saturn) or common knowledge.[65] He uses arguments commonly employed in the 1660s debate about comets—regardless of the side—and he criticizes Aristotle in order to advance an explanation grounded on "A new Conjecture concerning the Nature of Comets," as the title of Art. 15 states.[66] What is missing from Rohault's chapter on comets, besides the use of recent astronomical observations, is a mathematical approach to the phenomenon. Instead, Rohault quickly recapitulates the common points of the discussion and integrates the Cartesian theory of comets, without naming Descartes. It looks like he adds the chapter only because the context requires this. The same is true for chapter 27, "Of the Influence of the Stars, and of judicial Astrology," as most of the early modern discussions about comets were complemented with a section about the effects of such remarkable celestial phenomena.[67]

---

64 See PII.26.1; Rohault, *System of Natural Philosophy*, vol. 2, 80.

65 Galileo's name appears several times in the second part of the *Traité*: PII.10.15, PII.14.7, PII.14.8, PII.16.10, PII.16.11, PII.16.13, PII.16.16; see Rohault, *System of Natural Philosophy*, vol. 2, 31, 40, 44–45. Huygens's explanation of the astronomical observations of Saturn is praised in PII.16.16 Rohault, *System of Natural Philosophy*, vol. 2, 45.

66 That would be PII.26.15, Rohault, *System of Natural Philosophy*, vol. 2, 83. The reader finds that "this is indeed done by a late eminent Philosopher," undoubtedly Descartes: "he conjectures, that what we call a Comet, is nothing else but one of those Stars, which, being by Degrees covered with Spots all over, so as entirely to lose its Light, could no longer keep its Place amongst the other Stars, but was carried away by one of their Vortexes, which impressed a Motion upon it proportionable to its Bigness, and Solidity, by which means it may come very near the Heaven of *Saturn*, where the Light which it receives from the Sun may make it visible." Rohault, *System of Natural Philosophy*, vol. 2, 84.

67 See, for example, chapter VII ("De la cause finale de la Comete, & de ses effets") in Grandamy, *Le cours de la comète*, 21–23. See also chapter VII ("Des effets de la Comete") in Denis, *Discours sur les comètes*, 127–156.

The evasive tone of the first article, reproduced above, of Rohault's chapter on comets is symptomatic for his cosmology. Unfortunately, owing to limits on the scope of this chapter, I can only sketch the argument for the entire section on cosmology, while preserving the focus on the chapter about comets. The notes of the lawyer "F." for the section on comets include only two names: Aristotle and Descartes. If one examines the entire section on cosmology, a longer list of names will emerge from the *Physique nouvelle* (*1667*): Aristarchus, Aristotle, Copernicus, Ecphantes, Galileo, Hipparchus, Huygens, Johannes Kepler (1571–1630), Philolaus, Ptolemy, Seleucus, and Tycho. The corresponding section of the *Traité* will expand the list with several other names besides: Archimedes, Gian-Domenico Cassini (1625–1712), Plato, Pope Gregory, Pythagoras, Socrates, Thales. Kepler's name does not appear in the 1671 treatise.[68] What is peculiar about the names included in Rohault's writings is the reliance on the classical sources (and examples), and the little use made of modern authors. Rohault refers to post-Copernican authors only when their contributions are well established (Galileo's observations with the telescope, Huygens's account of Saturn, Cassini's explanation of the spots observed on Jupiter). But the name most glaringly omitted from Rohault's *Traité* is that of Descartes, despite his having framed his arguments upon Descartes's natural philosophical thought. What are we to make of this?

4 Conclusion

The first conclusion of this chapter is that Rohault's cosmology had not kept pace with the astronomical knowledge of the time. The manuscript evidence reveals little interest in cosmological debates—even the lecture recorded by lawyer "F." does not show any sign of engagement in the debate about comets—

---

68 While the erasure of Kepler's name might raise some concerns, it is the unofficial status of the *Physique nouvelle* that should restrain us from placing too much interpretative weight upon this text. For example, just as in the case of the notes taken by the lawyer "F." in the Ms. 2225 (Paris, Bibliothèque Sainte-Geneviève), one can speculate that some details were added (or even omitted) by the scribe. Alternatively, one can explain this reference through Rohault's sources, as far as can be reconstructed from the inventory of books in his library at the end of his life. Both Noël Durret (ca. 1590–ca. 1650) and Giovanni Battista Riccioli (1598–1671)—mentioned in n. 30 above—refer to Kepler. But both Durret and Riccioli discuss many other post-Copernican astronomers and natural philosophers not taken up by Rohault, which makes it difficult to explain Rohault's criteria for selecting the authorities in his *Traité*. To avoid too much emphasis on this topic, I should add that neither Durret nor Riccioli are mentioned by Rohault.

and the section on "le système du monde" of the *Physique nouvelle* focuses on the standard topic of presenting the available hypotheses on the issue of the centrality of the Earth. The same structure is preserved in the *Traité* of 1671, with enriched details in the discussion of the Copernican hypothesis.

A more thorough study of the development of Rohault's views during the 1660s is needed, but some provisional inferences can be made on the basis of the material discussed in this chapter. Above, I referred to the curious lack of references to cosmology in early modern accounts about Rohault's conferences and I conjectured that this could be explained in three different ways: the topic was not addressed in his public lectures, his explanations were not new and hence they lacked appeal, or his cosmological views were still being developed.

As documented above, Rohault lectured on some topics incorporated at the end of his published cosmology. The manuscript of the Bibliothèque Sainte-Geneviève includes a discourse on the tides and a lecture on comets. In this sense, the first of the three explanatory conjectures is not tenable. The second conjecture is more difficult to assess, as there is only one note about the views of Rohault's audience: the final remark by "F." in the Paris Ms. 2225, that one who is not familiar with Descartes's philosophy and illustrations will fail to grasp the argument of Rohault's lecture. Ironically, in the *Discours sur les comètes*, Jean-Baptiste Denis carefully tackled such worries: "[...] if anyone finds some difficulty, it will not come from the thing itself, but from that one who is not versed in the principles of this Philosophy [Descartes's]."[69] The connection with Descartes is explicit in Denis's conclusions, "But if someone is troubled by the principles we employed to derive the previous consequences, he can clarify [them] by reading the entire third part of Mr. Descartes's principles."[70] It was beyond the scope of this chapter to analyse the differences between Denis's treatise and that of Rohault, but is worth stressing the more polemical attitude of the former. Moreover, Denis uses a large variety of materials and sources in his Cartesian explanation of the nature and causes of comets. Rohault is almost excusing himself for having to deal with this topic (Art. 1 in his chapter on comets in the *Traité*), while early manuscript versions reveal either a lack of interest in this subject (*Physique nouvelle*) or a general lecture framed in terms of the old and the new explanations (Ms. 2225). It is for these reasons that the

---

69   See Denis, *Discours sur les comètes*, 78: "si quelqu'un y trouve quelque difficulté, elle ne proviendra pas tant de la chose mesme, que de celuy qui n'estant pas versé dans les principes de cette Philosophie [Descartes's]."

70   Denis, *Discours sur les comètes*, 126: "Mais si quelqu'un avoit de la peine pour les principes, d'où nous avons tiré les consequences precedents, il pourra s'en éclaircir en lisant toute la troisiéme partie des principes de Monsieur Descartes."

third conjecture seems to be the most tenable: unlike in other parts of his natural philosophy (pneumatic experiments, optics, magnets), where Rohault was actively involved in experimentation and ongoing debates, in the section on cosmology, he struggles to stay up to date with recent developments. Maybe he lacked the resources to devote himself to astronomical observations or the topics were not very interesting for Rohault. Be that as it may, in the end, as we have seen above, he was forced to tackle them, in the case of the comets, for example.

Rohault was definitely not part of the French debates about comets and his insertion of the Cartesian vortex model seems to have developed closer to the moment of the publication of the *Traité* in 1671. The caution exercised in this presentation of a Cartesian cosmology is revealed in the permit to publish the treatise, quoted above,—"a Treatise on physics, or of the Science of nature & that of Cosmography, seen by Mr. de Mezeray, our Counselor & Historiographer"—with the second part of the book validated by the Royal historiographer. Even if one would have worried about censorship, the content of Rohault's cosmology is far from being dangerous. Indeed, if we return to the starting point and the lessons Rohault gave to the Conti family in 1668 and 1669, we can conclude that the two globes commissioned to him, together with private lessons, offer the best framework in which his standard cosmological account should be interpreted.

Rohault's lessons must have been introductory, discussing general themes such as the competing hypotheses of the world system. Two illustrations were featured in the second section of the *Traité*, one reproducing the Ptolemaic system and the other the Copernican worldview. In the treatise, Rohault does not hesitate to defend the hypothesis of Nicolaus Copernicus (1473–1543), but the large space allowed for the discussion of Ptolemy's system in the (unofficial) *Physique nouvelle* suggests a late development of the Copernican section. Rohault's private lessons in cosmology seem to have relied too heavily on the traditional astronomical thinking to take full advantage of modern explanations, and this would justify the need to rework them in the *Traité*. A final word about the context which forced Rohault to tackle these topics: the central episode in the French debates about comets was the January 1665 conference at the Collège de Clermont in Paris. Members of the Conti family were among the audience and maybe Rohault was required to develop his views on cosmological topics in order to answer the questions that his famous employers might have put to him subsequent to the conference. Once more, this would explain the slow transition from the traditional account—covered undoubtedly in Rohault's private lessons—to the more encompassing explanation he presented in his published cosmology.

Acknowledgments

This work was supported by a grant of Ministry of Research and Innovation, CNCS—UEFISCDI, project number PN-III-P1–1.1-TE-2019–0841, within PNCDI III. I would like to thank Ovidiu Babeș, Ioana Bujor, Trevor McClaughlin, John Schuster, Grigore Vida, and the anonymous reviewers for their valuable comments and suggestions on the prior drafts.

CHAPTER 20

French Cartesianisms in the 1690s: The Textbooks of Regis and Pourchot

Tad M. Schmaltz

1 French Cartesian Textbooks

The project of presenting Cartesian views in a manner suitable for teaching started with Descartes himself.[1] This is clear from the history of his *Principles of Philosophy* (1644). Descartes announces this work to Mersenne in a 1640 letter as "un Cours de ma Philosophie" that serves as an alternative to the current philosophy texts used in the schools.[2] Indeed, he describes his plan to publish his work with an annotated copy of the "Cours de la Philosophie ordinaire" of the French Cistercian Eustachius a Sancto Paulo (Eustache Asseline) (1573–1640).[3] Descartes abandoned this combined-publication part of the project by 1641,[4] but claimed in his 1642 *Epistola ad Patrem Dinet* that he still intends in the *Principles* to present his philosophy in "a style more suited to the current practice in the Schools."[5] In particular, he modelled this text on the relatively new French genre of the philosophy *cursus*, for which Eustachius's textbook provided a notable model. During the seventeenth century, the teaching of philosophy in France increasingly moved from the Faculty of Arts to *collèges de plein exercice*.[6] Indeed, La Flèche, the Jesuit school where Descartes had been

---

1 Regarding Descartes's own intentions concerning the teaching of his philosophy, see Chapter 1, in this volume. In the paragraphs that follow, I borrow from the discussion in Schmaltz, *Early Modern Cartesianisms*, 84–85, 88.

2 Descartes to Mersenne, 11 November 1640, AT III 233, CSMK III 157. But alongside the desire to be introduced into the schools, Descartes also had a low opinion of academic philosophy and philosophers, on which, again, see Chapter 1, in this volume.

3 Namely, Eustachius a Sancto Paulo, *Summa Philosophiae quadripartita de rebus dialecticbus, moralibus, physicis, et metaphysicis* (Paris, 1609), which was republished several times throughout the seventeenth century.

4 See his comments in Descartes to Mersenne, 22 December 1641, AT III 470. In an earlier letter to Mersenne that same year, he indicated that since the recent death of Eustachius prevented him from receiving permission for the project, he was disinclined to continue to pursue it; Descartes to Mersenne, 21 January 1641, AT III 286.

5 Descartes, *Epistola ad P. Dinet*, AT VII 577, CSMK II 389.

6 This institution was a fifteenth-century development in Paris, but it greatly expanded in the

educated, was itself just such a *collège*. In these colleges, the *cursus* replaced the texts of Aristotle and commentaries thereon that had previously provided the basis for philosophical instruction. These new textbooks bypassed the need for laborious investigations of Aristotle's own works by presenting the basic elements of a properly Christianized Aristotelian position in a readily accessible form.[7] Descartes clearly intended his *Principles* to present his own philosophical system in the same way, and thus to promote its dissemination within the French academy.

The typical scholastic *cursus* divides philosophy into the practical sciences of logic and morals and the speculative sciences of physics and metaphysics, with the topics being taught in that order. With respect to the practical sciences, logic—as the science of right reason—concerns the human intellect, whereas ethics—as the science of action—concerns the human will. Since logic provides a basic preparation for the study of philosophy, it is to be taught prior to ethics. With respect to the speculative sciences, physics is the science of natural body, whereas metaphysics is the science of being as such. Physics, being less abstract than metaphysics and so closer to the senses, tended to be taught prior to metaphysics.

Unlike the traditional *cursus*, Descartes's *Principles of Philosophy* includes no explicit treatment of the practical sciences. At least partially in response to this absence, Descartes indicates in the preface to the French translation (1647) of the *Principles* that both logic and ethics can be appended to the parts of philosophy he presents in this text. He notes that the first step to studying philosophy is to study logic. He distinguishes the "true logic," which teaches us "to direct our reason with a view to discovering truths which are evident," from "the logic of the schools." The latter, by contrast, is only "a dialectic that teaches ways of expounding to others what one already knows or even of holding forth without judgment about things one does not know."[8] The true logic is to be restricted to the four rules of method Descartes introduced to the public in his *Discourse on Method* (1637).[9] In the aforementioned preface, Descartes

---

late-sixteenth and seventeenth centuries, largely due to its adoption as a primary means of education. These *collèges* typically combined secondary instruction in Latin and Greek grammar and rhetoric with at least two years of university-level work in philosophy.

7   For further discussion of this innovation in French teaching, see Brockliss, "Rapports de structure." On the evolution in the schools from dependence on the *corpus Aristotelicum* to the use of the new philosophy textbook, see Schmitt, "Rise of the Philosophical Textbook." On the nature of these textbooks in early modern scholasticism, see Ariew, *Descartes and the First Cartesians*, ch. 2.

8   Descartes, *Principes de la philosophie*, *Préface* AT IX-B 13, CSMK I 185–186.

9   Descartes, *Discours de la méthode* II, AT VI 18–19, CSMK I 120. The discussion of the rules is pre-

adds that the study of philosophy appropriately ends with the science of morals, which, as "the highest and most perfect moral system" that "presupposes a complete knowledge of the other sciences," is to be considered last.[10] But though the indication here is that logic and ethics can be included as parts of the Cartesian system (albeit with ethics coming last rather than second), it remains the case that these parts are not formally set out in Descartes's version of the *cursus*.

By contrast with logic and ethics, metaphysics and physics are front and centre in the *Principles*, though not in the same sequence as they occur in the scholastic order: metaphysics is presented prior to physics. In a famous passage in the preface to the French translation, Descartes indicates the centrality of these disciplines to his system when he speaks of metaphysics as the roots of the tree of philosophy, and of physics as its trunk.[11] Even so, the discussion of metaphysics in this text does not closely correspond to metaphysics in the scholastic *cursus*. The latter introduces metaphysics by focusing on the concept of being and the nature of 'transcendental' properties of being, namely, unity, truth and goodness. It focuses also on the relation of essence to existence and the division of being into uncreated and created substance. Descartes provides no systematic treatment of these issues.[12] Moreover, in the preface to the French translation, he acknowledges the deficiencies of his text with respect to its treatment of physics, noting that it fails to provide any systematic consideration of "the nature of all the particular bodies that exist on the earth, namely, minerals, plants, animals and, most importantly, man." After allowing that he most likely will never be in a position to provide such a consideration, Descartes requests that future generations "forgive me if from now on I give up working on their behalf."[13]

In fact, soon after Descartes's death, there were those who took up the task of providing a complete presentation of the Cartesian system in scholastic form. To my knowledge, the first to do this was Jacques du Roure (fl. 1653–

---

faced by a critique of scholastic logic similar to the one presented in the preface to the French translation of the *Principia philosophiae*; see *Discours de la méthode* II, AT VI 17, CSMK I 119.

10   Descartes, *Principes de la philosophie*, *Préface* AT IX-B 14, CSMK I 186.

11   Descartes, *Principes de la philosophie*, *Préface*, AT IX-B 14, CSMK I 186.

12   Perhaps most relevant to traditional metaphysics is the claim in the *Principles of Philosophy* that substance does not apply "univocally" to God and creatures (*Principia philosophiae* I 51, AT VIII-B 24, CSMK I 210). In this text, there is also a brief discussion of the metaphysical status of universals (*Principia philosophiae* I 55–59, AT VIII-B 26–28, CSMK I 212–213), an issue that was usually treated in the part of the *cursus* concerning logic.

13   Descartes, *Principes de la philosophie*, *Préface*, AT IX-B 16–17, CSMK I 187–188.

1683), a friend of the academic Cartesian Johannes Clauberg (1621–1665).[14] Du Roure's version of the Cartesian *cursus* is *La philosophie divisée en toutes ses parties* (1654), which—as its subtitle notes—promises to compare the philosophy "of the ancients and new authors, and principally of the Peripatetics and of Descartes."[15] Our focus in this chapter, however, is on two different attempts toward the end of the seventeenth century to provide a textbook suitable for the teaching of Cartesian views in France. The first is the *Système de philosophie* (1690) of Pierre-Sylvain Regis (Régis) (1632–1707). I argue in section 2 that this work is to be linked to the attempt to popularize the new Cartesian philosophy outside of the French universities. The second textbook is the *Institutio philosophica* (1695) of the Paris professor Edmond Pourchot (Edmundus Purchotius) (1651–1734). In contrast with Regis's *Système*, Pourchot's textbook is to be associated with an academic rather than a popular French Cartesianism. In particular, I argue in section 3 that there was an important connection between Pourchot's *Institutio* and disputes internal to the University of Paris that derived from the initial imposition of an anti-Cartesian formulary in 1691. I then conclude in section 4 with some brief reflections on the different agendas reflected in the textbooks of Regis and Pourchot.

2   Regis and Popular French Cartesianism

In his role as Permanent Secretary of the Académie des Sciences, Bernard le Bovier de Fontenelle (1657–1757) offered an official eulogy for Regis, who had been appointed as a member of the Académie in 1699.[16] In his memorial, Fontenelle reports that Regis was studying for an ecclesiastical career at the Sorbonne when the popular public presentations of Cartesian physics by Jacques Rohault (which had commenced by 1659) turned him away from this path by

---

14   For more on du Roure and his association with Clauberg, see Roux, "Premiers éléments." As Roux notes, du Roure probably met with Clauberg during the visit of the latter to Paris between 1646 and 1648. In a 1652 letter to Clauberg, he praises the Cartesians as *philosophes raisonnables*.

15   The full title of du Roure's text is *La philosophie divisée en toutes ses parties, établie sur des principes, evidens et expliquée en tables et par discours, ou particuliers, ou tirez des anciens, des nouveaux auteurs; & principalement des Peripaticiens, et de Descartes* (Paris, 1654). A revised second edition of this text was published in 1665 with the new title, *Abrégé de la vraye philosophie*.

16   The appointment of Regis was part of a *renouvellement* of the Académie that included the appointment of Malebranche and others who had a sympathy for Cartesian science. For the complete eulogy, see Fontenelle, *Éloges*, 96–107.

directing him instead to a study of the new Cartesian philosophy. In 1665, Fontenelle continues, Regis departed for Toulouse as a kind of missionary to spread the good news of Cartesianism. His conferences were reportedly attended by various notables, and "Ladies even made a part of the crowd."[17] According to Fontenelle, Regis presided over a discussion in French of "a thesis of pure Cartesianism" (to my knowledge, now lost) dedicated to "one of the leading Ladies of Toulouse" (to my knowledge, of unknown identity) who under Regis's influence became a "fort habile Cartésienne". After his time in Toulouse, Regis accompanied his patron the Marquis de Vardes, first to Aigues-Mortes and then to Montpellier, to continue his work of proselytization for Cartesianism. He returned to Paris in 1680, in part to revive Rohault's Cartesian conferences (Rohault having died in 1672 with no one to replace him as *conférencier*), and in part to publish a manuscript entitled *Système de philosophie*. As Fontenelle notes, in both of these enterprises Regis was thwarted by the Archbishop of Paris, François de Harlay de Champvallon (1625–1695), who suspended his conferences and refused him permission to publish.[18]

We can understand the reaction of Archbishop Harlay in the context of the controversies over Cartesianism that were triggered by a 1671 decree to the University of Paris, issued by Louis XIV through Harlay, then recently installed as Archbishop of Paris. This decree requires "that no other doctrine than the one conveyed in the rules and statutes of the University [be] taught in the universities." The stated reason for the decree was the king's desire,

> having learned that certain opinions that the faculty of theology had once censored and that the Parlement had prohibited from teaching and from publishing, are now being disseminated, not only in the University but also in the rest of this city and in certain parts of the kingdom, either by strangers, or also by people within [...] [to] prevent the course of this opinion that could bring some confusion in the explanation of our mysteries.[19]

---

17   Fontenelle, *Éloges*, 97: "Les dames même faisait partie de la foule." Cf. the report that "les Dames même tiennent souvent le premier rang" were found in the crowds at Rohault's conferences; from the preface to the second volume of Clerselier's edition of Descartes's *Lettres* (Paris, 1659), reproduced in AT V 758–759.

18   Fontenelle, *Éloges*, 97–99.

19   The full text of the decree (quoted in Bouillier, *Histoire*, vol. 1, 469): "Le roi ayant appris que certaines opinions que la Faculté de Théologie avait censurées autrefois et que le parlement avait défendu d'enseigner ni de publier, se répandent présentement, non seulement dans l'Université, mais aussi dans le reste de cette ville et quelques autres du royaume, soit par des étrangers, soit aussi par gens de dedans, voulant empêcher le cours de cette opin-

The most likely culprits as the sources of these "certain opinions" are works by Rohault and Dom Robert Desgabets, published around the time of this decree, that attempt to reconcile the Catholic doctrine of transubstantiation with Cartesian matter theory.[20] Though the king made no explicit mention of Descartes or Cartesianism, his decree prompted a campaign throughout the 1670s against the teaching of the new Cartesian philosophy in various French universities and religious organizations. The issue of the Eucharist continued to be prominent in French discussions of Cartesianism through 1680, as illustrated by the publication that year of the *Sentimens de Monsieur Descartes ... opposez à la Doctrine de l'Eglise, et conforme aux erreurs de Calvin sur le sujet de l'Eucharistie* of 'Louis de la Ville' (the Jesuit Louis le Valois). Since Cartesianism remained a theologically hot topic when Regis arrived in Paris, Harlay was simply not in a position to allow the public promotion of Cartesian views.

There is evidence that Regis revised the text of his *Système* after the refusal of permission to publish. In his discussion of 'the seat of the soul', for example, Regis rejects Descartes's identification of this with the pineal gland in favour of the conclusion reached in the *Neurographia universalis* (1684) of Raymond Vieussens (1641–1715) that the seat of the soul is to be identified rather with the 'centrale ovale' in the brain.[21] There is also a 1682 draft of the section of the *Système* on morals that reveals that Regis modified an earlier version of his text to avoid any suggestion of sympathy for Jansenism (which, as we will dis-

---

ion qui pourrait porter quelque confusion dans l'explication de nos mystères, poussé de son zèle et de sa piété ordinaire, il m'a commandé de vous dire ses intentions. Le roi vous exhorte, Messieurs, de faire en sorte que l'on n'enseigne point dans les universités d'autre doctrine que celle qui est portée par les règlements et les statuts de l'Université, et que l'on n'en mette rien dans les thèses, et laisse à votre prudence et à votre sage conduite de prendre les voies nécessaires pour cela." For further discussion of the 1671 decree and its context, see Schmaltz, *Early Modern Cartesianisms*, 28–35. The Paris condemnation was preceded by the Louvain condemnation of Cartesianism in 1662 and by the listing in 1663 of Descartes's works on the Roman *Index librorum prohibitorum*. On the context of the Louvain condemnation, see Chapter 15, in this volume.

20   The two works in question are Desgabets's anonymous *Considérations sur l'état present de la controverse touchant le Très Saint-Sacrement de l'autel, où il est traité en peu mots l'opin-ion qui enseigne que la matière du pain est change en celle du corps de Jésue-Christ par son union substantielle à son l'ame et à son personne divine* (s.l., 1671), published prior to the decree, and Rohault's *Entretiens sur la philosophie* (Paris, 1671), published after the decree but previewed in his open "Lettre à Monsieur Guyard," which was circulating prior to the decree.

21   For a discussion of the influence of Vieussens on Regis with respect to this point, see Schmaltz, "Pineal Gland."

cover in section 3, was particularly controversial in France around this time).[22] These modifications, along with a decrease in the temperature around public discussions of Cartesianism, no doubt allowed Harlay to grant the permission formerly withheld to publish the *Système* in 1688, but even then only on condition that Descartes's name be removed from the title.[23]

When the *Système* was finally published in 1690, it drew a relatively rapid response from scholastic and Cartesian critics alike. It was the target of the *Réflexions critiques sur le système cartesien de la philosophie de Mr. Regis* (1692) of the Paris scholastic Jean du Hamel, the *Réponse à Regis* (1693) of Nicolas Malebranche, and *La vraye et la fausse métaphysique* (1694) of the *malebranchiste* Henri de Lelevel.[24] During the 1690s, Regis also was the primary Cartesian respondent to the criticisms of Descartes's philosophy in the work of Pierre-Daniel Huet, as reflected in Huet's own—derisive—reference to him as "the Prince of the Cartesian philosophers" who is "today recognized in all the alleyways and among witty and accomplished ladies as the protector of subtle matter, Patron of Globules, and Defender of Vortices."[25] The publication of his *Système*, then, positioned Regis right in the middle of public discussions of Cartesianism in France in the last decade of the seventeenth century.

I have indicated that Regis's *Système* was intended to serve as a kind of scholastic *cursus*. However, it deviates from the traditional *cursus* in presenting the four parts of philosophy in the order specified by Descartes in his preface to

---

22   In particular, Regis modified the text to eliminate any suggestion of sympathy for Jansenist views of human freedom and the operation of grace. Relevant here is the fact that the decade-long *Paix de l'Église*—the truce in the official French campaign against Jansenism—came to an end in 1679. On the differences between the draft and the final text, see Kieft, "Morale rappelée à l'ordre?."

23   The full title of the first edition of this text is *Système de philosophie, contenant la logique, la metaphysique, la physique et la morale* (Paris, 1690). Descartes's name was restored to the title of the second edition of this text, published in Amsterdam in 1691: Regis, *Cours entiers de philosophie; ou Système generale selon les principes de M. Descartes concernant la logique, la metaphysique, la physique, et la morale.* [...] *Derniere Edition, enrichie d'un très-grand nombre des figures, et augmentee d'un discours sur la Philosophie ancienne et modern, où l'on fait en abregé l'histoire de cette Science.*

24   For Regis's exchanges with Du Hamel, Malebranche and Lelevel, see Schmaltz, *Radical Cartesianism*, 233–241, 245–251, and 251–256, respectively.

25   From the dedication in Huet's *Nouveaux mémoires* (published under the pseudonym M.G. de l'A): "À Monsieur Regis, prince des philosophes cartésiens," who is "aujourd'hui reconnu dans toutes les ruelles et parmi les Dames spirituelles et virtuoses pour protecteur de la matière subtile, Patron des Globules, et Defenseur des Tourbillons." The reference to the "les Dames" recalls Fontenelle's previously mentioned report that "les Dames" who attended Regis's conferences in Toulouse "faisait partie de la foule"; see note 17. On the exchange between Regis and Huet, see Schmaltz, *Radical Cartesianism*, 223–233.

the French translation of the *Principles of Philosophy*. Indeed, Regis takes his Cartesian ordering to indicate the sense in which the views in his text constitute a 'system'. In the preface to the *Système*, he protests that those who have attempted to provide a consideration of all the parts of philosophy have failed to show the relations among the parts: "because it is not sufficient to compose a body to join the several parts together; these parts must also have some relation among themselves, without which they would produce only a difform and monstrous body."[26] In contrast, Regis claims that his text serves to illustrate that

> Morals supposes Physics, Physics supposes Metaphysics, and Metaphysics Logic; and by this means all the parts of Philosophy have such a relation and such a connection together that I believe that all that results from their assembly can justly be called *the General System of Philosophy*.[27]

In Regis's *Système*, then, we have a system that follows Descartes's own conception of the relations obtaining among the various parts of philosophy.[28]

As we have seen, Regis's textbook continues the attempt of du Roure to complete Descartes's own project of producing a proper textbook of his philosophy.[29] However, the texts of du Roure and Regis differ from Descartes's *Principles of Philosophy* in being composed originally in French. The two Cartesians seem to have been concerned, not to produce a text for use in the schools, but to produce something for the educated public. In this respect, Regis's *Système* has less in common with Descartes's text than with early modern scholastic manuals such as Louis de Lesclache's *La philosophie divisée en cinq parties* (1648–1650).[30] These manuals were typically dedicated to people associated

---

26   Regis, *Système*, vol. 1, *Prèface*, n.p.: "Car il ne suffit pas pour faire de joindre plusieurs parties ensemble, il faut aussi que ces parties ayent de certains rapports enre ells, sans lesquels elle ne produisent qu'un corps difforme et monstrueux.".

27   Regis, *Système*, vol. 1, *Préface*, n.p.: "Ainsi la Morale suppose la Physique; la Physique suppose la Metaphysique; & la Metaphysique la Logique; & par ce moyen toutes les parties de la Philosophie ont un tel rapport, & une telle liaison ensemble, que j'ay cru le tout qui résultede leur assemblage, pouvoit justement este appellé *le Système Générale de la Philosophie*."

28   On the relative novelty of the use of the term '*système*' in early modern French philosophy, see Armogathe, "Préface."

29   However, whereas du Roure's text is in line with Descartes's original project of offering a comparison of the views of the ancients (Peripatetics) and moderns (Descartes), Regis's text is closer to the published version of Descartes's *Principia philosophiae* in largely foregoing such a comparison. Cf. note 60.

30   To the traditional four parts of philosophy, Lesclache adds a fifth part, on theology. Other examples of early modern scholastic manuals published in French include Boujou, *Corps*

with the court (as in the case of Regis's *Système*),[31] which suggests that there was a French market for this sort of text among members of the nobility who were eager for university-level learning that did not require a proficiency in Latin.[32] Du Roure and Regis were evidently attempting to tap into this market by presenting the new Cartesian philosophy in a familiar scholastic form.

Though the *Système* is standardly Cartesian with respect to the structure of the parts of philosophy, the connection of the content of Regis's text to Descartes's own views is less straightforward. We can understand the complexity of this relation in terms of the distinction made by the aforementioned Desgabets—who, we will find, has an important connection to Regis—between two different kinds of 'supplement' to Descartes's philosophy. Desgabets presents himself as offering a 'first supplement' to Descartes, "insofar as I try to rectify his own thoughts on matters where it seems to me that he has left the correct path that leads to the truth." The 'second supplement', by contrast, involves

> the new application that one would make of his incontestable principles to phenomena that he has not known, or to truths of which he has not spoken, [such as] MM. de Cordemoy, Rohault, de la Forge, Clauberg and others have produced in the excellent works they have given to the public, where one sees the manner we should extend our thoughts to equally excellent and useful matters.[33]

There is a sense in which Regis's *Système* clearly counts as a second supplement that extends Descartes's principles to phenomena Descartes himself had never considered. This is indicated by the comment of Simon Foucher (1644–1696) in a 1701 letter to Leibniz:

> You know that I think M. Regis has given the public a great system of philosophy in three quarto volumes with several figures. This work contains many important treatises, such as the one on percussion by M. Mariotte, chemistry by M. l'Emeri, medicine by M. Vieuxsang [i.e., Vieussens] and

---

de toute la philosophie (Paris, 1614); Scipion Dupleix's *Corps de philosophie, contenant la logique, l'ethique, la physique, et la metaphysique* (Geneva, 1627); René de Cerizier's *Le philosophe françois* (Paris, 1650). For more on Lesclache, see Chapter 18, in this volume.

31   Regis dedicated the *Système* to the abbé de Louvois, the son of the minister in charge of the Académie des Sciences at that time, Michel le Tellier, marquis de Louvois.

32   For this point, see Blair, "Natural Philosophy."

33   Desgabets, *Oeuvres*, vol. 5, 156.

M. d'Uvernai [i.e., du Verney]. He even speaks of my treatise on Hygro-
meters, although he does not name it. There is in it a good portion of the
physics of M. Rohault and he there refutes Father Malbranche [i.e., Maleb-
ranche], M. Perrault, M. Varignon—the first concerning ideas, the second
concerning weight, and the third, who has recently been received by the
Académie royale des Sciences, also on weight. The *Météores* of Father
Lamy also in part adorns his work, and the remainder is from Descartes.
M. Regis conducted himself rather skillfully in his system, especially in his
morals.[34]

The indication here is that Regis's *Système* offers an updating of Descartes that
takes into consideration recent advances in physics, chemistry, medicine and
meteorology (as well as the theory of ideas).

Even so, there is a significant portion of Regis's text that seems to count
as a first supplement in the sense intended by Desgabets. One of the cor-
rections of Descartes occurs in the section of the *Système* devoted to logic. I
have indicated that Descartes himself attempted to replace scholastic syllogistic logic with his own rules of method. However, later Cartesians recognized
that a discussion of these rules was no match for the much richer treatment
in scholastic manuals of the logical operations of mind. This is reflected, for
instance, in the *Logique, ou l'art de penser* (1662) of the Port-Royal *solitaires*
Antoine Arnauld and Pierre Nicole, a work which provided the basis for the
treatment of logic in Regis's *Système*.[35] In line with the standard arrangement
in the scholastic *cursus*, the discussion in the *Logique* is organized around the
three principal operations of mind, namely, conception, judgment and reas-

---

34   Foucher to Leibniz, 30 May 1691, in Leibniz, *Die philosophischen Schriften*, vol. 1, 398–400:
      "Vous savez comme je pense, que Mr Regis a donné au public un grand systeme de philo-
      sophie en 3 in quarto avec plusieurs figures. Cet ouvrage renferme plusieurs traitez de plus
      considérables comme de la percussion de Mr Mariotte, de chymie de M^r l'Emeri, de la
      medicine de Mr Vieuxsang et de M^r d'Uvernai. Il y parle mesme de mon traitté des Hygro-
      metres, quoyqu'il ne me nomme pas. La physique de Mr Rohault y a bonne part, il y refute
      le P. de Malbranche, Mr Perraut, Mr Varignon; le 1er touchant les idées, le 2e touchant la
      pesanteur, et le 3e, lequel a esté nouvellement receu de l'Académie royale des Sciences,
      touchant la pesanteur aussi. Les Methdores du Pere l'Ami font encore une partie des orne-
      mens de cet ouvrage, et le rests de Mr Descartes. Ce n'est pas que Mr Regis ne se soit conduit
      assez adroitement dans son système, surtout dans sa morale." On the final sentence of this
      passage, see note 48.

35   The *Port-Royal Logic* went through five editions, with the last appearing in 1683. On the
      connection of this text to the section on logic in Regis's *Système*, see Milani, "L'Art de
      penser." On the teaching of the *Logique*, see Chapter 17, in this volume.

oning, with an additional consideration given to ordering (method). In some cases, Cartesian elements are imported into the discussion, as in the consideration of conception in terms of "clear and distinct" or "confused and obscure" ideas, and in the appeal—in the discussion of ordering—to the axiom that "everything contained in the clear and distinct idea of a thing can truthfully be affirmed of it."[36] Nonetheless, the *Logique* also provides an extended discussion of the syllogism as part of the consideration of the mental operation of reasoning; there is just such a discussion also in the first part of Regis's *Système*. Following Arnauld and Nicole, then, Regis adopts (a Cartesianized version of) the very "logic of the schools" that Descartes consistently disparaged.[37]

However, in the metaphysics section of the *Système*, there are rather more dramatic deviations from Descartes that reflect Desgabets's influence on Regis. The importance of Desgabets for this text is indicated by the claim in a 1712 dictionary that "Regis had a great deal to do with Father Desgabets, and he profited greatly from his illuminations [*lumières*] in the three volumes of his philosophy that he published."[38] In his last published text, moreover, Regis calls Desgabets "one of the greatest metaphysicians of our age."[39] According to one distinctive metaphysical thesis in Desgabets that is particularly relevant to us here, Descartes's doctrine of the creation of eternal truths requires the "indefectibility of creatures," according to which the substances God creates have an atemporal existence that renders them unchangeable (indefectible). Desgabets distinguishes these substances from the 'modal beings' dependent on them, which, in contrast to the substances, have a temporal existence that is subject to change.[40] Regis's insistence in the *Système* on the need to distinguish the creation of the world from its conservation is clearly influenced by the

---

36   Arnauld and Nicole, *La logique ou l'art de penser*, 156–162, 381–382.

37   To discern the difference, one need only compare Descartes's remark in the *Discours de la méthode* II, AT VI 17, that syllogisms "servent plutôt a expliquer à autrui les choses qu'on sait, ou même, ... a parler, sans jugement, de celles qu'on ignore" to Regis's indication that this form of logic is an important species of reasoning (Regis, *Système*, vol. 1, 31–39). In the preface to his text, Regis (*Système*, vol. 1, *Préface*, n.p.) notes that his discussion of logic draws on "des réfléxions qui ont été proposes par l'Auteur de l'Art de penser, auxquelles il est difficile de rien ajoûter."

38   Moréri, *Le Grand dictionnaire historique, ou le mélange curieux de l'histoire sacré et profane* (Paris, 1712), vol. 2, 602.

39   Regis, *Usage*, 696.

40   See *Traité de l'indéfectibilité des creatures* (started in 1653), in Desgabets, *Oeuvres*, vol 2, 34. For further discussion of this position in Desgabets, see Schmaltz, *Radical Cartesianism*, ch. 2.

earlier distinction drawn by Desgabets. In Descartes, of course, God's conservation of the world is not really distinct from his initial creation of the world.[41] For Regis, however, creation involves God's immediate production of the 'permanent' being of substances, which is not subject to a changing temporal duration. Conservation, by contrast, involves God's mediate production of the 'successive' being of modes, which is subject to this kind of duration.[42] Here a central metaphysical position in Descartes is corrected in the light of the innovations of Desgabets.

In addition to the revised view of the nature of created substance received from Desgabets, in the *Système*, Regis draws a further difference from Descartes concerning the status of corporeal substance. Descartes had claimed that body is divisible by its very nature into really distinct parts, and indeed that this very divisibility serves to distinguish body as a thing that is extended from mind as a thing that thinks.[43] Regis argues, however, that since created corporeal substance is indefectible, it cannot be subject to changes due to division. Divisibility pertains not directly to corporeal substance, but rather to what Regis calls its *quantité*, that is, "body considered as such or such according to size."[44] Various quantities are the divisible, and thus mutable, modes of the one indivisible and immutable corporeal substance. In this way, Regis endorses a kind of 'monism' with respect to corporeal substance that other commentators have attributed to Descartes but that seems difficult to square with Descartes's own account of divisibility.[45]

The final section of the *Système*, on morals, is perhaps the least Cartesian of the four sections of this text.[46] For one thing, the basic structure of the section is not recognizably Cartesian. Regis divides morality into three main parts: natural, civil, and Christian, with the first following from natural reason alone, the second from the relation of the individual to a particular state, and the third from what divine revelation requires for morality. This schema can-

---

41   As indicated in Descartes, *Meditationes* III, AT VII 49, CSMK II 33.

42   Regis, *Système*, vol. 1, 107–109. The distance from Descartes is somewhat obscured by the fact that Regis, in one instance, explicitly endorses the very identification of conservation with creation that he is committed to rejecting; Regis, *Système*, vol. 1, 99. In a later text, however, he emphasizes that his distinction between creation and conservation serves to "correct the reasonings of a very considerable philosopher," identified in the margin as "M. Descartes"; Regis, *Usage*, 321–322.

43   See Descartes, *Meditationes* VI, AT VII 85–86, CSMK II 59.

44   Regis, *Système*, vol. 1, 279–280.

45   For discussion of the monist interpretation of Descartes's account of corporeal substance, as well as its difficulties, see Schmaltz, "Descartes on the Metaphysics."

46   For discussion of Cartesian textbooks on ethics, see Chapter 13, in this volume.

not be found in Descartes, but is familiar from the work of Thomas Hobbes (1588–1679) and Samuel Pufendorf (1632–1694).[47] Moreover, the influence specifically of Hobbes is evident in Regis's view of the formation of civil society by means of a contract that involves the surrender of the rights one has in a 'state of nature'.[48]

Given the—Desgabets-inspired—deviations from Descartes in the metaphysics section of the *Système*, as well as the importance of non-Cartesian sources for its section on morals, it might be difficult to understand how this work could be seen—as it clearly was at the time—as representative of a Cartesian position. Here, I think, Regis's notoriety among his contemporaries as a popularizer of Cartesianism—the true successor to Rohault—helps to explain this perception. Moreover, a part of the explanation must derive from the fact that 'Cartesianism' was a somewhat fluid notion in the early modern period. This is illustrated by the latter-day assessment that "there hardly was a doctrine, view, or argument that was advanced by everyone thought, and rightly thought, to be Cartesian."[49] Cartesianism was not a doctrine with a well-defined 'essence', but rather more akin to a biological species characterized by phenotypic diversity.[50] In the early modern period, an allegiance to Descartes could come with different approaches, strategies and positions. Which, to return to the first point, makes it all the more important for Regis's status as a 'Cartesian' that he was perceived as—and perceived himself to be— a primary promoter of Descartes's views (albeit in sometimes corrected and refined formulations).

Regis's *Système* clearly had an impact in the public sphere. But does it serve to explain the increasing importance of Cartesianism to the curriculum of French universities toward the end of the seventeenth century? I have already indicated that since it was written in French, this textbook would not have

---

47   In particular, the tripartite division of the *Système* is found in Hobbes's *De cive* (Paris, 1642) and Pufendorf's *De officio hominis et civis juxta legem naturalem* (Lund, 1673).

48   On the formation of civil society, see Regis, *Système*, vol. 3, 412–416, 447–449. Recognizably Hobbesian also is Regis's view that the sovereign is not subject to civil law; see Regis, *Système*, vol. 3, 461–463. There is a discussion of the relation of Regis's views to those of Hobbes in Canziani, "Entre Descartes et Hobbes." Incidentally, the fact that the section on morals in Regis's text reflects the views in non-Cartesian sources may well explain why, in the passage from his letter to Leibniz cited earlier, the Cartesian critic Foucher singled it out for praise.

49   From the editorial introduction in Lennon and Easton, eds., *The Cartesian Empiricism of François Bayle*, 1.

50   For further consideration of this suggestion, see Schmaltz, *Early Modern Cartesianisms*, 5–11.

been suitable for use at institutions such as the University of Paris. Nor, to my knowledge, is there any evidence that the *Système* was ever used in any French classroom.[51] However, it is possible that this text had an indirect effect on university teaching. The dissemination of Regis's Cartesian textbook, together with his and Rohault's public presentations of Cartesianism, could well have pushed French professors towards the teaching of the new ideas that were exciting their students outside of the classroom.[52]

Nevertheless, there is one example of an early modern French Cartesian textbook that was produced in Latin and that did not originate outside of French academia, but was rather a product thereof. What makes this work—namely, the *Institutio philosophica* of Pourchot—particularly interesting for our purposes is that it was one of the targets of an internal campaign against Cartesianism at the University of Paris during the 1690s and early-1700s. In the final section of this chapter, I focus on the significance of Pourchot and his *Institutio* for French academic disputes over Cartesianism.

3   Pourchot and Academic French Cartesianism

Paul Dibon has observed—correctly—that "the history of Cartesianism in the Netherlands is dominated by university controversies."[53] Descartes's philosophy gained a foothold in Dutch universities during his own lifetime, and battles over Cartesianism raged in the Netherlands throughout the seventeenth century.[54] French academics, by contrast, for the most part simply ignored Descartes during his lifetime, and Cartesianism seriously infiltrated the University of Paris only toward the very end of the seventeenth century. To be sure, there was the initial surge of interest connected to the royal decree in 1671. However, the issue of Cartesianism was not fully engaged within the university

---

51   Roger Ariew, "Descartes and the First Cartesians Revisited," 612, has argued, to the contrary, that "writing in the vernacular increased significantly throughout the seventeenth century," citing the fact that "the Oratorians even began to teach in French." However, Latin was still the language of instruction at the University of Paris into the eighteenth century. Moreover, I know of no evidence that the Oratorians, or anyone else, used Regis's manual for classroom instruction.

52   This is suggested as a reason for the change in the academic fortunes of Cartesianism in France, in Brockliss, "Aristotle, Descartes and the New Science," 66–67.

53   Dibon, *Regards sur la Hollande*, 606.

54   This explains the fact that a large portion of the chapters in this volume are devoted to the early modern teaching of Descartes in a Dutch context.

until 1691, with the start of a campaign directed against the new philosophy. Just as the temperature around public disputes over the Cartesian account of the Eucharist had cooled sufficiently for Archbishop Harlay to grant Regis permission to publish his *Système*, the controversy over Cartesianism started to heat up among professors at the University of Paris.

One Cartesian *Mémoire* provides an account of the events leading up to the 1691 controversy.[55] According to this report, in August of that year, a list of propositions "supposedly extracted from the writings of professors of philosophy at the University" was passed along to Harlay by a person motivated by "ill-will toward the University" (*mal intentionée pour l'Université*). Having examined various theses sponsored by these professors, the archbishop found nothing corresponding to the propositions, and that seemed to be the end of the matter. However, the following October, the same list of propositions was brought to the attention of the king. Always highly sensitive to any hint of disorder in his universities, Louis XIV instructed his archbishop to conduct a new examination of the philosophy professors. Not wanting to be delayed by protracted investigations, Harlay took the simpler route of having the rector of the University, Gentilhomme, require the signature of each member of the Faculty of Arts on a formulary denouncing the propositions.

The *Mémoire* reveals the identity neither of the person behind the 1691 formulary nor of its targets within the University. It does, however, relate an incident in 1699 involving a certain member of the Paris Faculty of Medicine, M. Desprez (that is, Jacques Desprez). Repeated complaints from this individual about deviations from the formulary were passed along to the king, apparently through Louis XIV's confessor, the Jesuit Père de la Chaize. The king again instructed his Archbishop of Paris, now Louis-Antoine de Noailles (1651–1729), to take measures that would restore peace within the University. This order resulted in a tribunal on the formulary, which ultimately recommended no further action. The author of the *Mémoire* claimed that what motivated Desprez's complaints was his desire to gain a certain benefice by showing his loyalty to the religion of the king; in this, he was thwarted by a report on his conduct to the king by "wise people [...] who have good will toward the University (*bien intentionées pour l'Université*)." Suspecting that Pourchot, then a syndic of the University, was a member of the group that had conspired against him, Desprez complained in 1700 to the Secretary of State, the comte de Pontchartrain, that Pourchot was in fact the principal proponent in the University of the doctrines condemned in the 1691 formulary. Once again, there was

---

55   The *Mémoire* is reproduced in Jourdain, *Histoire de l'université de Paris*, vol. 2, 127–128.

an investigation by Archbishop Noailles, which included a review of a forth-coming second edition of Pourchot's *Institutio*. Once again, no new action was taken.[56]

One cannot accept at face value the report's account of Desprez's motiva-tions. The *Mémoire* has a decidedly Cartesian slant, after all. However, there can be little doubt that by 1700, Pourchot had become a primary target of the campaign against Cartesianism in the University of Paris. This is confirmed in the famous "Arrêt burlesque" (1671) of Nicolas Boileau-Despréaux (1636–1711), which had originally been written in defence of "Cartistes et Gassendistes" in the controversies arising from Louis XIV's 1671 decree, but which in a 1700 edi-tion added to its list of heroes "Pourchotistes et Malebranchistes."[57] It would be reasonable to regard Pourchot as a main target of the campaign against Cartesianism, and even of the 1691 formulary, since he and his colleague Guil-laume Dagoumer were at the vanguard of the effort to smuggle the Cartesian philosophy into university courses by presenting it in a scholastic form.

While the 1671 edict was a vague decree directed against those who would introduce some sort of confusion into a discussion of the mysteries, the 1691 condemnation was constituted by a formulary rejecting a set of specific pro-positions that "His Majesty desires not to be held in the schools." In particular, the formulary rejects the following eleven propositions:

1. One must rid oneself of all kinds of prejudices, and doubt everything before being assured of any knowledge;
2. One must doubt whether there is a God until one has a clear and distinct knowledge of it;
3. We do not know whether God did not create us such that we are always deceived in the very things that appear the clearest;
4. In philosophy, one must not develop fully the unfortunate consequences that an opinion might have for faith, even when the opinion appears incompatible with faith; notwithstanding this, one must stop at that opin-ion, if it seems evident;
5. The matter of bodies is nothing other than their extension, and one can-not exist without the other;
6. One must reject all the reasons the theologians and the philosophers have used until now with Saint Thomas to demonstrate that there is a God;
7. Faith, hope, and charity and generally all the supernatural habits are noth-ing spiritual distinct from the soul, as the natural habits are nothing spir-itual distinct from mind and will;

56   Jourdain, *Histoire de l'université de Paris*, vol. 2, 128.
57   Boileau-Despréaux, *Oeuvres*, vol. 3, 144.

8. All the actions of the infidels are sins;

9. The state of pure nature is impossible;

10. The invincible ignorance of natural right does not excuse sin;

11. One is free, providing that one acts with judgment and with full knowledge, even when one acts necessarily.[58]

This formulary subsequently became the primary vehicle for addressing academic controversies over Cartesianism in Paris: further signings of it were required in 1693, 1704, 1705 and 1707.

We have seen good reason to suspect that Pourchot was a target of the 1691 formulary. Pourchot had a social status that clearly differs from that of Regis. Whereas Regis was a public intellectual, Pourchot devoted himself to an academic career.[59] At the tender age of twenty-six, he began teaching philosophy at the Collège des Grassins. By the time of the initial imposition of the formulary, he was a philosophy professor in the Collège Mazarin, where he remained until 1703, after which he devoted himself to teaching Hebrew in the Collège de Sainte Barbe, which he helped to establish. He had significant tours of administrative duty, serving as rector of the Sorbonne seven times, and as its syndic for forty years. In 1695, Pouchot published—anonymously—a collection of lecture notes from his long career of teaching philosophy: *Institutio philosophica ad faciliorem veterum ac recentiorum philosophorum intelligentiam comparatae.*[60]

---

58   "1. Il faut se défaire de toutes sortes de préjugés, et douter de tout avant que de s'assurer d'aucune connaissance; 2. Il faut douter s'il y a un Dieu, jusqu'à ce qu'on en ait une claire connaissance; 3. Nous ignorons si Dieu ne nous a pas voulu crêer de telle sorte que nous soyons toûjours trompés dans les choses mêmes qui paraissent les plus claires; 4. En philosophie, il ne faut pas se mettre en peine des conséquences fâcheuses qu'un sentiment peut avoir pour la foy, quand même il paroîtrait incompatible avec elle; nonobstant cela, il faut s'arrêter à cette opinion, si elle semble évidente; 5. La matière des corps n'est rien autre chose que leur étendue, et l'une ne peut être sans l'autre; 6. Il faut rejeter toutes les raisons dont les theologiens et les philosophes se sont servi jusqu'icy avec S. Thomas, pour démontrer qu'il y a un Dieu; 7. La Foy, l'Esperance, et la Charité et généralement les habitudes surnaturelles ne sont rien de spirituel distingués de l'âme, comme les habitudes naturelles ne sont rien de spirituel distingués de l'esprit et de la volonté; 8. Toutes les actiones des infidèles sont des pechés; 9. L'état de pure nature est impossible; 10. L'ignorance invincible du droit naturel n'excuse pas de pecher; 11. On est libre, pourvû qu'on agisse avec jugement et avec pleine connoissance, quand même on agiroit necessairement" (quoted in Ariew, "Condamnations," 5).

59   For biographical information on Pourchot, I am relying on Piaia, "Histories of Philosophy," 88–89.

60   Pourchot's *Institutio philosophica* (Paris, 1695) is closer to du Roure's Cartesian textbook than to Regis's *Système* in offering a comparison of 'old' and 'new' views; see note 29. An example of an earlier scholastic manual that offers such a comparison is provided by the *Philosophia vetus et nova ad usum scholae accommodata* (Paris, 1678) of Jean-Baptiste du

He continued to revise this text over several editions, some of which were published under his name and with the plural form of the title: *Institutiones philosophiae*.[61] Though the first edition was not published until four years after the formulary was first imposed, the work clearly covers material from lectures prior to that time. The first of this four-volume textbook is devoted to logic, metaphysics and the elements of geometry; the second to "general physics" (*physica generalis*), including a consideration of basic matter theory; the third to "special physics" (*physica specialis*), including a consideration of cosmology and the nature of living bodies; and the fourth to morals, with a set of *Exercitationes scholasticae* appended.[62]

As in the case of Regis's *Système*, there is the recognizably Cartesian ordering of the standard parts of philosophy: logic, metaphysics, physics, and morals. Another 'new' feature, not anticipated in Regis's *Système*, is the addition of a summary of basic Euclidean geometry to metaphysics in the foundations of physics.[63] Moreover, the section on general physics includes a defence of a Cartesian theory of the elements, whereas the section on special physics includes a presentation of Cartesian cosmology.[64] There are thus Cartesian features of the *Institutio* that would have been obvious to Pourchot's Paris colleagues.

With respect to the specific points to which Pourchot's critics took exception, the first four propositions of the 1691 formulary highlight the dangers of the Cartesian method of doubt, while the last four are drawn from various Church condemnations of the theological views of the Louvain theologian

---

Hamel (not to be confused with Regis's scholastic critic Jean du Hamel). On du Hamel's text, see Schmaltz, *Early Modern Cartesianisms*, 112–115.

61   Subsequent editions were published in both France and Italy: Paris, 1700; Lyon, 1711; Venice, 1715; Lyon, 1716–1717, 1733; Paris, 1733; Padua 1751; and Venice, 1755.

62   The *Exercitationes* was republished as a separate volume in 1700 and 1711, and an *Appendix ad Institutiones philosophicas* was published separately in 1733.

63   See Pourchot, *Institutio*, vol. 1, 481–536.

64   For the defence of the Cartesian theory of the elements, see Pourchot, *Institutio*, vol. 2, 87–88. For the presentation of Cartesian cosmology, identified as the "Copernico-Cartesian system" (*Systema Copernicii sive Cartesii*), see Pourchot, *Institutio*, vol. 3, 20–38. On scriptural grounds, Pourchot himself ultimately favours the geostatic Tychonic system over the Copernico-Cartesian posit of a moving Earth; Pourchot, *Institutio*, vol. 3, 42–46. Nonetheless, the latter is accepted as a useful 'hypothesis', in distinction from an asserted 'thesis'; Pourchot, *Institutio*, vol. 3, 31–32. To my knowledge, Pourchot never considered Descartes's (in)famous proposal in his *Principia philosophiae* that the Earth, strictly speaking, is at rest since it orbits the sun together with the celestial fluid that surrounds it; for this proposal, see Descartes, *Principia philosophiae* III 28–29, AT VIII-A 90–91, CSMK I 252.

Cornelius Jansenius (Jansen) (1585–1638) and his controversial precursor at
Louvain, Michael Baius (Michel de Bay) (1513–1589). In contrast, the middle
three propositions constitute something of a miscellany that nonetheless bear
the clearest connection to the views in Pourchot's *Institutio*. In attempting to
understand the motivations for the choice of the propositions in the formulary,
I will start with the middle set.[65]

In his textbook, Pourchot might seem to violate the injunction against the
fifth proposition in endorsing the doctrine in Descartes that the nature of body
consists in its being extended in length, breadth, and depth.[66] However, Pourchot was well aware of the theological minefield here, and was careful to stip-
ulate from the start that physics considers body only *quantus naturale est*, and
to note at one point that given the doctrine of the Eucharist, matter can be
said to be extended only 'naturally' and not 'essentially'.[67] Though these moves
allow him to assert sincerely that he has complied with the letter of the 1691
formulary, his subtleties were no doubt lost on the critics of Descartes within
the university.

The sixth proposition rejects traditional arguments drawn from Aquinas for
the existence of God. Clearly relevant here is the fact that Descartes offered his
own arguments as preferable to those of Aquinas.[68] In the section of his *Institutio* on metaphysics, Pourchot in fact highlights the 'metaphysical' arguments of
'Cartesius' drawn from Meditations III and V. To be sure, Pourchot also reports
the 'physical' arguments in Aquinas's *Summa Theologiae*, and thus can hardly
be said to reject "all the reasons" of the theologians and the philosophers that
derive from "Saint Thomas".[69] Even so, the spotlight in Pourchot's textbook is on

---

65   My discussion here of the middle propositions draws on the remarks in Schmaltz, "Tale of
      Two Condemnations," 215–217.

66   Pourchot, *Institutio*, vol. 2, 84–87. Pourchot follows Descartes in endorsing a plenist con-
      ception of matter, in opposition to Gassendi's atomism. For the arguments against atoms
      and void space in Descartes's *Principia philosophiae*, AT VIII-A 51–52, 49; CSM I 231–232,
      229–230.

67   Pourchot, *Institutio*, vol. 2, 8 and 29–30, respectively. Pourchot takes only the aptitude to
      be extended (what he calls *extensio radicitus*), and not actual extension (*extensio actualis*),
      to be essential to matter; see Pourchot, *Institutio*, vol. 2, 31.

68   For instance, when responding to the claim of his Dutch critic Caterus that his argument
      for the existence of God in *Meditation* III takes "exactly the same approach as that taken
      by St. Thomas" (AT VII 94, CSMK II 68), Descartes insists that his approach differs funda-
      mentally from Aquinas's insofar as he regards "the existence of God as much more evident
      than the existence of anything perceived by the senses," and does not require that any suc-
      cession of causes require a first cause; AT VII 106–107, CSMK II 77–78.

69   See Pourchot, *Institutio*, vol. 1, 284–299.

Descartes's arguments. One can understand how a sensitive critic would here see an affront to tradition, but also why, in his review of this text, Archbishop Noailles would have found nothing objectionable.

It might seem that, in contrast to the fifth and sixth propositions, the seventh proposition, concerning natural and supernatural habits, is not relevant to a consideration of Cartesianism. However, the discussion in the 1682 *Principes de la philosophie contre les nouveaux philosophes* of the Oratorian Jean-Baptiste de la Grange (ca. 1641—after 1680) indicates the relevance of the seventh proposition. In his text, La Grange considers the Cartesian axiom that anything other than substance is a mode that is inseparable from the substance it modifies. He concludes that this axiom conflicts not only with the Catholic doctrine that eucharistic accidents can exist apart from their subject, but also with the teaching of the Council of Trent that supernaturally induced states, such as charity and justifying grace, are causes that inhere in the soul but nonetheless are distinct from it.[70]

In his lectures on "pneumatology" (*pneumatologia*: a traditional subdiscipline of metaphysics devoted to spiritual substances), Pourchot insists that supernatural habits are distinct from our natural habits simply in virtue of the fact that the former are infused by God.[71] To be sure, he also is defiant in denying that any of these habits are—or that Church doctrine requires that they are—entities that can subsist apart from the soul. However, Pourchot is careful to allow for the position that spiritual habits differ in some sense from natural habits, and thus are distinct in some sense from natural features of the soul. Here he is helped somewhat by ambiguities in the requirement to deny the seventh proposition. It is unclear whether one was required to deny simply that spiritual habits are nothing distinct from the soul, or rather that spiritual habits are nothing distinct in the same way that natural habits are nothing distinct. It was therefore open to Pourchot to understand the requirement in this second, more restricted manner. Even though La Grange had argued at length against the claim that natural habits are nothing distinct from the soul, moreover, Pourchot could read the formulary as prohibiting only a (restricted) claim about spiritual habits.[72]

---

70   La Grange, *Principes de la philosophie* (Paris, 1675), 90–99. On the eucharistic accidents, see La Grange, *Principes de la philosophie*, 109–135. La Grange claims that the doctrine of the Council of Trent—that grace is an infused cause of meritorious action—requires that such grace cannot simply be a mode of the soul.

71   Pouchot, *Institutio*, vol. 1, 469, 475–476.

72   La Grange, *Principes de la philosophie*, 65–98. La Grange insists in particular that the moral virtues and science or knowledge are "Etres [qui] different de notre Ame."

To this point, we have discovered some connections between the middle propositions of the formulary and the discussion in Pourchot's *Institutio*. It seems that there can be no such connection to the first four propositions, which concern the Cartesian method of doubt. As with the *Logique* and Regis's *Système*, the discussion of logic in the *Institutio* is structured around the standard division among simple perceptions, judgments concerning propositions, and the combination of propositions in syllogistic reasoning.[73] Pourchot's discussion of logical 'method' provides what is, in comparison to the lengthy discussion of the other logical operations, a rather brief consideration of analysis and synthesis.[74] Again in line with previous Cartesian texts, Pourchot introduces the Cartesian terminology of clear/distinct and confused/obscure ideas, and he emphasizes the utility of doubting beliefs based on prejudice.[75] Yet he does not require that one "doubt everything" even to the extent of allowing that God could deceive, and he certainly never concedes that there could be clear and distinct ideas that are incompatible with faith.[76]

Nor is the *Institutio* directly implicated in the sort of theological issues concerning sin, grace and free will that are prominent in the last four propositions of the 1691 formulary. The eighth and ninth propositions were most likely drawn rather from a 1567 papal bull, *Ex omnibus afflictionibus*, that condemns as heretical a list of 79 theses purportedly drawn from the work of Baius. Indeed, the eighth proposition is taken verbatim from one of the theses quoted in this papal bull.[77] In affirming that all actions of the infidels are sins, Baius was setting himself in opposition to the view, common among the Jesuits, that even those who have fallen retain the power to turn away from sin. The papal bull also condemns a proposition that asserts, in effect, that God cannot create humans with a "pure nature" (*natura pura*) that is free from the effects both of

---

73   Pourchot, *Institutio*, vol. 1, 26–28.
74   Pourchot, *Institutio*, vol. 1, 181–193.
75   On the Cartesian terminology of clear/distinct and confused/obscure ideas, see Pourchot, *Institutio*, vol. 1, 59–65; on the utility of doubt, see Pourchot, *Institutio*, vol. 1, 104. I find nothing similar in Pourchot to Descartes's use of 'hyperbolic doubt' to justify the rule that clear and distinct ideas are true.
76   I have argued elsewhere that the likely source for the first four propositions is the *Censura philosophiae cartesianae* (Paris, 1689) of Huet, which includes a detailed critical examination of Descartes's method of doubt as well as the charge that Descartes accepts his account of matter on the basis of clear and distinct perception, even though he recognizes its incompatibility with what the Church teaches concerning the Eucharist; see Schmaltz, *Early Modern Cartesianisms*, 52–53.
77   Thesis 25 claims: "Omnia opera infidelium sunt peccata, et philosophorum virtutes sunt vitia" (Denzinger, *Enchiridion*, 431).

original sin and of supernatural additions such as grace.[78] Sixteenth-century Jesuits had introduced the notion of pure nature in order to consider humans simply in terms of their natural ends and powers. Baius opposed the very possibility of pure nature on the grounds that human nature cannot be conceived in abstraction from its dependence on divine grace. Jansenius took up the case against pure nature in his *Augustinus* (published posthumously, 1640), where he attempted to dodge the charge of heresy by claiming (somewhat dubiously) that the earlier bull had condemned merely the manner in which Baius rejected pure nature, and not the rejection itself.[79]

Jansenius further suggests in the *Augustinus*—in line with the tenth proposition of the formulary—that, in their fallen state, human beings may be incapable of the judgment required to perform good works, and are yet still responsible for their own sins.[80] Moreover, Baius was condemned in the 1567 bull for holding that an action can be free "even if necessary," and one of the five propositions that later bulls condemned and attributed to the *Augustinus* held that merit or demerit for action requires not freedom from the necessity of willing and acting, but merely freedom from external constraint.[81]

In condemning the last four propositions, then, the formulary was condemning deviations from Church tradition derived from the work of Baius and Jansenius. Here is a clear shift away from the earlier focus of French critics on the tensions between Cartesianism and the doctrine of the Eucharist. Why this change in strategy? One possible reason is that Cartesians such as Pourchot became adroit at sidestepping difficulties pertaining to the Eucharist. But more basically, I suspect that after two decades, these difficulties no longer retained their original force, as suggested by the fact that Harlay felt comfortable enough to grant Regis permission to publish the *Système* in 1688. There were the difficulties for the Cartesian method indicated in the first four propositions of the formulary, but these were only tenuously connected to university teach-

---

78   Thesis 55 claims: "Deus non potuisset ab initio talem creare hominem, qualis nunc nascitur" (Denzinger, *Enchiridion*, 434). Though there is no direct reference here to 'pure nature', the thesis was taken to assert the impossibility that God can create a human being in an initial state that does not include any supernatural aid.

79   See Jansenius, *Augustinus* (Louvain, 1640), vol. 2, 669–679, 975–980.

80   Jansenius, *Augustinus*, vol. 2, 305–310.

81   For Baius, see Thesis 39: "Quod voluntarie fit, etiam si necessario fiat, libere tamen fit" (Denzinger, *Enchiridion*, 433). The third of the condemned Jansenist propositions claims, "Ad merendum et demerendum in statu naturae lapsae non requiritur in homine libertas a necessitate, sed sufficit libertas a coactione" (Denzinger, *Enchiridion*, 445, 447). The basis for the attribution of this proposition to Jansenius is his discussion in the *Augustinius*, vol. 2, 657–664.

ing, and did not, in any case, concern a doctrine that the Church had explicitly declared to be heretical. Jansenism, on the other hand, was formally identified as heresy, and by the 1690s the defeat of Jansenism in France had become a matter of state. One significant offshoot of these countervailing tendencies—or so I would propose—was the Cartesio-Jansenism condemned in the 1691 formulary.

Pourchot himself was as insulated from the Jansenism of the last four propositions of the formulary as he was from the Cartesianism of its first four propositions. In the *Institutio*, for example, he explicitly affirms that human freedom (*libertas humana*) excludes both external and internal necessitation.[82] Moreover, the tactic in the formulary of linking Cartesianism to a theologically heretical and politically dangerous Jansenism ultimately proved unsuccessful. This is illustrated by the fact that in 1720, the University of Paris instituted new statutes that recommend the use in the classroom of the *Port-Royal Logic* in matters pertaining to logic, Descartes's *Meditations* in matters pertaining to metaphysics, and to draw "from old and recent" works (*ex veterum recentiorumque*) for treatments of physics.[83] The author of these statues? Reportedly, none other than our own Edmond Pourchot![84]

4   Teaching Cartesianism in the 1690s

Even this limited survey of Cartesian textbooks serves to reveal that there were very different strategies for disseminating Cartesian ideas in early modern France. In the case of Regis, we have an attempt to provide to the educated public a version of Descartes's system that has a complete (though modified) scholastic form and that accommodates scientific discoveries and philosophical insights that postdate Descartes himself. Even though Regis was prohibited from including Descartes's name in the title of the edition of the *Système* published in France, he was nonetheless able to claim in the preface to this edition that "all that I have said [must] be attributed to *Monsieur Descartes*, whose

---

82   Pourchot, *Institutio*, vol. 1, 462–464. There is an elaboration of this point, appropriately enough, in the section on morals; see Pourchot, *Institutio*, vol. 4, 80–83.

83   Jourdain, *Histoire de l'université de Paris*, vol. 2, 173.

84   Jourdain, *Histoire de l'université de Paris*, vol. 2, 169 n. 1: "Ce projet de statuts … est écrit de la main du Pourchot." No doubt the adoption of Pourchot's statutes was made possible by the death of Louis XIV in 1715, as well as by the subsequent policy of the regent, Philippe, duc d'Orléans, of letting the university conduct its own affairs. However, it is striking that the regent allowed the adoption of these statutes just when he was starting to return to Louis XIV's policy of enforcing public opposition to Jansenism.

Method and Principles I have followed even in the explanations that are different from his."[85] Regis's system can thus be attributed to Descartes, even in cases
where it departs from the letter of Descartes's own views. What Regis offers,
then, is a new and improved system of Descartes, suitable for the appraisal of
an educated public at the end of the seventeenth century.[86]

In contrast, we have in Pourchot's *Institutio* a document that, first of all, is directed to the classroom rather than to the public, and secondly, is not presented
as something to be attributed to—or seen as deriving from—Descartes. With
respect to the first point, Pourchot needed to be sensitive to the constraints on
teaching reflected in the 1691 formulary. To be sure, Regis also faced constraints,
as shown by his protracted negotiations with Archbishop Harlay over permission to publish the *Système*. Nonetheless, these negotiations allowed Regis to
attribute his system to Descartes (even if not in the title), something Pourchot
certainly would not have been in a position to do in his *Institutio*. Indeed, given
the need to avoid the acknowledgement of any debt to Descartes, one may wonder whether there is good reason to characterize Pourchot's text as 'Cartesian'
in the first place. To be sure, the *Institutio* follows, as does Regis's *Système*, a
Cartesian rather than a more traditional scholastic order with respect to its
various parts, but this seems to provide a rather slim basis for taking Pourchot's
text to be Cartesian. While he may not have attributed his views to Descartes,
however, Pourchot was a key participant in academic disputes over Cartesianism during the 1690s and early-1700s and did not need to explicitly ally himself with Descartes in order for his colleagues to discern such a connection.
Whatever external influence the explicitly Cartesian views in Regis's *Système*
may have had on the French academic reception of Descartes, it is the less
partisan presentation in Pourchot's *Institutio* that played a direct role in the
battles over—and the ultimate victory of—Cartesianism within the University
of Paris.

---

85   Regis, *Système*, vol. 1, *Préface*, n.p.: "Tout ce que j'ai dit, devant être attribué à *Monsieru
      Des-Cartes*, don't j'ai suivi le Méthode et les Principes dans les explications mêmes qui
      sont différentes des siennes." As indicated in note 23, the retitled second edition of the
      Système, published in Amsterdam in 1691, includes the claim that the "Système generale"
      of Regis is "selon les principes de M. Descartes".
86   Although I cannot argue the point here, it seems to me that in his later text *Usage*, Regis
      is more concerned to distance himself from Descartes.

Bibliography

Primary Literature

Manuscript Sources

Berlin, Preußischer Kulturbesitz, Geheimes Staatsarchiv

GR, I HA Rep.51, Nr. 94: *Acta des Königl. Geheimen Staats-Archivs betreffend der Philoso-phorum [zu Frankfurt a. O.] Irrungen unter sich* (1655–1659).

Brussels, Koninklijke Bibliotheek van België / Bibliothèque royale de Belgique

MS. 19376: Nicolaus Lamine and Joannes Van den Hove (professors) and Cornelius Van Riethoven (student), *Logica* (1689).

MS. 21127: anonymous (professor) and anonymous (student), *Physica* (1757).

MS. II 106: Joannes Baptista Wauchier (professor) and Leo Josephus Daco (student), *Physica* (1678).

MS. II 168: Joannes Baptista Scheppers (professor) and Franciscus Van der Meersche (student), *Physica* (1648–1649).

MS. II 737: anonymous (professor) and Arnoldus Meesters (student), *Physica* (1652).

MS. II 3214: Petrus Josephus Van Tieghem and Michael Bessemers (professors) and Petrus Josephus Brunin (student), *Physica* (1720–1721).

MS. II 3294: Petrus Josephus Heijlen (professor) and Joannes Baptista Eliart (student), *Physica* (1763).

MS. II 3352: Arnoldus Mennekens and Nicolas Du Bois (professors) and Adrianus Wustenraedt (student), *Physica* (1649–1650).

MS. II 3703: Philippus Van Billoen (professor) and Maximilianus Emmanuel Marij (student), *Physica* (1730–1731).

MS. II 4269: Joannes Lauvaux (professor) and Gisbertus Wauters (student), *Physica* (1755).

MS. II 4272: Antonius Cinck and Natalis Du Bois (professors) and Arnoldus Godefridus Dues (student), *Logica* (1714–1715).

MS. II 4352: Carolus Verlaine and Martinus Page (professors) and Joannes Josephus De Libotton (student), *Logica* (1743).

MS. II 4523: anonymous (professor) and Waltherus Van Beynen (student), *Physica* (1759).

MS. II 5342: Petrus Josephus Vrammout and Josephus Forgeur (professors) and Antonius Josephus Jonckers (student), *Physica* (1773).

MS. II 5412: Franciscus Josephus Engelbert and Hubertus Cornet (professors) and Guilielmus Borghs (student), *Logica* (1735).

MS. II 5444: Martinus Page (professor) and Joannes Van Nuffel (student), *Physica* (1739).

MS. II 5602: Joannes Franciscus Verberght and Gerardus Deckers (professors) and Dominicus Josephus Lodewijckx (student), *Physica* (1760–1761).

Copenhagen, KGL Bibliothek

E don. var. 145 kvart: Johannes De Raey, *Annotata ad* Principa philosophica *Rev. Des-Cartes, excerpta in collegio, habito sub Joh. de Raei, inchoato die 1. Maji 1658, finito die 20. Decembris* (1658).

Ghent, Universiteitsbibliotheek Gent

MS. 3564: Taillart (professor) and Hollanders (student), *Commentaria in Aristotelem et Johannes de Sacrobosco* (1626).

Gotha, Forschungsbibliothek

Chart. A 707, fols. 190^r–202^v: anonymous, *Kürtzliche Erleuterung etlicher Vorgaben von der Würkung des MagnetSteins nach Anleitung und gesetzten Grundstücken von Cartesio* (*post* 1644).

Göttingen, Niedersächsische Staats- und Universitätsbibliothek Göttingen

MS. Philol. 264: *Responsiones Renati Des Cartes ad quasdam difficultates* (*post* 1648).

Hamburg, Staats- und Universitätsbibliothek Hamburg Carl von Ossietzky

Cod. Philos. 273: Burchard de Volder, *Notulae quaedam in nobilissimi doctissimique viri Dni Renati Descartes* Principiorum philosophiae *partem primam-quartam* (*ante* 1673).

Cod. Philos. 274: Burchard de Volder, *Dictata in Carthesii* Principia philosophica (*post* 1673).

Leiden, University Library

MS. BPL 907: Johannes De Raey, *Analysis sive argumenta eorum, quae continentur in Cartesii* Dissertatione de Methodo *et* Principiis Philosophiae (*post* 1667).

MS. BPL 2841, fols. 1^r–32^r: Burchard de Volder, *Dictata in Carthesii* Principia philosophica (*post* 1673).

PAP 20: Petrus Wassenaer, *Album amicorum* (1648).

Liège, Bibliothèque de l'Université de Liège

MS. 1264: Adrianus De Neve (professor) and anonymous (student), *Logica* (1674).

MS. 2538: Leopoldus Iustinus Stam (professor) and anonymous (student), *Logica* (1673).

London, British Library

MS. Sloane 1216, fols. 75^r–128^v: Burchard de Volder, *Dictata in Carthesii* Principia philosophica (*post* 1673).

MS. Sloane 1274: Theodoor Craanen (professor) and Christopher Love Morley (student), *Dictata, ut videtur, in Theodori Craanen tractatum physico-medicum de homine* (1677–1679).

MS. Sloane 1292, fols. 78ʳ–141ʳ: Burchard de Volder (professor) and Christopher Love Morley (student), *Experimenta philosophica naturalia* (1676–1677).

Louvain, KU Leuven Libraries

MS. 202: Guilielmus Philippi (professor) and Alexander Van der Goos (student), *Metaphysica* (1639).

MS. 211: Egidius Muel and Deodatus Werici (professors) and Balthasar Cox (student), *Metaphysica & Physica* (1686–1687).

MS. 229: Jacobus Boudart (professor) and Vincentius Taijmont (student), *Physica & Metaphsyica* (1665–1666).

MS. 236: Andreas Van Horick and Joannes Franciscus Thijsbaert (professors) and Joannes Du Mont (student), *Logica* (1761).

MS. 241: Philippus Van Billoen and Laurentius Lilivelt (professors) and Joannes Ferdinandus Seiger (student), *Logica* (1737).

MS. 244: Hermannus Josephus Petit and Thomas Lambertus Ghenne (professors) and Michael Van Aerschot: *Logica* (1761–1763).

MS. 247: Joannes Josephus Mottin (professor) and Joannes Cyprianus Schruers (student), *Physica* (1772).

MS. 250: Robertus De Novilia and Dominicus Snellaerts (professors) and Georgius Jodoigne (student), *Logica* (1678).

MS. 257: Arnoldus Mennekens and Nicolas Du Bois (professors) and Petrus Watier (student), *Physica & Metaphsyica* (1652).

MS. 261: Joannes Stevenot (professor) and Michael Van den Biesche (student), *Physica* (1675–1676).

MS. 284: anonymous (professor) and Petrus Jacobus Franciscus Walkiers (student), *Physica* (1779–1780?).

MS. 302: Josephus Forgeur (professor) and Antonius Josephus Jonckers (student), *Physica* (1774).

MS. 308: anonymous (professor) and Jean-Joseph Havelange (student), *Physica* (1767).

MS. 326: anonymous (professor) and Joannes Franciscus Bertrand (student), *Physica* (1751).

MS. 359: Jacobus Van Waes (professor) and Henricus Dewael (student), *Physica* (1755).

MS. PM0326: Joannes Baptista Goffoul and Aegidius Franciscus Audenaert (professors) and Christophorus Devisscher (student), *Logica* (1723).

PRECA0021: Petrus Franciscus Van Audenrode (professor) and Lambertus Antonius Sterpin (student), *Physica* (1785).

Louvain-la-Neuve, Archives de l'Université catholique de Louvain

MS. C4: Joannes Baptista Bultot (professor) and Joannes Franciscus Bossart (student), *Physica* (1756).

MS. C28: Hubertus Cornet and Franciscus Antonius Graven (professors) and Judocus Henricus Grondel (student), *Philosophia & Metaphysica* (1746).

MS. C59: anonymous (professor) and Xavier Jacquelart (student), *Physica* (1785).

MS. C72: Michael Bessemers (professor) and Joannes Van Anderwerelt (student), *Physica* (1715).

MS. C75: Thomas Josephus Amand (professor) and Maximillianus Plischart (student), *Physica* (1739).

MS. C90: Ursmarus Narez (professor) and Jacobus Steijart (student), *Physica* (1708–1709).

MS. C96: Guilielmus Philippi and Sebastianus Stockmans (professors) and Philippus C. De Wauldret (student), *Physica & Metaphysica* (1640–1641).

MS. C97: Guilielmus Philippi and Sebastianus Stockmans (professors), Philippus C. De Wauldret (student), *Logica* (1640).

MS. C145: Martinus Page and Joannes Baptista Bultot (professors) and Joannes Josephus Lambillion (student), *Logica* (1757).

MS. C148: anonymous (professor) and Franciscus Josephus Lahaye (student), *Physica* (1777–1778).

MS. C158: Engelbertus Verheyden (professor) and Remigius Franciscus Staumont (student), *Physica* (1776).

MS. C163: Jan-Pieter Minckelers (professor) and Joannes Josephus Marselle (student), *Physica*, 1781.

MS. C165: Charles Léger De Decker and Gaspar Van Goirle (professors) and Albertus Boonen (student), *Physica & Metaphysica* (1680–1681).

MS. C174: Joannes Vandenhove and Gerardus Josephus Dominicus De Quareux (professors) and Abrahamus Wagemaeckers (student), *Logica* (1698).

MS. C202: anonymous (professor) and anonymous (student), *Physica* (1774).

MS. C210: anonymous (professor) and anonymous (student), *Physica* (1761).

Paris, Archives Nationales, Châtelet de Paris

MS. Y//184-Y//187: *Insinuations* (15 October 1644–2 March 1651).

Paris, Bibliothèque Sainte-Geneviève

MS. 2225: Jacques Rohault, *Conférences sur la physique, faites en 1660–1661, par Jacques Rohault, et recueillies par un de ses auditeurs qui, dit-il, y a ajouté du sien* (1660).

Pretoria, National Library of South Africa

MSD27: Burchard de Volder, *Dictata in Carthesii* Principia philosophica (*post* 1673).

Rome, Biblioteca Nazionale Centrale "Vittorio Emanuele II"

Fondo Gesuitico 1323: Niccolò Zucchi, *Philosophia magnetica per principia propria proposita et ad prima in suo genere promota* (completed *post* 1653).

The Hague, Koninklijke Bibliotheek

MS. 72 A 7: Burchard de Volder, *Notulae quaedam in nobilissimi doctissimique viri Dni Renati Descartes* Principiorum philosophiae *partem primam-quartam* and *Dictata in Carthesii* Principia philosophica (1690).

Uppsala, Universitetsbibliotek

A 209: Andreas Drossander, *Prolegomena in physicam Hoffwenii* (1683).

Utrecht, Utrecht University Library

MS. 715: Daniel Voet (professor) and Johannes Rauwers (student), *Dictata in Phisicam Senguerdi*[*i*] (1656 ca.).

Warsaw, Biblioteka Narodowa

MS. BN Rps 3365 II: Burchard de Volder, *Dictata in Carthesii* Principia philosophica (*post* 1673).

Washington, Burndy Library Manuscripts Collection (Smithsonian Libraries)

MSS 001296 B: Engelbertus Verheyden (professor) and anonymous (student), *Tractatus in quo de vero partiu*[*m*] *universi situ et motu, seu de vero mundi systemate inquiritur* (1770 ca.).

Printed Texts

Achrelius, Daniel Erici (praes.) and Petter Svensson Ulnerus (resp.): *Contemplationum mundi dissertatio decima tertia, de geocosmi semine, magnetismo rerum naturalium, tum qualitatibus veneni* (Abo, 1681).

Aemilius, Antonius: *Oratio in obitum clarissimi et praestantissimi viri Henrici Renerii* [...] *habita in templo maximo postridie exequiarum xv Kal. April. MDCXXXIX; accedit ejusdem carmen funebre* (Utrecht, 1639).

Aemilius, Antonius: *Orationes* (Utrecht, 1651).

Allinga, Petrus: *Sedige verdedigingh van de eer en leer der wijdt-beroemste Professoren en voornaemste Leeraren van Nederlandt tegen de schriftelijcke beschuldigingh ingestelt tot der selver verswaringh van den Eerwaerdigen, Godtsaligen Hoogh-geleerden D. Hermannus Witsius, Predikant to Leeuwarden, in sijn Boeck genaemt De twist des Heeren met sijn Wijngaert* (Amsterdam, 1672).

Alsted, Johann Heinrich: *Systema physicae harmonicae, quatuor libellis methodice propositum, in quorum I. Physica mosaica delineatur, II. Physica Hebraeorum, Rabbinica et Cabbalistica proponitur. III. Physica peripatetica ... plenius pertractatur. IV. Physica chemica perspicue et breviter adumbratur* (Herborn, 1612, 1616, 1642).

Andreas, Valerius: *Fasti academici studii generalis Lovaniensis* (Louvain, 1650).

Aristotle, *The Complete Works*, transl. Jonathan Barnes, 2 vols. (Princeton, NJ, 1984).

Arnauld, Antoine, and Pierre Nicole: *La Logique ou l'art de penser*, 5th edition (Paris, 1683), eds. Pierre Clair and François Girbail (Paris, 1981).

Bacon, Francis: *The Instauratio magna*, part II: *Novum Organum*, eds. Graham Rees and Maria Wakely, in *The Oxford Francis Bacon*, 16 vols. (Oxford, 2012—), vol. 11.

Bacon, Francis: *The Works of Francis Bacon*, eds. James Spedding, Robert L. Ellis, and D.D. Heath, 14 vols. (London, 1861–1879).

Baillet, Adrien: *Vie de Monsieur Descartes*, 2 vols. (Paris, 1691); repr. (Hildesheim, 1972).

Bartholin, Caspar: *Specimen compendii physicae: praecipua philosophiae naturalis capita* (Copenhagen, 1687).

Bassecour, Fabricius de la: *Defensio cartesiana, in duas distributa partes, succincte conscripta et amice adversus D. Johannem Schulerum* [...] *occasione utriusque libri ab ipso nuper editi in partem primam atque secundam* Principiorum philosophiae cartesianae (Amsterdam, 1671).

Bérigard, Claude Guillermet de: *Circulus pisanus* (Udine, 1643; Padua, 1661).

Beverwijck, Johan van, ed.: *Epistolicae quaestiones* (Rotterdam, 1644).

Bilberg, Johan (praes.) and Ericus E. Odhelius (resp.): *Specimen cogitationum de magnetismis rerum* (Stockholm, 1683).

Bilberg, Johan (praes.) and Andreas Plaan (resp.): *Disputatio physica de magnete* (Uppsala, 1687).

Birlens, Symon (praes.) and Michael Godefridus D'Overschie (resp.): *Theses philosophicae* (Louvain, 14 August 1660).

Bisterfeld, Johann Heinrich: *Elementorum logicorum libri tres:* [...] *Phosphorus catholicus* [...] *Consilium de studiis feliciter instituendis* (Leiden, 1657).

Bisterfeld, Johann Heinrich: *Philosophiae primae seminarium* [...] *editum ab Adriano Heereboord* [...] *qui dissertationem praemisit de philosophiae primae existentia et usu* (Leiden, 1657).

Bisterfeld, Johann Heinrich: *Scripturae sacrae divina eminentia et efficientia* [...] *denuo in duabus disputationibus. Accedit eiusdem authoris, ars concionandi* (Leiden, 1654).

Blaeu, Willem: *Institutio astronomica de usu globorum & sphaerarum caelestium ac terrestrium: duabus partibus adornata, una, secundum hypothesin Ptolomaei, per terram quiescentem. Altera, juxta mentem N. Copernici, per terram mobilem. Latinè reddita à M. Hortensio* (Amsterdam, 1634).

Boileau-Despréaux, Nicolas: *Oeuvres completes*, ed. Lèon Thiessé, 3 vols. (Paris, 1810).

Borel, Pierre: *Vitae Renati Cartesii summi philosophi compendium* (Paris, 1656).

Bouju, Théophraste: *Corps de toute la philosophie* (Paris, 1614).

Boyle, Robert: *A Defence of the Doctrine Touching the Spring and Weight of the Air* (London, 1662).

Boyle, Robert: *Hydrostatical Paradoxes, Made out by New Experiments (for the Most Part Physical and Easie)* (Oxford, 1666).

Boyle, Robert: *New Experiments Physico-Mechanicall, Touching the Spring of the Air and its Effects* (Oxford, 1660).

Boyle, Robert: *The Excellency of Theology Compared with Natural Philosophy* (London, 1674).

Boyle, Robert: "The History of Fluidity and Firmness," in Robert Boyle: *Certain Physiological Essays and Other Tracts* [...], *Increased by the Addition of a Discourse about the Absolute Rest in Bodies* (London, 1669; 1st ed. 1661), 163–206.

Brandt, Gerard: *Historie der Reformatie* (s.l., 1668; 1674).

Brandt, Gerard: *On-partydig chronyxken der voornaemste Nederlantsche geschiedenissen, zoo kerkelijk als politijcq* (Rotterdam, 1658).

Brandt, Gerard: *Stichtelijke gedichten* (s.l., 1665).

Brandt, Gerard: *Verhaal van de Reformatie, in en ontrent de Nederlanden* (Amsterdam, 1663).

Brucker, Johann Jakob: *Historia critica philosophiae a mundi incunabulis ad nostram usque aetatem deducta*, 5 vols. (Leipzig 1742–1744).

Burgersdijk, Franco: *Idea philosophiae moralis* (Leiden 1623, 1629, 1635, 1640, 1644).

Burgersdijk, Franco: *Idea philosophiae naturalis* (Leiden 1622, 1627, 1635, 1640, 1645, 1648, 1652; Amsterdam 1657, 1648, 1657).

Burgersdijk, Franco: *Institutionum logicarum libri duo ex Aristotelis, Keckermanni, aliorum praecipuorum logicorum praeceptis recensitis* (Leiden, 1626); 3rd ed. [...] *ex Aristotelis praeceptis nova methodo ac modo formati, atque editi* (Leiden, 1645).

Burgersdijk, Franco: *Institutionum metaphysicarum libri duo* (Leiden, 1640).

Burgersdijk, Franco: *Logica practica, oft Oeffening der reden-konst*, transl. Lodewijk Meyer (Amsterdam, 1648).

Burman, Frans: *Synopsis theologiae et speciatim oeconomiae foederum Dei*, 2 vols. (Amsterdam, 1699).

Cabeo, Niccolò: *Philosophia magnetica* (Ferrara, 1629).

Calamy, Edmund: *An Historical Account of My Own Life* (London, 1830).

Camphuysen, Dirck: *Stichtelyke rymen*, 20th edition (Amsterdam, 1685).

Camphuysen, Dirck: *Theologische werken* (Amsterdam, 1661); 3rd edition (Amsterdam, 1682).

Capivaccius, Hieronymus: *Medicina practica, seu methodus cognoscendorum et curandorum omnium humani corporis affectuum* (Venice, 1594).

Catalogus instructissimae bibliothecae Joannis Henrici Glazemaker (Amsterdam, 1683).

Catalogus [...] *librorum* [...] *Adriani Heereboord* (Leiden, 1661).

Catalogus van de boeken die gedrukt of in meerder getal te bekomen zijn (s.a., s.l.)

Clauberg, Johannes: *Defensio cartesiana adversus Jacobum Revium* (Amsterdam, 1652).

Clauberg, Johannes: *Logica vetus et nova quadripartita, modum inveniendae ac tradendae veritatis in Genesi simul et analysi facile methodo exhibens* (Amsterdam 1654).

Clauberg, Johannes: *Nadere uitbreiding wegens Renati Cartesii Bedenkingen van d'eerste Wijsbegeerte*, transl. Jan Hendrik Glazemaker (Amsterdam, 1683).

Clauberg, Johannes: *Opera omnia philosophica*, 2 vols. (Amsterdam, 1691).

Clauberg, Johannes: *Physica contracta in qua tota rerum universitas per clara & certa principia succincte & dilucide explicatur* (Leipzig, 1689).

Clauberg, Johannes: *Physica, quibus rerum corporearum vis & natura, mentis ad corpus relatae proprietates, denique corporis ac mentis arcta & admirabilis in homine conjunctio explicantur* (Amsterdam, 1664).

Coeman, Sibertus and Burchard de Volder: *Orationes duae, quarum altera inauguralis* (Leiden, 1679).

Colerus, John: *The Life of Benedict de Spinosa* [...] *Done out of French* (London, 1706).

Comenius, Jan Amos: *Comenius' självbiografi / Comenius about Himself*, eds. Stig Nordström and Wilhelm Sjöstrand (Uppsala, 1975).

Comenius, Jan Amos: *Continuatio admonitionis fraternae de temperando charitate zelo ad S. Maresium* (Amsterdam, 1669).

Comenius, Jan Amos: *Jana Amoso Komenského Korrespondence*, ed. Adolf Patera (Prague, 1892).

Conimbricenses: *In libros* Ethicorum *Aristotelis* ad Nicomachum, *aliquot Conimbricensis cursus disputationes* (Lisbon, 1593).

Conway, Anne: *Principiae philosophiae antiquissimae et recentissimae de Deo, Christo & creatura, id est de spiritu & materia in genere* (Amsterdam, 1690).

Conway, Anne: *The Principles of the Most Ancient and Modern Philosophy*, eds. and transl. Allison P. Coudert and Taylor Corse (Cambridge, 1996).

Craanen, Theodoor: *Opera omnia, nunc demum conjunctim edita: tomus prior est tractatus physico-medicus de homine, in quo status ejus tam naturalis, quam praernaturalis, quoad theoriam rationalem mechanice demonstratur: cum figuris aeneis & indicibus tam capitum, quam rerum & verborum locupletissimis. Tomus alter continens observationes, quibus emendatur & illustratur Henrici Regii* Praxis medica, *medicationum exemplis demonstrata* (Antwerp, 1689).

Craanen, Theodoor: *Tractatus physico-medicus de homine, in quo ejus tam naturalis, quam praeternaturalis, quoad theoriam rationalem mechanice demonstratur* (Leiden, 1689).

Craanen, Theodoor (praes.) and various (resp.): *Exercitationes philosophicae de natura hominis: de conjunctione animae cum corpore* (Nijmegen, 1663–1665).

Cyrano de Bergerac, Savinien: *Histoire comique, contenant les états et empires de la Lune* (Paris, 1657).

Dalrymple, James: *Physiologia nova experimentalis, in qua, generales notiones Aristotelis, Epicuri, & Cartesii supplentur* (Leiden, 1686).

De Bie, Alexander: *Disputatio logica de qualitate* (Amsterdam, 1679).

De Bie, Alexander (praes.) and Joannes Brandlight (resp.): *Disputatio mathematica de acus magneticae inconstanti deviatione* (Amsterdam, 1658).

De Bie, Alexander (praes.) and Sibertus Coeman (resp.): *Disputatio mathematica de acus magneticae deviatione* (Amsterdam, 1658).

De Bie, Alexander (praes.) and Burchard de Volder (resp.): *Disputatio philosophica inauguralis de simplicitate Dei, echo et republica* (Utrecht, 17 October 1660).

De Bie, Alexander (praes.) and Joannes du Pire (resp.): *Disputatio de magnete, quae est de ejus ὀρθοβορεοδείξει* (Amsterdam, 1658).

De Bruyn, Johannes: *Defensio doctrinae cartesianae, de dubitatione et dubitandi modo, ut & de idea Dei in nobis, deque existentiae ejus demontratione ex ea idea, adversus objectiones Reineri Vogelsangii* (Amsterdam, 1670).

De Bruyn, Johannes: *Epistola ad Isaacum Vossium de natura et proprietate lucis* (Amsterdam, 1663).

De Bruyn, Johannes (praes.) and Thomas Bolwerck (resp.): *Disputatio physica de specialibus quibusdam motus effectibus* (Utrecht, 7 June 1665).

De Bruyn, Johannes (praes.) and Gerard Bornius (resp.): *Disputatio physica de mundo* (Utrecht, 14 May 1653).

De Bruyn, Johannes (praes.) and Johannes Camp (resp.): *Disputationum physicarum septima de corporis loco et modo existendi in loco* (Utrecht, 14 December 1655).

De Bruyn, Johannes (praes.) and Esias Clemens (resp.): *Disputationis physicae de alitura pars prima* (Utrecht, 29 April 1654).

De Bruyn, Johannes (praes.) and Theodor Craanen (resp.): *Disputatio mechanico-mathematica de trochlea* (Utrecht, 21 June 1654).

De Bruyn, Johannes (praes.) and Johannes de Buisson (resp.): *Disputatio physica continens theses et problemata e physica generali pars prima* (Utrecht, 10 December 1653).

De Bruyn, Johannes (praes.) and Samuel Gelei (resp.): *Disputatio physica de coelo* (Utrecht, 19 April 1656).

De Bruyn, Johannes (praes.) and Michael Michielzon (resp.): *Disputatio physica de mari* (Utrecht, 14 May 1656).

De Bruyn, Johannes (praes.) and Caspar Scharbach (resp.): *Disputationum physicarum duodecima de motu pars tertia* (Utrecht, 29 October 1656).

De Bruyn, Johannes (praes.) and Nicolaus Shepheard (resp.): *Disputationis physicae de alitura pars quinta* (Utrecht, 22 September 1660).

De Bruyn, Johannes (praes.) and Aegidius Taeispil (resp.): *Disputationum physicarum prima de philosophia in genere* (Utrecht, 20 October 1652).

De Bruyn, Johannes (praes.) and Cornelis Van Ackersdijck (resp.): *Disputationis physicae de alitura pars septima* (Utrecht, 12 December 1663).

De Bruyn, Johannes (praes.) and Henricus Van Rhee (resp.): *Disputatio philosophiae aliquot nobilissimae materiae ex physica et mathesi selectae continens* (Utrecht, 11 December 1661).

De Bruyn, Johannes (praes.) and Johannes Van Swanevelt (resp.): *Disputatio philosophica de libero hominis arbitrio* (Utrecht, 25 April 1657).

De Bruyn, Johannes (praes.) and Johannes Van Swanevelt (resp.): *Disputatio philosophica miscellanea* (Utrecht, 2 July 1656).

De Bruyn, Johannes (praes.) and Johannes Van Swanevelt (resp.): *Disputationum physicarum quarta de corporis divisibilitate* (Utrecht, 13 October 1655).

De Bruyn, Johannes (praes.) and Everard Van Sypesteyn (resp.): *Diputatio physica de causis descensus gravium* (Utrecht, 12 October 1653).

De Bruyn, Johannes (praes.) and Johannes Verwey (resp.): *Disputationum philosophicarum de naturali Dei cognitione quarta* (Utrecht, 12 December 1666).

De Bruyn, Johannes (praes.) and Johannes Witkint (resp.): *Disputationum physicarum decima-quinta de motu pars sexta* (Utrecht, 6 November 1658).

De Ceriziers, René: *Le philosophe français* (Paris, 1643).

Dechales, Claudius Franciscus Milliet: *Cursus seu mundus mathematicus*, vol. 2 (Leiden, 1690).

De Courcelles, Étienne: *Opera Theologica* (Amsterdam, 1675; London, 1684).

De Courcelles, Étienne: *Synopsis ethices: tractatus lectu dignissimus*, 3rd ed. (Cambridge, 1702).

[De Decker, Charles Léger] I.T. Philosopho Lovaniensi: *Cartesius seipsum destruens, sive Dissertatio brevis in qua Cartesianae contradictiones, & hallucinationes variae aequivocationibus, aliisque illusionibus & artibus innixae: pluresque immoderatae adversus philosophiam communem expostulationes, panduntur & refelluntur* (Louvain, 1675).

De la Barre, François Poulain: *De l'égalité des deux sexes. De L'éducation des dames. Del'excellence des hommes*, ed. Marie-Frédérique Pellegrin (Paris, 2011).

De la Barre, François Poulain: *Three Cartesian Feminist Treatises*, transl. Marcelle M. Welch and Vivien Bosley (Chicago, IL, 2002).

Denis, Jean-Baptiste: *Discours sur les comètes, suivant les principes de M. Descartes. Où l'on fait voir combien peu solides & mal fondées sont les opinions de ceux, qui croyent que les comètes sont composées* (Paris, 1665).

De Raey, Johannes: *Clavis philosophiae naturalis, seu introductio ad naturae contemplationem, aristotelico-cartesiana* (Leiden, 1654); 2nd ed. (Leiden, 1677).

De Raey, Johannes: *Cogitata de interpretatione* (Amsterdam, 1692).

De Raey, Johannes: *Miscellanea philosophica* (Amsterdam, 1685).

De Raey, Johannes: *Oratio inauguralis de gradibus et vitiis notitiae vulgaris circa contemplationem naturae, et officio philosophi circa eandem* (Leiden, 1651).

De Raey, Jan (praes.) and Johannes Crooswyck (resp.): *Disputationum physicarum ad problemata Aristotelis, quintae de materia subtili pars posterior* (Leiden, 4 May 1653).

De Raey, Johannes (praes.) and Martinus Duyrkant (resp.): *Disputatio philosophica de dubitatione* (Leiden, 1665).

De Raey, Johannes (praes.) and Johannes Freherus (resp.): *Disputatio philosophica de constitutione logicae* (Leiden, 1668).

De Raey, Johannes (praes.) and Johannes von Flammerdinge (resp.): *Disputatio philosophica de loco* (Leiden, 1667).

Des Amorie Van der Hoeven, Abraham: "De Philippo a Limborch, theologo, dissertatio historico-theologica," in Abraham des Amorie Van der Hoeven: *De Joanne Clerico et Philippo a Limborch dissertationes duae* (Amsterdam, 1843), part 2.

Descartes, René: *Alle de werken van Renatus des Cartes*, various translators (s.l., 1692).

Descartes, René: *Bedenkingen van d'eerste wysbegeert*, transl. Jan Hendrik Glazemaker (Amsterdam, 1656).

Descartes, René: *Conversation with Burman*, ed. and transl. John Cottingham (Oxford, 1976).

Descartes, René: *De homine, figuris et latinitate donatus a Florentio Schuyl* (Leiden, 1662).

Descartes, René: *Étude du bon sens. La recherche de la vérité et autres écrits de jeunesse (1616–1631)*, eds. Gilles Olivo and Vincent Carraud (Paris, 2013).

Descartes, René: *Epistolae* (Amsterdam, 1668).

Descartes, René: *Ethice, in methodum et compendium, gratiâ studiosae juventutis, concinnata* (London, 1685; Cambridge, 1702).

Descartes, René: *Geometria à Renato Des Cartes anno 1637 gallice edita, nunc autem cum notis [...] Francisci à Schooten* (Leiden, 1649).

Descartes, René: *Gespräch mit Burman*, ed. and transl. Hans W. Arndt (Hamburg, 1982).

Descartes, René: *Le monde ou Traité de la Lumière* (Paris, 1664).

Descartes, René: *Les principes de la philosophie*, transl. Claude Picot (Paris, 1647).

Descartes, René: *Lettres*, ed. Claude Clerselier, 3 vols. (Paris, 1657, 1659, 1667).

Descartes, René: *Mediationes de prima philosophia, of Bedenkingen van d'eerste wysbegeerte*, transl. Willem Deurhoff (Amsterdam, 1687).

Descartes, René: *Opera philosophica* (Amsterdam, 1654–1656).

Descartes, René: *Passiones animae* (Amsterdam, 1685).

Descartes, René: *Principles of Philosophy*, transl. with explanatory notes by Valentine Rodger Miller and Reese P. Miller (Dordrecht-London, 1983).

Descartes, René: *Redenering van 't beleed om zijn reden wel te beleiden*, transl. Jan Hendrik Glazemaker (Amsterdam, 1656); 2nd ed., *Proeven der wijsbegeerte of Redenering van de middel om de reden wel te beleiden, en de waarheid in de wetenschappen te zoeken: de verregezichtkunde, verhevelingen, en meetkunst* (Amsterdam, 1659).

Descartes, René: *The Correspondence of René Descartes, 1643*, ed. Theo Verbeek, Erik-Jan Bos, and Jeroen Van de Ven (Utrecht, 2003).

Descartes, René: *The World and Other Writings*, ed. and transl. Stephen Gaukroger (Cambridge, 1998).

Descartes, René: *Tutte le lettere, 1619–1650*, ed. Giulia Belgioioso (Milan, 2005).

[Desgabets, Robert]: *Considérations sur l'état present de la controverse touchant le Très Saint-Sacrement de l'autel, où il est traité en peu mots l'opinion qui enseigne que la matière du pain est change en celle du corps de Jésue-Christ par son union substantielle à son l'ame et à son personne divine* (s.l., 1671).

Desgabets, Robert: *Oeuvres philosophiques inédites*, 7 vols., ed. Joseph Beaude (Amsterdam, 1983–1985).

Desmarets, Samuel (praes.) and Johannes Placentinus (resp.): *De duplici volumine Naturae et Scripturae, ex quo notitia Dei arcessenda est* (1652).

De Volder, Burchard (praes.) and Johannes Balthasar Helvetius (resp.): *Disputatio philosophica de magnete* (Leiden, 1677).

De Volder, Burchard (praes.) and various (resp.): *Disputatio philosophica de aëris gravitate prima [-quinta]* (Leiden, 1676–1678).

De Volder, Burchard (praes.) and various (resp.): *Disputatio philosophica de rerum naturalium principiis, prima [-quintadecima]* (Leiden, 1674–1676).

De Vries, Gerard: *De natura Dei et humanae mentis determinationes pneumatologicae: accedunt de catholicis rerum attributis ejusdem determinationes ontologicae* (Utrecht, 1687).

De Vries, Gerard: *De Renati Cartesii Meditationibus a Petro Gassendo impugnatis dissertatiuncula historico-philosophica* (Utrecht, 1691).

De Vries, Gerard: *Introductio historica ad Cartesii philosophiam* (Utrecht, 1683).

De Vries, Gerard: *Logica compendiosa* (Utrecht, 1684).

De Vries, Gerard: *Narrator confutatus, sive animadversiones in narrationem de controversiis nuperius in Academia Ultraiectina motis* (Utrecht, 1679).

De Vries, Gerard (praes.) and Jacobus Broedelet (resp.): *Disputatio philosophica, qua disquiritur, an philosophia cartesiana sit aristotelicae, prout hodie docetur, praeferenda* (Utrecht, 1676).

De Vries, Gerard (praes.) and Adolph Schröder (resp.): *Disputatio medica de catalepsi*, (Utrecht, 1692).

De Vries, Gerard (praes.) and Henricus Van Galen (resp.): *Exercitationis philosophicae ad Cartesii* Princip. Philos. *part. 1. art.* LXXVI. *de officio philosophi circa revelata, pars prima* (Utrecht, 1687).

De Vries, Gerard (praes.) and various (resp.): Exercitationes rationales de Deo, divinisque perfectionibus. Accedunt ejusdem dissertationes de infinito; nullibitate spirituum; [...]. In quibus passim quae de hisce philosophatur Cartesius cum rectae rationis dictamine conferuntur. (Utrecht, 1685).

De Vries, Gerard (praes.) and various (resp.): *Exercitationes rationales de Deo, divinisq[ue] perfectionibus. Nec non philosophemata miscellanea. Edito nova, ad quam, praeter alia, accedit diatriba singularis gemina altera, de cogitatione ipsa mente; altera, de ideis rerum innatis.* (Utrecht, 1695).

Dilly, Antoine: *De l'ame des betes* (Lyon, 1676).

Du Hamel, Jean: *Réflexions critiques sur le système cartesien de la philosophie de Mr. Regis* (Paris, 1692).

Du Hamel, Jean-Baptiste: *De meteoris et fossilibus libri duo: in priore libro mixta imperfecta, quaeque in sublimi aëre vel gignuntur, vel apparent, fusè, pertractantur: posterior liber mixta perfecta complectitur: ubi salium, bituminum, lapidum, gemmarum, & metallorum naturae causae, & usus inquiruntur* (Paris, 1660).

Du Hamel, Jean-Baptiste: *Philosophia vetus et nova ad usum scholae accomodata* (Paris, 1678).

Du Hamel, Jean-Baptiste: *Philosophia vetus et nova: tomus posterior qui physicam generalem et speciale tripartitam complectitur*, ed. Jacques Nicolas Colbert, vol. 2 (Nuremberg, 1682).

Du Hamel, Jean-Baptiste: *Philosophia vetus et nova, ad usum scholae accommodata, in Regia Burgundia olim pertractata*, 3rd ed., vol. 2 (Paris, 1684).

Duker, Gisbertus Wesselus: *Disputatio philosophica inauguralis de recta ratiocinatione* (Franeker, 1686).

Du Moulin, Pierre: *Les elemens de la philosophie morale* (Paris, 1624; Paris, 1643).

Dupleix, Scipion: *Corps de philosophie, contenant la logique, l'éthique, la physique et la métaphysique*, 2 vols. (Geneva, 1627).

Dupleix, Scipion: *L'ethyque* (Paris, 1610), ed. Roger Ariew (Paris, 1993).

Du Roure, Jacques: *La philosophie divisée en toutes ses parties, établie sur des principes, evidens et expliquée en tables et par discours, ou particuliers, ou tirez des anciens, des nouveaux auteurs; & principalement des Peripaticiens, et de Descartes* (Paris, 1654), 2nd ed. *Abrégé de la vraye philosophie* (Paris, 1665),

Durret, Natalis: *Nouvelle theorie des planetes, conforme aux observations de Ptolomée, Copernic, Tycho, Lansberge, & autre excellens astronomes, tant anciens que modernes. Avec les tables Richeliennes et Parisiennes, exactement calculées* (Paris, 1635).

[Essenius, Andreas] Modestinus Liberius Philosophus: *Specimen philosophiae cartesianae, thesibus aliquot expressum* (Utrecht, 1656).

Essenius, Andreas (praes.) and Thomas Bolwerck (resp.): *De conscientia pars quarta* (Utrecht, 1666).

Essenius, Andreas (praes.) and Justinus Danneels (resp.): *Disputationis practicae, de conscientia, pars quinta* (Utrecht, 10 November 1666).

Ethica, sive Summa moralis disciplinae, in tres partes divisa. Authore Fr. Eustachio a *s. Paulo ... Cui Accesserunt Steph. Curcellaei, et Renati Des-Cartes* Ethices (Cambridge, 1707).

Eustachius a Sancto Paulo: *Summa philosophiae quadripartita de rebus dialectibus, moralibus, physicis, et metaphysicis* (Paris, 1609; Cambridge, 1648).

Fabri, Honoré: *Physica, id est, scientia rerum corporearum, in decem tractatus distributa 4 continens tractatum septimum et octavum*, vol. 4 (Leiden, 1671).

Fénelon, François de Salignac de La Mothe: *Oeuvres*, ed. Jacques Le Brun, 2 vols. (Paris, 1983–1997).

Fénelon, François de Salignac de La Mothe: *Traité de l'éducation des filles* (Paris, 1687).

Fernel, Jean: *Universa medicina, tribus et viginti libris absoluta: editio postrema* (Geneva, 1577).

Fontenelle, Bernard le Bovier de: *Éloges des académiciens de l'Académie Royale des Sciences*, vol. I (The Hague, 1731).

Fontenelle, Bernard le Bovier: *Entretiens sur la Pluralité des Mondes* (s.l., 1686).

Fragment de physique (Paris, 1662).

Frassen, Claude: *Philosophia academica, quam ex selectissimis Aristotelis et doctoris subtilis Scoti rationibus ac sententiis* (Paris, 1657).

Freinsheim, Johann: *Orationes* (Frankfurt, 1655).

Froidmont [Fromondus], Libert: *Ant-Aristarchus sive orbis-terrae inmobilis liber unicus: in quo decretum S. Congregationis S.R.E. Cardinal. an. [1616] adversus pythagorico-copernicanos editum defenditur* (Antwerp, 1631).

Froidmont, Libert: *Labyrinthus, sive de compositione continui* (Antwerp, 1631).

Froidmont, Libert: *Meteorologicorum libri VI* (Antwerp, 1627).

Gabriel, Daniel: *Voiage du monde de Descartes* (Paris, 1690); transl. Gerard de Vries (?), *Iter per mundum Cartesii* (Amsterdam, 1694).

Galen: *In Hippocratis Aphorismos commentaria*, in *Claudii Galenii Opera Omnia*, ed. C.G. Kühn, 20 vols. (Leipzig, 1821–1833); repr. (Hildesheim, 1965), vol. 17.

Galen: *Aliquot opuscula nunc primum Venetorum opera inventa et excusa, quorum sequens tibi pagella catalogum indicabit* (Leiden, 1550).

Galen: *In Hippocratis praedictionum librum primum commentariorum libri tres* (Paris, 1535).

Galen: *On the Natural Faculties*, ed. and transl. Arthur John Brock (London-New York, NY, 1916).

Geulincx, Arnold: *Sämtliche Schriften*, ed. Herman. J. de Vleeschauwer (Stuttgart, 1968).

Geulincx, Arnold: *Saturnalia, seu (ut passim vocantur) Quaestiones quodlibeticae in utramque partem disputatae*, 2nd ed. (Leiden, 1665).

Goudin, Antoine: *Philosophia juxta inconcussa tutissimaque Divi Thomae dogmata* (Paris, 1726).

Grandamy, Jacques: *Le cours de la comète qui a paru sur la fin de l'année 1664 et au commencement de l'année 1665* (Paris, 1665).

Gronovius, Jakob: *Burcheri de Volder laudatio* (Leiden, 1709).

Hahn, Petrus Olai (praes.) and Gabriel Procopoeus (resp.): *Disputatio physica amicitiam magnetis cum ferro exhibens* (Abo, 1698).

Hartlieb, *The Hartlib Papers*, eds., Mark Greengrass, Michael Leslie and Michael Hannon, eds., (Sheffield, HRI Online Publications, 2013): https://www.dhi.ac.uk/projects/hartlib/.

Heereboord, Adriaan: *Collegium ethicum, in quo tota philosophia moralis aliquot disputationibus [...] explicatur* (Leiden, 1649).

Heereboord, Adriaan: *Collegium logicum: in quo tota philosophia rationalis [...] per theses proponitur* (Leiden, 1649).

Heereboord, Adriaan: *Collegium physicum: in quo tota philosophia naturalis aliquot disputationibus breviter & perspicue per theses proponitur* (Leiden, 1649).

Heereboord, Adriaan: *Meletemata philosophica maximam partem metaphysica* (Leiden, 1654); 2nd ed. (Leiden, 1659); 3rd ed. (Nijmegen, 1664–1665), 4th ed. (Amsterdam, 1665); 5th ed. (Amsterdam, 1680).

Heereboord, Adriaan: *Philosophia naturalis, cum commentariis peripateticis antehac edita: nunc vero hac posthuma editione mediam partem aucta, & novis commentariis, partim e Nob. D. Cartesio, Cl. Berigardo, H. Regio, aliisque praestantioribus philosophis, petitis, partim ex propria opinione dictatis, explicata* (Leiden, 1663).

Heereboord, Adriaan: *Sermo extemporaneus de ratione philosophice disputandi* (Leiden, 1648).

Heereboord, Adriaan (praes.) and Andreas Du Pré (resp.): *Disputationum ex philosophia selectarum decima-quinta de angelis* (Leiden, 1644).

Heereboord, Adriaan (praes.) and Petrus Georgii Goethals (resp.): *Disputationum ex philosophia selectarum trigesima-quarta quae est prima de formis* (Leiden, 1648).

Heereboord, Adriaan (praes.) and Ioannes Hardenbergh (resp.): *Disputationum ex philosophia selectarum trigesima-quinta quae est secunda de formis* (Leiden, 1648).

Heereboord, Adriaan (praes.) and Daniel Ouzeel (resp.): *Disputationum ex philosophia selectarum trigesima-sexta quae est tertia de formis* (Leiden, 1648).

Heereboord, Adriaan (praes.) and Guilelmus Caspari Wassembergh (resp.): *Disputatio metaphysica ad Aristotelis l. 3* Metaphys. (Leiden, 1646).

Heereboord, Adriaan (praes.) and Iacobus Caroli Zouterius (resp.): *Disputationum ex philosophia selectarum tertia de notitia Dei naturali* (Leiden, 25 March 1643).

Hegel, Georg Wilhelm Friedrich: *Vorlesungen über die Geschichte der Philosophie*, in *Werke in zwanzig Bänden*, eds. Eva Moldenhauer and Karl Markus Michel, vol. 20 (Frankfurt am Main, 1979).

Heidanus, Abraham, [Burchard de Volder and Christoph Wittich]: *Consideratien, over eenige saecken onlangs voorgevallen in de Universiteyt binnen Leyden* (Leiden, 1676).

Hentschel, Samuel (praes.) and Johannes Brewer (resp.): *De vita hominis disputatio VI* (Wittenberg, 1661).

Hentschel, Samuel (praes.) and Johannes Ludovicus Hartmannus (resp.): *De vita hominis disputatio II* (Wittenberg, 1659).

Hentschel, Samuel (praes.) and Georg Jaenisch (resp.): *De vita hominis disputatio I* (Wittenberg, 1659).

Hentschel, Samuel (praes.) and Matthaeus Lampert (resp.): *De vita hominis disputatio V* (Wittenberg, 1660).

Hentschel, Samuel (praes.) and Jacobus Rhetius (resp.): *De vita hominis disputatio III* (Wittenberg, 1660).

Hentschel, Samuel (praes.) and Johann Georg Wedel (resp.): *De vita hominis disputatio IV* (Wittenberg, 1660).

Hilleprand, Gerhard: *Examen doctrinae cartesianae ad regulam fidei et rationis: honoribus illustrissimorum, ac perillustrium dominorum, dominorum dum in antiquissima ac celeberrima universitate viennensi* (Vienna, 1707).

Hobbes, Thomas: *De cive* (Paris, 1642).

Hobbes, Thomas: *The Correspondence of Thomas Hobbes*, ed. Noel Malcolm, 2 vols. (Oxford, 1994).

Hoffwenius, *Synopsis physica, disputationibus aliquot Academicis comprehensa* (Stockholm, 1678); 2nd ed. (Stockholm, 1698); 3rd ed. (Pärnu, 1699); 4th ed. (Pärnu, 1700).

Huber, Ulricus: *De concursu rationis et sacrae Scripturae liber* (Franeker, 1687).

Huet, Pierre Daniel: *Censura philosophiae cartesianae* (Paris, 1689; Paris 1694).

[Huet, Pierre-Daniel] M.G. de l'A.: *Nouveaux mémoires pour servir à l'histoire du cartésianisme* (s.l., 1692); repr. (Paris, 1996).

Huygens, Constantijn: *De briefwisseling van Constantijn Huygens (1608–1687)*, ed. Jacob Adolf Worp, 6 vols. (The Hague, 1911–1917).

Huygens, Christiaan: *Oeuvres complètes*, eds., D. Bierens de Haan, Johannes Bosscha, Diederik Johannes Korteweg, Albertus Antonie Nijland, 22 vols. (The Hague, 1888–1950).

Huygens, Christiaan: *Systema saturnium, sive de causis mirandorum Saturni phaenomenôn, et comite ejus planeta novo* (The Hague, 1659).

Ibn Tufail, *Philosophus autodidactus sive Epistola Abi Jaafar, ebn Tophail de Hai ebn Yokdhan*, transl. Johannes Bouwmeester (Amsterdam, 1672).

Jansenius, Cornelius: *Augustinus, seu doctrina Sancti Augustini de humanae naturae sanitate, aegritudine medicina*, 3 vols. (Louvain, 1640).

Jansen van ter Goes, Anthony: *Zederymen, bestaande in zangen en gedigten* (Amsterdam, 1656).

Jelles, Jarig: *Belydenisse des algemeenen en christelijcken geloofs* (Amsterdam, 1684).

Kant, Immanuel: *Anthropologie in pragmatischer Hinsicht abgefasst* (Köningsberg, 1798).

Kant, Immanuel: *Gesammelte Schriften*, 29 vols. (Berlin, 1902—).

Kircher, Athanasius: *Magnes; sive, de arte magnetica opus tripartitum* (Rome, 1641); 3rd ed. (Rome, 1654).

Kirchmayer, Georg Kaspar: *Commentatio epistolica de phosphoris et natura lucis nec non de igne* (Wittenberg, 1680).

Kirchmayer, Georg Kaspar: *Institutiones metallicae, das ist, wahr- und klarer Unterricht vom edlen Bergwerck* (Wittenberg, 1687).

Kirchmayer, Georg Kaspar (praes.) and Arnoldus Lendericus (resp.): *Ex physicis disputatio publica de coelo* (Wittenberg, 1659).

Kirchmayer, Georg Kaspar (praes.) and Johannes Frid. Scharfius (resp.): *De origine vitae humanae et potioribus quibusdam philosophiae cartesianae speciminibus, additis, a fine, notis apologeticis, ad secunda responsa placentiniana* (Wittenberg, 1660).

Koerbagh, Adriaan [Vreederyk Warmond]: *Een Bloemhof van allerley lieflijkheyd* (Amsterdam, 1668).

Koerbagh, Adriaan [Vreederyk Warmond]: *Een ligt schijnende in duystere plaatsen* (Amsterdam, 1668).

Kók, A.L.: *Ont-werp der Neder-duitsche letter-konst*, ed. Gerardus Rutgerus Wilhelmus Dibbets (Amsterdam, 1649); repr. (Assen, 1981).

La Forge, Louis de: *Traité de l'esprit de l'homme, de ses facultez et fonctions, et de son union avec le corps, suivant les Principes de René Descartes* (Paris, 1666); transl. *Tractatus de mente humana, eiusque facultatibus et functionibus, nec non de eiusdem unione cum corpore, secundum principia R. Descartes* (Amsterdam, 1669).

La Grange, Jean-Baptiste de: *Les principes de la philosophie contre les nouveaux philosophes Descartes, Rohault, Regius, Gassendi, Le P. Maignan, &c.* (Paris, 1675).

Lana de Terzi, Francesco: *Magisterium naturae et artis, opus physico-mathematicum*, 3 vols. (Brescia-Parma, 1684–1692).

L'art de vivre heureux, formé sur les idées les plus claires de la raison et du sens commun et sur de très belles maximes de Monsieur Descartes (Paris, 1687), ed. Sébastien Charles (Paris, 2009).

Le Clerc, Jean: "Éloge de feu Mr. De Volder professeur en philosophie et aux mathematiques, dans l'academie de Leide," *Bibliothèque choisie*, 18 (1709), 346–354.

Le Clerc, Jean: *Physica, sive, de rebus corporeis libri quinque* (London, 1696).

Le Grand, Antoine: *Entire Body of Philosophy according to the Principles of the famous Renate Descartes*, 2 vols. (London, 1694).

Le Grand, Antoine: *Institutio philosophiae, secundum principia Renati Descartes ... ad usum juventutis academicae* (London, 1672); 2nd ed. (Nuremberg, 1683).

Le Grand, Antoine: *Philosophia veterum e mente Renati Descartes, more scholastico breviter digesta* (London, 1671).

Le Grand, Antoine: *Scydromedia seu sermo quem Alphnsus de la Vita habuit coram comite de Falmouth* (London, 1699),

Leibniz, Gottfried Wilhelm: *Die philosophischen Schriften von Gottfried Wilhelm Leibniz*, 7 vols., ed. Carl Immanuel Gerhardt (Berlin, 1875–1890); repr. (Hildesheim, 1960).

Lelevel, Henri de: *La vraye et la fausse métaphysique, ou l'on refute les sentimens de M. Regis, et de ses adversaires de cette matiere* (Paris, 1694)

Le Long, Isaac: *Hondertjaarige Jubelgedachtenisse der Akademie van Utrecht: bestaande in een beknopt verhaal, van derselver oorsprong* (Amsterdam, 1736).

Lesclache, Louis de: *La philosophie divisée en cinq parties* (Paris, 1648–1650).

Lesclache, Louis de: *Les avantages que les femmes peuvent recevoir de la philosophie, ou l'abrégé de cette science* (Paris, 1667).

[Le Valois, Louis] La Ville, Louis de: *Sentimens de Monsieur Descartes... opposez à la Doctrine de l'Eglise, et conforme aux erreurs de Calvin sur le sujet de l'Eucharistie* (Caen, 1680).

Lidén, Johan Henrik: *Catalogus disputationum in Academiis et Gymnasiis Sveciae, atque etiam, a Svecis, extra patriam habitarum, quotquot huc usque reperiri potuerunt*, 2 vols. (Uppsala, 1778–1780).

Limmer, Conrad (praes.) and Johann Christian Wolff (resp.): *Dissertatio philosophica, de magnete ejusque effectibus* (Zerbst, 1693).

Lipstorp, Daniel: *Copernicus redivivus, seu de vero mundi systemate, liber singularis* (Leiden, 1653).

Lipstorp, Daniel: *Specimina philosophiae Cartesianae: quibus accedit, ejusdem authoris, Copernicus redivivus* (Leiden, 1653).

Locke, John: *Some Thoughts Concerning Education* (London, 1693).

Locke, John: *The Correspondence*, ed. Esmond S. de Beer, 8 vols. (Oxford, 1978–1989).

Lufneu, Hermann: "Memoire communiqué [...] sur une expérience curieuse d'hydrostatique," *Nouvelles de la république des lettres* (April 1685), 381–389.

Lufneu, Hermann: "Réponse de Mr Lufneu à la difficulté que Mr Pujolas luy a faite dans les Nouvelles du mois de Janvier dernier," *Nouvelles de la république des lettres* (March 1687), 239–249.

Luyken, Caspar: *Ondersoeck over den inhoud van twee boecxkens* (Amsterdam, 1655).

Maccovius, Joannes: *Metaphysica: ad usum quaestionum in philosophia ac theologia adornata & applicata* (Leiden, 1645); 2nd ed., *accedit* [...] *tractatus de anima separata* (Leiden, 1650); 3rd ed., *tertium edita, et explicata, vindicata, refutata per Adrianum Heereboord* (Amsterdam, 1651; Leiden, 1658; Bremen, 1668).

Maignan, Emmanuel: *Pars secunda philosophiae naturae, cursus philosophicus, concinnatus ex notissimis cuiusque principiis* (Toulouse, 1653).

Maire, Joannes: *Recentiorum disceptationes de motu cordis, sanguinis, et chyli, in animalibus, quorum series post alteram paginam exhibetur* (Leiden, 1647).

Malebranche, Nicolas: *De la recherche de la vérité, où l'on traite de la nature de l'esprit de l'homme, & de l'usage qu'il en doit faire pour éviter l'erreur dans les sciences* (Paris, 1674–1675).

Maresius, Samuel: *De abusu philosophiae cartesianae* (Groningen, 1670).

Mersenne, Marin: *Correspondance du P. Marin Mersenne, Religieux Minime*, eds. Cornelis de Waard, René Pintard, Bernard Rochot and Armand Beaulieu, 17 vols. (Paris, 1933–1988).

Meyer, Lodewijk: *Nederlandsche Woordenschat* (Amsterdam, 1650).

Meyer, Lodewijk: *Philosophy as the Interpreter of Holy Scripture*, transl. Samuel Shirley (Milwaukee, WI, 2005).

Molhuysen, Philip Christiaan: *Bronnen tot de geschiedenis der Leidsche Universiteit*, 7 vols. (The Hague, 1913–1924).

Molière: *Oeuvres complètes*, ed. Robert Jouanny, 2 vols. (Paris, 1979).

Montaigne, Michel de: *Oeuvres complètes*, eds. Albert Thibaudet and Maurice Rat, 14 vols. (Paris, 1962).

More, Henry: *A Collection of Several Philosophical Writings* (London, 1662).

More, Henry: *Conjectura cabbalistica* (London, 1653).

More, Henry: *Enchiridion metaphysicum, sive de rebus incorporeis succincta et luculenta dissertatio* (London, 1671).

More, Henry: *Observations upon Anthroposophia Theomagica* (London, 1650).

More, Henry: *The Immortality of the Soul, so farre forth as it is Demonstrable from the Knowledge of Nature and the Light of Reason* (London, 1659).

Moréri, Louis: *Le grand dictionnaire historique, ou le mélange curieux de l'histoire sacré et profane*, 5 vols. (Paris, 1712).

Morhof, Daniel Georg: *Polyhistor, literarius philosophicus et practicus* (Lübeck, 1714).

Morley, Christopher Love: *Collectanea chymica* (Leiden, 1689).

Murhard, Friedrich Wilhelm August: *Versuch einer historisch-chronologischen Bibliographie des Magnetismus* (Kassel, 1797).

Newton, Isaac: *Certain Philosophical Questions: Newton's Trinity Notebook*, eds. and transl. James E. McGuire and Martin Tamy (Cambridge, 1983).

Niceron, Jean-Pierre: "Burcher de Volder," in *Mémoires pour servir á l'histoire des hommes illustres dans la république des lettres. Avec un catalogue raisonné de leurs ouvrages*, ed. Jean-Pierre Niceron, 43 vols. (Paris, 1733), vol. 22, 48–57.

North, Roger: *Lives of the Right Hon. Francis North […] together with the Autobiography of the Author*, ed. Augustus Jessop, 3 vols. (London, 1890).

Ode, Jacob: *Principia philosophiae naturalis, in usum scholarum privatarum conscripta, et captui studiosae juventutis accommodata* (Utrecht, 1727).

Oldenburg, Henry: *The Correspondence of Henry Oldenburg*, eds. and trans. A. Rupert Hall and Marmi Boas Hall, 13 vols. (London, 1965).

Paquot, Jean Noël: *Mémoires pour servir à l'histoire littéraire des dix-sept provinces des Pays-Bas, de la principauté de Liège et de quelques contrées voisines*, 18 vols. (Louvain, 1763–1770).

Petermann, Andreas: *Philosophiae cartesianae adversus censuram Petri Danielis Huetii vindicatio, in qua pleraque intricatiora Cartesii loca clare explanantur* (Leipzig, 1706).

Placentinus, Johannes [Jan Kołaczek]: *Dissertatio philosophica, exhibens modum praecavendi errorem in veritati philosophicae, imprimis naturalis inquisitione atque dijudicatione, iuxta principia Renati des Cartes* (Frankfurt an der Oder, 1660).

Placentinus, Johannes: *Dissertatio philosophica, probans, calorem, et motum membrorum naturalem, in humano corpore: adeoque vitam non procedere ab anima, quae unica est scilicet rationalis, sed a materia coelesti subtilissima, analoga elemento solis et stellarum fixarum* (Frankfurt an der Oder, 1659).

Placentinus, Johannes: *Geotomia sive terrae sectio exhibens* (Frankfurt an der Oder, 1657).

Placentinus, Johannes: *Renatus Des-Cartes triumphans, id est, principia philosophiae cartesianae in Alma Viadrina ventilata atque defensa* (Frankfurt an der Oder, 1655).

Placentinus, Johannes (praes.) and Samuel Kaldenbach (resp.): *Discussio mathematica erotematis, an terra moveatur?* (Frankfurt an der Oder, 24 February 1655).

Placentinus, Johannes (praes.) and Georgius Mellemanus (resp.): *Disputatio mathematica exhibens themata ex universa mathesi* (Frankfurt an der Oder, 3 December 1653).

Placentinus, Johannes (praes.) and Reinholdus Tinctorius (resp.): *De origine caloris et motus membrorum naturalis, in humano corpore, adeoque vitae brutae, quae nobis communis est cum bestiis* (Frankfurt an der Oder, 1660).

Placentinus, Johannes (praes.) and Johann Reinhold von Boyen (resp.): *Dissertatio physica, refutans propositam in Academia Wittenbergensi* (Frankfurt an der Oder, 1660).

Plemp, Vopiscus Fortunatus: *De fundamentis medicinae* (Louvain, 1638), 2nd ed., *Fundamenta medicinae ad scholae acribologiam aptata* (Louvain, 1644); 3rd ed. (Louvain, 1654); 4th ed. (Louvain, 1664).

Plemp, Vopiscus Fortunatus: *Ophthalmographia, sive tractatio de oculi fabrica, actione et usu* (Amsterdam, 1632); 2nd ed. (Louvain, 1648); 3rd ed. (Louvain, 1659).

Plemp, Vopiscus Fortunatus, Libertus Fromondus, Petrus Damasus De Coninck, Christianus Lupus, F. Joan Rivius and Henricus Vanden Nouwelandt, *Doctorum aliquot in Academia Lovaniensi virorum iudicia de philosophia cartesiana*, in Vopiscus Fortunatus Plemp, *Fundamenta medicinae*, 3rd ed. (Louvain, 1654), 375–387.

Poelenburgh, Arnoldus: *Epistola ad C.H. in qua liber octavus Summæ controversiarum Ioannis Hoornbeeckii* (Amsterdam, 1655).

Poiret, Pierre: *Cogitationes rationales de Deo, anima, et malo* (Amsterdam, 1685).

[Pourchot, Edmond]: *Institutio philosophica ad faciliorem veterum ac recentiorum philosophorum intelligentiam comparatae*, 4 vols. (Paris, 1695).

Pufendorf, Samuel: *De officio hominis et civis juxta legem naturalem* (Lund, 1673).

Ray, John: *Observations Topographical, Moral, & Physiological; Made in a Journey Through part of the Low-Countries, Germany, Italy, and France* (London, 1673).

Reael, Laurens and Kaspar Van Baerle: *Observatien of ondervindingen aen de magneetsteen: en de magnetische kracht der aerde* (Amsterdam, 1651).

Relaas van de beroertens op Parnassus: ontstaan over het drukken van de Beginselen der eerste wijsbegeerte van de heer Renatus Descartes (Amsterdam, 1690).

Regiomontanus, Johannes, *An Terra moveatur an quiescat* (Nuremberg, 1553).

Regis, Pierre-Sylvain: *L'usage de la raison et de la foy, ou l'accord de la foy et de la raison* (Paris, 1704), ed. Jean-Robert Armogathe (Paris, 1996).

Regis, Pierre-Sylvain: *Système de philosophie, contenant la logique, la metaphysique, la physique et la morale*, 3 vols. (Paris, 1690); 2nd ed. *Cours entiers de philosophie; ou système generale selon les principes de M. Descartes concernant la logique, la metaphysique, la physique, et la morale.* [...] *Derniere edition, enrichie d'un très-grand nombre des figures, et augmentee d'un discours sur la philosophie ancienne et modern, où l'on fait en abregé l'histoire de cette science* (Amsterdam, 1691).

Register van Holland en Westvriesland van de jaaren 1655 en 1656 (s.l., s.a.).

Regius, Henricus: *Brevis explicatio mentis humanae, sive animae rationalis* (Utrecht, 1657).

Regius, Henricus: *Disputatio medico-physiologica pro sanguinis circulatione* (Utrecht, 1640).

Regius, Henricus: *Fundamenta physices* (Amsterdam, 1646).

Regius, Henricus: *Physiologia sive cognitio sanitatis, tribus disputationibus in Academia Ultrajectina publice proposita* (Utrecht, 1641).

Regius, Henricus: *Praxis medica, medicationum exemplis demonstrata*, 2nd ed. (Utrecht, 1657).

Regius, Henricus: *Responsio sive notae in appendicem ad corollaria theologico-philosophica* (Utrecht, 1642).

Regius, Henricus: *Spongia, qua eluntur sordes animadversionum, quas Jacobus Primirosius, doctor medicus, adversus theses pro circulatione sanguinis* (Utrecht, 1640).

Regius, Henricus (praes.) and Cornelius Godefridus Elbarchs (resp.): *Medicatio viri catalepsi laborantis* (Utrecht, 1645).

Regius, Henricus (praes.) and Petrus a Wouw (resp.): *Disputatio medica inauguralis de catalepsi* (Utrecht, 1653).

Reneri, Henricus: "De lectionibus ac exercitiis philosophicis," in *Illustris Gymnasii Ultrajectini inauguratio una cum orationibus inauguralibus* (Utrecht, 1634).

Revius, Jacobus, *Kartesiomania, hoc est, furiosum nugamentum, quod Tobias Andreæ, sub titulo Assertionis methodi cartesianæ, orbi literato obstrusit, succincte ac solide confutatum* (Leiden, 1654).

Riccioli, Giovanni Battista: *Almagestum novum astronomiam veterem novamque complectens observationibus aliorum, et propriis novisque theorematibus, problematibus, ac tabulis promotam* (Bologna, 1651).

Riddermarck, Andreas (praes.) and Magnus S. Aulaenius (resp.): *Dissertatio philosophica de magnetis ac ferri amoribus & odiis, eorumq[ue] interitu* (Lund, 1692).

Rohault, Jacques: *A System of Natural Philosophy*, ed. L.L. Laudan (New York, NY, 1969).

Rohault, Jacques: "Discours de la fiévre," in René Descartes, *Le monde* (Paris, 1664).

Rohault, Jacques: *Entretiens sur la philosophie* (Paris, 1671).

Rohault, Jacques: *Oeuvres posthumes*, ed. Claude Clerselier (Paris, 1682).

Rohault, Jacques: *Physique nouvelle (1667)*, ed. Sylvain Matton (Paris-Milan, 2009).

Rohault, Jacques: *System of Natural Philosophy, Illustrated with Dr. Samuel Clarke's Notes, Taken Mostly Out of Sir Isaac Newton's Philosophy*, 2 vols. (London, 1723); repr. (New York, NY-London, 1987).

Rohault, Jacques: *Traité de physique* (Paris, 1671), ed. Simone Mazauric (Paris, 2014).

Rohault, Jacques: *Tractatus physicus*, transl. Théophile Bonet (Geneva, 1674).

Rohault, Jacques: *Tractatus physicus: Latine vertit, recensuit, & uberioribus jam adnotationibus, ex illustrissimi Isaaci Newtoni Philosophia maximam partem haustis, amplificavit & ornavit Samuel Clarke*, transl. Samuel Clarke (Amsterdam, 1708).

Scheid, Johann Valentin (praes.) and Joannes Joachimus Kast (resp.): *Quaestionum decades duae de magnete* (Strasbourg, 1683).

Scheiner, Christoph: *Oculus, hoc est fundamentum opticum* (Innsbruck, 1619).

Schenck, Johannes: *Observationes medicae de capite humano* (Basel, 1584).

Schertzer, Johannes Adam: *Breviarium Eustachianum* (Leipzig, 1663).

Scheurleer, Henry: *Bibliotheca Martiniana, ou catalogue de la bibliothèque de feu Monsr. David Martini* (The Hague, 1752).

Schoock, Martin: *Admiranda methodus novae philosophiae Renati Des Cartes* (Utrecht, 1643).

Schoock, Martin (praes.) and various (resp.): *Physica generalis* (Groningen, 1660).

Schott, Gaspar: *Magia universalis naturae et artis*, 4 vols. (Würzburg, 1657–1659).

Schott, Gaspar: *Mechanica hydraulico-pneumatica* (Würzburg, 1657).

Schuler, Johannes: *Examinis philosophiae Renati Des-cartes specimen, sive brevis et perspicua* Principiorum philosophiae *cartesianae refutatio* (Amsterdam, 1666; Cambridge, 1685).

Schuler, Johannes: *Exercitationes ad* Principiorum philosophiae *Renati Des-Cartes partem primam* (Utrecht, 1667; Cambridge, 1682).

Schuler, Johannes: *Philosophia nova methodo explicata. Cujus pars prior continet exercitationum philosophicarum libros VI, metaphysicam, theologicam naturalem, angelographiam, psychologiam, logicam, ethicam* [...] *Pars posterior: continens librum VII sive physicam*, 2 vols. (The Hague, 1663).

Schuyl, Florian: *De veritate scientiarum et artium academicarum: qua, demostrata mentis et Dei opt. max. existentia, omnium scientiarum et artium certitudo* [...] *deducitur* (Leiden, 1672).

Seneca, Lucius Annaeus: *Ad Lucilium epistulae morales*, transl. Richard M. Gummere, 3 vols. (London-New York, NY, 1925–1961).

Senguerd, Arnoldus: *Collegium physicum, in quo viginti disputationibus in illustri Amstelodamensium Gymnasio publice ventilatis, physica systematice proponitur* (Amsterdam, 1652).

Senguerd, Arnoldus: *Introductio ad physicam libri sex* (Utrecht, 1644); 2nd ed. (Amsterdam, 1653); 3rd ed. (Amsterdam, 1666).

Senguerd, Arnoldus: *Physicae exercitationes* (Amsterdam, 1658).

Senguerd, Arnoldus (praes.) and Petrus ab Eyndhoven (resp.): *Disputatio physica decima continens quaestiones de mineralibus in specie* (Utrecht, 1645).

Senguerd, Arnoldus (praes.) and Friedricus Renesse (resp.): *Disputatio physica nona continens quaestiones de mineralibus in genere* (Utrecht, 1645).

Senguerd, Wolferdus: *Philosophia naturalis, quatuor partibus primarias corporum species, affectiones, vicissitudines, et differentias, exhibens* (Leiden, 1681); 2nd ed. (Leiden, 1685).

Senguerd, Wolferdus (praes.) and Cornelius Udemans (resp.): *Disputationum physicarum selectarum septima decima; quae est de particulis subtilibus, secunda* (Leiden, 1679).

Sennert, Daniel: *Practica medicina, liber primus* (Wittenberg, 1636).

Siegfried, Caspar Esaias: "Curiöse Gedancken vom Magnete," in *Deliciarum manipulus, Das ist: Annehmliche und rare Discurse von mancherley nützlichen und curiosen Dingen*, transl. M.M., 2 vols. (Dresden, 1704), vol. 2, 65–144.

Siegfried, Caspar Esaias (praes.) and Ehrenfried Pfundt (resp.): *Disputatio physica de magnete* (Weissenfels, 1673).

Spinoza, Baruch: *Collected Works*, ed. and transl. Edwin Curley, 2 vols. (Princeton, NJ, 1985–2016).

Spinoza, Baruch: *Opera posthuma* (Amsterdam, 1677).

Spinoza, Baruch: *Tractatus theologico-politicus* (Amsterdam, 1670).

Steensen [Steno], Niels: *Discours de Monsieur Stenon sur l'anatomie du cerveau* (Paris, 1669).

Stevin, Simon: *Hypomnemata mathematica.* [...] *Tomus quartus* [...] *de statica* (Leiden, 1605).

Tatinghoff, Johann: *Clavis philosophiae naturalis antiquo-novae* (Leiden, 1655).

Tepelius, Johannes: *Historia philosophiae cartesianae* (Nuremberg, 1674).

Testimonium Academiae Ultrajectinae, et narratio historica, qua defensae, qua exterminatae novae philosophiae (Utrecht, 1643).

Vallerius, Harald (praes.) and Andreas Linnrot (resp.): *Disputatio physico mathematica de pyxide magnetica: sive, ut vocant, compasso nautico* (Stockholm, 1699).

Van Gutschoven, Gerard: *Regulae munitionum analogicae, earumque ex methodo Fritagii et Dogenii usus compendiosus* (Brussels, 1673).

Van Gutschoven, Gerard: *Usus quadrantis geometrici* (Brussels, 1674).

Van Gutschoven, Willem (praes.) and Lambertus Buntinx (resp.): *Philosophia* (Louvain, 18 August 1655).

Van Gutschoven, Willem (praes.) and Lambertus Lamberti (resp.): *Philosophia* (Louvain, 11–12 August 1655).

Van Gutschoven, Willem (praes.) and Egbertus Van der Gheest (resp.): *Philosophia* (Louvain, 6 August 1653).

Van Gutschoven, Willem (praes.) and Hilarius Van Veen (resp.): *Philosophia* (Louvain, 31 July 1651).

Van Gutschoven, Willem (praes.) and Hilarius Van Werm (resp.): *Philosophia* (Louvain, 31 July 1651).

Van Musschenbroek, Johannes Joosten: *Descriptio antliae pneumaticae et instrumentorum ad eam inprimis pertinentium* (s.l., 1700).

Van Sichen, Willem: *Integer cursus philosophicus brevi, clara et ad docendum discendumque facili methodo digestus* (Antwerp, 1666).

Van Velden, Martinus (praes.) and Carolus Allegambe (resp.): *Theses philosophicae* (Leiden, 1 July 1695).

Van Velthuysen, Lambert: *Nader bewys dat noch de leere van der sonne stilstant en des aertryx beweging, noch de gronden van de philosophie van Renatus Des Cartes strijdig sijn met Godts woort: gestelt tegen een tractaet van J. du Bois, predikant tot Leyden; genaemt schadelickheyt van de Cartesiaensche philosophie* (Utrecht, 1657); transl. *Luculentior probatio, quod neque doctrina de quiete Solis et motu Terrae, neque principia Philosophiae Renuti Descartis verbo divino contraria sunt*, in Lambert Van Velthuysen, *Opera omnia*, 2 vols. (Rotterdam, 1680), vol. 2, 1177–1183.

Van Velthuysen, Lambert: *Opera omnia*, 2 vols. (Rotterdam, 1680).

Van Velthuysen, Lambert: *Disputatio de finito et infinito, in qua defenditur sententia clarissimi Cartesii, de motu, spatio, et corpore* (Amsterdam, 1651).

Varenius, Bernard: *A Complete System of General Geography Explaining the Nature and Properties of the Earth* [...] *Originally written in Latin by Bernhard Varenius, M.D. Since improved and illustrated by Sir Isaac Newton and Dr. Jurin; and now translated into English* (London, 1733).

Varenius, Bernard: *Geographia generalis, in qua affectiones generales telluris explicantur* (Amsterdam, 1650).

Vesti, Justus (praes.) and Johann Andreas Fischer (resp.): *Disputatio physico-medica de magnetismo macro- et microcosmi* (Erfurt, 1687).

Vieussens, Raymond: *Neurographia universalis* (Lyon, 1684).

Vincentius, Joannes: *Discussio peripatetica, in qua philosophiae Cartesianae principia, per singula fere capita, seu articulos dilucide examinantur* (Toulouse, 1677).

Voet, Daniel: *Compendium pneumaticae, adjectis aliquot ejusdem argumenti disputationibus publice antehac propositis*, ed. Gisbertus Voetius (Utrecht, 1661).

Voet, Daniel: *Meletemata philosophica: in quibus antiquae & receptae philosophiae fulcra magis stabiliuntur: novatorum philosophandi methodus evertitur: adjectis selectis quibusdam disputationibus ante hac publice ab eodem dum viveret, habitis* (Utrecht, 1661).

Voet, Daniel: *Physiologia: adjectis aliquot ejusdem argumenti disputationibus publice antehac propositis*, ed. Paul Voet (Utrecht, 1661).

Voet, Daniel (praes.) and Petrus Arlebautius (resp.): *Disputatio philosophica de opera-*

tionibus animalibus et habitibus brutorum pars posterior (Utrecht, 24 March 1658).

Voet, Daniel (praes.) and Nicolaas Beets (resp.): *De ubi spirituum, pars quarta* (Utrecht, 9 December 1653); repr. in Voet, *Compendium pneumaticae*, 113–121.

Voet, Daniel (praes.) and Libertus Essenius A.F. (resp.): *Disputationum selectarum decima-quarta de errore, secunda* (Utrecht, 22 December 1656).

Voet, Daniel (praes.) and Thomas Essingius (resp.): *Disputationum selectarum secunda de specificatione scientiarum pars secunda* (Utrecht, 19 May 1655).

Voet, Daniel (praes.) and Lambertus Groen (resp.): *De ubi spirituum, pars quinta* (Utrecht, 14 December 1653); repr. in Voet, *Compendium pneumaticae*, 121–133.

Voet, Daniel (praes.) and Samuel Junius (resp.): *Disputationum selectarum undecima de quaestionibus de nomine pars prior* (Utrecht, 19 March 1656).

Voet, Daniel (praes.) and Adrianus Ketelaer (resp.): *Disputationum selectarum octava continens quaestiones duas de evidentia et certitudine pars tertia* (Utrecht, 23 February 1656).

Voet, Daniel (praes.) and Cornelius Mogge (resp.): *Disputatio philosophica de operationibus animalibus et habitibus brutorum pars prior* (Utrecht, 10 March 1658).

Voet, Daniel (praes.) and Arnoldus Niepoort (resp.): *De ubi spirituum* (Utrecht, 29 June 1653); repr. in Voet, *Compendium pneumaticae*, 87–95.

Voet, Daniel (praes.) and Johannes Rauwers (resp.): *Disputationis metaphysicae, de formali ratione libertatis pars quarta* (1657).

Voet, Daniel (praes.) and Justus Smetius (resp.): *Disputationum selectarum quarta de pluralitate motus specifica: appetitu innato* (Utrecht, 15 November 1654); repr. in Voet, *Physiologia*, 120–124.

Voet, Daniel (praes.) and Regnerus Van Mansvelt (resp.): *Disputatio philosophica continens quaestiones illustres philosophicas* (Utrecht, 1656).

Voet, Daniel (praes.) and Caspar Verstegen (resp.): *Disputatio philosophica de habitu pars secunda quae est de habitu corporis* (Utrecht, 24 May 1656).

Voet, Daniel (praes.) and Reinerus Vossius (resp.): *Disputationum selectarum sexta continens quaestiones duas de evidentia et certitudine pars prior* (Utrecht, 1 December 1655).

[Voet, Gijsbert] Suetonius Tranquillus, *Den overtuyghden Cartesiaen, ofte clare aenwysinge uyt de bedenckingen van Irenaevs Philalethius, dat de stellingen en allegatien in de nader openinge tot laste der Cartesianen, in saecken de theologie raeckende, nae waerheyt en ter goeder trouwe zijn by een gebracht* (Leiden, 1656).

Voet [Voetius], Gijsbert: *Selectae disputationes* (Utrecht, 1669).

Voet, Gijsbert: *Selectarum disputationum theologicarum pars prima*, 4 vols. (Utrecht, 1648).

Voet, Paul (praes.) and Leonardus Van Ryssen (resp.): *Disputatio metaphysica pro novitate essendi* (Utrecht, 15 November 1651).

Von Guericke, Otto: *Experimenta nova (ut vocantur) Magdeburgica de vacuo spatio* (Amsterdam, 1672).

Waldschmidt, Johann Jakob (praes.) and Christian Kursner (resp.): *Disputatio physica, de magnete* (Marburg, 1683).

Waterlaet, Lambertus van den: *Prodromus sive examen tutelare orthodoxae philosophiae principiorum: contra fictitium quoddam huius temporis eorum pentagonum*, 2 vols. (Leiden, 1642).

Wendelinus, Marcus Fredericus: *De christlijke ghódt-ghe-leertheidt. In 't Latyn beschreeven en nu ten besten van de lief-hebbers der ghódt-ghe-leertheidt ver-taalt door Alhardt L. Kók* (Amsterdam, 1655).

Wits, Herman: *De twist des Heeren met sijn wijngaert* (Leeuwarden, 1671).

Wittich, Christoph: *Annotationes ad Renati Des-Cartes* Meditationes (Dordrecht, 1688).

Wittich, Christoph: *Consensus veritatis in Scriptura divina et infallibili revelatae cum veritate philosophica a Renato Des Cartes detecta* (Nijmegen, 1659).

Wittich, Christoph: *Theologia pacifica* (Leiden, 1671).

Worthington, John: *The Diary and Correspondence of John Worthington*, ed. James Crossley and Richard C. Christie, 3 vols. (Manchester, 1847–1886).

Wotton, Henry: *Reliquiae Wottonianae, or, A Collection of Lives, Letters, Poems* (London, 1651).

Zwinger, Theodor: *Scrutinium magnetis physico-medicum* (Basel, 1697).

Zwinger, Theodor (praes.) and Jeremias Gemuseus (resp.): *Disquisitionum physicarum de magnete secunda* (Basel, 1686).

Zwinger, Theodor (praes.) and Hieronymus Gemuseus (resp.): *Disquisitionum physicarum de magnete quinta* (Basel, 1692).

Zwinger, Theodor (praes.) and Theodor Gernler (resp.): *Disquisitionum physicarum de magnete prima* (Basel, 1685).

Zwinger, Theodor (praes.) and Daniel Schönauer (resp.): *Disquisitionum physicarum de magnete septima* (Basel, 1697).

Secondary Literature

Aalderink, Mark: *Philosophy, Scientific Knowledge and Concept Formation in Geulincx and Descartes* (Utrecht, 2009).

Agostini, Igor: "Descartes and More on the Infinity of the World," *British Journal for the History of Philosophy*, 25 (2017), 878–896.

Agostini, Igor: "Quelques remarques sur l'epistola ad V.C. de Henry More," *Les études philosophiques*, 108 (2014), 7–35.

Agostini, Igor: *La démonstration de l'existence de Dieu: les conclusions des cinq voies de saint Thomas d'Aquin et la preuve a priori dans le thomisme du XVIIe siècle* (Turnhout, 2016).

Agostini, Siegrid: *Les lettres de Claude Clerselier (1646–1681)* (Turnout, 2021).

Ahsmann, Margreet: *Collegia en Colleges: Juridisch onderwijs aan de Leidse Universiteit 1575–1630 in het bijzonder het disputeren* (Groningen, 1990); transl. Irene Sagel-Grande, *Collegium und Kolleg: der juristische Unterricht an der Universität Leiden 1575–1630, unter besonderer Berücksichtigung der Disputationen* (Frankfurt am Main, 2000).

Aiton, Eric J.: "Descartes's Theory of the Tides," *Annals of Science*, 11 (1955), 337–348.

Akkerman, Fokke: "J.H. Glazemaker, an Early Translator of Spinoza," in *Spinoza's Political and Theological Thought*, ed. Cornelis de Deugd (Amsterdam, 1984), 23–29.

Akkerman, Fokke: *Studies in the Posthumous Works of Spinoza* (Groningen, 1980).

Album promotorum, qui inde ab anno MDCXXXVIo usque ad annum MDCCCXVum in Academia Rheno-Trajectina gradum doctoratus adepti sunt (Utrecht, 1936).

Album studiosorum Academiae Rheno-Traiectinae, 1636–1886: http://dspace.library.uu.nl/handle/1874/371445.

Alcover, Madeleine: *Cyrano relu et corrigè: Lettres, Estats du soleil, Fragment de physique* (Geneva, 1990).

Andersen, Claus A.: *Metaphysik im Barockscotismus: Untersuchungen zum Metaphysikwerk des Bartholomaeus Mastrius. Mit Dokumentation der Metaphysik in der scotistischen Tradition ca. 1620–1750* (Amsterdam-Philadelphia, PA, 2016).

Anfray, Jean Pascal: *"Partes extra partes*: etendue et impénétrabilité dans la correspondance entre Descartes et More," *Les études philosophiques*, 108 (2014), 37–59.

Antoine-Mahut, Delphine and Stephen Gaukroger, eds.: *Descartes' Treatise on Man and its Reception* (Dordrecht, 2016).

Allgemeine Deutsche Biographie, 56 vols. (Leipzig, 1875–1912); repr. (Berlin, 1967–1971).

Appold, Kenneth G.: "Academic Life and Teaching in Post-Reformation Lutheranism," in *Lutheran Ecclesiastical Culture, 1550–1675*, ed. Robert Kolb (Leiden, 2008), 65–116.

Ariew, Roger: "Censorship, Condemnations, and the Spread of Cartesianism," in *Cartesian Empiricisms*, eds. Mihnea Dobre and Tammy Nyden (Dordrecht, 2013), 25–46.

Ariew, Roger: *Descartes among the Scholastics* (Leiden, 2011).

Ariew, Roger: *Descartes and the First Cartesians* (Oxford, 2014).

Ariew, Roger: "Descartes and the First Cartesians Revisited," *Perspectives on Science*, 26 (2018), 599–617.

Ariew, Roger: "Descartes, les premiers cartésiens et la logique," *Revue de métaphysique et de morale*, 49 (2006), 55–71.

Ariew, Roger: "La *Logique de Port-Royal*, les premiers cartésiens et la scolastique tardive," *Archives de philosophie*, 78 (2015), 29–48.

Ariew, Roger: *"Le meilleur livre qui ait jamais été fait en cette matière*: Eustachius a Sancto Paulo and the Teaching of Philosophy in the Seventeenth Century," in *Teaching Philosophy in the Early Modern Europe: Text and Image*, eds. Susanna Berger and Daniel Garber (Dordrecht, 2021), 31–46.

Ariew, Roger: "Quelques condamnations du cartésianisme: 1662–1706," *Bulletin carté-sien*, XXII, *Archives de philosophie*, 57 (1994), 1–6.

Ariew, Roger: "The Infinite in Descartes' Conversation with Burman," *Archiv für Ge-schichte der Philosophie*, 69 (1987), 140–163.

Ariew, Roger: "The Nature of Cartesian Logic," *Perspectives on Science*, 29 (2021), 275–291.

Ariew, Roger: "Theory of Comets at Paris during the Seventeenth Century," *Journal of the History of Ideas*, 53 (1992), 355–372.

Ariew, Roger and Marjorie Grene: "Ideas, in and before Descartes," *Journal of the History of Ideas*, 56 (1995), 87–106.

Ariew, Roger and Marjorie Grene: "The Cartesian Destiny of Form and Matter," *Early Science and Medicine*, 2 (1997), 300–325.

Armogathe, Jean-Robert: "Early German Reactions to Huet's Censura," in *Skepticism in the Modern Age: Building on the Work of Richard Popkin*, eds. José R. Maia Neto, Gianni Paganini, and John Christian Laursen (Leiden-Boston, MA, 2009), 297–308.

Armoganthe, J.R., "Eustachius a Sancto Paulo", in The Cambridge History of Seven-teenth-Century Philosophy, ed. Daniel Garber and Michael Ayers, 2 vols (Cam-bridge, 2000), 1424–1425.

Armogathe, Jean-Robert: "Préface," in Pierre-Sylvain Régis, *La morale ou les devoirs de l'homme raisonnable, de l'homme civil et de l'homme chrestien (1682)*, ed. Sylvain Mat-ton (Paris-Milan, 2015), v–xiv.

Armogathe, Jean-Robert: Theologia cartesiana: *L'explication physique de l'Eucharistie chez Descartes et Dom Desgabets* (The Hague, 1977).

Armogathe, Jean-Robert and Vincent Carraud: "La première condamnation des *Oeuvres* de Descartes, d'après des documents inédits aux Archives du Saint-Office," *Nouvelles de la République des Lettres*, 2 (2001), 103–137.

Armogathe, Jean-Robert and Vincent Carraud: "The First Condemnation of Descartes's *Oeuvres*: Some Unpublished Documents from the Vatican Archives," *Oxford Studies in Early Modern Philosophy*, 1 (2005), 67–110.

Ashworth, Earline J.: "Petrus Fonseca on Objective Concepts and the Analogy of Being," in *Logic and the Workings of the Mind: The Logic of Ideas and Faculty Psychology in Early Modern Philosophy*, ed. Patricia A. Easton (Atascadero, CA, 1997), 46–63.

Aucante, Vincent: *La philosophie médicale de Descartes* (Paris, 2006).

Ayers, Michael: "Ideas and Objective Being," in *The Cambridge History of Seventeenth-Century Philosophy*, eds. Daniel Garber and Michael Ayers (Cambridge, 1998), vol. 2, 1062–1107.

Baldwin, Martha: "Athanasius Kircher and the Magnetic Philosophy" (PhD diss., Uni-versity of Chicago, 1987).

Balz, Albert G.A.: "Clerselier (1614–1684) and Rohault (1620–1675)," *The Philosophical Review*, 39 (1930), 445–458.

Barth, Christian: "Leibnizian *Conscientia* and its Cartesian Roots," *Studia Leibnitiana*, 43 (2011), 216–236.

Bauer, Barbara, ed.: *Melanchthon und die Marburger Professoren (1527–1627)* (Marburg, 1999).

Beck, Andreas J.: *Gisbertus Voetius (1589–1676): sein Theologieverständnis und seine Gotteslehre* (Göttingen, 2007).

Beck, Leslie J.: *The Method of Descartes: A Study of the* Regulae (Oxford, 1952).

Belgioioso, Giulia: "Descartes, Elisabeth et le cercle cartésien de La Hague," in *Elisabeth de Bohême face à Descartes: deux philosophes?*, eds. D. Kolesnik-Antoine and Marie-Frédérique Pellegrin (Paris, 2014), 13–44.

Bellis, Delphine: "Empiricism without Metaphysics: Regius' Cartesian Natural Philosophy," in *Cartesian Empiricism*, eds. Mihnea Dobre and Tammy Nyden (Dordrecht, 2013), 151–183.

Bellis, Delphine: "The Perception of Spatial Depth in Kepler's and Descartes' Optics: A Study of an Epistemological Reversal," in *Boundaries, Extents and Circulations: Space and Spatiality in the Early Modern Natural Philosophy*, eds. Koen Vermeir and Jonathan Regier (Dordrecht, 2016), 125–152.

Berger, Susanna: *The Art of Philosophy: Visual Thinking in Europe from the Late Renaissance to the Early Enlightenment* (Princeton, NJ-Oxford, 2017).

Berkel, Klaas van: "Descartes' Debt to Beeckman: Inspiration, Cooperation, Conflict," in *Descartes' Natural Philosophy*, eds. Stephen Gaukroger, John A. Schuster and John Sutton (London-New York, NY, 2000), 48–59.

Bertoloni Meli, Domenico: *Mechanicism: A Visual, Lexical, and Conceptual History* (Pittsburgh, PA, 2019).

Biagioli, Mario: *Galileo, Courtier: The Practice of Science in the Culture of Absolutism* (Chicago, IL-London, 1993).

Biagioli, Mario: "Le prince et les savants: La civilté scientifique au 17e siècle," *Annales: historie, sciences sociales*, 6 (1995), 1417–1453.

Biagioli, Mario: "The Social Status of Italian Mathematicians, 1450–1600," *History of Science*, 27 (1989), 41–95.

Bianchi, Luca: *Le verità dissonanti: Aristotele alla fine del Medioevo* (Rome-Bari, 1990).

Blair, Ann: "Mosaic Physics and the Search for a Pious Natural Philosophy in the Late Renaissance," *Isis*, 91 (2000), 32–58.

Blair, Ann: "Natural Philosophy," in *The Cambridge History of Science*, vol. 3, ed. Katharine Park and Lorraine Daston (Cambridge, 2006), 365–406.

Blay, Michel: "Introduction," in Jacques Rohault, *Physique Nouvelle (1667)*, ed. Sylvain Matton (Paris-Milan, 2009), v–xix.

Blum, Paul Richard: *"In fugam vacui*—Avoiding the Void in Baroque Thought," *Quaestio*, 17 (2017), 426–460.

Blum, Paul Richard: *Studies on Early Modern Aristotelianism* (Leiden-Boston, MA, 2012).

Bohatec, Josef: *Die cartesianische Scholastik in der Philosophie und reformierten Dogmatik des 17. Jahrhunderts* (Leipzig, 1912); repr. (Hildesheim, 1966).

Bordoli, Roberto: *Dio, ragione, verità: le polemiche su Descartes e su Spinoza presso l'Universita di Franeker (1686–1719)* (Macerata, 2009).

Borghero, Carlo: "Discussioni sullo scetticismo di Descartes (1650–1712)," *Giornale critico della filosofia italiana*, 77 (1998), 1–25.

Borghero, Carlo: "Osservazioni conclusive," in *Eredità cartesiane nella cultura britannica*, eds. Brunello Lotti and Paola Dessì (Florence, 2011), 253–268.

Bos, Egbert P., and Henri A. Krop: *Franco Burgersdijk (1590–1635): Neo-Aristotelian in Leiden* (Leiden, 1993).

Bos, Erik-Jan: "Descartes and Regius on the Pineal Gland and Animal Spirits, and a Letter of Regius on the True Seat of the Soul," in *Descartes and Cartesianism: Essays in Honour of Desmond Clarke*, eds. Stephen Gaukroger and Catherine Wilson (Oxford, 2017), 95–111.

Bos, Erik-Jan: "Een kleine geschiedenis van een cartesiaans versje," in *Bijzonder onderzoek: een ontdekkingsreis door de Bijzondere Collecties van de Universiteitsbibliotheek Utrecht*, eds. Bart Jaski, Hans Mulder, and Marco van Egmond (Utrecht, 2009), 244–251.

Bos, Erik-Jan: "Henricus Regius et les limites de la philosophie cartésienne," in *Qu'est-ce qu'être cartésien?*, ed. Delphine Kolesnik-Antoine (Lyon, 2013), 53–68.

Bos, Erik-Jan: *The Correspondence between Henricus Regius and Descartes* (Utrecht, 2002).

Bots, Hans: "Témoignages sur l'ancienne université de Nimègue," *Lias*, 19 (1992), 232–234.

Bouillier, Francisque: *Histoire de la philosophie cartésienne*, 2 vols. (Paris, 1868; Geneva, 1970).

Briquet, Fortunée: *Dictionnaire historique des Françaises connues par leurs écrits* (Strasbourg, 2016).

Brockliss, Lawrence: "Aristotle, Descartes and the New Science: Natural Philosophy at the University of Paris, 1600–1740," *Annals of Science*, 38 (1981), 33–69.

Brockliss, Laurence: *French Higher Education in the Seventeenth and Eighteenth Centuries: A Cultural History* (Oxford-New York, NY, 1987).

Brockliss, Lawrence: "Rapports de structure et de contenu entre les *Principia* et les cours de philosophie des colleges," in *Descartes*: Principia Philosophiae (*1644–1994*), eds. Jean-Robert Armogathe and Giulia Belgioioso (Naples, 1996), 491–516.

Brown, Harcourt: *Scientific Organizations in Seventeenth Century France (1620–1680)* (Baltimore, MD, 1934).

Buickerood, James G.: "The Natural History of the Understanding: Locke and the Rise of Facultative Logic in the Eighteenth Century," *History and Philosophy of Logic*, 6 (1985), 157–190.

Buning, Robin O.: *Henricus Reneri (1593–1639): Descartes' Quartermaster in Aristotelian Territory* (Utrecht, 2013).

Buys, Ruben: *Sparks of Reason: Vernacular Rationalism in the Low Countries (1550–1670)* (Hilversum, 2015).

Canziani, Guido: "Entre Descartes et Hobbes: la morale dans le *Système* de Pierre-Sylvain Régis," in Pierre-Sylvain Régis, *La morale ou les devoirs de l'homme raisonnable, de l'homme civil et de l'homme chrestien (1682)*, ed. Sylvain Matton (Paris-Milan, 2015), 59–136.

Capozzi, Mirella and Gino Roncaglia: "Logic and Philosophy of Logic from Humanism to Kant," in *The Development of Modern Logic*, ed. Leila Haaparanta (Oxford, 2009), 78–158.

Caps, Geraldine: *Les "médicins cartésiens": héritage et diffusion de la représentation mécaniste du corps humain (1646–1696)* (Hildesheim, 2010). s

Cassan, Elodie: "La *Logique* de Port-Royal: une logique cartésienne?," in *Qu'est-ce qu'être cartésien?*, ed. Delphine Kolesnik-Antoine (Lyon, 2013), 159–177.

Cassirer, Ernst: *Das Erkenntnisproblem in der Philosophie und Wissenschaft der neueren Zeit*, vol. II (Berlin, 1907).

Cassirer, Ernst: *Descartes: doctrine, personnalité, influence*, transl. Philippe Guilbert (Paris, 2008).

Cellamare, Davide: "A Theologian Teaching Descartes at the Academy of Nijmegen (1655–1679): Class Notes on Christoph Wittich's Course on the *Meditations on First Philosophy*," *Intellectual History Review*, 30 (2020), 585–613.

Cellamare, Davide: "Confessional Science and Organisation of Disciplines: Anatomy, Psychology, and Anthropology Between the Sixteenth and the Seventeenth Centuries," *Quaestio*, 17 (2017), 461–480.

Cellamare, Davide: "History of a Productive Failure: The Use of Andreas Vesalius's *De humani corporis fabrica* in Some Sixteenth-century Books on the Soul," in *Towards the Authority of Vesalius: Studies on Medicine and the Human Body*, eds. Erika Gielen and Michèle Goyens (Turnhout, 2018), 403–436.

Cellamare, Davide: "Soul, The," in *Encyclopedia of Early Modern Philosophy and the Sciences*, eds. Dana Jalobeanu and Charles T. Wolfe (Springer online, 2020), https://link.springer.com/referenceworkentry/10.1007%2F978--3--319--20791--9_209--1.

Cellamare, Davide and Paul Bakker: "Libertus Fromondus' Christian Psychology: Medicine and Natural Philosophy in the *Philosophia christiana de anima* (1649)," *Lias*, 42 (2015), 37–66.

Chalmers, Alan: *One Hundred Years of Pressure: Hydrostatics from Stevin to Newton* (Dordrecht, 2017).

Chang, Kevin: "From Oral Disputation to Written Text: The Transformation of the Dissertation in Early Modern Europe," *History of Universities*, 19 (2004), 129–187.

Chen-Morris, Raz and Ofer Gal: "Baroque Optics and the Disappearance of the Observer: From Kepler's Optics to Descartes' Doubt," *Journal of the History of Ideas*, 71 (2010), 191–217.

Clair, Pierre: "De Gueudreville interlocuteur de Rohault," *Recherches sur le XVIIe siècle*, 4 (1980), 47–52.

Clair, Pierre: *Jacques Rohault (1618–1672): Bio-Bibliographie* (Paris, 1978).

Clark, William: *Academic Charisma and the Origins of the Research University* (Chicago, IL-London, 2006).

Clarke, Desmond M.: *Descartes: A Biography* (Cambridge, 2006).

Clarke, Desmond M.: *Descartes's Theory of Mind* (Oxford, 2003).

Clarke, Desmond M.: *Occult Powers and Hypotheses: Cartesian Natural Philosophy under Louis XIV* (Oxford-New York, NY, 1989).

Clarke, Desmond M.: "The Physics and Metaphysics of the Mind: Descartes and Regius," in *Mind, Method, and Morality: Essays in Honour of Anthony Kenny*, eds. John Cottingham and Peter Hacker (Oxford, 2010), 187–207.

Clarke, Desmond M., and Erik-Jan Bos: "Henricus Regius," *The Stanford Encyclopedia of Philosophy* (Spring 2020 Edition), ed. Edward N. Zalta, https://plato.stanford.edu/archives/spr2020/entries/henricus-regius/.

Clericuzio, Antonio: *Elements, Principles and Corpuscles: A Study of Atomism and Chemistry in the Seventeenth Century* (Dordrecht, 2000).

Coesemans, Steven: "Faculties of the Mind: The Rise of Facultative Logic at the University of Louvain" (PhD diss., KU Leuven, 2019).

Coffey, Justin M.: "Catatonia in Adults: Epidemiology, Clinical Features, Assessment, and Diagnosis," *Up-to-Date*, www.uptodate.com/contents/catatonia-in-adults-epidemiology-clinical-features-assessment-and-diagnosis.

Colie, Rosalie: *Light and Enlightenment: A Study of The Cambridge Platonists and the Dutch Arminians* (Cambridge, 1957).

Collacciani, Domenico: "La censure du cartésianisme à Louvain: les thèses de médecine du 29 août 1662," *Bulletin cartésien*, XLVI, *Archives de philosophie*, 80 (2017), 162–165.

Collacciani, Domenico: "Manuscrits cartésiens à la Kongelige Bibliotek de Copenhague," *Bulletin cartésien*, XLIV, *Archives de philosophie*, 78 (2015), 168–172.

Collacciani, Domenico: "The Reception of *L'homme* Among the Leuven Physicians: The Condemnation of 1662 and the Origins of Occasionalism," in *Descartes' Treatise on Man and its Reception*, eds. Delphine Antoine-Mahut and Stephen Gaukroger (Dordrecht, 2016), 103–125.

Collacciani, Domenico and Sophie Roux: "La querelle optique de Bourdin et de Descartes à la lumière des thèses mathématiques soutenues au collège de Clermont," in *Chemins du cartésianisme*, eds. Antonella Del Prete and Raffaele Carbone (Paris, 2017), 51–84.

Collacciani, Domenico and Sophie Roux: "The Mathematics Theses Defended at Collège de Clermont (1637–1682): How to Guard a Fortress in Times of War," in *Teaching*

Philosophy in Early Modern Europe: Text and Image, eds. Susanna Berger and Daniel Garber (Dordrecht, 2021), 79–138.

Collet, Dominik: *Die Welt in der Stube: Begegnungen mit Außereuropa in Kunstkammern der Frühen Neuzeit* (Göttingen, 2007).

Cook, Harold J.: "Princess Elisabeth's Cautions and Descartes' Suppression of the *Traité de l'Homme*," *Early Science and Medicine* 26 (2021), 289–313.

Cook, Harold J.: "The New Philosophy in the Low Countries," in *The Scientific Revolution in National Context*, ed. Roy Porter (Cambridge, 2012), 115–149.

Cooper, Alix: "Placing Plants on Paper: Lists, Herbaria, and Tables as Experiments with Territorial Inventory at the Mid-Seventeenth-Century Gotha Curt," *History of Science*, 56 (2018), 257–277.

Cottingham, John: "A Brute to the Brutes? Descartes' Treatment of Animals," *Philosophy*, 53 (1978), 551–559.

Cousin, Victoir: *Fragments philosophiques*, in Victoir Cousin, *Oeuvres* II (Brussels, 1841), 1–273.

Crapulli, Giovanni: *Le note marginali latine nelle versione olandesi di opere di Descartes di J.H. Glazemaker* (Rome, 1969).

Darge, Rolf, Emmanuel J. Bauer, Günter Frank, eds.: *Der Aristotelismus an den europäischen Universitäten der Frühen Neuzeit* (Stuttgart, 2009).

De Angelis, Simone: *Anthropologien: Genese und Konfiguration einer 'Wissenschaft vom Menschen' in der Frühen Neuzeit* (Berlin, 2010).

Dear, Peter: "Method and the Study of Nature," in *The Cambridge History of Seventeenth-Century Philosophy*, eds. Daniel Garber and Michael Ayers, 2 vols. (Cambridge, 1998), vol. 1, 147–177.

De Baar, Maria P.A.: *'Ik moet spreken': Het spiritueel leiderschap van Antoinette Bourignon* (Zutphen, 2004).

De Buzon, Fréderic: *La science cartésienne et son objet* (Paris, 2013).

Dechange, Klaus: "Die frühe Naturphilosophie des Henricus Regius (Utrecht, 1641)" (Inaugural diss., Münster, Westfälische Wilhelms-Universität Münster, Medizinische Fakultät, 1966).

De Clercq, Peter: "Exporting Scientific Instruments around 1700: The Musschenbroek Documents in Marburg," *Tractrix*, 3 (1991), 79–120.

De Clercq, Peter: *The Leiden Cabinet of Physics: A Descriptive Catalogue* (Leiden, 1997).

De Dijn, Herman: "Adriaan Heereboord en het Nederlands Cartesianisme," *Algemeen Nederlands Tijdschrift voor Wijsbegeerte*, 75 (1983), 56–69.

De Haan, Adrianus. A.M.: *Het wijsgerig onderwijs aan het Gymnasium Illustre en de Hogeschool te Harderwijk, 1599–1811* (Harderwijk, 1960).

De Hoog, Adriaan Cornelis: "Some Currents of Thought in Dutch Natural Philosophy, 1675–1720" (PhD diss., University of Oxford, 1974).

Del Prete, Antonella: "*Duplex intellectus et sermo duplex*: Method and the Separation of Disciplines in Johannes De Raey," in *Physics and Metaphysics in Descartes and his Reception*, eds. Delphine Antoine-Mahut and Sophie Roux (New York, NY-London, 2019), 161–174.

Del Prete, Antonella: "Y-a-t-il une interprétation cartésienne de la Bible? Le cas de Christoph Wittich," in *Qu'est-ce qu'être cartésien?*, ed. Delphine Kolesnik-Antoine (Lyon, 2013), 117–142.

De Mûelenaere, Gwendoline: *Early Modern Thesis Prints in the Southern Netherlands: An Iconological Analysis of the Relationships between Art, Science and Power* (Leiden, 2022).

Denzinger, Heinrich: *Enchiridion symbolarum: definitionum et declarationum de rebus fidei et morum* (Barcelona, 1963).

De Pater, Cornelis: "Experimental Physics," in *Leiden University in the Seventeenth Century: An Exchange of Learning*, eds. Theodoor Herman Lunsingh Scheurleer and Guillaume Henri Marie Posthumus Meyjes (Leiden, 1975), 309–327.

Des Chene, Dennis: "Cartesian Science: Régis and Rohault," in *A Companion to Early Modern Philosophy*, ed. Steven Nadler (Malden, MA, 2002), 183–196.

Des Chene, Dennis, "*Cartesiomania*: Early Receptions of Descartes," *Perspectives on Science*, 3 (1995), 534–581.

Des Chene, Dennis: *Life's Form: Late Aristotelian Conceptions of the Soul* (Ithaca, NY-London, 2000).

Des Chene, Dennis: *Spirits and Clocks: Machine and Organism in Descartes* (Ithaca, NY-London, 2001).

De Vregille, Pierre: "Les jesuites et l'étude du magnetisme terrestre," *Études*, 104 (1905), 492–510.

De Vrijer, Marinus Johannes Antonie: *Henricus Regius: een "cartesiaansch" hoogleraar aan de Utrechtsche hoogeschool* (The Hague, 1917).

Dibbets, Gerardus Rutgerus Wilhelmus, "Kóks Burgersdijk vertalingen en de Nederlandse Woordenschat," *Geschiedenis van de wijsbegeerte in Nederland*, 2 (1991), 13–35.

Dibbets, Gerardus Rutgerus Wilhelmus, "Inleiding," in A.L. Kók, *Ont-werp der Neder-duitsche letter-konst*, ed. Gerardus Rutgerus Wilhelmus Dibbets (Assen, 1981), xiii–xxvii.

Dibon, Paul: "Der Philosophieunterricht in den Niederlanden," in *Grundriss der Geschichte der Philosophie*, founded by Friedrich Ueberweg: *Die Philosophie des 17. Jahrhunderts*, ed. Jean-Pierre Schobinger, 4 vols. (Basel, 1988–2001), vol. 2 (1993): *Frankreich und Niederlande* (1993), 33–70, 76–86.

Dibon, Paul: "Histoire des idées au XVIIe siècle," *École pratique des hautes études. IVe section, Sciences historiques et philologiques. Annuaire* 1966–1967 (1966), 405–414.

Dibon, Paul: *La philosophie néerlandaise au siècle d'or*, vol. 1: *L'Enseignement philosophique dans les universités a l'époque précartésienne (1575–1650)* (Paris, 1954).

Dibon, Paul: *Regards sur la Hollande du Siècle d'Or* (Naples, 1990).

Dibon, Paul: "Scepticisme et orthodoxie reformée dans la Hollande du Siecle d'Or," in *Scepticism from the Renaissance to the Enlightenment*, eds. Richard Henry Popkin and Charles B. Schmitt (Wiesbaden, 1987), 55–81.

Dieckhöfer, Klemens: *Der niederländische Arzt Daniel Voet: ein Neoaristoteliker des 17. Jahrhunderts* (Münster, 1970).

Dijksterhuis, Eduard J.: *Archimedes* (Groningen, 1938).

Dijksterhuis, Eduard J.: *Simon Stevin* (The Hague, 1943).

Dijksterhuis, Eduard J., Cornelia Serruirer and Paul Dibon, eds.: *Descartes et le Cartésianisme hollandais* (Paris, 1950–1951).

Dobre, Mihnea: "Depicting Cartesian Cosmology in the Seventeenth Century," in *Cartesian Images: Picturing Philosophy in the Early Modern Age*, eds. Mattia Mantovani and Davide Cellamare (Leiden, 2023; forthcoming).

Dobre, Mihnea: *Descartes and Early French Cartesianism: Between Metaphysics and Physics* (Bucharest, 2017).

Dobre, Mihnea: "Jacques Rohault," in *Cambridge Descartes Lexicon*, ed. Lawrence Nolan (Cambridge, 2016), 657–659.

Dobre, Mihnea: "Jacques Rohault and Cartesian Experimentalism," in *Oxford Handbook of Descartes and Cartesianism*, eds. Steven Nadler, Delphine Antoine-Mahut and Tad Schmaltz (Oxford, 2019), 388–401.

Dobre, Mihnea: "Jacques Rohault on Medicine," in *Descartes and Medicine: A System with Obscurities and Lights, and its Reception*, ed. Fabrizio Baldassarri (forthcoming).

Dobre, Mihnea: "Jacques Rohault's Mathematical Physics," *HOPOS: The Journal of the International Society for the History of Philosophy of Science*, 10 (2020), 414–439.

Dobre, Mihnea: "Rohault, Jacques," in *Encyclopedia of Early Modern Philosophy and the Sciences*, eds. Dana Jalobeanu and Charles T. Wolfe (Cham, 2019), 1–6.

Dobre, Mihnea: "Rohault's Cartesian Physics," in *Cartesian Empiricisms*, eds. Mihnea Dobre and Tammy Nyden-Bullock (Dordrecht, 2013), 203–226.

Dobre, Mihnea: "Rohault's *Traité de physique* and Its Newtonian Reception," in *The Circulation of Science and Technology: Proceedings of the 4th International Conference of the ESHS, Barcelona, 18–20 November 2010*, ed. A. Roca-Rosell (Barcelona, 2012), 389–394.

Dobre, Mihnea, Ovidiu Babeş and Ioana Bujor, eds.: *Cartesian Cosmological Illustrations*: *A Digital Approach*: https://cartesian.unibuc.ro/s/cosmologicalillustrations/page/welcome.

Dobre, Mihnea and Tammy Nyden, eds.: *Cartesian Empiricisms* (Dordrecht, 2013).

Donini, Pierluigi: "The History of the Concept of Eclecticism," in *Commentary and Tradition: Aristotelianism, Platonism, and Post-Hellenistic Philosophy*, ed. by Mauro Bonazzi (Berlin-New York, NY, 2011), 197–209.

Duker, Arnoldus Cornelius: *School-gezag en eigen-onderzoek: historisch-kritische studie van den strijd tusschen Voetius en Descartes* (Leiden, 1861).

Dunér, David: *The Natural Philosophy of Emanuel Swedenborg: A Study in the Conceptual Metaphors of the Mechanistic World-View* (Dordrecht-New York, NY, 2013).

Dunin-Borkowsky, Stanislaus von: *Der junge Spinoza* (Münster, 1910).

Dupré, Sven: "The Return of the *Species*: Jesuit Responses to Kepler's New Theory of Images," in *Religion and the Senses in Early Modern Europe*, eds. Wietse De Boer and Christine Göttler (Leiden, 2012), 473–487.

Du Rieu, Nicolaas, ed.: *Album studiosorum academiae lugduno batavae MDLXXV–MDCCCLXXV; accedunt nomina curatorum et professorum per eadem secula* (The Hague, 1875).

Ebbersmeyer, Sabrina: "Elisabeth of Bohemia," in *Encyclopedia of Early Modern Philosophy and the Sciences*, eds. Dana Jalobeanu and Charles Wolfe (Springer online, 2022), https://doi.org/10.1007/978--3--319--20791--9_409--1.

Eberhardt, Kai-Ole: *Christoph Wittich (1625–1687): Reformierte Theologie unter dem Einfluss von René Descartes* (Göttingen, 2018).

Elias, Norbert: *Die höfische Gesellschaft: Untersuchungen zur Soziologie des Königtums und der höfischen Aristokratie* (Frankfurt am Main, 2002).

English Short Title Catalogue: http://estc.bl.uk.

Eriksson, Gunnar: "Framstegstanken i de Cartesianska Stridernas Uppsala. Kring D-ebatten Om Naturens Konstans Och Vetenskapernas Tillväxt," *Lychnos* (1967), 137–185.

Feingold, Mordechai: "Aristotle and the English Universities in the Seventeenth Century: A Re-evaluation," in *European Universities in the Age of Reformation and Counter-Reformation*, ed. Helga Robinson-Hammerstein (Dublin, 1998), 135–148.

Feingold, Mordechai: "The Humanities," in *Seventeenth-Century Oxford*, ed. Nicholas Tyack (Oxford, 1997), *sub voce*.

Feingold, Mordechai: "The Mathematical Sciences and New Philosophies," in *Seventeenth-Century Oxford*, ed. Nicholas Tyack (Oxford, 1997), *sub voce*.

Feingold, Mordechai and Andrea Sangiacomo, eds.: *Reshaping Natural Philosophy: Tradition and Innovation in the Early Modern Academic Milieu*, special issue of *History of Universities*, 33.2 (2020).

Finocchiaro, Maurice: "The *Port-Royal Logic*'s Theory of Argument," *Argumentation*, 11 (1997), 393–410.

Foisneau, Luc: *Dictionnaire des philosophes français du XVIIe siècle* (Paris, 2015).

Forlivesi, Marco: "La distinction entre concept formel et concept objectif: Suárez, Pasqualigo, Mastri," *Les études philosophiques*, 60 (2002), 3–30.

Foucault, Michel: "Qu'est-ce que la critique?," *Bulletin de la société française de philosophie*, 27 (1978), 35–63.

Fournier, W.J.: "Vries, Gerardus De," in *Biografische Lexicon voor de Geschiedenis van het Nederlandse Protestantisme*, eds. Doede Nauta, Aart de Groot, Johannes van den Berg, O.J. de Jong (Kampen, 1983), part 1, 417.

Freedman, Joseph S.: *European Academic Philosophy in the Late Sixteenth and Early Seventeenth Centuries: The Life, Significance, and Philosophy of Clemens Timpler (1563/4–1624)*, 2 vols. (Zürich, 1988).

Freedman, Joseph S.: "The Career and Writings of Bartholomew Keckermann," *Proceedings of the American Philosophical Society*, 141 (1997), 305–364.

Friedenthal, Meelis, Hanspeter Marti and Robert Seidel, eds.: *Early Modern Disputations and Dissertations in an Interdisciplinary and European Context* (Leiden, 2020).

Friedrich, Arnd: *Die Gelehrtenschulen in Marburg, Kassel und Korbach zwischen Melanchthonianismus und Ramismus* (Darmstadt-Marburg, 1983).

Frijhoff, Willem: "Pieter de la Court and Comenius's Third Visit to Holland (1642)," *Acta Comeniana*, 8 (1986), 183–192.

Fuchs, Thomas: *The Mechanization of the Heart: Harvey and Descartes* (Rochester, NY, 2001).

Gabbey, Alan: "Anne Conway et Henry More: Lettres sur Descartes (1650–1651)," *Archives de philosophie*, 40 (1977), 388–404.

Gabbey, Alan: "*Philosophia cartesiana triumphata*: Henry More (1646–1671)," in *Problems in Cartesianism*, eds. Thomas. M. Lennon, John. M. Nicholas and John. W. Davis (Kingston-Montreal, QC, 1982), 171–250.

Gabbey, Alan: "Spinoza's Natural Science and Methodology," in *The Cambridge Companion to Spinoza*, ed. Don Garrett (Cambridge, 1996), 142–191.

Garber, Daniel: "Descartes and the Scientific Revolution: Some Kuhnian Reflections," *Perspectives on Science*, 9 (2001), 405–422.

Garber, Daniel: *Descartes' Metaphysical Physics* (Chicago, IL-London, 1992).

Garber, Daniel: "Descartes, the Aristotelians, and the Revolution that did not Happen in 1637," *The Monist*, 71 (1988), 471–486.

Garber, Daniel: "Novatores," in *The Cambridge History of the Scientific Revolution*, eds. David Marshall Miller and Dana Jalobeanu (Cambridge, 2022), 35–57.

Garber, Daniel and Michael Ayers, eds.: *The Cambridge History of Seventeenth-Century Philosophy*, 2 vols. (Cambridge, 2000).

Garber, Daniel and Lesley Cohen: "A Point of Order: Analysis, Synthesis and Descartes's *Principles*," *Archiv für Geschichte der Philosophie*, 64 (1982), 136–147.

Gardies, Jean-Louis: "La *Logique de Port-Royal*: Esquisse d'un bilan," *Revue des sciences philosophiques et théologiques*, 84 (2000), 83–92.

Gascoigne, John: "A Reappraisal of the Role of the Universities in the Scientific Revolution," in *Reappraisals of the Scientific Revolution*, eds. David C. Lindberg and Robert S. Westman (Cambridge, 1991), 207–260.

Gascoigne, John: *Cambridge in the Age of the Enlightenment: Science, Religion and Politics from the Restoration to the French Revolution* (Cambridge, 1989).

Gascoigne, John: "Isaac Barrow's Academic Milieu," in *Before Newton: The Life and Times of Isaac Barrow*, ed. Mordechai Feingold (Cambridge 1990), 250–290.

Gaukroger, Stephen: *Cartesian Logic: An Essay on Descartes's Conception of Inference* (Oxford, 1989).

Gaukroger, Stephen: *Descartes: An Intellectual Biography* (Oxford, 1995).

Gaukroger, Stephen: *Descartes' System of Natural Philosophy* (Cambridge, 2002).

Gellera, Giovanni: "A 'Calvinist' Theory of Matter? Burgersdijk and Descartes on *res extensa*," *Intellectual History Review*, 28 (2018), 255–270.

Gellera, Giovanni, "The Scottish Faculties of Arts and Cartesianism (1650–1700)," *History of Universities*, 29 (2017), 166–187.

Geudens, Christophe and Steven Coesemans: "Introduction to 'Studies in Post-Medieval Logic'," *History and Philosophy of Logic*, 41 (2020), 305–308.

Geudens, Christophe and Jan Papy: "The Teaching of Logic at Leuven University (1425–1797): Perpetually Peripatetic? A First Survey of a Research Project on Student Notebooks and their European Context," *Neulateinisches Jahrbuch: Journal of Neo-Latin Language and Literature*, 17 (2015), 360–378.

Giannini, Giulia and Mordechai Feingold, eds.: *The Institutionalization of Science in Early Modern Europe* (Leiden, 2019).

Gilby, Emma: *Descartes's Fictions: Reading Philosophy with Poetics* (Cambridge, 2019).

Golvers, Noël, "Note on Descartes's Entrance at the Coimbra Colégio das Artes (mid-17th Century)," *Revista filosófica de Coimbra*, 29 (2020), 485–494.

Gootjes, Albert: "The *Collegie der Sçavanten*: A Seventeenth-Century Cartesian Society in Utrecht," in *Enlighted Religion: From Confessional Churches to Polite Piety in the Dutch Republic*, eds. Joke Spaans and Jetze Touber (Leiden-Boston, MA, 2019), 156–182.

Goudriaan, Aza: *Reformed Orthodoxy and Philosophy, 1625–1750: Gisbertus Voetius, Petrus van Mastricht, and Anthonius Driessen* (Leiden-Boston, MA, 2006).

Goudriaan, Aza: "Ulrik Huber (1636–1694) and John Calvin: The Franeker Debate on Human Reason and the Bible (1686–1687)," *Church History and Religious Culture*, 91 (2011), 165–178.

Goudriaan, Aza, ed. and transl.: *Jacobus Revius, A Theological Examination of Cartesian Philosophy: Early Criticisms (1647)* (Leiden, 2002).

Granada, Miguel Ángel: *Sfere solide e cielo fluido: momenti del dibattito cosmologico nella seconda metà del Cinquecento* (Milan, 2002).

Grant, Edward: *Much Ado about Nothing: Theories of Space and Vacuum from the Middle Ages to the Scientific Revolution* (Cambridge, 1981).

Grant, Edward: *Planets, Stars, and Orbs: The Medieval Cosmos, 1200–1687* (Cambridge, 1994).

Grendler, Paul F.: *The Jesuits and Italian Universities, 1548–1773* (Washington, D.C., 2017).

Grene, Marjorie: "The Heart and Blood: Descartes, Plemp and Harvey," in *Essays on the Philosophy and Science of René Descartes*, ed. Stephen Voss (New York, NY, 1993), 324–336.

Groenhuis, Gerrit: "De predikanten: De sociale positie van gereformeerde predikanten in de Republiek der Verenigde Nederlanden voor ± 1700" (PhD diss., Universiteit Utrecht, 1977).

Gueroult, Martial: *Descartes selon l'ordre des raisons*, 2 vols. (Paris, 1954).

Hall, A. Rupert: "Further Newton Correspondence," *Notes and Records of the Royal Society of London*, 37 (1982), 7–34.

Hamid, Nabeel, "Domesticating Descartes, Renovating Scholasticism: Johann Clauberg and the German Reception of Cartesianism," *History of Universities*, 33 (2020), 57–84.

Hamou, Philippe: *La Mutation du visible: essai sur la portée épistémologique des instruments d'optique au XVIIe siècle*, volume 1: *Du* Sidereus Nuncius *de Galilée à la* Dioptrique *cartésienne* (Villeneuve d'Ascq, 1999).

Harrison, Peter: "Animal Souls, Metempsychosis, and Theodicy in Seventeenth-Century English Thought," *Journal of the History of Philosophy*, 31 (1993), 519–544.

Harth, Erica: *Cartesian Women: Versions and Subversions of Rational Discourse in the Old Regime* (Ithaca, NY, 1992).

Hatfield, Gary: "Descartes' Physiology and its Relation to his Psychology," in *The Cambridge Companion to Descartes*, ed. John Cottingham (Cambridge, 1992), 335–370.

Hatfield, Gary: "*L'Homme / De homine*: Images as Interpretations," in *Cartesian Images: Picturing Philosophy in the Early Modern Age*, eds. Mattia Mantovani and Davide Cellamare (Leiden, 2023; forthcoming).

Hellyer, Marcus: *Catholic Physics: Jesuit Natural Philosophy in Early Modern Germany* (Notre Dame, IN, 2005).

Henry, John: " 'Mathematics Made No Contribution to the Public Weal': Why Jean Fernel Became a Physician," *Centaurus*, 53 (2011), 193–220.

Henry, John: "The Reception of Cartesianism," in *The Oxford Handbook of British Philosophy in the Seventeenth Century*, ed. Peter Anstey (Oxford, 2013), 116–143.

Het Utrechts Archief: https://hetutrechtsarchief.nl.

Heydt, Colin: *Moral Philosophy in Eighteenth-Century Britain: God, Self, and Other* (Cambridge, 2017).

Hirai, Hiro: "Il calore cosmico in Telesio fra il *De generatione animalium* di Aristotele e il *De carnibus* di Ippocrate," in *Bernardino Telesio tra filosofia naturale e scienza moderna*, eds. Giuliana Mocchi, Sandra Plastina and Emilio Sergio (Pisa, 2012).

Hirai, Hiro: *Le concept de semence dans les théories de la matière à la Renaissance: De Marsile Ficin à Pierre Gassendi* (Turnhout, 2005).

Hirai, Hiro: *Medical Humanism and Natural Philosophy: Renaissance Debates on Matter, Life and the Soul* (Leiden, 2011).

Holdsworth, Richard: "Directions for a Student in the Universities," in *The Intellectual Development of John Milton*, ed. Harris Francis Fletcher, 2 vols. (Champaign, IL, 1956–1961), 623–655.

Hoogendoorn, Klaas: *Bibliography of the Exact Sciences in the Low Countries from ca. 1470 to the Golden Age (1700)* (Leiden-Boston, MA, 2018).

Hoskin, Michael: "'Mining All Within': Clarke's Notes to Rohault's *Traité de physique*," *The Thomist*, 24 (1961), 353–363.

Hotson, Howard: *Commonplace Learning: Ramism and its German Ramifications, 1543–1630* (Oxford, 2007).

Hotson, Howard: *Johann Heinrich Alsted (1588–1638): Between Renaissance, Reformation and Universal Reform* (Oxford, 2000).

Hotson, Howard: *The Reformation of Common Learning: Post-Ramist Method and the Reception of the New Philosophy, 1618–c. 1670* (Oxford, 2020).

Hübener, Wolfgang: "Descartes-Zitate bei Clauberg zum Quellenwert frühcartesianischer Kontroversliteratur für die Descartesforschung," *Studia Leibnitiana*, 5 (1973), 233–239.

Hutchison, Keith: "What Happened to Occult Qualities in the Scientific Revolution?," *Isis*, 73 (1982), 233–253.

Hutton, Sarah: *Anne Conway: A Woman Philosopher* (Cambridge, 2004).

Hutton, Sarah: *British Philosophy in the Seventeenth Century* (Oxford, 2015).

Hutton, Sarah: "Cartesianism in Britain," in *Oxford Handbook of Descartes and Cartesianism*, eds. Delphine Kolesnik Mahut, Stephen Nadler and Tad Schmalz (Oxford, 2019), 496–513.

Hutton, Sarah: "Henry More's *Epistola H. Mori ad V.C.* and the Cartesian Context of Newton's Early Cambridge Years," in *Collected Wisdom of the Early Modern Scholar: Essays in Honor of Mordechai Feingold*, eds. Gideon Manning and Anna Marie Roos (forthcoming).

Hutton, Sarah: "Henry More et Descartes: une copie manuscrite de leur correspondance dans le *Notebook* de Thomas Clarke (1654)," transl. Dan Abib, *Bulletin cartésien*, XLIX, *Archives de philosophie*, 83 (2020), 162–167.

Israel, Jonathan: *The Dutch Republic: Its Rise, Greatness, and Fall, 1477–1806* (New York, NY, 1995).

Israel, Jonathan: "The Early Dutch and German Reaction to the *Tractatus theologico-politicus*: Foreshadowing the Enlightenment's More General Spinoza Reception?," in *Spinoza's* Theological-Political Treatise: *A Critical Guide*, eds. Yitzhak Melamed and Michael A. Rosenthal (Cambridge, 2010), 72–100.

Israel, Jonathan: *Radical Enlightenment: Philosophy and the Making of Modernity, 1650–1750* (Oxford, 2001).

Jalobeanu, Dana: "The Cartesians of the Royal Society: The Debate over Collisions and the Nature of Body (1668–1670)," in *Vanishing Matter and the Laws of Motion*: *Descartes and Beyond*, eds. Peter Anstey and Dana Jalobeanu (London, 2011), 103–129.

Janiak, Andrew: *Newton* (Chichester, 2015).

Janiak, Andrew: *Newton as Philosopher* (Cambridge, 2008).

Janowski, Zbigniew: *Cartesian Theodicy: Descartes' Quest for Certitude* (Dordrecht, 2000).

Jesseph, Douglas M.: "Fromondus, Libertus," in *The Historical Dictionary of Descartes and Cartesian Philosophy*, eds. Roger Ariew, Dennis Des Chene, Douglas M. Jesseph, Tad M. Schmaltz and Theo Verbeek (Lanham, 2015), 146–147.

Jourdain, Charles: *Histoire de l'université de Paris au XVIIe et au XVIIIe siècle*, 2 vols. (Paris, 1862–1866).

Kallinen, Maija: *Change and Stability: Natural Philosophy at the Academy of Turku (1640–1713)* (Helsinki, 1995).

Kallinen, Maija: "Kartesiolaisuus Turun akatemian meteorologisissa väitöskirjoissa, 1678–1702," *Opusculum*, 13 (1993), 67–98.

Kallinen, Maija: "Naturens hemliga krafter: Daniel Achrelius' *Contemplationes mundi*," *Historisk tidskrift för Finland*, 76 (1991), 317–346.

Kathe, Heinz: *Die Wittenberger philosophische Fakultät, 1502–1817* (Vienna-Cologne-Weimar, 2002).

Kato, Yoshi and Kuni Sakamoto: "Between Cartesianism and Orthodoxy: God and the Problem of Indifference in Christoph Wittich's Anti-Spinoza," *Intellectual History Review*, 32 (2022), 239–257.

Kearney, Hugh F.: *Scholars and Gentlemen: Universities and Society in Pre-Industrial Britain, 1500–1700* (Ithaca, NY, 1970).

Kernkamp, Gerhard Wilhelm: *Acta et decreta Senatus; Vroedschapsresolutiën en andere bescheiden betreffende de Utrechtse Academie* (Utrecht, 1936).

Kernkamp, Gerhard Wilhelm: *De Utrechtse Academie 1636–1815* (Utrecht, 1936).

Keyser, Marja, ed.: *Glazemaker 1682–1982: Catalogus bij een tentoonstelling over de vertaler Jan Hendriksz Glazemaker in de Universiteitsbibliotheek van Amsterdam* (Amsterdam, 1982).

Kieft, Xavier: "Conversation with Burman," in *The Cambridge Descartes Lexicon*, ed. Lawrence Nolan (Cambridge, 2016), *sub voce*.

Kieft, Xavier: "Une morale rappelée à l'ordre? Régis et l'invention du cartésianisme autorisé," in Pierre-Sylvain Régis, *La morale ou les devoirs de l'homme raisonnable, de l'homme civil et de l'homme chrestien (1682)*, ed. Sylvain Matton (Paris-Milan, 2015), 19–57.

Klever, Wim: "Burchardus de Volder (1643–1709): A Crypto-Spinozist on a Leiden Cathedra," *Lias*, 15 (1988), 191–241.

Klever, Wim: "Jan Hendricksz Glazemaker: The Addressee of Letter 84?," *North American Spinoza Society, NASS Monograph*, 13 (2007), 25–31.

Knuttel, William, ed.: *Acta der particuliere synoden van Zuid-Holland 1621–1700*, 6 vols. (The Hague, 1908–1916).

Kolesnik-Antoine, Delphine: "Le rôle des expériences dans la physiologie d'Henricus

Regius: les 'pierres lydiennes' cartésianisme," *Journal of Early Modern Studies*, 2 (2013), 125–145.

Kolesnik-Antoine, Delphine, ed.: *Qu'est-ce qu'être cartésien?* (Lyon, 2013).

Kolesnik-Antoine, Delphine and Marie-Frédérique Pellegrin, eds.: *Elisabeth de Bohême face à Descartes: deux philosophes?* (Paris, 2014).

Kremer, Elmar J.: "Arnauld on the Nature of Ideas as a Topic in Logic: The Port-Royal Logic and On True and False Ideas," in *Logic and the Workings of the Mind: The Logic of Ideas and Faculty Psychology in Early Modern Philosophy*, ed. Patricia A. Easton (Atascadero, CA, 1997), 64–82.

Krop, Henri: "Medicine and Philosophy in Leiden around 1700: Continuity or Rupture?," in *The Early Enlightenment in the Dutch Republic, 1650–1750*, ed. Wiep Van Bunge (Leiden-Boston, MA, 2003) 173–196.

Krop, Henri: "*Scholam naturae ingrediamur*: Adrianus Heereboord als geschiedschrijver van de filosofie," *Geschiedenis van de wijsbegeerte in Nederland: documentatieblad van de Werkgroep "Sassen,"* vol. 4.1 (1993), 3–13.

Krop, Henri: "Schulerus, Johannes (1619–1674)," in *The Dictionary of Seventeenth and Eighteenth-Century Dutch Philosophers*, eds. Wiep Van Bunge, Henri Krop, Bart Leeuwenburgh, Han Van Ruler, Paul Schuurman and Michiel Wielema, 2 vols. (Bristol, 2003), vol. 2, 899–902.

Krop, Henri: "The *Philosophia S. Scripturae interpres* between Humanist Scholarship and Cartesian Science: Lodewijk Meyer and the Emancipatory Power of Philology," in *The Dutch Legacy: Radical Thinkers of the 17th Century and the Enlightenment*, eds. Sonja Lavaert and Winfried Schöder (Leiden, 2017), 90–120.

Landucci, Sergio: *La mente in Cartesio* (Milan, 2002).

Larmore, Charles: "Descartes' Psychologistic Theory of Assent," *History of Philosophy Quarterly*, 1 (1984), 61–74.

Lennon, Thomas M. and Patricia A. Easton, eds.: *The Cartesian Empiricism of François Bayle* (New York, NY, 1992).

Lerner, Michel-Pierre: *Le monde des spheres: Genèse et triomphe d'une représentation cosmique* (Paris, 2008).

Lind, Gunter: *Physik im Lehrbuch, 1700–1850: zur Geschichte der Physik und ihrer Didaktik in Deutschland* (Berlin-New York, NY, 1992).

Lindborg, Rolf: *Descartes i Uppsala: striderna om "Nya filosofien" 1663–1689* (Stockholm, 1965).

Lindeboom, Gerrit Arie: *Florentius Schuyl (1619–1669) en zijn betekenis voor het cartesianisme in de geneeskunde* (The Hague, 1974).

Lodge, Paul: "Burchard de Volder: Crypto-Spinozist or Disenchanted Cartesian?," in *Receptions of Descartes: Cartesianism and Anti-Cartesianism in Early Modern Europe*, ed. Tad Schmaltz (London-New York, NY, 2005), 128–145.

Lodge, Paul, ed.: *The Leibniz-De Volder Correspondence: With Selections from the Correspondence between Leibniz and Johann Bernoulli* (New Haven, CT, 2013).

Lolordo, Antonia: "'Descartes' One Rule of Logic': Gassendi's Critique of Clear and Distinct Perception," *British Journal for the History of Philosophy*, 13 (2005), 51–72.

Lolordo, Antonia: *Pierre Gassendi and the Birth of Modern Philosophy* (Cambridge, 2006).

Lotze, Karl-Heinz and Sascha Salatowsky, eds.: *Himmelsspektakel: Astronomie im Protestantismus der Frühen Neuzeit; Katalog zur Ausstellung der Universitäts- und Forschungsbibliothek Erfurt/Gotha in Zusammenarbeit mit der Physikalisch-Astronomischen Fakultät der Friedrich-Schiller-Universität Jena vom 12. April bis 21. Juni 2015* (Gotha, 2015).

Luard, Henry.R.: *A Catalogue of the Manuscripts Preserved in the Library of the University of Cambridge* (Cambridge, 1867); repr. (Cambridge, 2011).

Lüthy, Christoph: *David Gorlaeus (1591–1612): An Enigmatic Figure in the History of Philosophy and Science* (Amsterdam, 2012).

Lüthy, Christoph: "Perplexing Particles: The Logic and Impact of Descartes's Illustrated Microworld," in *Cartesian Images: Picturing Philosophy in the Early Modern Age*, eds. Mattia Mantovani and Davide Cellamare (Leiden, 2023; forthcoming).

Lüthy, Christoph: "Where Logical Necessity Turns into Visual Persuasion: Descartes' Clear and Distinct Illustrations," in *Transmitting Knowledge: Words, Images and Instruments in Early Modern Europe*, eds. Sachiko Kusukawa and Ian Maclean (Oxford, 2006), 97–133.

Luyendijk-Elshout, Antonie M.: "*Oeconomia animalis*, Pores and Particles," in *Leiden University in the Seventeenth Century: An Exchange of Learning*, eds. Theodoor Herman Lunsingh Scheurleer and Guillaume Henri Marie Posthumus Meyjes (Leiden, 1975), 295–307.

Machamer, Peter and J.E. McGuire: *Descartes's Changing Mind* (Princeton, NJ, 2009).

Madan, Falconer: *Oxford Books*, 3 vols. (Oxford, 1895–1931).

Magister Dixit Collection: http://lectio.ghum.kuleuven.be/lectio/magister-dixit-collection.

Manning, Gideon, "Descartes and the Bologna Affair," *British Journal for the History of Science*, 47 (2014), 1–13.

Mantovani, Mattia: "A Priori Proofs of God's Existence in 17th-Century Scholastics," *Quaestio*, 19 (2019), 492–497.

Mantovani, Mattia: "A Tale of Two Images: Picturing the Human Machine among Cartesians," in *Cartesian Images: Picturing Philosophy in the Early Modern Age*, eds. Mattia Mantovani and Davide Cellamare (Leiden, 2023; forthcoming).

Mantovani, Mattia: "Descartes' Man under Construction: The *Circulatory Statue* of Salomon Reisel, 1680," *Early Science and Medicine*, 25 (2020), 101–134.

Mantovani, Mattia: "Herbert of Cherbury, Descartes and Locke on Innate Ideas and Universal Consent," *Journal of Early Modern Studies*, 8 (2019), 83–115.

Mantovani, Mattia: "The Eye and the Ideas: Descartes on the Nature of Bodies" (PhD diss., Humboldt-Universität zu Berlin 2018).

Mantovani, Mattia: "The First of All Natural Sciences: Roger Bacon on *perspectiva* and Human Knowledge," *Vivarium*, 59 (2021), 186–214.

Mantovani, Mattia: "The Institution of Nature: Descartes on Human and Animal Perception," *Oxford Studies in Early Modern Philosophy*, 12 (2022), 1–30.

Mantovani, Mattia: "*Visio per sillogismum*: Sensation and Cognition in 13th-Century Theories of Vision," in *Medieval Perceptual Puzzles: Theories of Sense Perception in the 13th and 14th Centuries*, ed. Elena Băltuţă (Leiden, 2020), 111–152.

Mantovani, Mattia and Davide Cellamare, eds.: *Cartesian Images: Picturing Philosophy in the Early Modern Age Century* (Leiden, 2023; forthcoming).

Manusov-Verhage, C.G.: "Jan Rieuwertsz, marchand libraire et éditeur de Spinoza," in *Spinoza to the Letter*, eds. Fokke Akkerman and Piet Steenbakkers (Leiden, 2005), 237–250.

Marin, Louis: *La critique du discours: sur la* Logique de Port-Royal *et les* Pensées *de Pascal* (Paris, 1975).

Martin, Craig: *Subverting Aristotle: Religion, History, and Philosophy in Early Modern Science* (Baltimore, MD, 2014).

Martin, John N.: "Distributive Terms, Truth, and the Port Royal Logic," *History and Philosophy of Logic*, 34 (2013), 133–154.

Martin, John N.: "The Structure of Ideas in the Port Royal Logic," *Journal of Applied Logic*, 19 (2016), 1–19.

Matriculation Catalogues of the University of Duisburg: https://www.uni-due.de/collca rt/matrikel/stud-alph-c.htm.

Matton, Sylvain: "Remarques sur le manuscrit de la *Physique nouvelle*," in Jacques Rohault, *Physique Nouvelle (1667)*, ed. Sylvain Matton (Paris-Milan, 2009), lxxiii–xc.

Mazauric, Simone: "Préface," in Jacques Rohault, *Traité de physique*, ed. Simone Mazauric (Paris, 2014), vii–xxxvi.

McClaughlin, Trevor: "Censorship and Defenders of the Cartesian Faith in Mid-Seventeenth Century France," *Journal of the History of Ideas*, 40 (1979), 563–581.

McClaughlin, Trevor: "Claude Clerselier's Attestation of Descartes's Religious Orthodoxy," *Journal of Religious Studies*, 20 (1980), 136–146.

McClaughlin, Trevor: "Descartes, Experiments, and a First Generation Cartesian, Jacques Rohault," in *Descartes' Natural Philosophy*, eds. Stephen Gaukroger, John Schuster and John Sutton (London, 2000), 330–346.

McClaughlin, Trevor: "Jacques Rohault and the Natural Sciences" (PhD diss., University of Cambridge, 1972).

McClaughlin, Trevor: "Quelques mots sur Rohault et Molière," *Europe*, 54 (1976), 178–183.

McClaughlin, Trevor: "Sur les rapports entre la Compagnie de Thévenot et l'Académie royale des Sciences," *Revue d'histoire des sciences*, 28 (1975), 235–242.

McClaughlin, Trevor: "Was There an Empirical Movement in Mid-Seventeenth Century France? Experiments in Jacques Rohault's *Traité de physique*," *Revue d'histoire des Sciences*, 49 (1996), 459–481.

McClaughlin, Trevor and Guy Picolet: "Un exemple d'utilisation du Minutier central de Paris: la bibliothèque et les instruments scientifiques du physicien Jacques Rohault selon son inventaire après décès," *Revue d'histoire des sciences*, 29 (1976), 3–20.

McGahagan, Thomas A.: "Cartesianism in the Netherlands, 1639–1676: the New Science and the Calvinist Counter-Reformation" (PhD diss., University of Pennsylvania, 1976).

McLaughlin, Peter: "Descartes on Mind-Body Interaction and the Conservation of Motion," *The Philosophical Review*, 102 (1993), 155–182.

McVaugh, Michael Rogers: "Losing Ground: The Disappearance of Attraction from the Kidneys," in *Blood, Sweat and Tears: The Changing Concepts of Physiology from Antiquity into Early Modern Europe*, eds. Claus Zittel, Manfred Horstmanshoff and Helen King (Leiden-Boston, MA, 2012), 103–137.

Menk, Gerhard: *Die Hohe Schule Herborn in ihrer Frühzeit: ein Beitrag zum Hochschulwesen des deutschen Kalvinismus im Zeitalter der Gegenreformation* (Wiesbaden, 1981).

Menn, Stephen: "The Great Stumbling Block: Descartes' Denial of Real Qualities," in *Descartes and His Contemporaries: Meditations, Objections, and Replies*, eds. Roger Ariew and Marjorie Grene (Chicago, IL-London, 1995), 182–207.

Meschini, Franco Aurelio: *Indice dei* Principia philosophiae *di René Descartes: indici lemmatizzati, frequenze, distribuzione dei lemmi* (Florence, 1996).

Milani, Nausicaa E.: "L'*Art de penser* nella logica del *Système* di Régis," *Annali della Scuola Normale Superiore di Pisa*, 5 (2012), 517–555.

Milhaud, Gérard: "Rohault et Molière," *Europe*, 50 (1972), 37–49.

Mirguet, Françoise and Françoise Hiraux: *Collection de cours manuscrits de l'Université de Louvain, 1425–1797: catalogue analytique* (Louvain-la-Neuve, 2003).

Molhuysen, Philip C.: *Bronnen tot de geschiedenis der Leidsche universiteit 1574–1811*, 7 vols. (The Hague, 1913–1924).

Monchamp, Georges: *Galilée et la Belgique: essai historique sur les vicissitudes du système de Copernic en Belgique* (Brussels-Paris, 1892).

Monchamp, Georges: *Histoire du cartésianisme en Belgique* (Brussels, 1886).

Monna, Adriaan: "Gedichten en andere uitgaven ter gelegenheid van de opening vande Utrechtse Academie in 1636," in *Over beesten en boeken*, eds. Koert Van der Horst, Peter A. Koolmees and Adrian Monna (Rotterdam, 1995), 241–260.

Moran, Bruce: "German Prince-Practitioners: Aspects in the Development of Courtly Science, Technology, and Procedures in the Renaissance," *Technology and Culture*, 22 (1981), 253–274.

Moran, Bruce: *Patronage and Institutions: Science, Technology and Medicine at the European Court, 1500–1750* (New York, NY, 1991).

Morison, Samuel Eliot: *Harvard College in the Seventeenth Century*, 2 vols. (Cambridge, MA, 1936).

Mouy, Paul: *Le développement de la physique cartésienne, 1646–1712* (Paris, 1934).

Nadler, Steven: *Arnauld and the Cartesian Philosophy of Ideas* (Princeton, NJ, 1989).

Nadler, Steven, Tad M. Schmaltz and Antoine-Mahut, Delphine, eds.: *The Oxford Handbook of Descartes and Cartesianism* (Oxford, 2019).

Newman, Lex: "Descartes on the Method of Analysis," in *The Oxford Handbook of Descartes and Cartesianism*, eds. Steven Nadler, Tad M. Schmaltz and Delphine Antoine-Mahut (Oxford, 2019), 65–88.

Nicolson, Marjorie and Sarah Hutton, eds.: *The Conway Letters*: *The Correspondence of Anne, Viscountess Conway, Henry More and their Friends, 1642–1684* (Oxford, 1992).

Nilsson, Alb.: "Johan Bilberg," in *Svenskt Biografiskt Lexikon* (Stockholm, 1924), *sub voce*.

"Notice des manuscrits des docteurs en médecine Van der Belen, Plempius, Peeters, Rega, etc.," *Annuaire de l'Université catholique de Louvain*, 27 (1863), 306–325.

Nuchelmans, Gabriel: "Proposition and Judgement," in *The Cambridge History of Seventeenth-Century Philosophy*, eds. Daniel Garber and Michael Ayers, 2 vols. (Cambridge, 1998), vol. 1, 118–131.

Nyden, Tammy: "De Volder's Cartesian Physics and Experimental Pedagogy," in *Cartesian Empiricisms*, eds. Mihnea Dobre and Tammy Nyden (Dordrecht, 2013), 227–249.

Omodeo, Pietro Daniel: *"An vita hominis procedat ab materia coelesti subtilissima?* Cartesian Astrobiology and Scholastic Responses in Seventeenth-Century Protestant Germany," in Philosophia vetus et nova: *Aristotelismus, neuere Philosophie und protestantische Gelehrsamkeit in Helmstedt und Mitteleuropa, 1600–1700*, eds. Bernd Roling, Benjamin Wallura and Sinem Kiliç (forthcoming).

Omodeo, Pietro Daniel: "Asymmetries of Symbolic Capital in Seventeenth-Century Scientific Transactions: Placentinus's Cometary Correspondence with Hevelius and Lubieniecki," in *Institutionalization of Science in Early Modern Europe*, eds. Giulia Giannini and Mordechai Feingold (Leiden, 2019), 52–80.

Omodeo, Pietro Daniel: "Central European Polemics over Descartes: Johannes Placentinus and His Academic Opponents at Frankfurt on Oder (1653–1656)," *History of Universities*, 29 (2016), 29–64.

Omodeo, Pietro Daniel: *Defending Descartes in Brandenburg-Prussia: The University of Frankfurt an der Oder in the Seventeenth Century* (Dordrecht, 2022).

Omodeo, Pietro Daniel and Alberto Bardi: "The Disputational Culture of Renaissance Astronomy: Johannes Regiomontanus's *An terra moveatur an quiescat*," in *Early Modern Disputations and Dissertations in an Interdisciplinary and European Context*, eds. Meelis Friedenthal, Hanspeter Marti and Robert Seidel (Leiden, 2020), 233–254.

Omodeo, Pietro Daniel and Jürgen Renn: *Science in Court Society: Giovanni Battista*

Benedetti's Diversarum speculationum mathematicarum et physicarum liber (*Turin, 1585*) (Berlin, 2019).

Omodeo, Pietro Daniel and Volkhard Wels, eds.: *Natural Knowledge and Aristotelianism at Early Modern Protestant Universities* (Wiesbaden, 2019).

Ong, Walter J.: *Ramus, Method and the Decay of Dialogue* (Cambridge, MA, 1958).

Osler, Margaret J.: *Divine Will and the Mechanical Philosophy* (Cambridge, 1994).

Oxford English Dictionary: https://www.oed.com/.

Palmerino, Carla Rita: "'Ils se trompaient d'ailleurs tous deux': Galileo's and Descartes' Representations of Ebb and Flow," in *Cartesian Images: Picturing Philosophy in the Early Modern Age*, eds. Mattia Mantovani and Davide Cellamare (Leiden, 2023; forthcoming).

Palmerino, Carla Rita: "Libertus Fromondus' Escape from the *Labyrinth of the Continuum* (1631)," *Lias*, 42 (2015), 3–36.

Pantin, Isabelle: "*Simulachrum, Species, Forma, Imago*: What Was Transported by Light into the Camera Obscura? Divergent Conceptions of Realism Revealed by Lexical Ambiguities at the Beginning of the Seventeenth Century," *Early Science and Medicine*, 13 (2008), 245–269.

Papy, Jan: "Logicacursussen aan de Oude Leuvense Universiteit: Scholastieke traditie en innovatie?," in Ex cathedra: *Leuvense collegedictaten van de 16de tot de 18de eeuw*, eds. Geert Vanpaemel, Katharina Smeyers, An Smets and Diewer van der Meijden (Louvain, 2012), 107–124.

Parageau, Sandrine: *Les ruses de l'ignorance* (Paris, 2010).

Pariente, Jean-Claude: *L'analyse du langage à Port-Royal* (Paris, 1985).

Park, Woosuk: "On Animal Cognition: Before and After the Beast-Machine Controversy," in *Philosophy and Cognitive Science: Western and Eastern Studies*, eds. Lorenzo Magnani and Ping Li (Berlin-Heidelberg, 2012), 53–74.

Pasnau, Robert: *Metaphysical Themes, 1274–1671* (Oxford, 2011).

Patrick, J. Max: "*Scydromedia*, a Forgotten Utopia of the 17th Century," *Philological Quarterly*, 23 (1944), 273–278.

Pearce, Kenneth L.: "Arnauld's Verbal Distinction between Ideas and Perceptions," *History and Philosophy of Logic*, 37 (2016), 375–390.

Peeters, Frank: "Leven en bedrijf van Timotheus ten Hoorn (1644–1715)," *Mededelingen van de stichting Jacob Campo Weyerman*, 25 (2002), 20–29.

Peeters, Frank: "Timotheus ten Hoorn Sells a Copy of Spinoza's *Posthumous Works*," *Quaerendo*, 12 (1982), 242–244.

Pellegrin, Marie-Frédérique: "Cartesianism and Feminism," in *The Oxford Handbook of Descartes and Cartesianism*, eds. Steven Nadler, Tad Schmaltz and Delphine Antoine-Mahut (Oxford, 2019), 565–579.

Pellegrin, Marie-Frédérique: "La science parfaite: savants et savantes chez Poulain de la Barre," *Revue philosophique de la France et de l'étranger*, 3 (2013), 377–392.

Pellegrin, Marie-Frédérique: "Martin, André," in Luc Foisneau, ed., *Dictionnaire des philosophes français du XVIIe siècle*, 2 vols. (Paris, 2015), vol. 1, 1169–1173.

Pellegrin, Marie-Frédérique: "The Feminine Body in the Correspondence between Descartes and Elisabeth," in *Elisabeth of Bohemia (1618–1680): A Philosopher in Historical Context*, eds. Sabrine Ebbersmeyer and Sarah Hutton (Cham, 2021), 193–204.

Perler, Dominik: "Suárez on Intellectual Cognition and Occasional Causation," in *Causation and Cognition in Early Modern Philosophy*, eds. Dominik Perler and Sebastian Bender (London, 2020), 18–38.

Petrescu, Lucian: "Descartes on the Heartbeat: The Louvain Affair," *Perspectives on Science*, 21 (2013), 397–429.

Piaia, Gregorio: "The Histories of Philosophy in France in the Age of Descartes," in *Models of the History of Philosophy*, vol. 2: *From the Cartesian Age to Brucker*, eds. Gregorio Piaia and Giovani Santinello (Dordrecht, 2011), 3–91.

Pollmann, Judith: "The Bond of Christian Piety: The Individual Practice of Tolerance and Intolerance in the Dutch Republic," in *Calvinism and Religious Toleration in the Dutch Golden Age*, eds. Ronnie Po-chia Hsia and Henk Van Nierop (Cambridge-New York, NY, 2002), 53–71.

Pousseur, Jean-Marie: "Bacon et la *Logique de Port-Royal*," *Revue des sciences philosophiques et théologiques*, 84 (2000), 23–40.

Pumfrey, Stephen: "Neo-Aristotelianism and the Magnetic Philosophy," in *New Perspectives on Renaissance Thought: Essays in the History of Science, Education and Philosophy, In Memory of Charles B. Schmitt*, eds. John Henry and Sarah Hutton (London, 1990), 177–189.

Pumfrey, Stephen: "William Gilbert's Magnetic Philosophy, 1580–1684: The Creation and Dissolution of a Discipline" (PhD diss., University of London, 1987).

Radelet-de Grave, Patricia: "Les Jésuites, l'Université de Louvain et les recherches en physique," in *Sedes scientiae: l'émergence de la recherche à l'université: contributions au séminaire d'histore des sciences 2000–2001*, eds. Patricia Radelet-de Grave and Brigitte Van Tiggelen (Louvain-la-Neuve-Turnhout, 2003), 72–125.

Reid, Jasper: *The Metaphysics of Henry More* (Dordrecht, 2012).

Renz, Ursula and Han Van Ruler: "Okkasionalismus," in *Enzyklopädie Philosophie* (Hamburg, 2010); *sub voce*.

Reusens, Edmond: *Documents relatifs à l'histoire de l'Université de Louvain (1425–1797)* (Louvain, 1881–1903).

Rey, Anne-Lise: "L'ambivalence de la notion d'action dans la dynamique de Leibniz: La correspondance entre Leibniz et de Volder (1e Partie)," *Studia Leibnitiana*, 41 (2009), 47–66.

Rey, Anne-Lise: "L'ambivalence de la notion d'action dans la dynamique de Leibniz: La correspondance entre Leibniz et de Volder (11e Partie)," *Studia Leibnitiana*, 41 (2009), 157–182.

Robinet, André: "Leibniz et la *Logique de Port-Royal*," *Revue des sciences philosophiques et théologiques*, 84 (2000), 69–81.

Roegiers, Jan: "Het Leuvense artesonderwijs 1426–1797: leerstof en onderwijsmethode," in Ex cathedra: *Leuvense collegedictaten van de 16de tot de 18de eeuw*, eds. Geert Vanpaemel, Katharina Smeyers, An Smets and Diewer van der Meijden (Louvain, 2012), 23–45.

Roling, Bernd: Odins Imperium: *der Rudbeckianismus als Paradigma an den skandinavischen Universitäten, 1680–1860*, 2 vols. (Leiden-Boston, MA, 2020).

Roux, Sophie: "A French Partition of the Empire of Natural Philosophy (1670–1690)," in *The Mechanization of Natural Philosophy*, eds. Daniel Garber and Sophie Roux (Dordrecht, 2013), 55–98.

Roux, Sophie: "Descartes atomiste?," in *Atomismo e continuo nel XVII secolo*, eds. Romano Gatto and Egidio Festa (Naples, 2000), 211–274.

Roux, Sophie: "La philosophie mécanique (1630–1690)" (PhD diss., École des Hautes Études en Sciences Sociales, Paris, 1996).

Roux, Sophie: "Logique et méthode au XVIIe siècle," *Cahiers philosophiques de Strasbourg*, 32 (2012), 21–46.

Roux, Sophie: "Pour une conception polémique du cartésianisme: Ignace-Gaston Pardies and Antoine Dilly dans la querelle de l'âme des bêtes," in *Qu'est-ce qu'être cartésien?*, ed. Delphine Kolesnik-Antoine (Lyon, 2013), 315–337.

Roux, Sophie: "Premiers éléments d'une enquête sur Jacques du Roure," *Bulletin cartésien*, XLIX, *Archives de philosophie*, 83 (2020), 168–180.

Roux, Sophie: "The Condemnations of Cartesian Natural Philosophy under Louis XIV (1661–1691)," in *The Oxford Handbook of Descartes and Cartesianism*, eds. Steven Nadler, Tad Schmaltz and Delphine Antoine-Mahut (Oxford, 2019), 755–779.

Roux, Sophie: "The Two Comets of 1664–1665: A Dispersive Prism for French Natural Philosophy Principles," in *The Idea of Principles in Early Modern Thought*, ed. Peter Anstey (New York, NY-London, 2017), 98–146.

Roux, Sophie: "Was There a Cartesian Experimentalism in 1660s France?," in *Cartesian Empiricisms*, eds. Mihnea Dobre and Tammy Nyden (Dordrecht, 2013), 47–88.

Ruestow, Edward G.: *Physics at Seventeenth and Eighteenth-Century Leiden*: *Philosophy and the New Science in the University* (The Hague, 1973).

Russell, Leonard James: "The Correspondence between Leibniz and De Volder," *Proceedings of the Aristotelian Society*, 28 (1927–1928), 155–176.

Salminen, Seppo J.: "Barokin filosofis-teologisen synteesin hajoaminen maassamme," *Suomen Kirkkohistoriallisen Seuran vuosikirja*, 73 (1983), 52–84.

Sander, Christoph: Magnes: *der Magnetstein und der Magnetismus in den Wissenschaften der Frühen Neuzeit* (Leiden-Boston, MA, 2020).

Sander, Christoph: "Magnetism in an Aristotelian World (1550–1700)," in *Aristoteles und die Naturphilosophie an den mitteleuropäischen Universitäten der Frühen Neuzeit, 1600–1700*, eds. Bernd Roling, Sinem Kılıç and Benjamin Wallura (forthcoming).

Sander, Christoph: "Nutrition and Magnetism: An Ancient Idea Fleshed out in Early Modern Natural Philosophy, Medicine and Alchemy," in *Nutrition and Nutritive Soul in Aristotle and Aristotelianism*, eds. Roberto Lo Presti and Georgia-Maria Korobili (Berlin, 2021), 285–318.

Sander, Christoph: "Tempering Occult Qualities: Magnetism and Complexio in Early Modern Medical Thought," in Complexio: *Across Times and Disciplines*, eds. Chiara Beneduce and Paul J.J.M. Bakker (forthcoming).

Sander, Christoph: "*Terra AB*: Descartes's Imagery on Magnetism and Its Legacy," in *Cartesian Images: Picturing Philosophy in the Early Modern Age*, eds. Mattia Mantovani and Davide Cellamare (Leiden, 2023; forthcoming).

Sangiacomo, Andrea, "Defect of Knowledge and Practice of Virtue in Geulincx's Occasionalism," *Studia Leibnitiana*, 46 (2014), 46–63.

Sassen, Ferdinand, "Adriaan Heereboord (1614–1661): De opkomst van het Cartesianisme te Leiden," *Algemeen Nederlands Tijdschrift voor Wijsbegeerte*, 36 (1942), 12–22.

Sassen, Ferdinand: *Het wijsgerig onderwijs aan de Illustre School te Breda* (Amsterdam, 1963).

Sassen, Ferdinand: *Studenten van de Illustre School te 's-Hertogenbosch, 1638–1810* (Amsterdam, 1970).

Sassen, Saskia: "The Intellectual Climate in Leiden in Boerhaave's Time," in *Boerhaave and His Time*, ed. Gerrit Arie Lindeboom (Leiden, 1970), 1–16.

Savini, Massimiliano: *Johannes Clauberg*: Methodus cartesiana *et ontologie* (Paris, 2011).

Savini, Massimiliano: "La critique des arguments cartésiens dans l'*Admiranda Methodus* de Martin Schoock," in *Il Seicento e Descartes*, ed. Antonella Del Prete (Florence, 2004), 168–197.

Schaap, R.C.C.M.M.: "Het fonds van de Amsterdamse uitgever-boekhandelaar Jan Rieuwertsz (1644–1687): een poging tot reconstructie" (MA thesis, Amsterdam, 1978).

Schlegelmilch, Sabine: "The Scientific Revolution in Marburg," in *Early Modern Disputations and Dissertations in an Interdisciplinary and European Context*, eds. Meelis Friedenthal, Hanspeter Marti and Robert Seidel, transl. Ulrike Nichols (Leiden-Boston, MA, 2020), 288–311.

Schmaltz, Tad: "A Tale of Two Condemnations: Two Cartesian Condemnations in 17th-Century France," in *Descartes e i suoi avversari*, ed. Antonella Del Prete (Florence, 2004), 203–221.

Schmaltz, Tad: "Cartesianism in Crisis: The Case of the Eucharist," in *Theology and Early Modern Philosophy (1550–1750)*, eds. Simo Knutila and Risto Saarinen (Helsinki, 2010), 119–139.

Schmaltz, Tad: "Descartes on the Metaphysics of the Material World," *Philosophical Review*, 127 (2018), 1–40.

Schmaltz, Tad: *Early Modern Cartesianisms: Dutch and French Constructions* (Oxford, 2017).

Schmaltz, Tad: "Galileo and Descartes on Copernicanism and the Cause of the Tides," *Studies in History and Philosophy of Science*, 51 (2015), 70–81.

Schmaltz, Tad: *Radical Cartesianism: The French Reception of Descartes* (Cambridge, 2002).

Schmaltz, Tad, ed.: *Receptions of Descartes: Cartesianism and Anti-Cartesianism in Early Modern Europe* (London, 2005).

Schmaltz, Tad: "Review Essay: *Descartes on Forms and Mechanisms*, by Helen Hattab, and *Descartes's Changing Mind*, by Peter Machamer and J.E. McGuire," *Oxford Studies in Early Modern Philosophy*, 6 (2012), 349–372.

Schmaltz, Tad M.: "The Curious Case of Henricus Regius," in *The Oxford Handbook of Descartes and Cartesianism*, eds. Steven Nadler, Tad M. Schmaltz and Delphine Antoine-Mahut (Oxford, 2019).

Schmaltz, Tad: "The Pineal Gland in Cartesianism" (forthcoming).

Schmidt, Robert W.: *The Domain of Logic According to Saint Thomas Aquinas* (The Hague, 1966).

Schmitt, Charles: *Studies in Renaissance Philosophy and Science* (London, 1981).

Schmitt, Charles: "The Rise of the Philosophical Textbook," in *The Cambridge History of Renaissance Philosophy*, eds. Charles B. Schmitt and Quentin Skinner (Cambridge, 1988), 792–804.

Schouten, J. and Dietlinde Goltz: "James Primrose und sein Kampf gegen die Theorie vom Blutkreislauf," *Sudhoffs Archiv*, 61 (1977), 331–352.

Schüller, Volkmar: "Samuel Clarke's Annotations in Jacques Rohault's *Traité de physique*, and How They Contributed to Popularising Newton's Physics," in *Between Leibniz, Newton, and Kant: Philosophy and Science in the 18th Century*, ed. Wolfgang Lefèvre (Dordrecht, 2001), 95–110.

Schuster, John: *Descartes-Agonistes: Physico-mathematics, Method & Corpuscular-Mechanism 1618–1633* (Dordrecht-New York, NY, 2012).

Schuurman, Paul: "Continuity and Change in the Empiricism of John Locke and Gerardus de Vries (1648–1705)," *History of European Ideas*, 33 (2007), 292–304.

Schuurman, Paul: *Ideas, Mental Faculties and Method: The Logic of Ideas of Descartes and Locke and Its Reception in the Dutch Republic, 1630–1750* (Leiden-Boston, MA, 2004).

Schuurman, Paul: "Raey, Johannes de (1621–1702)," in *The Dictionary of Seventeenth-Century Dutch Philosophers*, eds. Wiep Van Bunge, Henri Krop, Bart Leeuwenburgh, Han Van Ruler, Paul Schuurman and Michiel Wielema, 2 vols. (Bristol, 2003), vol. 2, 813–816.

Scribano, Emanuela: *Angeli e beati: Modelli di conoscenza da Tommaso a Spinoza* (Bari, 2006).

Scribano, Emanuela: *Da Descartes a Spinoza: Percorsi della teologia razionale nel Seicento* (Milan, 1988).

Sellin, Paul R.: *Daniel Heinsius and Stuart England: With a Short-Title Checklist of the Works of Daniel Heinsius* (Leiden, 1968).

Sgarbi, Marco: "What Does a Renaissance Aristotelian Look Like? From Petrarch to Galilei," *Hopos*, 7 (2017), 226–245.

Shapiro, Lisa: "Descartes' Pineal Gland Reconsidered," *Midwest Studies in Philosophy*, 35 (2011), 259–286.

Shapiro, Lisa, ed.: *The Correspondence between Princess Elisabeth of Bohemia and René Descartes* (Chicago, IL, 2007).

Short Title Catalogue Netherlands: https://picarta.oclc.org/psi/DB=3.11/.

Simon, Gérard: *Kepler, rénovateur de l'optique* (Paris, 2017).

Simon, Gérard: "La théorie cartésienne de la vision, réponse à Kepler et rupture avec la problematique médiévale," in *Descartes et le Moyen Âge*, eds. Joël Biard and Roshdi Rashed (Paris, 1997), 107–118.

Skalnik, James Veazie: *Ramus and Reform: Church and University at the End of the Renaissance in France* (Kirksville, MO, 2002).

Smith, A. Mark: *From Sight to Light: The Passage from Ancient to Modern Optics* (Chicago, IL-London, 2014).

Smith, Justin: "Heat, Action, Perception: Models of Living Beings in German Medical Cartesianism," in *Cartesian Empiricism*, eds. Mihnea Dobre and Tammy Nyden (Dordrecht, 2013), 105–124.

Sortais, Gaston: "Descartes et la Compagnie de Jésus: menaces et avances, 1640–1646," *Estudios*, 57 (1937), 441–468.

Sortais, Gaston: "Le cartésianisme chez les Jésuites français au XVIIe et XVIIIe siècle," *Archives de philosophie* 6 (1929), 4–24.

Spink, Aaron: "The Experimental Physics of Jacques Rohault," *British Journal for the History of Philosophy*, 26 (2018), 850–870.

Steubing, Johann Hermann: *Geschichte der Hohen Schule Herborn* (Hadamar, 1823); repr. (Kreutzl, 1983).

Stévart, Armand: *Procès de Martin-Etienne Van Velden, Professeur à l'université de Louvain* (Brussels, 1871).

Stevenson, Leslie: "Freedom of Judgement in Descartes, Hume, Spinoza and Kant," *British Journal for the History of Philosophy*, 12 (2004), 223–246.

Strauch, Solveig: *Veit Ludwig von Seckendorff (1626–1692): Reformationsgeschichtsschreibung—Reformation des Lebens—Selbstbestimmung zwischen lutherischer Orthodoxie, Pietismus und Frühaufklärung* (Münster, 2005).

Strauss, Leo: *Persecution and the Art of Writing* (Chicago, IL, 1952).

Strazzoni, Andrea: *Burchard de Volder and the Age of the Scientific Revolution* (Cham, 2019).

Strazzoni, Andrea: *Dutch Cartesianism and the Birth of Philosophy of Science: From Regius to 's Gravesande* (Berlin-Boston, MA, 2019).

Strazzoni, Andrea: "How Did Regius Become Regius? The Early Doctrinal Evolution of a Heterodox Cartesian," *Early Science and Medicine*, 23 (2018), 362–412.

Strazzoni, Andrea: "La filosofia aristotelico cartesiana di Johannes de Raei," *Giornale critico della filosofia italiana*, 7 (2011), 107–132.

Strazzoni, Andrea: "On Three Unpublished Letters of Johannes De Raey to Johannes Clauberg," *Noctua*, 1 (2014), 66–103.

Strazzoni, Andrea: "The Cartesian Philosophy of Language of Johannes De Raey," *Lias*, 42 (2015), 89–120.

Strazzoni, Andrea: "The Dutch Fates of Bacon's Philosophy: *libertas philosophandi*, Cartesian Logic and Newtonianism," *Annali della Scuola Normale Superiore di Pisa. Classe di Lettere e Filosofia*, serie 5, vol. 4.1 (2012), 251–281.

Strazzoni, Andrea: "The Medical Cartesianism of Henricus Regius: Disciplinary Partitions, Mechanical Reductionism and Methodological Aspects," *Galilaeana: Journal of Galilean Studies*, 15 (2018), 181–220.

Summers, David: *The Judgment of Sense: Renaissance Naturalism and the Rise of Aesthetics* (Cambridge, 1987).

Sytsma, David S.: "Calvin, Daneau, and *physica mosaica*: Neglected Continuities at the Origins of an Early Modern Tradition," *Church History and Religious Culture*, 95 (2015), 457–476.

Tachau, Katherine H.: *Vision and Certitude in the Age of Ockham: Optics, Epistemology, and the Foundations of Semantics, 1250–1345* (Leiden, 1988).

Tegnér, Elof and Martin J. Weibull: *Lunds Universitets historia: 1668–1868*, 2 vols. (Lund, 1868).

Thijssen-Schoute, C. Louise, *Nederlands Cartesianisme* (Amsterdam, 1954).

Thijssen-Schoute, C. Louise: *Uit de Republiek der Letteren: Elf studiën op het gebied der ideeëngeschiedenis van de Gouden Eeuw* (The Hague, 1967).

Timmermans, Linda: *L'accès des femmes à la culture (1598–1715)* (Paris, 1993).

Trevisani, Francesco: *Descartes in Germania: la ricezione del cartesianesimo nella Facoltà filosofica e medica di Duisburg (1652–1703)* (Rome, 1992); revised edition, transl. Hans-Peter Kröner, *Descartes in Deutschland: die Rezeption des Cartesianismus in den Hochschulen Nordwestdeutschlands* (Vienna, 2011).

Vallinkoski, Jorma: *Turun Akatemian väitöskirjat, 1642–1828*, 2 vols. (Helsinki, 1962).

Van Asselt, Willem J.: *The Federal Theology of Johannes Coccejus* (Leiden, 2001).

Van Bunge, Wiep: *From Stevin to Spinoza: An Essay on Philosophy in the Seventeenth-Century Dutch Republic* (Leiden-Boston, MA-Cologne, 2001).

Van Bunge, Wiep: "Philosophy," in *1650: Hard-Won Unity*, eds. Willem Frijhoff and Marijke Spies (Assen and Basingstoke, 2004), 280–346.

Van Bunge, Wiep: "Vries, Gerard de (1628–1705)," in *The Dictionary of Seventeenth and Eighteenth-Century Dutch Philosophers*, eds. Wiep Van Bunge, Henri Krop, Bart Leeuwenburgh, Han Van Ruler, Paul Schuurman and Michiel Wielema, 2 vols. (Bristol, 2003), vol. 2, *sub voce*.

Van Bunge, Wiep, Henri Krop, Bart Leeuwenburgh, Han Van Ruler, Paul Schuurman and Michiel Wielema, eds.: *The Dictionary of Seventeenth and Eighteenth-Century Dutch Philosophers*, 2 vols. (Bristol, 2003).

Van der Horst, Koert: *Catalogus van de collectie collegedictaten van de Utrechtse Universiteitsbibliotheek* (Utrecht, 1994).

Van der Horst, Koert: "Collegedictaten als bron van kennis voor het universitaire onderwijs," in Hans C. Teitler, ed., *Studeren en promoveren in Utrecht*, special issue of *Utrechtse Historische Cahiers*, 17 (1996), 14–21.

Van der Horst, Koert: "De twee vroegste *Series Lectionum* van de Utrechtse Universiteit: 1656 en 1672," in *The Dictionary of Seventeenth and Eighteenth-Century Dutch Philosophers*, eds. Wiep Van Bunge, Henri Krop, Bart Leeuwenburgh, Han Van Ruler, Paul Schuurman and Michiel Wielema (Rotterdam, 1995), vol. 2, 261–282.

Van der Horst, Koert, Peter Koolmees and Adriaan Monna, eds.: *Over beesten en boeken* (Rotterdam, 1995).

Van Hardeveld, Ike: *Lodewijk Meijer (1629–1681) als lexicograaf* (Utrecht, 2000).

van Meerkerk, Edwin: "Een filosofische lessenserie aan de Kwartierlijke Academie, 1663–1665," *Jaarboek Numaga*, 45 (1998), 43–61.

van Meerkerk, Edwin: "The Right Use of Reason: The Only Philosophical Thesis at the Former University of Nijmegen 1668," *Lias*, 26 (1999), 77–102.

Van Miert, Dirk: *Humanism in an Age of Science: the Amsterdam Athenaeum in the Golden Age, 1632–1704*, transl. Michiel Wielema (Leiden-Boston, MA, 2009).

Van Miert, Dirk: "The Disputation Hall in the Seventeenth-Century Dutch Republic: An Urban Location of Knowledge," in *Locations of Knowledge in Dutch Contexts*, eds. Fokko Jan Dijksterhuis, Andreas Weber and Huib Zuidervaart (Leiden-Boston, MA, 2019), 211–231.

Van Otegem, Matthijs: *A Bibliography of the Works of Descartes (1637–1704)*, 2 vols., (Utrecht, 2002).

Vanpaemel, Geert: "Cartesianism in the Southern Netherlands: The Role of the Louvain Faculty of Arts," *Bulletin de La Societe Royale Des Sciences de Liege*, 55 (1986), 221–230.

Vanpaemel, Geert: "De eerste microscoop aan de Leuvense universiteit," *Scientiarum Historia*, 25 (1999), 13–25.

Vanpaemel, Geert: "De verspreiding van het cartesianisme," in *Geschiedenis van de wetenschappen in België van de Oudheid tot 1815*, eds. Robert Halleux, Carmélia Opsomer and Jan Vandersmissen (Brussels, 1998), 259–272.

Vanpaemel, Geert: *Echo's van een wetenschappelijke revolutie: de mechanistische natuurwetenschap aan de Leuvense Artesfaculteit (1650–1797)* (Brussels, 1986).

Vanpaemel, Geert: "Kerk en wetenschap: de strijd tegen het cartesianisme aan de Leuvense universiteit," *De zeventiende eeuw*, 5 (1989), 182–189.

Vanpaemel, Geert: "Rohault's *Traité de physique* and the Teaching of Cartesian Physics," *Janus*, 71 (1984), 31–40.

Vanpaemel, Geert: "*Terra autem in aeternum stat*: Het kosmologie-debat te Leuven," *De Zeventiende Eeuw*, 2 (1986), 101–115.

Vanpaemel, Geert: "The Louvain Printers and the Establishment of the Cartesian Curriculum," *Studium: Tijdschrift voor Wetenschaps- en Universiteitsgeschiedenis*, 4 (2011), 241–254.

Vanpaemel, Geert, Katharina Smeyers and An Smet, Diewer van der Meijden, eds.: Ex cathedra: *Leuvense collegedictaten van de 16de tot de 18de eeuw* (Louvain, 2012).

Van Ruler, Han: "Arnold Geulincx (1624–1669)," *The Internet Encyclopedia of Philosophy*, www.iep.utm.edu/geulincx.

Van Ruler, Han: "Minds, Forms, and Spirits: The Nature of Cartesian Disenchantment," *Journal of the History of Ideas*, 61 (2000), 381–395.

Van Ruler, Han: "Substituting Aristotle: Platonic Themes in Dutch Cartesianism," in *Platonism at the Origins of Modernity*, eds. Douglas Hedley and Sarah Hutton (Dordrecht, 2007), 159–175.

Van Ruler, Han: *The Crisis of Causality: Voetius and Descartes on God, Nature, and Change* (Leiden-New York, NY, 1995).

Van Ruler, Han: "The *Philosophia Christi*, its Echoes and its Repercussions on Virtue and Nobility," in *Christian Humanism: Essays in Honour of Arjo Vanderjagt*, eds. Alasdair A. MacDonald, Zweder R.W.M. von Martels and Jan R. Veenstra (Leiden, 2009), 235–263.

Van Ruler, Han and Anthony Uhlmann, Martin A. Wilson eds.: *Arnold Geulincx, Ethics. With Samuel Beckett's Notes* (Leiden, 2006).

Van Vloten, Johannes: *Ysselkout* (Deventer, 1855).

Vasoli, Cesare: *La dialettica e la retorica dell'umanesimo: invenzione e metodo nella cultura del XV e XVI secolo* (Milan, 1968).

Venn, John and John Archibald Venn, eds.: *Alumni cantabrigienses*, 2 vols. (Cambridge, 1922–1954).

Verbeek, Theo: "Bio-Bibliographical Sketch," in *Johannes Clauberg (1622–1665)*, ed. Theo Verbeek (Dordrecht, 1999), 181–200.

Verbeek, Theo: *Descartes and the Dutch: Early Reactions to Cartesian Philosophy, 1637–1650* (Carbondale-Edwardsville, IL, 1992).

Verbeek, Theo: "Descartes's *Letter to Voetius*: The Method of Philosophy and Morality," *Church History and Religious Culture*, 100 (2020), 289–300.

Verbeek, Theo: "Dutch Cartesian Philosophy," in *A Companion to Early Modern Philosophy*, ed. Steven Nadler (Oxford, 2002), 167–182.

Verbeek, Theo: "From 'Learned Ignorance' to Scepticism: Descartes and Calvinist Orthodoxy," in *Scepticism and Irreligion in the Seventeenth and Eighteenth Centuries*, eds. Arjo Vanderjagt and Richard H. Popkin (Leiden-New York, NY, 1993), 31–45.

Verbeek, Theo: "Gassendi et les Pays-Bas," in *Gassendi et l'Europe, 1592–1792: actes du colloque international de Paris, 'Gassendi et sa postérité (1592–1792)*, Sorbonne, 6–10 octobre 1992, ed. Sylvia Murr (Paris, 1997), 263–273.

Verbeek, Theo: "Generosity," in *Emotional Minds: The Passions and the Limits of Pure Enquiry in Early Modern Philosophy*, ed. Sabrina Ebbersmeyer (Berlin, 2012), 19–30.

Verbeek, Theo: "Heereboord, Adriaan (1614–1661)," in *The Dictionary of Seventeenth and Eighteenth-Century Dutch Philosophers*, eds. Wiep Van Bunge, Henri Krop, Bart Leeuwenburgh, Han Van Ruler, Paul Schuurman and Michiel Wielema (Bristol, 2003), vol. 1, 395–397.

Verbeek, Theo, ed.: *Johannes Clauberg (1622–1665) and Cartesian Philosophy in the Seventeenth Century* (Dordrecht, 1999).

Verbeek, Theo, ed.: *La querelle d'Utrecht: René Descartes et Martin Schoock* (Paris, 1988).

Verbeek, Theo: "Le contexte historique des *Notae in programma quoddam*," in *Descartes et Regius, Autour de l'*Explication de l'esprit humain, ed. Theo Verbeek (Amsterdam, 1993), 1–33.

Verbeek, Theo: "Les *Principia* dans la culture néerlandaise du XVIIe siècle," in *Descartes: Principia Philosophiae*, eds. Jean-Robert Armogathe and Giulia Belgioioso (Naples, 1996), 701–712.

Verbeek, Theo: "Regius's *Fundamenta physices*," *Journal of the History of Ideas*, 55 (1994), 533–551.

Verbeek, Theo: "The Invention of Nature: Descartes and Regius," in *Descartes' Natural Philosophy*, eds. Stephen Gaukroger, John Shuster and John Sutton (London-New York, NY, 2000), 149–167.

Verbeek, Theo: "Tradition and Novelty: Descartes and Some Cartesians," in *The Rise of Modern Philosophy*, ed. Tom Sorell (Oxford, 1993), 167–196.

Verbeek, Theo, ed.: *Descartes et Regius: Autour de l'*Explication de l'esprit humain (Amsterdam, 1993).

Vermeir, Koen: "Mechanical Philosophy in an Enchanted World: Cartesian Empiricism in Balthasar Bekker's Radical Reformation," in *Cartesian Empiricisms*, eds. Mihnea Dobre and Tammy Nyden-Bullock (Dordrecht, 2013), 275–306.

Vermeulen, Corinna: "René Descartes, *Specimina philosophiae*: Introduction and Critical Edition" (PhD diss., Universiteit Utrecht, 2007).

Vermij, Rink: *The Calvinist Copernicans: The Reception of the New Astronomy in the Dutch Republic, 1575–1750* (Amsterdam, 2002).

Viskolcz, Noémi: *Johann Heinrich Bisterfeld (1605–1655): Bibliográfia* (Budapest-Szeged, 2003).

Visser, Piet: "Blasphemous and Pernicious: the Role of Printers and Booksellers in the Spread of Dissident Religious and Philosophical Ideas in the Netherlands in the Second Half of the Seventeenth Century," *Quaerendo*, 26 (1996), 303–326.

Weijers, Olga: *A Scholar's Paradise: Teaching and Debating in Medieval Paris* (Turnhout, 2015).

Weill-Parot, Nicolas: "Astrology, Astral Influences, and Occult Properties in the Thirteenth and Fourteenth Centuries," *Traditio*, 65 (2010), 201–230.

Westfall, Richard S.: *Never at Rest: A Biography of Isaac Newton* (Cambridge, 1980).

Wiesenfeldt, Gerhard: "Academic Writings and the Rituals of Early Modern Universities," *Intellectual History Review*, 26 (2016), 1–14.

Wiesenfeldt, Gerhard: *Leerer Raum in Minervas Haus: Experimentelle Naturlehre an der Universität Leiden, 1675–1715* (Amsterdam, 2002).

Wijnne, Johan A.: *Resolutiën, genomen bij de Vroedschap van Utrecht, betreffende de Illustre School en de Akademie in hare stad, van de jaren 1632–1693* (Utrecht, 1888).

Wilson, Catherine: *Epicureanism at the Origins of Modernity* (Oxford, 2008).

Young, Robert: *Comenius in England* (Oxford, 1932).

Zittel, Claus: "Conflicting Pictures: Illustrating Descartes's *Traité de l'homme*," in *Silent Messengers: The Circulation of Material Objects of Knowledge in the Early Modern Low Countries*, eds. Sven Dupré and Christoph Lüthy (Münster, 2011), 217–260.

Zittel, Claus: Theatrum philosophicum: *Erfahrungsmodi und Formen der Wissensrepräsentation bei Descartes* (Berlin, 2007).

Index